DEPOSITIONAL SYSTEMS

DEPOSITIONAL SYSTEMS

A Genetic Approach to Sedimentary Geology

RICHARD A. DAVIS, JR.

University of South Florida

PRENTICE-HALL, INC., Englewood Cliffs, New Jersey 07632

Library of Congress Cataloging in Publication Data

Davis, Richard A. (Richard Albert), (date)
   Depositional systems.

   Bibliography: p.
   Includes index.
   1. Sedimentation and deposition. I. Title.
QE581.D38 1983   551.3   82-24125
ISBN 0-13-198960-X

Editorial/production supervision:
  Zita de Schauensee
Cover design: Photo Plus Art
Manufacturing buyer: John Hall
Cover photograph: Ona Beach,
  Oregon, by William T. Fox

10 9 8 7 6 5 4 3

ISBN 0-13-198960-X

Prentice-Hall International, Inc., *London*
Prentice-Hall of Australia Pty. Limited, *Sydney*
Editora Prentice-Hall do Brasil, Ltda., *Rio de Janeiro*
Prentice-Hall Canada Inc., *Toronto*
Prentice-Hall of India Private Limited, *New Delhi*
Prentice-Hall of Japan, Inc., *Tokyo*
Prentice-Hall of Southeast Asia Pte. Ltd., *Singapore*
Whitehall Books Limited, *Wellington, New Zealand*

TO DICK AND DOLLY

Contents

Contents

10 Intertidal Environments   325

11 Coastal Bays: Estuaries and Lagoons   364

Contents

12   The Barrier Island System   403

Contents   xiii

Preface

Sedimentation and stratigraphy have long been an integral part of the undergraduate geology curriculum. Some curricula contain a one-term combined course, others have separate courses in each. Generally, laboratory and/or field activities are also included in some fashion. Over the past several decades there have been surprisingly few textbooks written for first-level courses on these topics, but there are several advanced or reference books emanating from various sources, especially topical symposia.

The excellent and popular books of the 1950s, such as *Stratigraphy and Sedimentation* by Krumbein and Sloss, *Principles of Stratigraphy* by Dunbar and Rodgers, and *Sedimentary Rocks* by Pettijohn, placed emphasis on the description of sediments, sedimentary rocks, and stratigraphic units. Basic principles of stratigraphy were emphasized but little was presented on the genesis of these stratigraphic units and the rocks contained therein.

Throughout the 1950s and 1960s much research emphasis was placed on modern sedimentary environments, with the major petroleum research laboratories at the forefront. Originally, the shallow marine and fluvial environments were emphasized, but other continental environments, and more recently the deep marine areas, have also received much attention.

Formulation of the term *depositional systems* is commonly attributed to the group at The University of Texas–Austin headed by L. F. Brown, W. L. Fisher, and

A. J. Scott. These people developed a course in the 1960s under that title, the primary objectives being to convey to the students an understanding and appreciation for modern depositional environments and their analogs in the rock record. This course was taught to second- or third-year undergraduates as a first soft-rock course beyond the introductory courses (physical and historical geology). Over the past decade or two, similar courses have been developed at many institutions and, similarly, traditional sedimentation-stratigraphy courses have shifted emphasis toward a depositional systems approach.

This book has resulted from many of the basic ideas of the UT-Austin group and is based on the course in Depositional Systems (GLY 4550) as it is taught at the University of South Florida. The overall philosophy is to provide all undergraduate geology students with a broad-brush treatment of sedimentary geology. Principles are covered in the first four chapters to the extent that they are necessary to understand the remainder of the text. Undoubtedly, some instructors will feel that certain environments are stressed too much and others are treated too briefly. Such are the problems of any textbook. The examples of environments of deposition chosen from the rock record are intended to span a range in geologic time and are for the most part the well-known, classic studies. Most people will probably choose to add their own favorite examples and delete some which are presented here.

The reader will encounter numerous reference citations within the text, more than are typical for a book aimed at this level. It is important for students to realize that textbooks are simply compilations of work originally done by hundreds of people. In the opinion of this writer, such citations are an integral part of the literature. All complete references are compiled at the end of the book rather than after each chapter. This arrangement facilitates their location as compared to searching for them at the end of a given chapter. Selected annotated references on each major topic are available after the chapters. A glossary of terms is also provided.

Acknowledgments

All books benefit greatly from the input of others, and this one is no exception. Ideas, illustrations, and reviews were obtained from numerous people. The cooperation and assistance of my colleagues and librarians here at the University of South Florida are greatly appreciated.

Reviewers provided much input into the text. These include: Robert E. Carver, Bruce W. Selleck (Chap. 1); P. D. Komar, George F. Oertel (Chap. 2); H. Edward Clifton, Robert H. Dott, Jr. (Chap. 3); Alan H. Coogan, Donald E. Owen (Chap. 4); M. Dane Picard, C. C. Reeves, Jr. (Chap. 5); Brain J. Bluck, Ralph E. Hunter (Chap. 6); Gail M. Ashley, Thomas E. Gustavson (Chap. 7); Norman D. Smith (Chap. 8); L. F. Brown, L. D. Wright (Chap. 9); Franz E. Anderson, George deVries Klein (Chap. 10); Robert B. Biggs (Chap. 11); John C. Kraft (Chap. 12); Robert L. Brenner (Chap. 13); Henry S. Chafetz, Albert C. Hine (Chap. 14); Donn S. Gorsline, Reinhard F. Hesse (Chap. 15); John B. Anderson, Arnold H. Bouma (Chap. 16). All or substantial parts of the manuscript were reviewed for the publisher by Patrick L. Abbott, Menno G. Dickelman, William T. Fox, Lee C. Gerhard, Raymond V. Ingersoll, George deVries Klein, Gerard V. Middleton, and A. M. Thompson.

Editors Logan Campbell and Doug Humphrey, production editor Zita de Schauensee, and their colleagues at Prentice-Hall, Inc., were most cooperative throughout the project. Wanda McClelland typed various drafts of the manuscript. Mary Haney and Iris Rose assisted in various clerical activities. Joyce Bland did the photographic reproduction. Lastly, I thank my wife and children for enduring my messy study and numerous periods of frustration during the writing effort.

Tampa, Florida   Richard A. Davis, Jr.

Introduction

This book treats sedimentation and stratigraphy from a genetic point of view. This approach differs significantly from that taken by most earlier textbooks on the subject, which relied largely on descriptive aspects of sediments, sedimentary rocks, and rock strata. Extensive and intensive studies on modern sedimentary environments have now provided much data on sedimentary processes and modern depositional facies. As a result, numerous models have been developed to characterize the sediment accumulations in various environments of deposition.

A **depositional system** is an assemblage of process-related sedimentary facies (Scott and Fisher in Fisher et al., 1969). Within each system there may be numerous depositional environments, each of which is characterized by its own sediments, fauna, and flora as well as processes to which they are subjected. The fundamental building blocks of each depositional system are the facies. These, of course, represent specific environments of deposition. Each depositional system is therefore comprised of a group of these genetically related building blocks (facies).

Most of the chapters in Parts II through IV are devoted to an entire depositional system. However, a few treat broad environments that may be an important part of more than one depositional system. For example, the lacustrine, fluvial, and deltaic systems are rather easy to define and to discuss as distinct depositional systems, even though streams feed into lakes and may develop a delta at their mouths. Fisher and Brown (1972) consider a barrier bar–strandplain system, a lagoon, bay, estuarine,

1

tidal flat system, and an eolian system. One could argue that a barrier island complex can include all of these.

In order to maintain as much genetic organization as possible, the chapters are related first by gross geographic location: continental, coastal, and open marine. In the continental regime all are easily recognizable and definable, except perhaps the desert system. It includes some aspects of fluvial (fans), lacustrine (playas), and eolian (dunes) environments. These environments do commonly occur adjacent to one another both at present and in the geologic record. They do in fact represent genetically related facies and therefore a depositional system.

In the coastal areas of the world there are numerous environments of sediment accumulation which relate to one another in a great variety of geographic arrangements. Various types of coastal bays, including estuaries and lagoons, tidal environments, and beaches, may be juxtaposed geographically and stratigraphically. For discussion purposes the coastal bays and tidal environments are considered separately and are separated from the barrier island system, even though they may be elements of this system. The open marine region is also fairly readily subdivided into systems, even on shelves and in shallow seas, where distinct processes tend to separate the carbonate system from the terrigenous system.

There is a certain level of assumed knowledge in the sections that treat the depositional systems themselves. A section on principles contains appropriate background information for students who have little exposure to sedimentary geology or for those who wish to review. The coverage of the many diverse topics in these chapters is broad and brief. Comprehensive treatment of the various topics considered must be obtained from the numerous cited references or from lecture material.

Part One

PRINCIPLES

1 Sediments and Sedimentary Rocks

In beginning courses in geology, one of the fundamental concepts which is considered is the rock cycle. In the rock cycle there are numerous stages or places where earth materials interact with earth processes. As a consequence of some of these processes, rocks are formed and some processes result in the breakdown of rocks, by physical means, by chemical means, or as is often the case, by a combination of both. The result is particles of sediment which may be mineral grains or rock fragments and liquids from which minerals may precipitate at or near surface pressure and temperature conditions.

These weathering products will be transported on or near the earth's surface, eventually coming to rest as an accumulation of particles which either settle due to a change in surface conditions, or which are precipitated from solution (physicochemical) or by an organism (biochemical). Eventually, some of these particles become buried, then cemented into sedimentary rocks and preserved as part of the geologic record in the relatively thin outer skin of the earth's crust.

In this chapter both the textural and compositional aspects of sediments are discussed, together with the formation of sedimentary rocks. These subjects are treated only in enough depth to make the last three sections of the book easily understood and meaningful. In-depth studies of sediments and sedimentary rocks can be pursued through the several comprehensive texts on sedimentary petrology. Among these are Blatt et al. (1980), Pettijohn (1975), Bathurst (1975), Pettijohn et al. (1972), and Friedman and Sanders (1978).

TEXTURE

The production of sedimentary particles during erosion can be related to many geologic phenomena. Uplift increases relief and exposes rocks, lowering of the sea level may do the same, and climate or organisms can cause erosion. Regardless of the mechanism, sediment particles are the product. In this discussion no attention will be paid to the chemistry or mineralogy of the particles. Emphasis is on the size, shape, and distribution of these particles as they accumulate in their environment of deposition.

Grain Size

There is actually a variety of ways by which one might assign a size to a sediment particle. Two come to mind immediately: the diameter and the volume of a particle. The latter is rather easy to deal with because shape is not a factor and it can easily be related to specific gravity. However, measurement of the volume of large numbers of grains is both difficult and time consuming. The diameter of a particle is far easier to measure, although there is an infinite range of particle shapes that may be exhibited. Although relatively few perfect spheres are produced as erosion products, most particles exhibit a rough approximation to a sphere.

The range in grain size of sedimentary particles is almost infinite, from less than a micrometer to particles several meters in diameter. Obviously, several techniques must be utilized for measuring such a wide spectrum of grain sizes. Virtually all grain size analyses involve extremely large numbers of grains, which places further constraints on size analysis techniques. Commonly, grains larger than a centimeter or so in diameter are measured individually with calipers or by a similar method. Grains ranging from a centimeter down to about 50 μm in diameter are measured by either sieving them through several sieves which separate the grains into rather narrow size classes or by allowing the grains to settle through a column of a fluid, usually water (Zeigler et al., 1960). The rate at which particles settle is proportional to their size, shape, and density, although certain assumptions are made. Such settling tubes have become the standard in recent years and may be quite sophisticated, with direct connection to computers for rapid data processing. Particles smaller than 50 μm may also be measured by settling rates, although such methods are time consuming and not very accurate. Recently, instruments utilizing photoelectric sensors, lasers, or x-ray beams have been developed for measuring these small grains.

Because sedimentary particles display such a wide range of sizes, it is practical to utilize a geometric or logarithmic scale for size classification. Udden (1898) devised a scale based on a factor of 2 such that as one moves from unity there is a multiplier or divisor of 2. Later modifications by Wentworth (1922) resulted in the grain size scale as it is used today (Table 1–1). Each size class possesses a name as well as a size range.

TABLE 1-1 WENTWORTH GRAIN SIZE SCALE

Limiting particle diameter

mm	ϕ units		Size class		
2048	− 11		Very large		
1024	− 10		Large	Boulders	
512	− 9		Medium		
256	− 8		Small		G
128	− 7		Large	Cobbles	R
64	− 6		Small		A
32	− 5		Very coarse		V
16	− 4		Coarse		E
8	− 3		Medium	Pebbles	L
4	− 2		Fine		
2	− 1		Very fine	Granules	
1	0		Very coarse		
$1/2$	+ 1	μm 500	Coarse		
$1/4$	+ 2	250	Medium	Sand	
$1/8$	+ 3	125	Fine		
$1/16$	+ 4	62	Very fine		
$1/32$	+ 5	31	Very coarse		
$1/64$	+ 6	16	Coarse		M
$1/128$	+ 7	8	Medium	Silt	U
$1/256$	+ 8	4	Fine		D
$1/512$	+ 9	2	Very fine		
				Clay	

The most recent significant modification of the Wentworth scheme was the logarithmic transformation proposed by Krumbein (1934). The phi (ϕ) scale is based on the negative log to the base 2:

$$\phi = -\log_2 d \qquad (1-1)$$

where d is the diameter in millimeters. This notation allows one to deal with simple whole numbers rather than large numbers and messy fractions (Table 1-1). Such a transformation simplifies statistical calculations and graphic plotting of grain size data.

Statistical analysis of grain size data

After sediment has been subjected to one of the standard methods for measuring grain size, these data are then used to calculate various statistical parameters which describe the nature of the population of particles in the sediment sample. Data ob-

(a)

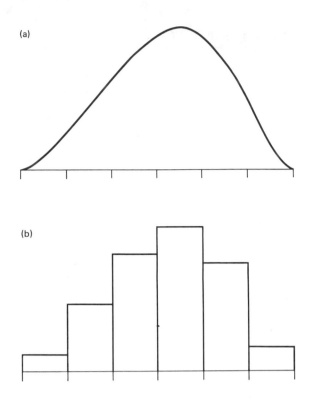

(b)

Figure 1-1 (a) Smoothed frequency curve and (b) the corresponding histogram for a hypothetical set of data.

tained from the size analysis are usually in the form of weight percentages for each size class, which may be at intervals of $\frac{1}{2}\phi$ or $\frac{1}{4}\phi$. These data may be plotted as a histogram or as a frequency curve (Figure 1-1).

The distribution of grain sizes in a sediment population typically follows or approaches a **log-normal distribution.** The resulting normal or bell-shaped curve provides data from which the four statistical **moments** are calculated. In a normal curve the two points of inflection are at 16% and 84%. The 68% included between these points lies within one **standard deviation** of the mean or average grain size.

The geologist generally plots grain size data as a cumulative frequency curve or as a probability plot (Figure 1-2). Cumulative curves or plots are advantageous because the percentiles can be read directly from the graph. They do have the disadvantage of being more difficult to interpret visually than the histogram or frequency curve.

Four statistical parameters are generally used to describe grain size distribution: mean, standard deviation, skewness, and kurtosis. These are most commonly determined graphically, although they can be calculated by moment statistics (McBride, 1971). In addition, the median and mode are of value in sediment analysis. The **median** is the grain size in the middle of the population: the size of the 50th percentile (Figure 1-2). The **mode** is the most frequently occurring grain size. Median values are easily determined from either a frequency curve or a cumulative curve, whereas the mode cannot be accurately determined from a cumulative plot (Figure 1-2).

The mean (first moment) is simply the statistical average expressed in ϕ units, usually to the nearest hundredth of a unit. In a normal distribution the median, mode,

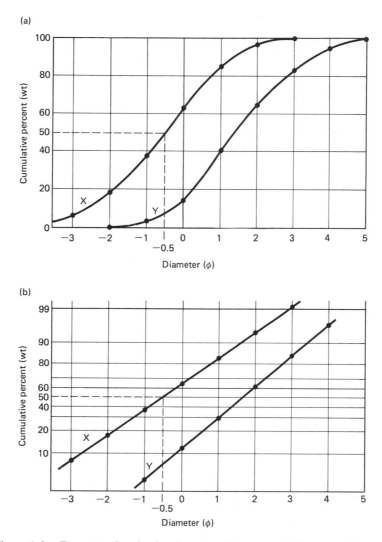

Figure 1-2 Two sets of grain size data plotted as a cumulative curve (a) on arithmetic paper and (b) on probability paper. The probability plot tends to straighten the tails of the frequency curves. The median (50th percentile) value for sample X is -0.5ϕ.

and mean are coincident [Figure 1-3(a)]; however, they typically show different size values for asymmetric curves [Figure 1-3(b)]. The graphic mean (Table 1-2) is calculated arithmetically from grain size values at the 16th, 50th, and 84th percentiles on a cumulative curve.

The uniformity in grain size within a sediment sample is measured by the standard deviation (second moment), which is called the **sorting value** in sedimentology. It measures the spread in grain size and is a valuable parameter because it commonly reflects conditions during deposition of the sediment. Sorting can be measured using

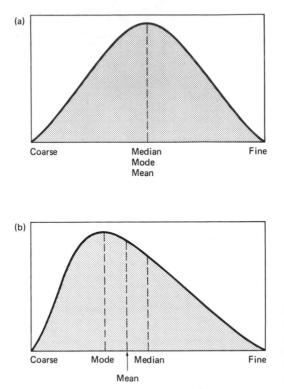

Figure 1-3 Frequency curves showing (a) perfectly symmetrical distribution where median, mode, and mean are coincident, and (b) asymmetrical distribution, where median, mode, and mean each have different grain size values.

TABLE 1-2 COMMON GRAPHIC STATISTICAL CALCULATIONS USED IN GRAIN SIZE ANALYSIS

Statistic		Formula
Graphic mean	M_z =	$\dfrac{\phi_{16} + \phi_{50} + \phi_{84}}{3}$
Graphic standard deviation	Σ_G =	$\dfrac{\phi_{84} - \phi_{16}}{2}$
Inclusive graphic standard deviation	Σ_I =	$\dfrac{\phi_{84} - \phi_{16}}{4} + \dfrac{\phi_{95} - \phi_5}{6.6}$
Graphic skewness	Sk_G =	$\dfrac{\phi_{16} + \phi_{84} - 2\phi_{50}}{\phi_{84} - \phi_{16}}$
Inclusive graphic skewness	Sk_I =	$\dfrac{\phi_{16} + \phi_{84} - 2\phi_{50}}{2(\phi_{84} - \phi_{16})} + \dfrac{\phi_5 + \phi_{95} - 2\phi_{50}}{2(\phi_{95} - \phi_5)}$
Graphic kurtosis	K_G =	$\dfrac{\phi_{95} - \phi_5}{2.44(\phi_{75} - \phi_{25})}$

SOURCE: After Folk, 1974, pp. 45–48.

TABLE 1-3 SORTING CLASSES BASED ON STANDARD DEVIATION

Ranges of values of sorting (φ units)	Sorting class	Environments of sands
<0.35	Very well sorted	Coastal and lake dunes; many beaches (foreshore); common on shallow marine shelf
0.35–0.50	Well sorted	Most beaches (foreshore); shallow marine shelf; many inland dunes
0.50–0.80	Moderately well sorted	Most inland dunes; most rivers; most lagoons; distal marine shelf
0.80–1.40	Moderately sorted	Many glaciofluvial settings; many rivers; some lagoons; some distal marine shelf
1.40–2.00	Poorly sorted	Many glaciofluvial settings
2.00–2.60	Very poorly sorted	Many glaciofluvial settings
>2.60	Extremely poorly sorted	Some glaciofluvial settings

SOURCE: Friedman and Sanders, 1978, p. 73.

either the graphic or inclusive graphic standard deviation (Table 1–2). Steep cumulative curves indicate good sorting and broad flat ones indicate poor sorting. Sorting values are also expressed in φ units (Table 1–3).

Skewness (third moment) is a measure of the asymmetry of the grain size distribution in a sediment sample. It can be easily recognized from a frequency curve but less so from a cumulative frequency curve. The frequency curve is asymmetric or skewed with the mode shifted to the left of the distribution for samples with abundant fine particles and to the right for abundant coarse particles (Figure 1–4). Skewness may be calculated from graphic data (Table 1–2) yielding a dimensionless value which may be positive (fine secondary population) or negative (coarse secondary population).

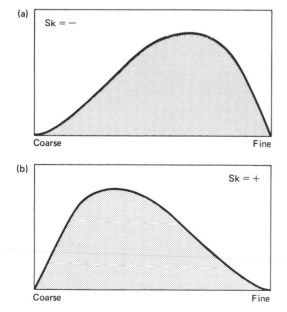

Figure 1-4 Frequency curves showing (a) negatively skewed grain size distribution, in which there is an abundance of coarse particles, and (b) positively skewed distribution, in which there is an abundance of particles on the fine end of the distribution.

Kurtosis (fourth moment) is commonly related to peakedness of a frequency curve. It may be calculated from graphic data and measures the relationship between the sorting in the central portion of the curve to the portion in the "tails" (Folk, 1974). Relatively peaked curves (> 1.00) are leptokurtic and flat curves (< 1.00) are platykurtic.

There are a few broad and oversimplified generalities that can be made to summarize the significance of the moments of grain size data. The mean grain size tends to reflect the competency of the transporting medium which brought the sediment to its environment of deposition. Sorting is the process that modifies the sediment after it has been moved to its site of deposition (Folk, 1974). It is an indication of variability in processes and also their rigor. Too little or too much rigor produces poor sorting. Intense and short-duration processes such as storms or breaking waves create much sediment movement and last for too short a time for sorting to take place. Preferential transport of certain size fractions gives rise to skewed sediments. Kurtosis does not appear to be of any geologic significance except as a means of characterizing a sediment population.

Grain Shape

Sediment particles display a great variety of geometric forms. This variation is due to a combination of the internal structure plus the origin and history of the particle. Some particles are simple and symmetrical, whereas others are extremely complex.

The **roundness** of a particle refers to the sharpness or smoothness of the edges and corners of a grain. Both physical abrasion and chemical reactions contribute to this characteristic, although abrasion is generally the most important of the two. Roundness can be measured by dividing the average radius of corners and edges by the radius of the maximum inscribed circle (Wentworth, 1919). This is typically accomplished by measuring cross sections or projected images of particles. A standard for comparison has been provided by Powers (1953), who developed a six-stage hierarchy comprised of descriptive names ranging from "very angular" to "well rounded" (Figure 1–5). Folk (1955) proposed a logarithmic conversion called the rho (ρ) scale, which ranges from 0 to 6, with the units corresponding to Powers' descriptive categories (Figure 1–5).

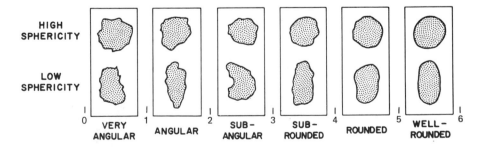

Figure 1–5   Outlines of classes in Powers' roundness scale, showing both high- and low-sphericity shaped particles. Numbers between classes represent the rho (ρ) scale of Folk (1955). (Modified from Powers, 1953, p. 118.)

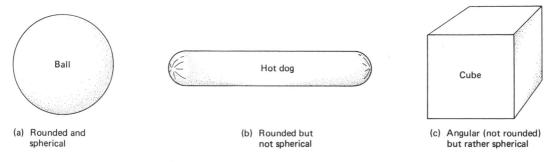

(a) Rounded and spherical

(b) Rounded but not spherical

(c) Angular (not rounded) but rather spherical

Figure 1-6 Examples of common objects which show various combinations of roundness and sphericity.

The term **sphericity** refers to the degree to which a particle approaches a sphere. Although many ways of determining sphericity are available, it is most common to compare the lengths of three mutually perpendicular axes. As this ratio approaches unity the particle is becoming more spherical. Sphericity can be measured by determining the volume of the particle and that of a sphere of equal volume, then comparing that to the volume of a circumscribing sphere (Wadell, 1932). Another sphericity determination is made by comparing the long, intermediate, and short diameters—in essence, a comparison of the minimum to the maximum cross-sectional area (Folk and Ward, 1957).

Sphericity is more strongly influenced by the origin of the particle than is roundness. Some grains are inherently elongate because of their crystallographic or biogenic makeup. Examples are such minerals as rutile or mica, which are rarely spherical, and many shell types, such as bivalves or branching corals. In addition, bedding or cleavage may create planes of weakness which yield on impact and prevent attainment of a spherical shape during transport.

It should be observed that roundness and sphericity may not be related in a given particle. That is, some perfectly rounded objects may show a low sphericity (e.g., a hot dog) and some rather spherical objects may be very angular (e.g., a cube). Of course, many things may be both well rounded and spherical (e.g., a Ping-Pong ball, quartz grain) (Figure 1-6).

Classification of shapes

Although roundness and sphericity are quite important and valuable in characterizing particle shape, they do not completely describe the particle. A simple but useful classification suggested by Zingg (1935) utilizes the ratios of the three mutually perpendicular diameters. The result is four primary shape classes (Figure 1-7): oblate (disk), equant, bladed, and prolate (roller). The classification is achieved by comparing the ratio of the intermediate and long diameter on one axis against the short versus intermediate diameters on the other axis.

By superimposing Wadell's sphericity values on the Zingg diagram it is apparent that different shapes may yield the same sphericity values. The importance of the Zingg classification is in its application to sediment transport. Although a disk and a roller could possess the same sphericity, their shapes are different and this would be

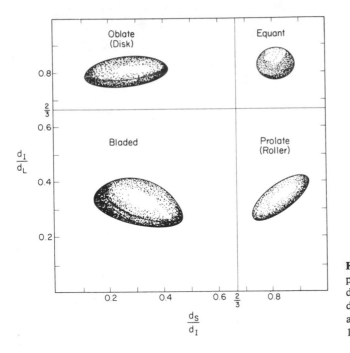

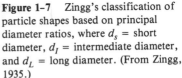

Figure 1-7 Zingg's classification of particle shapes based on principal diameter ratios, where d_s = short diameter, d_I = intermediate diameter, and d_L = long diameter. (From Zingg, 1935.)

reflected in their rate and mode of transport. Shape could also be a factor in the orientation of a particle when it comes to rest.

Surface Texture

The surface texture of sand and gravel size particles may be affected by physical and chemical phenomena, both during transport and in situ during diagenesis. The major cause of surface markings on particles is the impact of one particle with another. Considerable recent research has been undertaken to try to establish characteristic surficial markings as indicators of particular environments of deposition. If successful, such criteria would be quite valuable as a tool for the stratigrapher or sedimentologist working with ancient terrigenous sedimentary deposits.

Pebbles, cobbles, and boulders commonly exhibit relatively large surficial markings which can generally be related to sediment transport phenomena. Linear scratches, grooves, or striations are especially common on gravel particles from glacial drift. These features result from the particle moving over bedrock or in relation to an adjacent particle. Crescent-shaped cracks, sometimes called chatter marks or percussion marks (Klein, 1963b; Campbell, 1963), are also found on large sediment particles (Figure 1-8). They result from sudden and intense impact between particles such as might occur in a rapidly flowing stream or in the surf zone.

The surface textural elements described above are visible to the unaided eye. There is a wide variety of surficial features present on sand-size particles that must be studied with the light microscope or with the scanning electron microscope (SEM). The relatively recent application of the SEM to such studies has revealed surface tex-

Part One   Principles

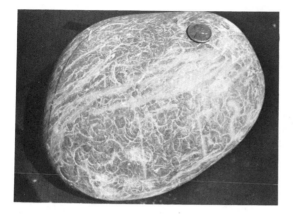

Figure 1-8 Photograph showing crescentic chatter marks on a sediment particle. Scale is a penny. (From Lindsay, 1972, Figure A.)

tures heretofore unknown (Krinsley and Doornkamp, 1973). The frosted surface that can be seen with a hand lens or light microscope becomes a delicate pattern of surface markings when viewed with the SEM (Figure 1-9). Many of these patterns and markings have been related to depositional environments, although there is no widespread agreement as to the significance of particular types of markings.

When viewing sediment particles from the rock record, another factor must be considered regarding surface textures: the diagenetic phenomena, which might create surface markings through both physical and chemical means. Great pressures may cause grains to display crescentic or concoidal fracture patterns. Chemical reaction between percolating groundwater or other fluids is the most important diagenetic phenomenon in producing surface textures.

As a result of the movement of sediment particles from one depositional environ-

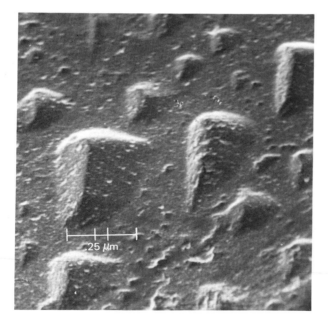

25 μm

Figure 1-9 Scanning electron micrograph showing delicate surface structures on sand-size quartz grains. (Courtesy of P. Scholle.)

ment to another, coupled with the markings produced diagenetically, it is very difficult or even impossible to ascertain an environment of deposition by examination of individual particles from the rock record.

Mass Properties

Sediment accumulations display characteristics of the entire deposit which may or may not be related to the size and shape of the constituent particles. The mass properties of significance to this discussion are fabric, porosity, and permeability.

Fabric

The arrangement of sedimentary particles in space is the **fabric** of a deposit. The shape of the individual sediment particles is the most important factor in determining the fabric. For example, it is common for nonspherical particles such as disks or rollers to be arranged preferentially with their long axes parallel. On the other hand, spheres are equidimensional and therefore do not display any particular spatial orientation.

Fabric results from despositional processes, some of which give rise to preferred orientation of sediment particles and some of which yield completely random particle orientations. All sediment particles that have one or two of their major axes elongate with respect to the others are likely to exhibit preferred orientation. Fluid transport of sediment may cause long axes to be aligned in prolate particles (rollers). Examples include many biogenic remains, especially high-spired gastropods and elongate mineral

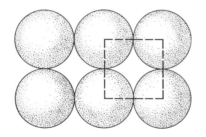

(a) Cubic packing

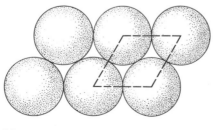

(b) Rhombic packing

Figure 1-10 Extremes of packing showing (a) least stable (cubic) and (b) most stable (rhombic) types. In nature, sediments typically fall somewhere between but toward the rhombic end of the spectrum.

Part One   Principles

grains such as rutile, zircon, or tourmaline. Disk- or blade-shaped particles may also show preferred orientation.

One aspect of fabric that is not necessarily shape dependent is **packing**, the spatial density of particles in a sediment accumulation. Consider a sediment composed entirely of sorted spherical particles. By arranging the particles in rows such that each particle rests directly above and below other particles, we have the least stable situation, commonly called loose or cubic packing [Figure 1–10(a)]. The most stable arrangement would be to nest an overlying particle between the two particles below, giving tight or rhombic packing [Figure 1–10(b)]. In nature most accumulations exhibit conditions that fall between the extremes, but usually closer to rhombic packing than to cubic.

The examples used above illustrate a grain-supported or particle-supported fabric due to the absence of fine-grained particles which might keep the particles separated [Figure 1–11(a)]. Many sediment accumulations have a combination of mud and coarser material, causing the coarse particles to be spatially separated [Figure 1–11(b)]. This fabric is called mud supported, with the coarse particles essentially floating like nuts in a pudding.

Porosity

Figure 1–10, which illustrates the extremes of packing, also serves as a good example of the range in **porosity** or open space in a sediment or rock. Porosity is defined as the percent of void space in the total volume. There is a great range in porosity among natural sediments and rocks. Important variables include sorting and the amount of matrix or cement. Porosity of the spherical, well-sorted particles shown in cubic packing (Figure 1–10) is 47.6%, whereas for rhombic packing it is reduced to 26% (Blatt et al., 1980).

A sediment comprised of clay mineral particles may have as much as 90% porosity, due to the way these flake-shaped particles accumulate in a "house-of-cards" arrangement; this is an example of very open packing. Compaction can greatly reduce this porosity by creating a preferred orientation for the flake-shaped particles.

Poorly sorted sediments that combine mud with sand or gravel display low porosity values, for obvious reasons. The pore spaces from Figure 1–10 are now occupied by fine particles.

Permeability

The size of the pore spaces and the degree to which they are interconnected represent two important factors in controlling **permeability,** the ability of a material to transmit fluids. The viscosity and pressure gradient on the fluid are also of significance.

It is rather easy to see how permeability is related to packing, porosity, and to grain size. For example, open packing produces increased porosity and therefore greater permeability than does closed packing. In general, coarse sediments are more permeable than fine sediments. Although clays may contain up to 90% porosity, they are quite impermeable because of the small size of the pore spaces and the general lack of connection between them. Sand and gravel, on the other hand, display high permeability values.

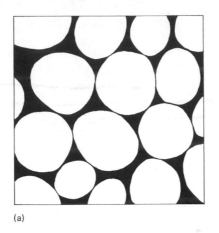

(a)

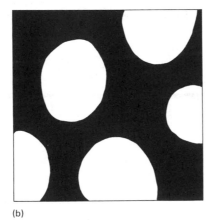

(b)

Figure 1-11   Contrast between (a) grain-supported and (b) mud-supported fabric. Remember that the flat or two-dimensional view may not show all grains in contact, but they do touch in the three-dimensional framework.

Permeability of a sediment may change drastically through time, not only due to compaction and the resulting loss of pore space but also due to precipitation of cements that occupy the pore space. This topic is discussed further later in the chapter.

Textural maturity

The concept of textural maturity was developed by Folk (1951) to integrate various textural properties of sediments. This combination of properties provides a sequence of stages that is the result of energy input during deposition. It is a valuable tool in the interpretation of the physical nature of the environment in which the sediment accumulated.

There are four stages in Folk's textural maturity classification: immature, submature, mature, and supermature (Table 1–4). Immature sediments contain a significant amount of fines, are poorly sorted, and are angular. These represent deposition in an environment where there is little or no mechanical energy to separate the particles. Submature sediments lack significant mud but are poorly sorted and angular. Mature sediments indicate winnowing of grains and display well-sorted but angular particles.

TABLE 1-4 TEXTURAL MATURITY SCHEME OF FOLK

Type	Description
Immature	> 5% clay, poor sorting, angular grains
Submature	Little or no clay, poor sorting, angular grains
Mature	Good sorting, angular grains
Supermature	Good sorting, well-rounded grains

SOURCE: Folk, 1951, p. 128.

Supermature sediments arc both well sorted and rounded. They indicate a high level of rigor and sediment transport. Textural maturity is a guide to depositional conditions; therefore, diagenetic phenomena are ignored (Folk, 1974).

Some exceptions to this scheme occur and are labeled as **textural inversions.** These are the result of mixing sediment populations from two environments or sources. One example is bimodal grain sizes, a mixture of well-rounded sand-size particles with mud or a mixture of rounded and angular particles. Such sediments should be considered in the lowest appropriate maturity level. Textural inversions are valuable in interpreting depositional environments. A number of examples is discussed in succeeding chapters.

MINERALOGY

Although the crust of the earth is composed of a wide variety of minerals, relatively few are present in great abundance. Those that are include the feldspars, iron-magnesium silicates such as pyroxenes and hornblendes, and quartz, with feldspars being most abundant. Sedimentary rocks are present as only a relatively thin skin near the surface of this crust. Although a large number of minerals is present in these rocks, they contain primarily four mineral types: feldspars, quartz, clay minerals, and carbonates, feldspars being the least abundant of the four.

This difference in abundance between the crust as a whole (mostly igneous) and sedimentary rocks is largely a reflection of the relative chemical stability of the minerals involved. Feldspars are comparatively unstable, especially the plagioclases, and they weather to produce clay minerals. Quartz is quite stable chemically and is physically very durable. Thus, during the cycle of weathering and transport, quartz tends to persist. Carbonate minerals are dominantly the result of physicochemical or biochemical precipitation from solution at or near the surface of the earth.

The result is that sediments and sedimentary rocks are comprised largely of a few minerals which are the relatively stable products of erosion. Those particles that are derived from preexisting rocks are called **terrigenous.**

This discussion will consider the mineral composition of sediments and sedimentary rocks. Emphasis is placed on the gross compositional attributes and on field and hand-specimen scale identification. Microscopic analysis and petrography will be mentioned only in cases of extreme diagnostic properties. The reader should consult standard texts on sedimentary petrology for in-depth discussions of this topic (Blatt et al., 1980; Blatt, 1982; Pettijohn et al., 1972; Bathurst, 1975).

Terrigenous Sediments

Four types of particles comprise the bulk of terrigenous sediments: feldspar, quartz, clay minerals, and rock fragments. The latter are simply particles derived from previously existing rocks which retain the character and composition of their source. Abundance of terrigenous particles is dependent on their availability, durability, and chemical stability (Folk, 1974). Only those minerals or weathering products that can be produced from a given source rock are available. Soft or easily cleaved minerals are subjected to greater size reduction (calcite) during transport than are hard minerals without cleavage (quartz). Similarly, chemical stability to weathering processes facilitates the incorporation of a mineral into a terrigenous sediment accumulation.

Types of terrigenous sediments

Quartz. The overall dominant mineral in terrigenous sediments is quartz, although there are some sediments which contain little or even no quartz. To the field geologist, virtually all quartz looks alike except for color. Under the petrographic microscope, however, it is apparent that there are many varieties. Some particles are single crystals, some are polycrystalline. The shape and arrangement of crystals as well as their optical properties may provide clues to the nature of the source rock from which the quartz was derived (Folk, 1974), although some authors disagree as to the criteria used (Blatt and Christie, 1963; Blatt, 1967).

Feldspar. Both potassium feldspar and plagioclase feldspars are common in some sediments. Because of its relative chemical instability, plagioclase is less common and is present in large amounts only under specific circumstances of supply and source rock proximity. Potassium feldspar, usually orthoclase or microcline, is abundant in many sediments and easy to recognize in sand-size or coarser particles. Some important environmental and tectonic interpretations are possible through close inspection of the feldspar particles in a sediment or sedimentary rock. For example, fresh feldspar, which is small compared to associated quartz grains, is generally interpreted as resulting from arid conditions, whereas altered feldspar is interpreted to suggest humid conditions. Although detailed examination of these characteristics requires the petrographic microscope, much can be accomplished with the hand lens.

Rock fragments. By far the most diverse of the primary constituents of terrigenous sediments is this category. Three major types of rock fragments exist; metamorphic, volcanic, and sedimentary (Folk, 1974). It is also possible to observe plutonic rock fragment particles, although they are uncommon and usually coarser than sand size. Within the metamorphic category, phyllite and schist particles are most common in sand-size particles, whereas gneissic fragments are generally large and relatively rare in sand but common in gravel. Sedimentary rock fragments occur in a variety of types, with shale, chert, and carbonate fragments most common. Recognition of rock fragment type is based on the same criteria as those used for recognizing their larger source rock types.

Clay minerals. The mud content of terrigenous sediment is dominated by clay minerals, although quartz, feldspar, and other constituents may also be present. The common minerals present are illite, kaolinite, chlorite, and montmorillonite (smectite). Determination of the particular mineral species present may be important in interpretation of its origin; for example, illite is commonly a weathering product of potassium feldspar and smectite is often an alteration product of volcanic material. Such determinations require x-ray analysis.

Accessory minerals. In addition to the aforementioned abundant particle types, it is not uncommon to find small percentages of other minerals included in terrigenous sediments. These frequently include **heavy minerals,** which are those minerals possessing a specific gravity greater than 2.85, and also micas and chert. Some of the heavy minerals, such as zircon, rutile, and tourmaline, are quite stable chemically. Chert is also quite stable both physically and chemically.

Provenance

Probably the most valuable role that terrigenous particles play in the interpretation of ancient geologic conditions is in their indication of source rock composition—their **provenance.** Certain minerals or suites of minerals can provide clues not only to the mineral composition of the source rock but also to its proximity to the site of deposition in which the particles accumulate. Most detailed provenance analyses must be carried out using the petrographic microscope.

The dominant mineral constituents or the accessory minerals present may be critical to establishing the provenance of a sediment. For example, a sediment rich in potassium feldspar undoubtedly was derived from a granitic source. The presence of kyanite or staurolite is indicative of a metamorphic source rock. Detrital chert may be significant in that it is commonly derived from a carbonate source. The carbonates are soft and relatively soluble, so that only the chert can withstand the rigors of weathering and transportation with eventual accumulation in a terrigenous sediment.

Chemical and Miscellaneous Sediments

Sediments that are not derived from the breaking up of previously existing rocks are considered in this section. This includes not only the material that precipitates directly from aqueous solution but also **biogenic** skeletal material, plant material, and volcanic ejecta, called **pyroclastic** material. Of the types of sediments named above, all but some of the minerals that precipitate directly from solution physicochemically are typically accumulated as transported sediment particles. Many evaporite minerals are quite soluble and do not withstand even local transport.

Physicochemical precipitates

This category includes a rather broad spectrum of sediments. Evaporites, non-skeletal carbonates, chert, phosphates, zeolites, and a variety of iron-rich sediments fall within this category. All are important in the rock record and most are valuable as indicators of specific environments of deposition.

Evaporites. These are minerals that precipitate from brines, as evaporation causes concentration of ions in solution. A great variety of evaporite minerals occurs in nature; however, only gypsum ($CaSO_4 \cdot 2H_2O$), anhydrite ($CaSO_4$), and halite (NaCl) are abundant. In the evaporite sequence developed by Usiglio in the nineteenth century, gypsum is the first common evaporite mineral to precipitate. This usually occurs near a salinity of 300‰. Because of the extreme environmental conditions necessary to produce gypsum or anhydrite, they are valuable in interpreting ancient environments of deposition. One problem that must be solved in using such a criterion, however, is that calcium sulfate may form diagenetically in the sediment and would then not be useful in the interpretation of depositional environments. Bedded gypsum or anhydrite is the result of precipitation at the sediment-brine interface, whereas scattered crystals suggest a diagenetic precipitate within the sediment.

Nonskeletal carbonates. Included are lime mud, intraclasts, pellets, and ooids. **Lime mud** or **micrite** is typically composed of aragonite ($CaCO_3$) needles a few micrometers in length and less than 1 μm in thickness (Figure 1-12). It is widespread on shallow banks or platforms in warm waters of the low latitudes such as the Bahama Platform and the Persian Gulf.

Intraclasts are disk-shaped particles that are typically aggregates of lime mud which have been torn up and transported locally, penecontemporaneous with accumulation (Folk, 1959). Some intraclasts may include other particle types incorporated into them. Intraclasts are sand size or larger and usually well rounded, testifying to their semisoft state during transport. The term "intraclast" should not be confused with **lithiclast,** which is reserved for carbonate fragments that were lithified prior to transport and may or may not be angular. These particles are a terrigenous rock fragment type.

Pellets are sand-size aggregates of lime mud which are spherical to ellipsoidal in shape. They are the fecal products of a great variety of mud ingesting organisms, including deposit feeders and filter feeders. The quantity of pellets and their distribution, both in modern environments and the rock record, has probably been underestimated. Superficial examination of lime mud or its rock equivalent (micrite) indicates homogeneous and structureless texture. Close examination commonly enables the observer to recognize the pellet shape and to establish this as the true nature of these sediments or rocks. Compaction commonly destroys their definition, so that under such circumstances the pellet form is obliterated.

Figure 1-12 Scanning electron micrograph of aragonite needles, western Andros Island, Bahamas. (Courtesy of R. P. Steinen.)

Ooids are typically spherical or nearly spherical in shape, are comprised of concentric layers of calcium carbonate around a nucleus, and are generally in the medium to coarse sand-size range. The nucleus may consist of almost any detrital particle, although quartz grains, biogenic fragments, or pellets are the most common. There is some disagreement about the details of the origin of ooids. They are commonly associated with shallow to intertidal environments, where there is at least moderate agitation. This agitation is provided by waves, wave-generated currents, and tidal currents.

Chert. Detrital chert occurs as a stable product of weathering and transport of carbonate rocks and from bedded cherts. There are two primary varieties: microcrystalline and cavity-filling (fibroradiating) chert (Folk and Weaver, 1952). Identification of the two types must be made with the petrographic microscope.

Phosphates. Varieties of calcium phosphate occur in a wide range of forms in several sedimentary environments. Ooids, pellets, and clasts of skeletal particles may be replaced by phosphate minerals and phosphate may precipitate directly as muds or nodules. Deep-sea deposits of phosphate nodules were first located by the Challenger Expedition in the 1870s. Extensive deposits have been mined for agricultural fertilizers from the Permian of the Rocky Mountains (Phosphoria Formation) and also in the Tertiary of central Florida.

Zeolites. These minerals are hydrous aluminosilicates with compositions similar to feldspars. They occur in association with detrital volcanic particles, in saline alkaline lakes, and in the deep ocean environment (Blatt et al., 1980). Zeolites are rare as terrigenous particles. They are probably most useful as paleoenvironmental indicators of lake deposits (Hay, 1966; High and Picard, 1965; Picard and High, 1972) but are rare in pre-Mesozoic rocks.

Iron-rich sediments. Much of the world's iron supply comes from sedimentary rocks. The most common occurrences are as banded, cherty Precambrian deposits (taconite) and as oxides that have replaced ooids. Some sedimentary iron ores are apparently bog or marsh deposits, some are related to volcanic activity, and others are at least partly a replacement product.

Skeletal particles

The vast majority of all skeletal material that contributes to sediments is calcium carbonate in composition. This may be in the form of aragonite, calcite, or high-magnesium calcite (at least 10 mole % $MgCO_3$). In addition, some siliceous, and phosphatic skeletal material is also contributed. Skeletal material may be the whole skeleton, disarticulated pieces, or broken fragments. Many small organisms such as foraminifers, small gastropods, pelecypods, or brachiopods may persist as whole skeletons (Figure 1–13). Echinoderms and trilobites tend to disarticulate into plates, spines, and so on. There are many ways whereby skeletal material is physically broken. It can occur through transport by wave and current motion but also by the activities of other organisms. The burrowing activities of infaunal varieties and feeding of fish or carnivorous invertebrates may cause considerable size reduction in large amounts of skeletal material. An excellent example is the parrotfish, which feeds by scraping and crushing the framework organisms on reefs. The fish utilizes the digestible portion of these organisms but passes great quantities of sediment through its digestive system.

Figure 1-13 Sedimentary particles which are complete shells of invertebrates. Lens cap is 5 cm in diameter.

The result is that much biogenic sand, silt, and clay is produced from a solid skeletal framework (Hoskin, 1963).

Sediment derived from the hard parts of organisms commonly implies an animal origin but it should not. A great variety of algae produce hard calcium carbonate particles which may form a significant contribution to sediment. These are largely in the green algae (Chlorophyta) and the red algae (Rhodophyta). Such taxa as *Halimeda* and *Penicillus* contribute a variety of particle sizes and shapes (Figure 1-14). In fact, the needle-shaped particles from *Penicillus* are nearly identical with the physiocochemically precipitated aragonite needles mentioned previously.

Diatoms, radiolarians, and some sponges are the primary contributors of siliceous skeletal material. These microscopic organisms have delicate and ornate tests (skeletons) and are an important constituent of deep-sea sediments. Some vertebrate skeletal material, composed of calcium phosphate, is also a contributor to sediments, although it is not volumetrically significant. Sharks' teeth are probably among the most common.

Plant debris

Some environments of deposition are characterized by great quantities of plant tissue accumulating as sediment. Climate plays an important role, as do the geomorphology, hydrodynamic characteristics, and the chemistry of the environment. Plant debris may consist of leaves, twigs, pollen, spores, or large tree trunks. In some places, especially deltas, plant debris may be carried in from distant parts of the drainage basin. It is incorporated with mineral particles and prevented from oxidation by rapid burial. Environments where luxuriant growth takes place, such as marshes, may contain sediment accumulations which are dominated by plant material. Here the great supply aids in preservation and prohibits or retards oxidation. In general, well-oxygenated environments where rates of sediment accumulation are low will not be favorable to incorporation of plant material into the sediment; it will be completely oxidized.

Part One   Principles

(a)

(b)

Figure 1-14   Green algae (a) *Halimeda* (b) *Penecillus,* which contribute calcium carbonate particles to the sediment.

Pyroclastic material

Explosive ejecta from volcanic eruption represent a borderline type of material, neither truly igneous nor sedimentary. Although the material originates in the igneous realm, it falls to earth as a solid and is thus sediment as well. The term **tephra** is used for all such volcanically derived material. Much tephra is in the form of glass, which occurs as small irregular particles that are quite angular and delicately structured. Some tephra consists of small crystals, usually plagioclase or iron-magnesium minerals, which can be identified by the petrographic microscope. Tephra is commonly well stratified as the result of its accumulation as an ash fall.

FORMATION OF SEDIMENTARY ROCKS

Once a sediment accumulates at its site of deposition, a great variety of postdepositional changes begins to take place. The term **diagenesis** is used to include all these changes, physical and chemical, up to metamorphism. Some aspects of packing are included and would be early diagenetic processes. The most important diagenetic phenomenon is cementation or lithification, whereby sediments become rocks.

Details of the various processes and characteristics of the lithification processes are the task of the sedimentary petrologist. The reader is referred to excellent discussions in other texts for such details (Blatt et al., 1980; Hatch et al., 1965; Pettijohn et al., 1972; Pettijohn, 1975; Bathurst, 1975). The discussions included here will be aimed primarily at providing a general appreciation of the lithification processes and some of the major postlithification phenomena. The primary purpose is to enable the reader to comprehend significant changes in the characteristics of sedimentary deposits that are the result of diagenesis. Some diagenetic processes alter the textural and compositional nature of these deposits and may therefore be important when interpretations of depositional environment are made. The geologist who can readily recognize and understand such diagenetic processes will be able to separate such phenomena in time from the depositional environment of the sediment body in question.

Siliceous Cementation

Because cement may constitute as much as one-fourth to one-third of the volume of a sedimentary rock, it is an important constituent. In general, silica precipitation in the form of cement takes place rather long after deposition and fairly deep below the surface, at least below the water table. The silica may be provided by percolating groundwater moving through the cement or it may come from local solution and reprecipitation. Most siliceous cement occurs in terrigenous sediments, where it occurs in a variety of forms. It may occur as a mosaic of crystals between grains, as microcrystalline chert, fibrous chert (chalcedony), in a hydrous form (opal), or as **overgrowths** on existing quartz grains. In the latter case there is secondary enlargement of the quartz grain which is optically continuous with the previously existing crystal. Overgrowths and pressure solution cementation are the most common forms of siliceous cementation. Pressure causes solution and reprecipitation along grain contacts and frequently distorts the crystals. Both forms of cementation result in a texture that is diagenetically produced and commonly different from the depositional texture of the sediment.

Carbonate Cementation

Calcium carbonate is generally both abundant and available in the diagenetic environment and occurs as cement in a wide variety of forms and rock types. It is a common cementing agent in both terrigenous and carbonate sediments. Until recently it was thought that carbonate cement was a late diagenetic phenomenon restricted to the subsurface environment. Modern cementation has been recognized in beachrock, reefs, and other carbonate environments.

Most carbonate cement is provided by groundwater passing through terrigenous or fragmental carbonate sediments. The cement may take the form of small needle-shaped crystals perpendicular to the host grain, it may be a relatively coarse mosaic of calcium carbonate, or it may result from pressure solution similar to the situation in siliceous cementing of quartz.

Although most carbonate cement is calcium carbonate, the iron carbonate, siderite, may act as a cement, especially in terrigenous sediments.

Miscellaneous Cements

Iron oxide (hematite) is probably the third most common cementing agent in sedimentary rocks. It is, however, far less abundant than siliceous or carbonate cements. The iron is provided through the weathering of hornblende, biotite, and other iron-rich silicates. Many of the rocks, called **red beds,** have at least some hematite cement.

Clay minerals frequently act as binding agents in terrigenous rocks. Some pressure solution may mobilize silica and permit it to reprecipitate as submicroscopic particles of cement.

CLASSIFICATION OF SEDIMENTARY ROCKS

Classifications in general are primarily for communication purposes. Virtually everything is classified. All classifications have slots into which items are placed depending on how their characteristics fit the framework on which the classification is based. Many of the items that are a product of human activity, such as manufactured goods, are easily placed in one category or another. On occasion it may be difficult to place something in a particular category. For example, consider vehicles; anyone can distinguish a car from a truck. What about the situation where the back of a car has been modified so that the car can be used like a truck. Is the resulting vehicle a car, a truck or perhaps a ''caruck''?

The above example is somewhat ludicrous; however it points out that borderline situations exist in classifications. This is particularly true in nature, where continua may be the rule. Such is the case with rocks in general, and certainly with sedimentary rocks. A classification cannot cover all possibilities nor can it suit all people. When a classification is too comprehensive, it becomes so complicated that it is not used. The point of this introduction is to make the reader aware of the problems that exist with rock classification and to point out that there is not a single right classification for each group of rocks. Basically, one selects a classification that is useful and is known by the people in the field.

In general, rock classifications come in three types: those that are descriptive, those that are genetic, and those combining these attributes. The latter is preferred because it tends to convey the greatest amounts of information; however, it is not always correct. The classifications described below for terrigenous and carbonate sedimentary rocks seem, at least to this author, to have the most to offer the practicing geologist. They are rather uncomplicated and widely known, they combine descriptive and genetic attributes, and nearly all possible sedimentary rocks fit into them.

Terrigenous Rocks

The classification of terrigenous rocks has been subjected to a rather long history of styles, major constituents, and terminology. Most of the classifications commonly found in the literature of the past half-century have been summarized by Klein (1963a) and McBride (1963).

The four primary terrigenous constituents of sediments or rocks (quartz, feldspar, rock fragments, and clay minerals) have been utilized in most of the classification systems developed, although clay minerals (matrix) have been excluded from some. The classification presented here is one developed by R. L. Folk and has evolved from previous classifications by Krynine (1948), Folk (1951, 1954, 1956, 1960), and McBride (1963). It has both descriptive and genetic attributes and is a useful, yet relatively simple classification.

This classification (Folk, 1974) is essentially designed for sandstones, although it is easily adapted for coarser-grained rocks and also for siltstones using the petrographic microscope. There are three major constituents: quartz, feldspar, and rock fragments. A rock composed entirely of quartz would fall on the point designated for that constituent on the ternary diagram, one equally divided between the three major constituents would be in the center, and so forth. These sand-size particles serve as the framework grains; clay minerals are present as matrix and are included in the textural part of the rock name. Those particles, in addition to the three end-member types, are added to the major rock name as modifiers.

The triangular diagram with the three end members is subdivided into seven categories, with quartz arenite, litharenite, and arkose as the rock names for each of the three poles (Figure 1–15). Appropriate modifiers are used for those rocks that fall

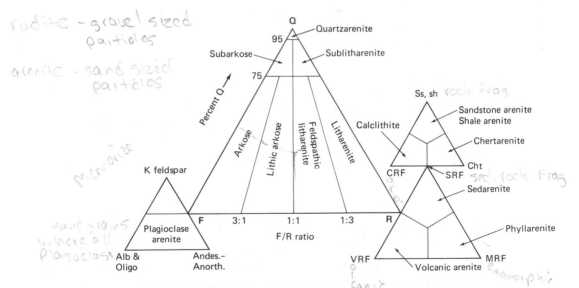

Figure 1–15   Sandstone classification of R. L. Folk. Note that triangles are not to scale and percentage subdivisions in the central triangle are out of proportion. (From Folk, 1974, p. 127.)

within other compositional categories; for example, lithic arkose indicates less than 75% quartz and a feldspar/rock fragment ratio between 3:1 and 1:1. Obviously, there are various feldspars or rock fragments that may be present (Figure 1–15). When either pole is dominated by a particular particle type it should be added to the name: for example, plagioclase arenite (plagioclase arkose), phyllarenite, volcanic arenite. In any rock where a modifier helps to define the major rock type more closely, it should be used.

The terminology described above serves as the fundamental basis for the terrigenous rock classification. To define the rock more precisely, various modifiers and additions to this terminology are necessary to complete the name. They include a grain size designation (e.g., medium sandstone), textural maturity (e.g., mature), accessory minerals (e.g., micaceous), and cement (e.g., calcite cemented). Thus there are five parts to the complete rock name (Folk, 1974). In some situations it may not be possible or practical to include all of these, especially in the field.

After becoming familiar with the classification, it is possible to recognize how source rock type and tectonics are related to the scheme. This gives the genetic aspect to the classification.

Mudstones

The most common sedimentary rocks are the fine-grained terrigenous mudstones, which include shales, claystones, and siltstones. This group of rocks has received little attention from sedimentologists partly because of its apparent homogeneity but probably more so because of the difficulty in conducting detailed petrologic studies, especially using thin sections. A recent book by Potter et al. (1980) has provided an excellent summary of the present knowledge of these fine-grained rocks.

The mineralogy of mudstones is dominated by clay minerals with abundant quartz. Other minerals are minor constituents overall but may be abundant locally. The fine clay- and silt-size particles that comprise mudstones are derived from chemical weathering of various source rock types and from physical abrasion of other rocks. Because it is virtually impossible to determine mineralogy of mudstones in the field or in hand specimens, any classification of these rocks must be based on general texture, stratification character, and induration. Potter et al. (1980) have devised such a classification (Table 1–5) that can be used with additional modifiers to name the rock in more detail. For example, known minerals, color, bedding style, and so on, can be added to the general rock names.

Carbonate Rocks

The classification of carbonate rocks does not have a long history. Until Folk's (1959) classification was published, there really was no comprehensive organization of carbonate rocks. Terms such as "coarse-grained, fossiliferous limestone" or "fine-grained limestone" were generally applied throughout the literature. In many respects the modern study of carbonates was born with Folk's landmark paper. Since that time numerous classifications have appeared, many of which can be found in references by Ham (1962), Pray and Murray (1965), Milliman (1974), and Bathurst (1975).

TABLE 1-5 CLASSIFICATION OF MUDSTONES (MORE THAN 50% OF GRAINS LESS THAN 0.062 mm)

Percentage clay-size constituents			0–32	33–65	66–100
	Field adjective		Gritty	Loamy	Fat or slick
Nonindurated	Beds	Greater than 10 mm	Bedded silt	Bedded mud	Bedded claymud
Nonindurated	Laminae	Less than 10 mm	Laminated silt	Laminated mud	Laminated claymud
Indurated	Beds	Greater than 10 mm	Bedded siltstone	Mudstone	Claystone
Indurated	Laminae	Less than 10 mm	Laminated siltstone	Mudshale	Clayshale
Metamorphosed		Degree of metamorphism Low	Quartz argillite	Argillite	
Metamorphosed		↓	Quartz slate	Slate	
Metamorphosed		High	Phyllite and/or mica schist		

SOURCE: Potter et al., 1980.

This discussion will consider two complementary classifications: one based largely on composition (Folk, 1959) and one based on depositional texture (Dunham, 1962). By using the two classifications in combination, it is possible to approximate a parallelism with the terrigenous rock classification described previously.

Dunham's classification

In the classification by Dunham there are two basic textural distinctions: those particles or components not bound together during deposition and those that are. The latter category includes in situ reef framework, algal stromalolites, and other similar deposits. Components that are not bound together include grains and lime mud

CLASSIFICATION ACCORDING TO DEPOSITIONAL TEXTURE

DEPOSITIONAL			TEXTURE	
Original components not bound together during deposition				Original components were bound together during deposition ... as shown by intergrown skeletal matter, lamination contrary to gravity, or sediment-floored cavities that are roofed over by organic or questionably organic matter and are too large to be interstices.
Contains mud *aragonite* (particles of clay and fine silt size)		Lacks mud and is grain supported		
Mud supported		Grain supported		
Less than 10 per cent grains	More than 10 per cent grains			
<u>Mud</u>stone	<u>Wacke</u>stone	<u>Pack</u>stone	<u>Grain</u>stone	<u>Bound</u>stone

a fosil mud grainstone

Figure 1-16 Carbonate classification based on depositional texture. (From Dunham, 1962, p. 117.)

(micrite). The relative percentages and shapes of these constituents provide for the other four categories in this classification. Grain-supported textures are, as the name implies, those sediments or rocks whose textural fabric is based on the contact between adjacent grains. One category lacks any micrite (grainstone) and the other has mud (packstone) (Figure 1-16). The mud-supported types include wackestones ($>10\%$ grains) and mudstone ($<10\%$ grains).

It should be observed that the nature of the grains involved is important in determining whether or not a grain-supported texture is produced. Certain biogenic particles, such as bivalve shells or branching coral or algae, may comprise only 30 to 40% of the volume but still produce a grain-supported texture. By contrast, spherical grains would have to comprise about two-thirds of the volume to produce a grain-supported texture (Figure 1-17).

Folk's classification

Folk's classification of carbonates (Folk, 1959, 1962) is based primarily on composition, but some textural aspects are also included. There are three types of constituents: micrite and sparite (cement), which are termed **orthochemical** constituents (**orthochem**), and particles or grains, called **allochemical** constituents (**allochem**). The particles include fossils (biogenic grains), ooids, pellets, and intraclasts. By combining

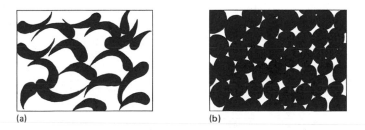

(a)   (b)

Figure 1-17 Two types of grainstones: (a) with little grain bulk; (b) with considerable grain bulk. The difference is due to the size and shape of the grains.

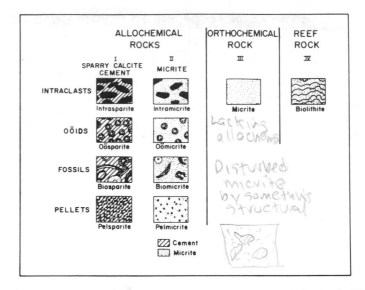

Figure 1-18 Classification of carbonate rocks as developed by R. L. Folk. (Modified from Folk, 1962, p. 71.)

these four particle types with the appropriate orthochemical constituent, eight rock types result (Figure 1-18). The primary allochemical constituent is coupled with the appropriate orthochem to give the rock name (e.g., oosparite, intramicite, or biomicrite).

It is possible to have a rock or sediment which contains only the orthochem micrite (Figure 1-18); however, it would not be possible to have a pure sparite with no grains present. Some rocks may have both micrite and sparite between the grains, in which case the term "poorly washed" is used as a modifier (e.g., poorly washed biosparite). The term "biolithite" is used for rocks where organisms grow in situ and is essentially equivalent to "boundstone" in Dunham's (1962) classification.

It is important to add modifiers to rock names in Folk's classification to give information on the major biogenic constituents or particle size. Examples could be a bryozoan biosparite or an intramicrudite, indicating coarse particles. Although the names and prefix combinations sound weird and hard to manage, the classification becomes easy to use with only a little practice.

The combination of the Dunham and Folk classifications gives a simple, yet comprehensive terminology for carbonate rocks. Such names as grainstone:oosparite, wackestone:pelmicrite, and boundstone:algal biolithite convey considerably more meaning than either of the classifications used separately.

ADDITIONAL READING

BATHURST, R. G. C., 1975. *Carbonate Sediments and Their Diagenesis,* 2nd ed. Elsevier, Amsterdam, 660 p. (Chapters 1 and 2). This is probably the best single book currently available on carbonate sediments. Good discussions and illustrations on the various particle types that comprise carbonate sediments. The book is primarily concerned with modern and Holocene deposits.

BLATT, H., 1982. *Sedimentary Petrology.* W. H. Freeman, San Francisco, 555 p. Excellent treatment of the subject. The best book on petrology of sedimentary rocks currently available.

BLATT, H., MIDDLETON, G. V., AND MURRAY, R. C., 1980. *Origin of Sedimentary Rock,* 2nd ed. Prentice-Hall, Englewood Cliffs, N.J., 782 p. (Chapters 2, 8, 9, and 12–19). In some respects this book is a companion volume to the present book. The suggested chapters expand considerably on the subjects treated in the preceding chapter of this book.

FOLK, R. L., 1974. *Petrology of Sedimentary Rocks.* Hemphills, Austin, Tex., 170 p. Designed initially as a comprehensive syllabus for courses taught by Folk at the University of Texas, this book has evolved over 20 years into a superb reference on sediments and sedimentary rocks. It contains little on carbonates and some illustrations are difficult to interpret, but this book is a must for the serious student of sedimentology.

FRIEDMAN, G. M., AND SANDERS, J. E., 1978. *Principles of Sedimentology.* John Wiley, New York, 792 p. (Chapters 2, 3, 6, and 7). This book is perhaps the most comprehensive single volume on sedimentology available. It contains nearly 2000 references.

HAM, W. E. (ED.), 1962, *Classification of Carbonate Rocks.* Amer. Assoc. Petroleum Geologists, Mem. No. 1, Tulsa, Okla., 279 p. Most of the modern classifications of carbonate rocks are described and discussed in detail. Excellent photomicrographs of carbonates are presented.

PETTIJOHN, F. J., 1975. *Sedimentary Rocks,* 3rd ed. Harper & Row, New York, 628 p. (Chapters 2, 3, 7, and 10). This third edition of Pettijohn's classic text is considerably expanded from earlier editions. Most of the discussion is descriptive.

PETTIJOHN, F. J., POTTER, P. E., AND SIEVER, R., 1972. *Sand and Sandstone.* Springer-Verlag, New York, 628 p. (Chapters 2, 3, 5, 6, and 10). Extensive discussion on petrography of sandstones and their diagenesis is presented. Excellent photomicrographs of terrigenous sedimentary rocks.

2 Sedimentary Processes

Chapter 1 considered the nature of sediment particles which are to be eroded, transported, and deposited in various sedimentary environments. In this chapter the discussion covers the various mechanisms that produce the erosion, transportation, and deposition. It should be observed that the application of physics to sedimentology is a relatively recent aspect of the discipline. In fact, modern sedimentologists owe a great deal to the pioneering efforts of numerous engineers who conducted research on sediment transport by the atmosphere and by water (Bagnold, 1941; Simons and Richardson, 1961, 1966).

During the first decades of this century most research on sediments by geologists was purely descriptive. That is, sediments were analyzed mineralogically and texturally, but no effort was expended toward understanding the nature of how or why sediment was carried from one location to another. When geologists studying sedimentary rocks from the geologic record tried to make interpretations about the environments of deposition of these rocks it became necessary to conduct studies of modern sedimentary environments in order to determine how these rocks originated. Such studies eventually led to the study of sedimentary processes as an important aspect of sedimentology.

Initially, much of the effort toward understanding sedimentary processes was conducted in wind tunnels, flumes, and wave tanks in the laboratory. Although such investigations are still important, the effort has expanded to all modern sedimentary environments not only on the earth but on other celestial bodies as well (e.g., the moon, Mars).

Sedimentary processes can conveniently be placed into three categories: biological, chemical, and physical. The biological processes involve various interactions between organisms and sediments. Chemical processes include diagenetic phenomena such as dissolution and the precipitation of minerals. These chemical processes are not critical to the primary objectives of this book and are not discussed. Most sedimentary processes are physical, such as sediment transport by waves, currents, wind, or by gravity alone. The following discussion will emphasize this aspect of sedimentation. The resulting sedimentary structures commonly generated by both biological and physical processes are discussed in Chapter 3.

BIOLOGICAL PROCESSES

This topic is quite broad. Organisms may be involved in such diverse activities as the production of sediment particles, both directly and indirectly, or they may cause changes in the arrangement or size of particles through physical activity. Only those biological processes that are evident in the sediment record and likely to be preserved in the rock record will be treated in this discussion. Such processes as the precipitation of skeletal material and chemical changes in the environment caused by organisms but which lead to precipitation of mineral particles, are excluded.

Organisms may cause both erosion and accumulation of unconsolidated sediment and may directly erode rock. Volumetrically, the most significant biological processes and the ones that may be preserved in the geologic record, are those which involve the rearrangement of sediment particles.

Degradation

As discussed in Chapter 1, a significant amount of modern marine sediment is **biogenic** in nature; that is, it is derived from organisms, typically as skeletal material. This biogenic debris ranges widely in particle size, shape, and origin. Some is generated by physical processes such as breakage during transport by currents, waves, and so on, and some is provided during the natural breakdown of tissues after an organism dies. There is, however, a great deal of degradation of skeletal material caused by the activities of other organisms. These include predators that consume organisms and those that break down biogenic particles through activities such as boring, burrowing, and searching for food.

Degradation caused by consumption

Both herbivorous and carnivorous organisms consume other organisms which contain skeletal components. Although some biochemical reactions may result in degradation, most of the breakdown of skeletal material is due to the crushing or scraping caused by the feeding and ingestion of predators.

Both vertebrate and invertebrate organisms are involved in this process of "making little ones out of big ones." An excellent example of a voracious predator is the parrotfish. These rather large fish break off pieces of coral with their heavy jaws and crush the skeletal material (Figure 2–1). Soft tissues present are digested and the skeletal fragments are excreted. The net effect of this activity on a reef is considerable, with great quantities of biogenic sediment of various particle sizes being produced (Hoskin, 1963).

Invertebrate organisms that ingest sediment include sea urchins, holothurians, and some gastropods. Considerable size reduction of particles may take place from ingestion to excretion. At least one researcher (Cloud, 1959) has suggested that chemical effects may also be important in this process. It is his opinion that some size reduction should take place with elimination of the smallest particles, resulting in coarse skewed sediments.

Degradation caused by burrowing

Many benthic organisms, especially invertebrates, are burrowers that cause degradation of sediment particles. These organisms burrow into unconsolidated substrate for shelter and/or for feeding. During these activities the more delicate particles are broken and the more durable ones may be abraded, causing size reduction and increased roundness.

A widespread type of burrowing is that caused by organisms which plow through the sediment in search of food. Others bury themselves for protection or create small excavations on the surface in their search for food. In all these cases the most common degradational process is the breaking of fragile biogenic particles. There is typically no rounding of the particles involved, and they are characterized by sharp breaks. The common end product is a fine sand or mud which contains freshly broken biogenic particles. Such a sediment is characteristic of bioturbation by organisms.

Degradation caused by boring

Erosion of a hard substrate by boring organisms is a widespread process throughout the marine environment. A great variety of organisms is involved in **bioerosion** (Neumann, 1966), ranging from bacteria to mollusks and echinoderms, and including a few plants. Some bore into shells of other organisms and some into rocky substrates. There are also some borers that prefer wood. Most borers carry on such activities to seek shelter (Warme, 1975), but a few do so to obtain food.

Exactly how many of the borers accomplish their excavation is not known. Many types apparently use purely mechanical means, whereas some combine mechanical boring with chemical secretions to aid in the process. A few organisms are suspected of using purely chemical means, but this has not yet been satisfactorily demonstrated (Warme, 1975).

Although some organisms are not true borers, they do cause bioerosion of hard substrates in their search for food. These include organisms that scrape and rasp the substrate, such as some echinoids, amphineurans, and gastropods that graze over rocks.

In general, borers do not cause a volumetrically significant amount of new sediment to be introduced into the environment. Borers and their borings are quite im-

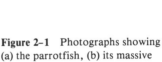

(a)

(b)

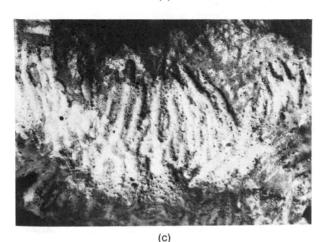

(c)

Figure 2–1 Photographs showing
(a) the parrotfish, (b) its massive
jaws which crush coral, and
(c) the scraping marks on coral.
(From Cloud, 1959, Plate 130.)

portant in stratigraphic analysis and reconstruction of ancient environments. These aspects are discussed in succeeding chapters. The interested reader is referred to an excellent article by Warme (1975) which summarizes borings and bioerosion.

Pelletization

Many organisms that ingest sediment in their feeding process pack it together in the form of pellets which are then excreted. They include filter feeders, deposit feeders, and some of those organisms mentioned above that break up material upon ingestion. The shapes of pellets vary greatly and range from less than a millimeter to more than a centimeter in diameter. Large pellets are composed of rather coarse and noncohesive material, resulting in rapid disaggregation upon excretion. There are, however, numerous organisms that excrete pellets of cohesive clay- and silt-size particles that may maintain their pelletized form.

Organisms such as holothurians (sea cucumbers) and some gastropods secrete mucus-bound pellets which disaggregate rapidly and are not important as sediment aggregates (Hoskin, 1963). Many of the organisms that ingest mud-size particles during feeding, excrete sand-size pellets, which then act as sand-size particles during physical processes of sedimentation. Included in this group are many varieties of worms, mollusks, burrowing shrimp, and some echinoids. The pellets produced are typically high in organic matter. The size and shape of the pellets varies with the taxon that produces them.

It is common for pellets produced in carbonate environments to be preserved in the rock record, due largely to relatively rapid cementation. In terrigenous muds, compaction generally proceeds more rapidly than cementation, and pellets are commonly destroyed. By contrast, sand-size terrigenous rocks may contain well-defined pellets because the pellets are protected from compaction by the sand-size particles. **Glauconite** pellets, such as those in Cambrian sandstones in the Upper Mississippi Valley, are thought to be biogenic (Berg, 1954).

Organism-Enhanced Sedimentation

The preceding sections have considered biological processes that were the result of direct modification brought about by the activity of the organisms in obtaining food or shelter. Another important biological process takes place when organisms modify the physical processes and cause sediment to accumulate. This occurs by two distinct methods: baffling and trapping of sediment.

Sediment baffles

A variety of sessile benthonic organisms causes modification to wave and current processes that results in decreased energy at or near the substrate, creating a baffle effect. This allows sediment particles in transport to come to rest or to be trapped by these organisms. The best examples of sediment baffles are dense stands of marine grass such as turtle grass (*Thalassia*) and exposed marsh grass (*Spartina*) (Figure 2–2). These plants form a carpet over the substrate and may physically trap particles or slow currents to the degree that their competence decreases and entrained sediment particles come to rest. In shallow marine areas turtle grass causes sediment to accumulate relatively rapidly in contrast to adjacent unvegetated areas. Similar but less efficient baffling effects are produced by digitate or branching sessile invertebrates such as corals, sea fans, and sea whips. These organisms cause sediment accumulations on reef environments. In addition to the enhancement of sediment accumulation, sediment is prevented from being carried away by being trapped in the roots and branches of the benthic organisms.

The various grasses in the salt marsh act in a similar but perhaps more efficient manner. When marshes are flooded with sediment-laden water, the fine suspended particles are trapped and the currents are diminished, allowing particles to settle into the dense network of the grass. Even under rather intense wave and current conditions, these fine particles are prevented from being removed by this efficient baffling and trapping mechanism (Frey and Basan, 1978).

Sediment trappers

Although some aspects of the baffling effects described above may be considered as trapping, there are certain organisms that contain a sticky mucilaginous film on their surface which facilitates direct trapping. Filamentous blue-green algae

(a)

(b)

Figure 2-2 Photographs of dense growths of (a) subtidal turtle grass and (b) intertidal marsh grass, which act as sediment baffles.

trap sediment particles on this coating like flies on flypaper. As currents carry sediment over the matty surface of these algae, particles stick to the surface in the form of a thin layer. The blue-green algae then grow up through the sediment layer and produce another sticky surface, upon which sediment particles accumulate. The result is a sequence of thin layers of bioaccumulated sediment particles. Such accumulations of sediment brought about by blue-green algae are called **stromatolites.** In the rock record these features straddle the boundary between fossils and sedimentary structures; actually, they are both.

Stromatolites may form in carbonate or terrigenous environments. Because of cementation rates, their chances for preservation in the rock record are high in carbonate environments and almost nonexistent in terrigenous environments. They can develop as flat layers (algal mats) or as cabbage-head shaped forms. Cabbage- and columnar-shaped varieties are typically associated with tidal current environments such as in Shark Bay, Western Australia.

PHYSICAL PROCESSES

Virtually all sediment particles are subject to physical processes of transport at one or more times during the period between their formation and their eventual incorporation into sedimentary rocks. To obtain a general understanding of the various interactions between these solid particles and their transporting media (typically fluids), it is necessary to consider some of the basic principles of physics as they apply to sedimentation. The treatment of these topics in this discussion will utilize only simple algebra. More details of the theory involved should be obtained from the several excellent references available on the subject (Graf, 1971; Yalin, 1977; Simons and Senturk, 1977). Excellent but less mathematical treatments of the subject designed specifically for geologists are found in Allen (1970a) and Middleton and Southard (1977), from which much of the following is summarized.

Fluid Flow

The motion of fluids, either water or air, provides the principal mechanisms for transport of sediment particles. In nature, sediment particles are typically subjected to force per unit area (pressure), which is normal to the surface, and a stress (shear stress) component, which is parallel to the surface. One must always consider the force of gravity as well. Properties of the fluid itself include its density and viscosity, both of which are important in fluid motion. Viscosity is essentially the ability of a fluid to resist shear or flow (Allen, 1970a).

In moving fluids there is some relative motion within the fluid as well as between the fluid and sediment particles which are being transported or passed over by the fluid. This is the **shear stress,** which is the shear force per unit area at some point in the fluid. In an open channel, flow is **uniform** if the average depth is unchanged, and **steady** if flow conditions remain unchanged. If steady uniform flow of a fluid over a plane surface is considered, it is possible to relate shear stress within the fluid to the velocity of flow. Shear stress (τ) attains its maximum value at the base of the fluid column and decreases to zero at the free surface, whereas the velocity of the fluid (u) is highest at the free surface and decreases to zero at the fluid–solid interface under laminar conditions (Figure 2–3).

Fluids move in two basic fashions: **laminar flow,** in which the fluid molecules move in linear paths, and **turbulent flow,** characterized by irregular and sinuous motion (Figure 2–4). Dimensionless terms are used in modeling or characterizing fluid flow. These terms are typically utilized in scaling the action of fluids to gain an understanding of natural processes in comparison to scaled-down laboratory models. One such dimensionless term is the **Reynolds number,** named after the designer of the experiment shown in Figure 2–4. The Reynolds number, Re, is the combination of

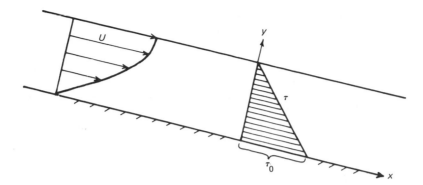

Figure 2-3 Distribution of velocity (u) and shear stress (τ) in steady uniform laminar flow down an inclined plane. (From Middleton and Southard, 1977, p. 1.7.)

the velocity, density, and viscosity of the fluid and some reference length measurement. It is expressed as

$$\text{Re} = \frac{UL\varrho}{\mu} \tag{2-1}$$

where U is the velocity, L the reference length, ϱ the density, and μ the viscosity. It is possible to determine the nature of flow (laminar or turbulent) by calculating the Reynolds number for a given set of conditions. For example, in a pipe where L is the diameter, an Re of >2000 denotes turbulent flow, whereas <2000 represents laminar conditions. A towed or settling sphere with Re < 1 is under laminar flow. Separation of flow begins at an Re of about 24 and causes eddies in the wake.

The second important dimensionless coefficient is the **Froude number (Fr)**, which considers the ratio between inertial and gravity forces. It is expressed as

$$\text{Fr} = \frac{U}{\sqrt{gD}} \tag{2-2}$$

with U the velocity, g the gravitational acceleration, and D the depth of the water. Froude numbers of <1.0 are called tranquil or subcritical and >1.0 are considered

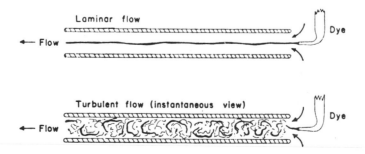

Figure 2-4 Reynold's experiment showing laminar and turbulent flow as dye is passed through a tube. (From Allen, 1970a, p. 33.)

supercritical. Application of Froude numbers has become widespread in sedimentology and is discussed later in the chapter.

Boundary layers

When fluids flow there is typically a velocity gradient normal to the surface over which flow takes place and a shear stress is created where this gradient exists. This zone of shear stress is the **boundary layer** of the flow. A typical example of describing boundary layer conditions is to place a thin, flat plate in the flow (Allen, 1970a; Middleton and Southard, 1977). The boundary layer thickens downstream from the leading edge of the plate and takes on two characteristics. It is laminar near the leading edge of the plate but abruptly becomes turbulent and thickens significantly (Figure 2–5). Turbulence is achieved when the distance x_{crit} is such that inertial forces on the boundary layer are large relative to viscous ones.

Flow separation

When the solid boundary or surface over which fluid flow is moving changes shape abruptly, the boundary layer separates from this surface, creating flow separation. The separation occurs tangentially with respect to the boundary surface and at, or slightly downstream from, the beginning of the expansion (Middleton and Southard, 1977). It is possible to locate the point of separation of the streamline and the point of attachment in this system (Figure 2–6). Note also that a slight backflow is created downstream from the flow separation. This phenomenon occurs over the crests of bedforms and is an important concept in sedimentation.

Settling of Particles

The settling of sediment particles through a fluid column is an important aspect of sedimentation. When mineral particles greater than colloidal size ($> 1\mu$m) are released in a still fluid, they accelerate due to gravitational forces until the fluid resistance force exerted is equal and opposite to the force of gravity on the grain.

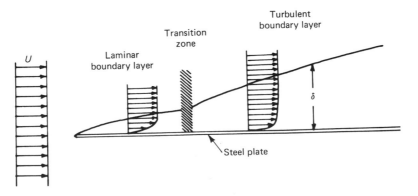

Figure 2–5 Development and structure of a boundary layer under both laminar and turbulent conditions, where U is velocity and δ is the boundary layer thickness. (From Allen, 1970a, p. 37.)

Part One   Principles

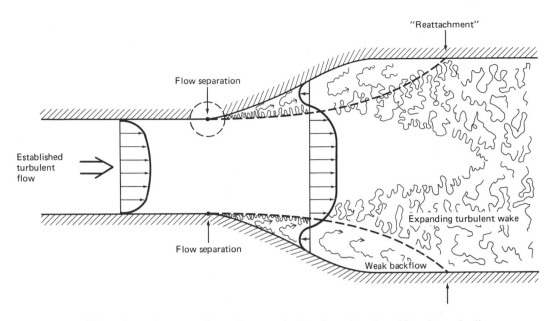

Figure 2-6 Flow separation as it occurs in flow through a pipe with a change in diameter. (From Middleton and Southard, 1977, p. 3.24.)

After that point the fall velocity is constant (Middleton and Southard, 1977). **Stokes' law** of settling relates the variables involved and states that

$$Ws = \left[\frac{(Ps - P)g}{18\mu} \right] d^2 \qquad (2\text{-}3)$$

where Ws is the settling velocity, d the particle diameter, $P_S - P$ the density difference between the particle and the field, and μ the viscosity of the fluid. It should be noted that there are assumptions and factors that limit the application of Stokes' law. In the strict sense, Stokes' law is valid only for a single particle; concentrations of sediment particles tend to retard settling. Figure 2–7 shows the relationship between Ws and d.

The effect of shape on particles moving through a fluid can be pronounced. Factors of shape variation complicate the settling of nonspherical particles because (1) resistance to settling varies with orientation of the particle, (2) during settling the sequence of orientations of the particle changes because of inertial effects, and (3) the settling path of the particle is neither straight nor vertical. To compensate somewhat for these situations, sedimentologists utilize the concept of **equivalent diameter.** This states that for a grain of given size, shape, and density that is settling in a given fluid, there is a sphere of the same density whose average vertical settling velocity is the same as that grain. For example, for its size, a pelecypod shell fragment settles rather slowly, due to its shape, which is rather flat or tabular. There is a sphere of the same density that settles at the same rate as the shell fragment. Its diameter is the equivalent diameter for the shell fragment.

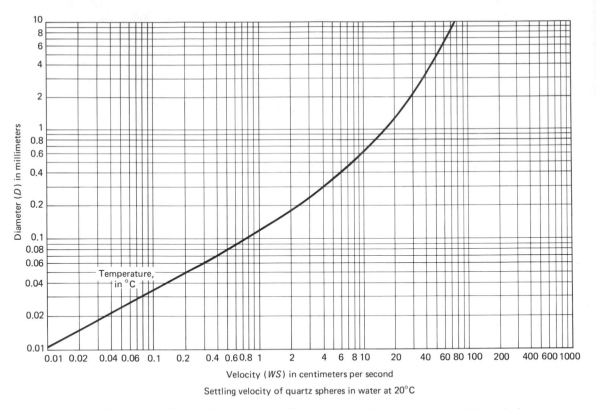

Figure 2-7 Plot of Stoke's relationship between grain diameter and settling velocity. (Modified from Rouse, 1937.)

Sediment Entrainment

As a fluid flows over a bed of sediment with gradually increasing velocity, there is a stage at which grains start to move on the bed. The velocity at which movement begins is the **threshold velocity** for that particle size and density in that fluid. The critical shear stress at which this motion first occurs varies not only with the size, shape, and density of the sediment particles, but also depends on whether particles are cohesive or cohesionless. Quartz sand, for example, is cohesionless, whereas muds are typically quite cohesive. A German engineer named Shields has developed a predictable relationship to determine the competence of flow. Using his dimensionless expression, it is possible to predict the initial movement of grains of various sizes.

Probably the most commonly cited relationship between flow velocity and grain size is that developed by Hjulstrom and later modified by Sundborg (1956). Although it does not predict the relationship between velocity of flow and grain size as accurately as the Shields relationship, it is a useful diagram (Figure 2-8). It should be observed that in the clay- and silt-size region, neither relationship appears to be very accurate. One point of interest on Hjulstrom's diagram is the difference in the

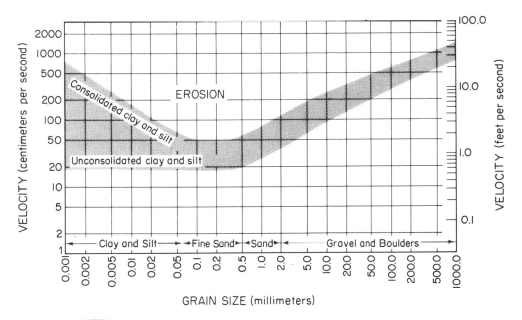

Figure 2-8 Hjulstrom's diagram showing the relationship between velocity, grain size, and particle movement, as modified by Sundborg (1956).

threshold velocity between cohesive and cohesionless sediments. In general, it requires a greater velocity to initiate motion in a cohesive mud than in a fine sand (Figure 2–8).

Grain movement

The motion of grains is caused primarily by two mechanisms: the fluid drag on grains in contact with the moving fluid and the hydrodynamic lift force. Considerable effort has been directed toward the understanding of sediment particle motion in the laboratory using flumes and wind tunnels. Among the more prominent efforts are those of Bagnold (1941), who conducted pioneering work on sand transport in the atmosphere.

Sediment particles tend to move in a variety of ways due to fluid transport, and this movement is neither continuous nor uniform. Grains may slide or roll over the substrate, movement to which the term **traction** has been applied. Grains that hop or bounce along the substrate are said to be in **saltation.** Both types of grain movement comprise the **bed load** of the fluid. Grains that move significant distances without contact with the bottom and in the main flow of the current comprise the **suspended load** or **wash load.** There is no distinct boundary between these types of sediment movement and commonly varying conditions cause grains to change from bed load to suspended load, or vice versa. Transport of sediment grains occurs during turbulent fluid motion.

Grain movement by fluid flow is markedly affected by shape and differs from effects of shape on settling velocity. The grains orient themselves to present the least resistance to flow or the most stable position against other grains or the hard surface on which they rest. Orientation is not necessarily with the maximum area against the flow; commonly, it is the opposite. Elongate grains (rollers) move and orient themselves differently with respect to the flow than do bladed or tabular grains. The size of the particle is also a factor, although it is not fully understood. Sand-size grains and gravel-size grains with the same shape typically achieve different orientations when subjected to similar flow conditions (Middleton and Southard, 1977).

Hydraulics and Sediment Textures

Much, although not all, of the textural characteristics of a sediment can be attributed to hydraulic sorting. In addition, the inherited characteristics of grains and effects of abrasion must be considered. In nature, conditions in a fluid flow system change through both space and time, causing erosion, deposition, or both. Such changing conditions create the possibility of hydraulic sorting. Among the most prominent causes are changes in the **competence** or **capacity** of a stream, which causes sediment particles to come to rest. It may therefore be expected that a sediment population will show characteristics determined by the hydraulic system in which the sediment accumulated (Middleton and Southard, 1977).

It has been postulated that it is possible to recognize three textural populations (Figure 2–9) in the accumulated sediment, each representing a mode of transport and having essentially a log-normal distribution (Visher, 1969). The traction population represents the coarsest particles and may be absent [Figure 2–9(a), (c), and (d)].

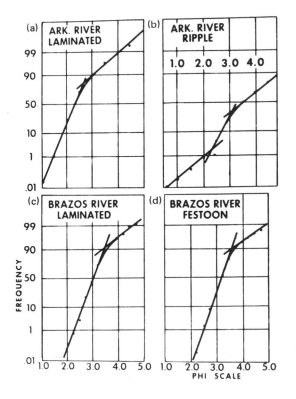

Figure 2–9 Grain size distribution on a logarithmic scale, showing distinct breaks which have been interpreted as representing change from one mode of transport to another. (After Visher, 1965, p. 126.)

Part One   Principles

Generally, about 1% of a sediment sample is necessary to define such a transport mode. The amount of sediment in the traction population is more likely due to unavailability of particles rather than to low fluid velocities (Middleton and Southard, 1977).

The saltation population comprises the bulk of the particles in a sediment sample and is well sorted. The latter is largely due to the efficiency of continued suspension and redeposition as a hydraulic sorting agent while grains bounce along. Visher (1969) has postulated that some samples may show two saltation populations, such as on a beach where swash and backwash act as separate hydraulic agents. The suspension portion of a sediment sample shows considerable variation due to both the intensity of turbulence (Visher, 1969) and the characteristics of source sediments, such as cohesion and flocculation (Middleton and Southard, 1977). Sorting within this population is poor and more than one mode may be present.

Generation of Bedforms

The interactions between flow and bed (sediments) take on a variety of scales, which can be generalized into three levels. The smallest scale is the interaction between the turbulence of flow and the grains themselves; this has been considered in earlier portions of the chapter. Intermediate are the interactions that occur on a scale larger than the grains but smaller than the flow itself (Harms et al., 1975). This gives rise to the various bed configurations or **bedforms** discussed below. The largest scale is that of the flow itself and includes the entire specific environment, such as the channel in which flow occurs.

Transport of grains over a loose bed of sediment particles produces various geometric patterns on the bed surface which are collectively called bedforms. Such features occur in various sizes and shapes, and they may migrate upstream or downstream. This migration is slow compared to the flow velocity and is the result of simultaneous erosion and deposition at various sites on the bed. The development of these bedforms is complicated and not well understood. Numerous empirical data are available, however, and the system of process-response mechanisms can be characterized. It is evident that as bedforms develop, there is considerable influence on flow characteristics, and in turn there is feedback to the bedforms themselves.

Primary emphasis here is put on the relationship between bedforms and flow velocity, a concept that was not synthesized until the work of Simons and Richardson (1961). Since that time numerous laboratory and field studies on bedform development and migration have been conducted by geologists and engineers.

As fluid flow is increased, the shear stress on bed material also increases, and eventually bed load movement is achieved. Almost immediately, small ripples are formed and these migrate in the direction of fluid flow. Under continued increase in flow velocity, the ripples may enlarge, change shape, or both. Eventually, large-scale bedforms called **megaripples** (or dunes) are formed. These may be destroyed at higher flow rates, resulting in **plane beds** and eventually in **antidunes.** This sequence of ripples, megaripples (dunes), plane beds, and antidunes represents the bedforms one would expect to result from the interaction of fluid flow and bed material as flow is increased (Figure 2–10). In fact, the foregoing sequence of bedforms is in response

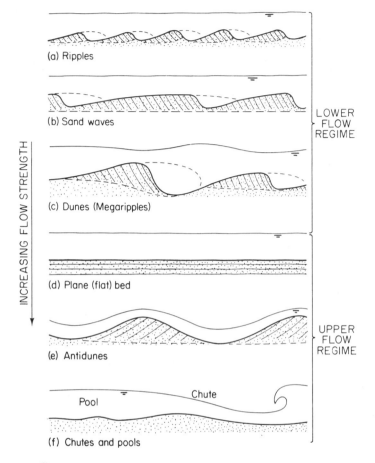

Figure 2-10 Flow-regime sequence of bedforms. (From Simons et al., 1965, p. 36.)

to flow power, not velocity only. Flow power is the product of the velocity (V) and boundary shear stress (τ_0) (Harms et al., 1975).

Flow regime concept

In the scheme of bedforms and related flow strength, Simons and Richardson (1961, 1966) subdivided the system into two **flow regimes.** The flow regime represents the sum of the relationships between fluid flow, bed configurations, dissipation of energy and the phase relationships between the bed surface and the water surface (Friedman and Sanders, 1978). A lower flow regime and an upper flow regime have been designated with a transition stage between. The lower flow regime is characterized by low Froude numbers (< 1.00) and the upper flow regime is the higher-energy portion of the scheme.

In the lower part of the lower flow regime, ripples form and the water surface

TABLE 2-1 CLASSIFICATION OF FLOW REGIME

Flow regime	Bedform	Bed material concentrations (ppm)	Mode of sediment transport	Type of roughness	Phase relation between bed and water surface
Lower regime	Ripples	10–200	Discrete steps	Form roughness predominates	Out of phase
	Ripples on dunes	100–1200			
	Dunes	200–2000			
Transition	Washed out dunes	1000–3000		Variable	
Upper regime	Plane beds	2000–6000	Continuous	Grain roughness predominates	In phase
	Antidunes	2000→			
	Chutes and pools	2000→			

SOURCE: Simons et al., 1965, p. 36.

displays few, if any irregularities. Sediment particles are transported intermittently and the concentration of sediment in transport is low (Table 2–1). Increased flow strength causes the development of ripples on dunes (megaripples), and then dunes as the upper part of the lower flow regime is achieved (Figure 2–10). Sediment is still transported intermittently and the concentration of bed material increases. The water surface shows distinct undulations which are out of phase with the bedforms. The highest portion of the lower flow regime is reached during the transition phase, when the dunes or megaripples are washed out (Simons et al., 1965).

In the upper flow regime the lower portion is characterized by plane beds with a flat and parallel water surface. Sediment concentration is high and transport is continuous, with grain roughness predominating over bed roughness (Figure 2–10). The upper part of the upper flow regime develops antidunes with standing and breaking waves; the water surface is in phase with bedforms. In the case of breaking waves, sediment transport again becomes intermittent (Harms and Fahnestock, 1965). The highest-energy phase causes the development of chutes and pools.

It should be remembered that the relationships described above were first recognized in flume studies utilizing uniform, unidirectional flow and well-sorted, cohesionless sediment. Such conditions do not exist in natural streams; flow strength varies with time and space (across the channel) and there are changes in morphology along the stream bed. Nevertheless, these relationships have been demonstrated as valid not only for streams, but also in estuaries, inlets, within the surf zone, and in virtually all environments where at least a significant unidirectional flow component is present. This concept is without question one of the significant contributions to sedimentology during the past century.

Role of grain size

Using numerous laboratory flume experiments, it is possible to see the general relationships between grain size and bedform using a constant depth and constant flow conditions. Because flat beds sometimes occur at two distinctly different posi-

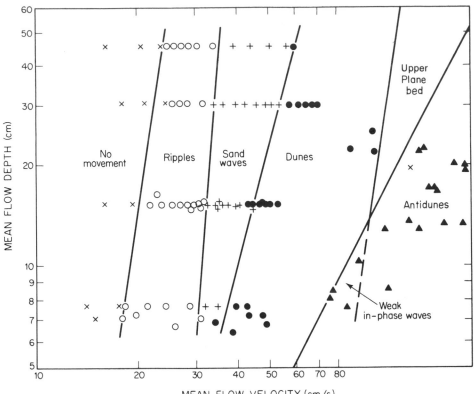

Figure 2–11 Depth-velocity diagram for flume runs, showing successive development of bedforms as flow velocity increases. (After Harms et al., 1975, pp. 18–20; Blatt et al., 1980, Figure 4–5.)

tions in the flow velocity spectrum, the terms **lower flat bed** and **upper flat bed** are used to separate the two types (Harms et al., 1975).

The depth-velocity diagram for flume studies for medium to coarse sand (about 1.0ϕ) shows the successive development of bedform types as velocity increases (Figure 2–11). A similar diagram for very fine sand (3.3ϕ) shows a sequence of no movement, ripples and upper flat beds as velocity increases. Large-scale bedforms are absent, which is consistent with a general absence of large-scale cross-stratification in fine sands (Harms, et al., 1975).

Similar conditions applied to medium sand (1.45 to 2.0ϕ) show that large-scale bedforms are present in the expected position between ripples and upper flat beds (plane beds). By way of contrast, very coarse sand (-0.20 to -0.45ϕ) does not develop ripples but shows a low-flat-bed phase where grain transport takes place in the low-energy portion of the diagram. Increasing velocities then give rise to large-scale bedforms (Figure 2–12).

Simons et al. (1965) have shown the stability fields for various bedforms in a similar relationship (Figure 2–13). Such a diagram is a generalization and was developed from alluvial channels, but it provides a fine tool for use in other environ-

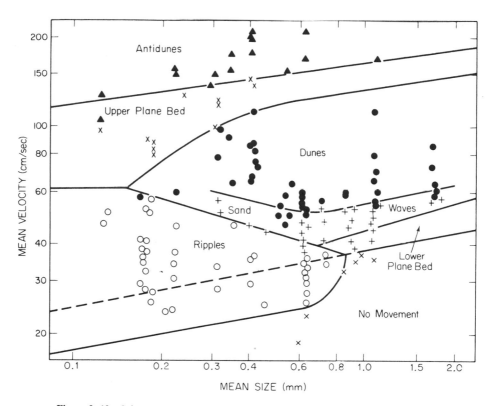

Figure 2-12 Schematic size-velocity diagram for a flow depth of about 40 cm. Note that dunes do not form in fine sediment and ripples do not form in coarse sediment. (After Harms et al., 1975, p. 21; Blatt et al., 1980, Figure 5-46.)

ments or for determining flow conditions for ancient environments now recorded in the rock record.

Sediment Gravity Processes

The title of this section refers to the flow of sediment or sediment-fluid mixtures under the direct action of gravity (Middleton and Hampton, 1973). Although such processes may be subaerial, most geological interest is in subaqueous flow. Sediment gravity flow is distinguished from fluid gravity flow which comprised most of the previous portion of this chapter. In fluid gravity flow (e.g., streams) the fluid moves due to gravity and causes the sediment to be carried along with the flow. In sediment gravity flow the reverse is true; the sediment is moved by gravity and causes the fluid to move with it.

There are five main types of sediment gravity processes (Figure 2-14):

1. **Turbidity currents,** in which fluid turbulence causes sediment to become suspended

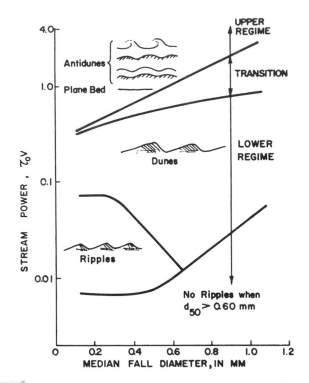

Figure 2–13 Relationships between stream power and median fall diameter to form of bed roughness. (After Simons et al., 1965, p. 52.)

2. **Liquefied sediment flows,** where sediment is supported by upward-flowing fluid as grains settle

3. **Grain flows,** where grain-to-grain interactions support sediment

4. **Debris flows,** where a mixture of fine sediment and fluid supports larger grains (Middleton and Hampton, 1973).

5. **Slump,** where masses of sediment move along shear planes

These flow types are forms of **density currents** which contain a density difference between the mass of the sediment-fluid mixture and the ambient fluid. In these sediment gravity flow types, especially in turbidity currents, it is necessary for some event to initiate the flow and cause sediment to become suspended. Once sediment is suspended, the sediment-fluid mixture flows laterally because it is characterized by a higher density than the surrounding fluid. Dropping a pebble into a shallow pond with a soft mud bottom produces such a phenomenon. The muddy water flows because suspension of sediment is maintained due to turbulence. This condition of a balance between turbulence and suspension is called **autosuspension** (Bagnold, 1962). Because of it, the turbidity current will flow in a steady uniform state until the slope changes or the supply of the suspension is exhausted (Middleton and Southard, 1977).

Turbidity currents

Although the concept of turbidity currents dates back almost a century (Forel, 1885), it has only been in the last three decades that their study has undergone careful scrutiny (Kuenen and Migliorini, 1950; Walker, 1965; Middleton, 1970). Turbidity currents are probably less widespread now than in numerous periods in the geologic past because the wide continental shelves separate the place of sediment influx at the coast from the slope, which is the site of most turbidity current activity.

Turbidity currents are short-lived pulses of sediment-laden water initiated by an event such as an earthquake, volcanic eruption, great and sudden influx of sediment, or any other phenomenon that would disturb a soft muddy bed. Unfortunately, turbidity currents have rarely been observed in nature; most of our information has been provided by laboratory experiments.

The surging turbidity current flow tends to develop four parts: head, neck, body, and tail (Figure 2–15) (Middleton and Hampton, 1973). The head is rather well defined, is typically the thickest part of the flow, and contains the coarsest sediment. Friction with the overlying fluid body causes fluid and sediment to sweep up and back into the less well defined neck. Most erosion or modification of the underlying surface over which the currents flows takes place at the head. Almost immediately afterward, rapid burial takes place from the body of the turbidity current.

Although rates of deposition by turbidity currents are quite variable, some generalities are appropriate. Along the margins it may be high, due to rapid loss of turbulence, and therefore be the mechanism for suspension. Decrease in gradient also causes rapid deposition. Some traction features, such as lamination or cross-stratification, may develop as sediment accumulates.

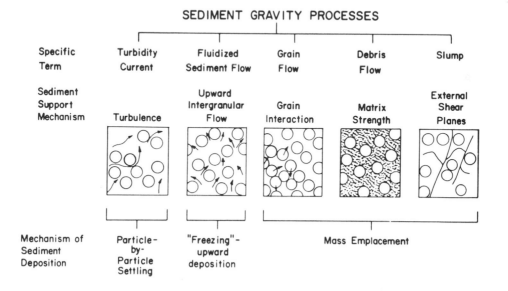

Figure 2-14 Classification of sediment gravity processes. (Modified from Middleton and Hampton, 1973, p. 3; after Klein, 1980, p. 121.)

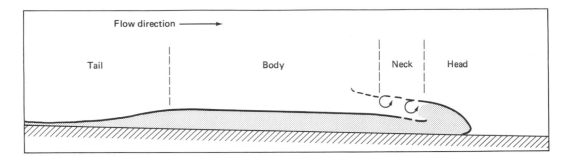

Figure 2–15 Schematic subdivision of a turbidity current into head, neck, body, and tail. (After Middleton and Hampton, 1973, p. 6.)

Liquefied sediment flows

Some subaqueous sands may have their fabric destroyed as a result of a sudden event which causes liquefaction. Once this fabric is destroyed, the grains are partially supported by the pore fluids and the sand is "quick." During such conditions the sand acts like a viscous fluid. This phenomenon can be observed on the low-tide beach or sand flats by tapping the surface with one's foot. Soon the firm surface becomes quick and the observer begins to sink into the quick sand.

This liquefaction enables sediment to flow in response to gravity. Because of a rapid loss of the excess pore pressure that causes the quick condition, transport of sediment may be short (Middleton, 1969). Frictional effects are large and flow may stop almost instantaneously due to the combination of friction and the loss of pore pressure (Middleton and Hampton, 1973).

Grain flows

Oversteepening of dry sand and the resulting sediment flow can readily be observed in eolian dunes. Bagnold (1954) postulated that this phenomenon of grain flow was the result of grain-to-grain interaction as opposed to fluid turbulence, thus creating what he termed a dispersive pressure. A similar flow occurs subaqueously, with the interstitial fluid being water or a mixture of water and fine sediment particles. The slope required to initiate grain flow varies depending on the nature of the interstitial material. In general, interstitial mud requires less slope to initiate flow than does interstitial water alone (Middleton and Hampton, 1973).

Deposition from grain flow is a mass emplacement phenomenon (Middleton and Hampton, 1973) and is caused by slope reduction or consolidation of flow. This brings about the dominance of the yield strength of the sediment over gravity stress.

Debris flows

Debris flow may take place in the subaerial or subaqueous environment. It is the slow downslope transport of a mixture of various-size sediment particles and water in response to gravity (Hampton, 1972). Large particles are supported in the flow by the mixture of mud and water in a buoyant condition. The strength and density of the mud-fluid mixture determines the competency of the flow. This phenomenon is very common in spring when thawing takes place and water is made available.

Debris flows may also be generated as the result of extreme rainfall and have caused catastrophic damage in some urban areas (Jahns, 1969).

Slump

Large masses of sediment may move downslope due to shear within the mass. These bodies retain their internal coherence, including stratification; however, deformation in the form of convolutions and fractures may occur (Klein, 1980). These large masses occur in a variety of submarine and subaerial settings where slopes are fairly steep and triggering mechanisms are present. Earthquakes, oversteepened slopes, and excess pore pressure in the sediments may cause slumping or sliding.

Such mass movement has been recognized for a long time in the subaerial environment, but only recently has it been depicted in the deep ocean. The cable breaks in the North Atlantic reported by Heezen and Ewing (1952) were attributed to submarine slumping, but the features were not properly defined until detailed seismic profiles were described in the 1970s (Carlson and Molnia, 1977; Hampton and Bouma, 1977). The size of individual slumps covers a broad range and may exceed 1000 km² (Moore, 1977).

Tides and Tidal Currents

The regular and predictable daily rise and fall of the water level along the coast constitutes the tides. Water-level changes and tidal currents generated by this phenomenon are important processes in sedimentation.

Sir Isaac Newton has provided an explanation for the tides through his law of gravitation, which states that any object in the universe attracts all other objects with a force that is proportional to the product of their masses and inversely proportional to the square of distance between them. Mathematically, it states that

$$F = G \; \frac{m_1 m_2}{r^2} \qquad (2\text{–}4)$$

where m_1 and m_2 are the masses of the objects in question, r the distance between them, and G the gravitational constant. This attraction between large celestial bodies may be great and gives rise to our regular sea-level changes (tides) in the form of **forced waves**. The mass attraction between the sun, moon, and earth causes distortions in the surface of the oceans, forming bulges (Figure 2–16). As the earth rotates on its axis, these bulges move across the earth as high tides and the intermediate and opposite effect is in the form of low tides. These deformations in the oceans vary depending on the relative positions of the sun, moon, and earth such that when the three celestial bodies are aligned there are maximum tides. These tides are called **spring tides** and occur during the new moon and full moon (Figure 2–16). Conversely, when the moon and sun are at right angles to the earth, they work independently of each other, producing minimal tidal ranges or **neap tides**. This condition exists during the first and third quarters (Figure 2–16).

The land masses of the earth interfere with the movement of these forced waves over the earth's surface and, together with irregularities in the coastal configuration,

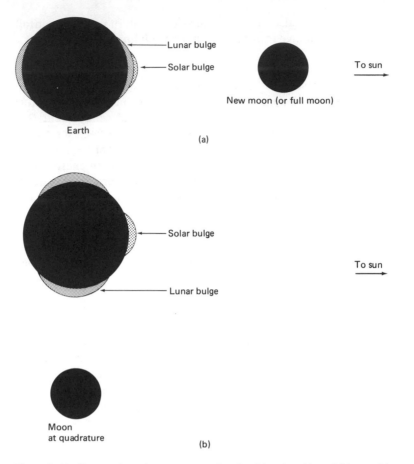

Figure 2-16 Sun, earth, and moon system, showing (a) spring tide and (b) neap tide.

cause considerable variation in the tides from one place to another. Constrictions and embayments in the coastline commonly cause water to be funneled into them, resulting in relatively large tidal ranges. The Bay of Fundy in Nova Scotia (Canada) represents an extreme case where spring tides exceed 15 m. Davies (1964, 1980) has classified coasts based on tidal range (Figure 2-17) as microtidal coasts (< 2 m), mesotidal coasts (2 to 4m), and macrotidal coasts (> 4 m). The development of sedimentary environments along the coast is influenced to a large extent by the tidal range (see Chapters 10 and 11).

Types of tides

There are three different tidal varieties that occur during a tidal day (24 hours and 50 minutes): semidiurnal, with two high and two low tides each day; diurnal, with one high and one low tide; and mixed tides, where they may be semidiurnal during part of the 28-day lunar cycle and diurnal during the other part (Figure 2-18). The reason for the different types of tides lies in the interaction of natural period oscillations of the water basin with tide-generating forces (Komar, 1976).

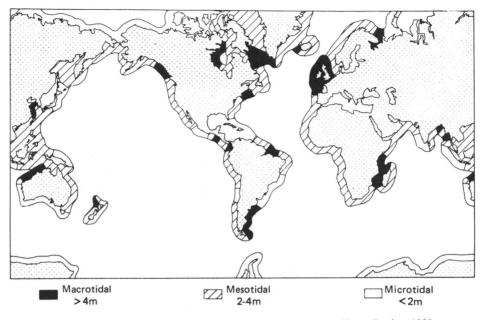

Figure 2-17 Worldwide distribution of spring tidal range. (From Davies, 1980, p. 51.)

There may be considerable variation in the high- and low-tide levels within the semidiurnal and mixed tides. For example, the tides at San Francisco show a marked inequality between the adjacent highs and lows throughout the lunar cycle (Figure 2-18), but they are not common. This inequality is caused by the variation in the moon's position with respect to the earth. When the moon is on the equator, equal tidal ranges result; however, when the moon is in the midlatitudes, the position of the bulge causes diurnal inequality as rotation occurs.

Tidal currents

Currents generated by the forced tidal waves may transport much sediment. Like the great geographic variation in tidal range, there is also great variety in the spatial distribution of tidal currents. In addition, there is a temporal variation to these currents.

In general, there is an increase in the strength of tidal currents as one approaches the coast up to a certain point—after this there is a decrease again. Although measurable tidal currents are generated as the tidal wave moves to and across the continental shelf, these currents are not important as sediment transporting agents. In fact, tidal currents are not important along open coast such as a smooth barrier island because their relatively low velocities are overshadowed by waves and longshore currents in the surf zone. Estuaries and other coastal embayments, inlets in barriers, and deltas represent the most important environments as far as tidal influence on sediments is concerned.

To understand and monitor sediment transport by tidal currents, it is first

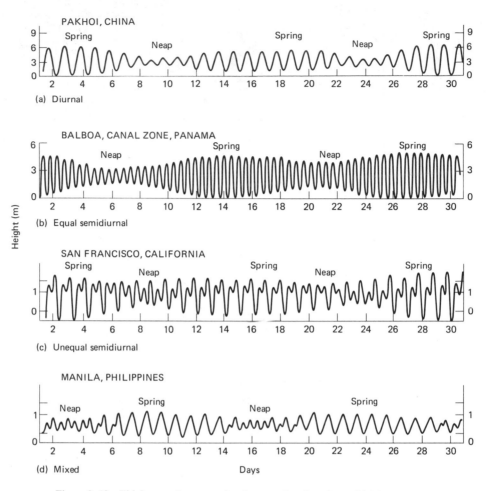

Figure 2–18 Tidal curves for a complete lunar cycle at locations which have diurnal, semidiurnal, and mixed tides.

necessary to measure tidal current speed throughout the tidal cycle. One might expect that a comparison between tide stage or level and tidal current speed would show that high tide would coincide with **flood slack** and low tide with **ebb slack.** Similarly, it might be expected that maximum tidal current speed would coincide with midflood and midebb tidal stages. In fact such relationships do not take place, which gives rise to the phenomenon of **time-velocity asymmetry** (Postma, 1967). Maximum flood currents are typically after midtide and maximum ebb currents occur near low tide (Figure 2–19). Similarly, flooding tidal currents do not equal ebbing currents with respect to time, velocity, and/or water volume. One may exceed the other; thus an inlet may be **flood dominated** or **ebb dominated.** In locations where runoff is significant, such as in deltas or mouths of estuaries, ebb-dominated conditions would prevail from the standpoint of both time and current speed.

Insofar as sediment transport and generation of bedforms is concerned, tidal currents act like currents in streams except that there is a rather regular and predic-

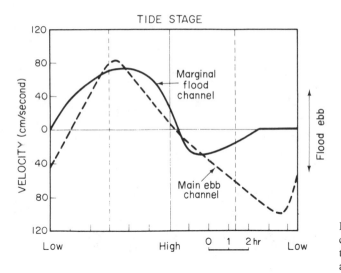

Figure 2-19 Tidal current velocity curves for an ebb-tidal delta illustrating time-velocity asymmetry. (From Hayes and Kana, 1976, p. I-59.)

table change in speed and direction during each tidal cycle. However, conditions are virtually always changing and there is no situation where equilibrium conditions exist. Sediment is nearly always in motion, bedforms are changing, and sediment bodies are moving or being modified.

Waves

A wave is a surface disturbance of a fluid medium. This commonly takes place between a liquid and a gas (the water and the atmosphere) but may also occur between two gaseous or two liquid media. Some surface water waves are the result of a disturbance to the container holding the water (the solid earth), whereas most are the result of the wind acting directly on the water surface.

Surface waves are commonly called **gravity waves** because they are controlled primarily by gravitational forces. Although the energy transfer between the wind and the water surface is complicated, it can be simplified by stating that the size of waves is a function of three factors:

$$\text{height, period} = f(W,\ F,\ D) \tag{2-5}$$

where W is the wind velocity, F the fetch or distance of the water surface over which the wind blows, and D the duration of the wind. An increase in any factor or combination of factors within limits causes waves to increase in size.

One can conveniently consider wind-generated waves in two categories: sea and swell. **Sea waves** are steep, short, and generally irregular waves actively being blown by the wind; **swell waves** are not under direct wind influence because the wind has ceased or the waves have moved beyond the area of active wind. As surface gravity waves reach the shallow coast they break, forming **surf.** Most breaking waves are **plunging breakers,** which form as the waves steepen, curl over, and break with an instantaneous crash of water, or **spilling breakers,** which break gradually over some

(a)

(b)

Figure 2–20 Photographs showing (a) spilling breakers on the Texas coast and (b) plunging breaker forming from swell on the Massachusetts coast.

distance (Figure 2–20) (Wiegel, 1964). Plunging breakers typically result from swell approaching the coast or from steep beach slopes, and spilling breakers are most commonly generated from sea waves or very gentle beach slopes (Komar, 1976).

Waves are described by their size and speed of movement. Principal morphologic components in a wave include the crest, typically somewhat peaked compared to the broad, concave-upward trough (Figure 2–21). The wave length (L) is the horizontal distance between successive crests or troughs, and the wave height (H) is the vertical distance from crest to trough. The dynamic nature of waves is such that measuring their wave length directly is difficult; therefore, the wave period (T) is invoked. This is the interval of time for successive crests (or troughs) to pass a fixed

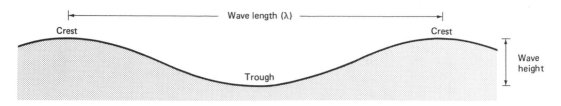

Figure 2-21   Ideal sinusoidal wave showing basic components.

point. The velocity of propagation of the wave (or celerity) is expressed as C, and it is related to the wave length and period such that

$$C = \frac{L}{T} \tag{2-6}$$

Water motion in waves

Although the wave form moves rapidly across the water surface, the water particles themselves follow a strikingly different path. Primary motion is orbital in nature, with a decrease in the size of the orbits with depth. The effective depth of this orbital motion is about equal to $\frac{1}{2}L$. The motion of water particles on any given wave is such that on the crest, movement is forward (in the direction of propagation), and in the trough, it is the opposite (Figure 2-22).

This orbital water motion is the process that interacts with the bed and causes sediment motion. When a propagating wave approaches shallow water, the orbital motion of the wave is modified at the bottom. The wave begins to steepen as the velocity of water particles in the crest increases. The crest becomes unstable when the inclusive angle becomes 120° and the wave breaks.

Although there is a pattern of sediment particle motion under breaking waves (Figure 2-23), the role of waves is more important in sedimentation as a means of suspending sediment particles and making them available for transport by shallow currents (see the following section). Whereas it is not generally analogous to a stream, research along the surf zone has shown that the concept of flow regime is also applicable in this environment (Clifton et al., 1971). Offshore, where swell waves are nearly sinusoidal in shape, the bed is covered with asymmetric ripples. As waves are affected by the bottom and begin to steepen, megaripples are generated. The intense energy associated with the breaking waves and the **bore** causes plane beds (Figure 2-24). In effect, this progression is comparable to moving through the lower part of the upper flow regime. Actually, the upper part of the upper flow regime is also represented in some cases because backwash in the swash zone may generate antidunes (Davis, 1978).

Nearshore Currents

Much sediment transport takes place in the beach and nearshore zone, partially as the result of wave action and also as the result of currents which are directly or indirectly the result of waves. These currents, in concert with waves, cause essentially all sediment transport along the open coast.

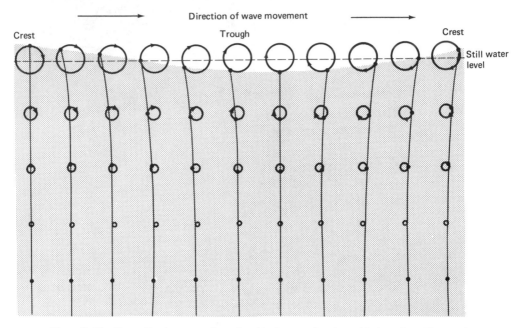

Direction of wave movement

Crest   Trough   Crest

Still water
level

Figure 2-22 Generalized cross section of an ideal wave, showing orbital motion. Observe that the orbital motion of water is essentially absent at depth of one-half wave length.

Longshore currents

As waves bend when they approach the coast and enter shallow water, there is a component of energy that causes water to move parallel to shore away from the acute angle made by the approaching wave and the shoreline. This is the **longshore current** (Figure 2–25) and is the dominant sediment transport process along the beach. The speed at which such currents travel varies considerably both with respect to location and wave conditions. Longshore current speed is dependent on the size and angle of the approaching waves and is typically greatest during storm conditions. Speeds in excess of 1 m/sec have been observed (Davis and Fox, 1972).

When long-period swell waves approach the coast, there is near total refraction with waves becoming parallel to the coast. This results in no significant longshore

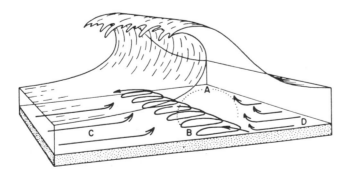

Figure 2-23 Diagram of grain motion associated with a breaking wave. Arrows show paths of largest grains (B), suspended particles (A), and those both landward (C) and seaward (D) of the wave. (From Ingle, 1966, p. 53.)

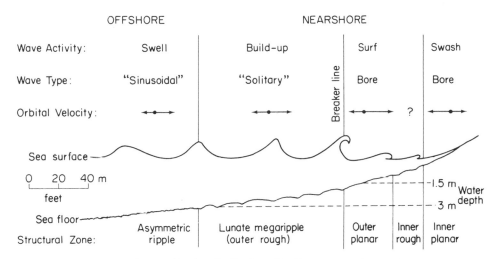

Figure 2-24 Diagram showing distribution of bedforms across the nearshore zone in a nonbarred coast. (After Clifton et al., 1971, p. 661.)

current. By contrast, the sea waves generated by storm conditions may enter the breaker zone at angles up to 30° or more, creating rapid longshore currents and much longshore transport of sediment or littoral drift.

The longshore current system has been called the "river of sand" because it behaves somewhat like a stream. The banks of this "stream" are the shore on one side and the outer limit of the breaker zone on the other. Nearly all longshore sediment transport occurs within this narrow zone. As currents move parallel to the coast, bedforms are generated in the same fashion and order as occurs in other unidirectional flow environments. This author has observed ripples, megaripples, and plane beds as flow increases. Thus the surf zone is being subjected to the entire spectrum of lower-flow-regime conditions as well as the lower part of the upper flow regime (Davidson-Arnott and Greenwood, 1976). Such high-energy conditions are relatively short lived, but much sediment transport can be achieved during this short period of time.

Rip currents

Water tends to "pile up" in the surf zone as the result of breaking waves. To compensate for this imbalance in water level, narrow, seaward-flowing currents called **rip currents** are generated. These currents have also been called undertow or rip tides and have been responsible for numerous swimming accidents.

Rip currents are fed by longshore currents and develop over saddles or low areas in the sandbars common to most shorelines (Figure 2–26). The rip current is narrow as it passes over this saddle, then expands into a plume of sediment-laden water (Shepard et al., 1941; Shepard and Inman, 1950). Although not as effective as longshore currents as a mechanism for sediment transport, rip currents do play a significant role in the dispersal of surf zone particles.

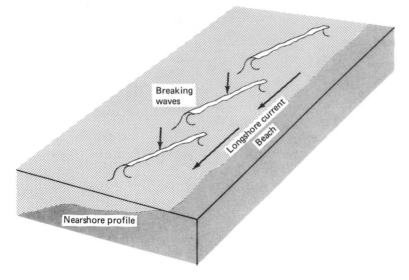

(a)

(b)

Figure 2–25 (a) Diagram and (b) photo showing generation of longshore current by approach of waves at an angle to the beach.

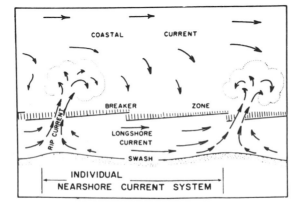

Figure 2–26 Diagram of coastal circulation system with rip currents. (From Komar, 1971, p. 2644.)

SUMMARY

It is obvious from the preceding discussion of the various processes to which sediment is subjected that the natural history of any given particle is very complicated. From the time a particle is produced by a source rock or generated by biological or chemical means until it is deposited in its final resting place, there may be several processes that impact on the particle. Most of these are either directly or indirectly a result of the force of gravity. As the reader covers the chapters on various depositional environments, it is important to recall these processes and how each contributes to the resulting attributes of the sediments and the environment as a whole. With these data in mind it is then possible to make reasonable interpretations of the conditions under which strata in the rock record were accumulated.

ADDITIONAL READING

ALLEN, J. R. L., 1970. *Physical Processes of Sedimentation: An Introduction.* Allen & Unwin, London, 248 p. Good elementary treatment of fluid mechanics and sediment transport. Excellent reading for the undergraduate student.

GRAF, W. H., 1971. *Hydraulics of Sediment Transport.* McGraw-Hill, New York, 513 p. Good comprehensive treatment of subject emphasizing details of fluid mechanics. Designed for the advanced student of engineering or a geologist with a good background of applied mathematics.

HARMS, J. C., SOUTHARD, J. B., SPEARING, D. R., AND WALKER, R. G., 1975. *Depositional Environments As Interpreted from Primary Sedimentary Structures and Stratification Sequences.* Soc. Econ. Paleontologists and Mineralogists, Short Course No. 2, Lecture Notes, Tulsa, Okla., 161 p. Excellent summaries of state of the art on sedimentary structures, their formation and stratigraphic significance in the rock record. A must for all serious students of depositional environments.

MIDDLETON, G. V. (ED.), 1965. *Primary Sedimentary Structures and Their Hydrodynamic Interpretation—A Symposium.* Soc. Econ. Paleontologists and Mineralogists, Spec. Publ. No. 12, Tulsa, Okla., 265 p. The first volume concerning geological studies of bedform generation. This publication represents a classic group of papers in the recent emphasis on processes in sedimentary environments.

MIDDLETON, G. V., AND SOUTHARD, J. B., 1977. *Mechanics of Sediment Transport.* Soc. Econ. Paleontologists and Mineralogists, Short Course No. 3, Lecture Notes, Tulsa, Okla. Superb and easily understandable treatment of fluid mechanics as applied to sediment transport. Can be handled by the advanced undergraduate student.

YALIN, M. S., 1972. *Mechanics of Sediment Transport.* Pergamon Press, Oxford, 290 p. Engineering approach to sediment transport with emphasis on fluid mechanics theory and application. Similar in treatment to the book by Graf.

3 Sedimentary Structures

One of the most useful tools available to the geologists studying depositional systems and their interpretation in the rock record is the spectrum of features called sedimentary structures, which are internal or surficial, megascopic, three-dimensional features of sediment or sedimentary rocks (Selley, 1970). The structure of a sediment, like its texture and composition, is a fundamental property (Pettijohn and Potter, 1964). It is concerned with the manner in which the sediment or sedimentary rock is put together and is generally large enough that the structure is best examined in at least hand specimen size but more commonly in the outcrop.

Under some circumstances it is difficult to separate a textural property from structure such as the imbrication of sediment particles or grading of the particle size stratigraphically. The fabric of a sediment or sedimentary rock is therefore commonly considered to be a sedimentary structure. At the opposite and large end of the spectrum are the huge bedforms such as aeolian dunes or subaqueous sand waves containing a rather simple and uniform internal organization. Such features may be considered as a sedimentary structure, although they may show smaller-scale structures incorporated within them.

Proposals for the classification of sedimentary structures have been numerous, generally incomplete, and without widespread adoption. In many respects there is no real need for a comprehensive and universal classification of sedimentary structures.

The individual structures and their mode of origin are of primary concern to the geologist.

Those classifications that have been proposed generally are characterized by at least some general implications of the genesis of the structures. For example, the two major categories of a classification by Pettijohn (1975) are inorganic and organic structures; some authors refer to these categories as physical and biogenic, respectively. Another means of classifying structures is based on the time of formation: primary structures formed during or shortly after accumulation of the sediment (at least pre-lithification), and secondary structures (often called chemical) formed after lithification.

In the following discussion no attempt is made to adhere to any of the existing classifications, nor is a new classification proposed. The organization is based on a combination of the genesis of the structures and their functional utilization by the geologist who will be interpreting environments in which the structures were developed.

The great diversity of sedimentary structures precludes complete treatment within this chapter. Comprehensive treatment of structures can be found in the excellent books by Pettijohn and Potter (1964), Conybeare and Crook (1968), in Part I of Reineck and Singh (1980), and Collinson and Thompson (1982).

PHYSICAL (INORGANIC) STRUCTURES

Most sedimentary structures are formed by physical processes without the direct influence of organisms at or shortly after sediment accumulation. These structures have been called mechanical or primary (Pettijohn, 1975) and exclude late diagenetic or chemical structures as well as biogenic structures.

Bedding Structures

In virtually all descriptions of outcrop, cores, or hand specimens of sedimentary rocks, some effort is devoted to describing the bedding or layering characteristics of the sequence. In the past this was done largely as a means of characterizing the stratigraphic sequence and in some cases was used for correlation purposes. Presently, the bedding characteristics are studied closely to aid the geologist in interpreting the environmental conditions that persisted during accumulation of the sediment in question. Most sedimentary structures have some expression in the bedding characteristics of the sedimentary sequence being examined.

In many discussions of sedimentary structures, surface features and internal bedding characteristics are discussed separately (Pettijohn and Potter, 1964). The basic purpose of this book is to provide the reader with the background necessary to interpret depositional environments from an examination of the sedimentary rock record. The primary tool for doing so is a comparison with analogous modern sedimentary environments. In the modern environment one typically views surface sedimentary structures (bedding planes) such as bedforms, whereas in the ancient record the typical view is normal to bedding. To relate the bedforms that can be observed on a modern tidal flat to the rock record, it is necessary to consider the stratigraphic view of these bedforms. This discussion combines the surface and internal characteristics of the various structures to allow the reader to understand the translation from the modern surface expression of a structure to the bedding that is displayed on the outcrop or in a core.

Horizontal bedding

Planar and horizontal or nearly horizontal sediment surfaces that exist in many places, such as on lake bottoms, gently sloping beaches, or in the deep marine environment, accumulate parallel and relatively continuous stratification. Although such bedding features may exhibit a rather uniform and monotonous appearance, they may originate in a variety of fashions and may display considerable textural and compositional variety.

There are various ways in which these parallel and essentially horizontal beds may originate, including sediment particles descending through a fluid and coming to rest in the absence of currents, or by being transported along the sediment-fluid interface by currents. The former are typically fine-grained accumulations and the latter generally are relatively coarse. Regardless of the method of transport and accumulation, the presence of the bedding itself signifies some type of change in depositional conditions. The term **sedimentation unit** is used to describe the sediment which was deposited under essentially constant conditions (Otto, 1938). Therefore, an individual bed is a sedimentation unit.

Rhythmites. Bedding that is a repetitious sequence of generally thin and alternating types of sediment particles may be called rhythmites (Bramlette, 1946). The cyclic nature of rhythmites may be caused by seasonal changes such as in **varves,** which produce repetitive thin beds of fine-grained sediment deposited in lakes (Figure 3–1). The summer season is represented by a thicker and coarser layer than that of the organic-rich winter layer. During the summer there is relatively high runoff from adjacent land areas. In winter there is a combination of low runoff and accumulation of organic material, producing a thin and generally carbonaceous layer. Each varve is a couplet and represents one year of sediment accumulation from suspension. Some varved sequences contain several thousand of these couplets and represent at least an equivalent period of time. Tidal conditions may produce rhythmites, which are the result of a combination of suspended sediment accumulation (mud) and bed load-transported sediment (fine sand). These interbedded sand and mud sequences (tidal bedding) represent accumulations from alternating flood and ebb tidal currents (Reineck, 1967).

Rhythmites are characterized by alternating layers of sediment texture, composition, or both. They reflect accumulation under alternating physical or chemical conditions. In the latter situation it may be possible to have an alternation of fine terrigenous sediment with a chemical precipitate such as calcite or an evaporite mineral.

Some conditions of sedimentation are such that although there is an alternation of processes, the net accumulation of sediment is in the form of uniform sedimentation units. Beach lamination is a rhythmite that may display such characteristics, although there is commonly some textural gradation within individual sedimentation

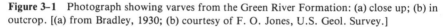

(a)   (b)

Figure 3–1 Photograph showing varves from the Green River Formation: (a) close up; (b) in outcrop. [(a) from Bradley, 1930; (b) courtesy of F. O. Jones, U.S. Geol. Survey.]

units (Clifton, 1969). Sediment is deposited during uprush and is reworked during backwash with upper-flow-regime plane bed conditions giving rise to the parallel bedding. Beach rhythmites are produced by alternations in grain size, composition, or a combination of both (Figure 3–2).

Other horizontal bedding. Some sedimentological conditions do not lend themselves to uniformly alternating conditions and therefore to rhythmites. Whereas most rhythmites are composed of mud or sand, the nonrhythmic horizontal beds may include the entire grain size spectrum. Such flat layers typically represent upper-flow-regime plane bed conditions, although some may develop under flat bed conditions prior to the initiation of ripples (Harms et al., 1975). Horizontal beds may be continuous over only a few meters or they may extend for a kilometer. These beds typically are deposited under the high-velocity conditions that exist at the beginning of the upper flow regime. Many of the various fluvial environments, such as braided streams, alluvial fans, and glacial outwash, represent examples of sand and gravel beds which may appear horizontal locally but which typically are not continuous throughout even a single exposure. By contrast, individual turbidites may extend for relatively great distances.

Bedforms and their internal stratification

The definition and general nature of bedforms were presented in Chapter 2. This discussion considers the various scales and geometries exhibited by the bed-

Figure 3–2 Photograph of foreshore beach stratification. These rhythmites are due to uprush and backwash on the beachface.

forms. Perhaps even more important is their internal organization of bedding, because it is this attribute that the geologist must be able to recognize when dealing with the rock record.

Current-generated structures (lower flow regime). Most bedforms that are generated in sedimentary environments are those in the lower flow regime of Simons and Richardson (1963; see also page 48 of this book). Although the general terminology associated with this sequence of bedforms is somewhat standard, there is not universal compliance, especially regarding spacing. This discussion will utilize the characteristics as presented by Harms et al. (1975) and modified by Boothroyd (1978) (Table 3–1). Ripples and megaripples (dunes of Simons and Richardson, 1963; Harms et al., 1975) are the most commonly observed bedforms in sedimentary environments. Sand waves are less common and are intermediate in terms of flow velocity strength.

In a general way the internal organization of the current-generated bedforms is similar regardless of scale and profile configuration. As sediment-laden fluids pass over the bedform, there is a distinct pattern to the flow and sediment movement (Figure 3–3), with three distinct zones down-current (lee side) of the crest: (1) the zone

TABLE 3-1 CHARACTERISTICS OF LOWER-FLOW-REGIME BEDFORMS

	Ripples	Megaripples	Sand waves
Spacing	< 60 cm	60 cm–6 m	> 6 m
Height/spacing ratio	Variable	Relatively large	Relatively small
Geometry	Highly variable	Highly variable	Straight to sinuous
Typical flow velocity	Low (~25–50 cm/sec)	High (~70–150 cm/sec)	Moderate (~40–80 cm/sec)

SOURCE: After Boothroyd, 1978.

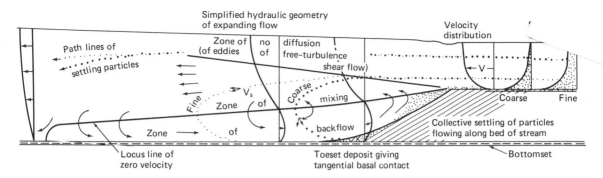

Figure 3-3 Flow pattern and sediment movement over bedforms. (After Jopling, 1967.)

of no diffusion; (2) the zone of mixing, where maximum turbulence exists; and (3) the zone of backflow, which is below the flow separation surface (Jopling, 1967). This flow pattern produces foreset stratification parallel to the slip face of the bedform as the major internal feature of the structure (Figure 3-3). Variations in the size and shape of the bedforms will cause corresponding changes in the cross-stratification.

The shape of the foreset stratification may also be influenced by the sediment texture, velocity, and bed shear stress and by the depth ratio (Jopling, 1963, 1965). The latter refers to the ratio of stream flow to the depth of the basin (i.e., deep water would have a high depth ratio). The resulting shapes of foreset strata cover a broad spectrum, including planar, tangential, concave, and convex (Figure 3-4).

The plan view of current-generated structures varies widely and is related to flow strength (Figure 3-5). As flow increases, the bedform changes from a linear shape (two-dimensional) through various three-dimensional configurations. These variations in bedform geometry have a profound effect on the cross-stratification generated by the migrating bedforms. Because most bedforms are in fact three-dimensional in nature, it is important to consider the three-dimensional appearance of the internal stratification of the structure. It may not be possible to view the three-dimensional nature of the stratification in the rock record, but if one has a thorough

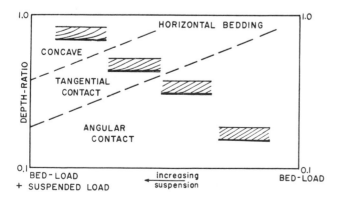

Figure 3-4 Changes of shape in foreset laminae as affected by various process factors. (After Jopling, 1965.)

Chapter 3 Sedimentary Structures

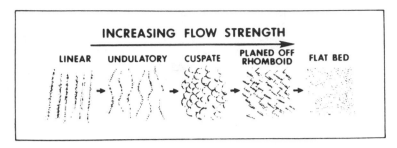

Figure 3-5 Relationships between shape of bedforms and flow strength. (From Hayes and Kana, 1976, p. I-10.)

knowledge of the bedform and its cross-stratification elements, it should be possible to reconstruct the bedform shape rather accurately.

Linear ripples and linear megaripples represent the simplest bedform configuration and cross-stratification pattern. Foreset stratification predominates normal to the bedform crests, with horizontal bedding parallel to the crests (Figure 3-6). The scale of the foreset beds or cross-set units can be used to determine whether ripples or megaripples generated the cross-strata.

As flow increases there is typically an accompanying change from linear crests to sinuous and crescentic bedforms (Figure 3-5). This change from two-dimensional to three-dimensional bedforms is reflected in the cross-stratification geometry, producing lenticular cross-sets both parallel to flow and normal to flow (Figure 3-7).

The term **trough cross-stratification** is used for the lenticular and trough-

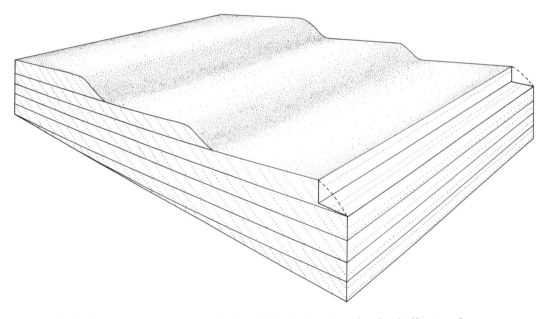

Figure 3-6 Block diagram showing the relationship between migrating bedforms and configuration of cross-stratification for straight bedforms. (From Reineck and Singh, 1980, p. 38.)

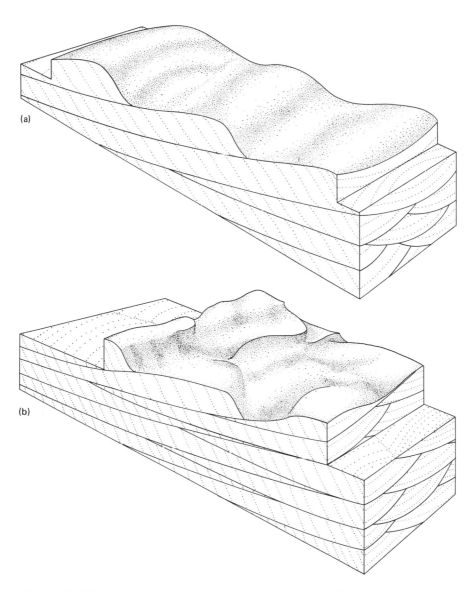

(a)

(b)

Figure 3-7 Relationships between (a) sinuous and (b) crescentic bedforms and configuration of cross-stratification. (From Reineck and Singh, 1980, pp. 39, 40.)

shaped sets produced by three-dimensional bedforms and **tabular cross-stratification** is used to describe those produced by linear bedforms. Another feature associated with generally linear bedforms is the **reactivation surface** (Boersma, 1967; Collinson, 1970). This is a locally horizontal to sloping surface that separates otherwise conformable cross-strata (Figure 3-8). Such surfaces record an interruption in the migration of a single bedform and are surfaces of some erosion. They can be caused by changes in flow rates, tidal stage, or tidal current direction (Harms et al., 1975). In

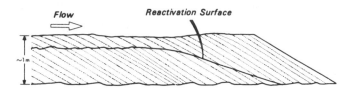

Flow

Reactivation Surface

~1 m

Figure 3-8 Diagram showing position and configuration of reactivation surface in a bedform sequence. (From Harms et al., 1975, p. 51.)

Figure 3-8, the area below the reactivation surface, represents a migrating bedform which was then partially eroded by a reversal of current direction followed by another change in flow, producing the overlying bedform.

When there is significant introduction of sediment during flow conditions, aggradation takes place. This condition gives rise to **climbing-ripple cross-stratification** (Figure 3-9), which is distinct from previously described types, although the surface bedform is the same. During only modest sediment influx and therefore low rate of aggradation, the ripple cross-strata are in-drift and lenticular parallel to flow

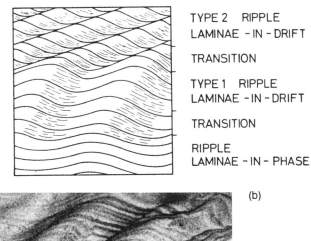

(a)

TYPE 2   RIPPLE
LAMINAE – IN – DRIFT

TRANSITION

TYPE 1   RIPPLE
LAMINAE – IN – DRIFT

TRANSITION

RIPPLE
LAMINAE – IN – PHASE

(b)

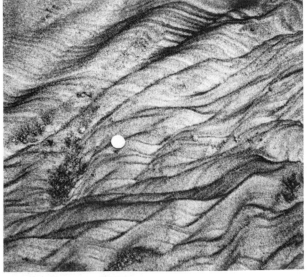

Figure 3-9 (a) Diagram and (b) photograph showing climbing ripple cross-lamination. White circle is U.S. 10 cent piece. (Courtesy of Joe R. Wadsworth).

(McKee, 1965; Jopling and Walker, 1968). Much sediment influx causes the ripple stratification to be undulatory and parallel; in other words, they are in phase (Figure 3–9). What happens in the in-drift situation is that the ripples are migrating both upward and with the flow, whereas in the in-phase type, aggradation is too rapid to produce migration with flow. Ripple-drift cross-stratification is common in ripples but is difficult to perceive in larger bedforms, although it has been reported from the Ganges-Brahmaputra River (Coleman, 1969).

Low-angle and gently undulating cross-stratification (Figure 3–10) with dips of 3 to 6 degrees has been termed **hummocky cross-stratification** (Harms et al., 1975). The sets may be concave or convex upward, with wavelengths of up to a few meters

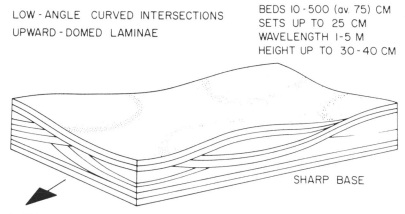

LOW - ANGLE CURVED INTERSECTIONS
UPWARD - DOMED LAMINAE

BEDS 10 - 500 (av. 75) CM
SETS UP TO 25 CM
WAVELENGTH 1-5 M
HEIGHT UP TO 30 - 40 CM

SHARP BASE

DIRECTIONAL SOLE MARKS

(a)

(b)

Figure 3–10 Illustrations of hummocky cross-stratification as shown (a) diagrammatically and (b) in the Cambrian of Wisconsin. [(a) from Walker, 1979, p. 85.]

and wave heights of several centimeters (Walker, 1979). Such features are presently interpreted as being generated by storm waves on the shelf environment.

Current-generated structures (upper flow regime). Bedforms generated by upper-flow-regime conditions are relatively uncommon in nature if we exclude upper plane beds. Antidunes can be observed in streams, tidal inlet systems, and on gently sloping beaches. Except for the foreshore beach (Figure 3–11), they are rarely exposed for study and antidunes have a low preservation potential for incorporation into the rock record. This is due to their dependence on high-velocity conditions, which may be local or of short duration. When these conditions wane, the bedforms adjust and antidunes are destroyed.

The surface expression of antidunes is characterized by low relief, only slightly asymmetric profile, and linear crests; they are two-dimensional (Reineck and Singh, 1980). Internally, antidunes display gently inclined cross-strata which dip upstream (Figure 2–10). Internal structure is typically faint and is usually destroyed by waning currents as flow decreases (Middleton, 1965). Cross-strata interpreted as representing antidunes have been recognized from the rock record (Hand et al., 1969).

Wave-generated structures. Oscillatory flow produced by waves may interact with the sediment substrate to produce bedforms. Actually, there is a transition from unidirectional flow, current ripples to wave-generated ripples, with combined-flow ripples as an intermediate type (Harms et al., 1975).

Wave ripples are typically symmetrical or only slightly asymmetric in profile, with peaked crests and rounded troughs (Figure 3–12). Ripple crests are straight or only slightly sinuous, with common bifurcations. According to Reineck and Singh (1980), the presence of bifurcating ripples is a good indication of wave genesis; such patterns are less common in current-generated ripples or megaripples. The size of wave-formed ripples is related to wave size and the grain size of the sediment (Inman,

Figure 3–11 Photograph of antidunes generated by backwash on beach at Jekyll Island, Georgia.

Part One   Principles

(a)

(b)

Figure 3–12 Photograph of wave-formed ripples in the shallow nearshore zone: (a) in the modern environment; (b) from the Ordovician New Richmond Sandstone (Wisconsin).

1957; Davis, 1965; Harms, 1969) as well as to wave period and the depth of water (Clifton, 1976).

Internally, wave-generated ripples may display a variety of chevronlike cross-stratification (Newton, 1968). Newton stressed the dominance of foreset stratification, and combined flow ripples show similar configuration and internal structure. In fact, there is a distinct asymmetry in the crest profile of most ripples whether they are wave-formed, combined-flow, or current-generated in origin.

In the rock record, one has great difficulty in separating combined-flow ripples from current-generated ripples. Because of this problem, effort has been made to use various ripple parameters to distinguish between these two types (Tanner, 1967). Two parameters are commonly employed: the ripple index (ripple length/ripple height) and the ripple symmetry index (length of horizontal projection of stoss side/length of horizontal projection of lee side) (Figure 3–13). In comparing wave-current-formed ripples, there is considerable overlap using the ripple index, whereas the ripple symmetry index shows relatively good separation (Tanner, 1967).

Other bedding types

Some bedding styles which may or may not be related to previously discussed bedforms or to horizontal stratification are not only distinctive in appearance but also denote specific modes of accumulation and are therefore useful to the geologist in interpreting environments of deposition. Some of these bedding types may have expression on the modern sediment surface, and some do not.

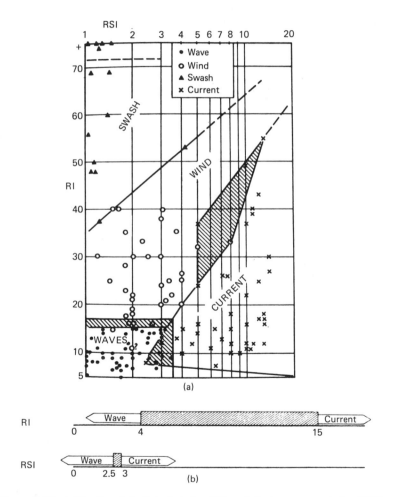

Figure 3–13 (a) Scatter plot of ripple index (RI) and ripple symmetry index (RSI) data, showing comparison between swash, wind, current, and wave-generated ripples. (After Tanner, 1967, Figure 1.) (b) Summary graph of these data. Hachured areas indicate zones of overlap. (From Reineck and Singh, 1980, Figure 35.)

Flaser and lenticular bedding. Cross-stratified ripple bedding commonly contains thin streaks of mud in the ripple troughs. This bedding type, called **flaser bedding,** may grade continuously into a sequence of mud with discontinuous, cross-stratified sand lenses called **lenticular bedding** (Reineck, 1960; Reineck and Wunderlich, 1968a). This spectrum of bedding types is dependent on two primary conditions: availability of both sand and mud, and alternation of relatively high energy and quiescent conditions. These bedding types may be produced in tidal environments, streams, or inlets, or on the intertidal beach—generally in any environment where ripples may form.

Flaser bedding develops when low-energy conditions follow ripple generation and suspended mud accumulates in a thin veneer over the ripples or perhaps only in the troughs (Figure 3–14). The advent of higher-energy conditions and ripple generation will remove the mud from the ripple crests, resulting in thin, discontinuous mud streaks in the rippled sequence (Reineck and Singh, 1980). Fecal pellets may accumulate in a similar fashion, due to their hydraulic differences in comparison with the sand in the ripples. The succeeding period of ripple formation covers the mud veneer and buries it within the cross-stratified ripple sequence (Figure 3–15). Actually, much of the flasers is of fecal origin, but compaction almost immediately destroys the identity of the pellets. X-radiography may enhance pellet definition and permit recognition.

As the availability of mud increases, there becomes a point where the mud is incorporated in continuous fashion in the rippled sequence. This is called **wavy bedding** and is characterized by undulating thin mud layers between rippled sand. As the mud increases, this undulating character is diminished.

Further increase in mud accumulation and decrease in rippled sand gives rise to lenticular bedding (Figure 3–16). The rippled sand somewhat resembles boudins; some are connected and some are not. Generalized diagrams of the typical styles from flaser bedding through lenticular bedding are presented in Figure 3–17.

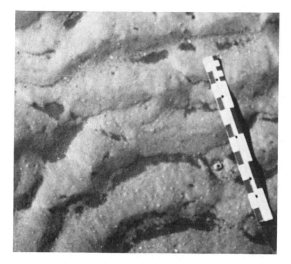

Figure 3-14 Mud accumulating in the troughs of ripples; incipient flaser bedding.

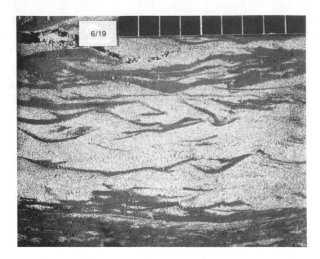

Figure 3–15 Photograph showing flaser bedding. Scale is in centimeters. (Courtesy of H.-E. Reineck.)

Graded bedding. Sedimentation units that are characterized by vertical gradation in particle size are said to be graded (Figure 3–18). The term graded bedding has been applied to such units for decades, although some authors (Friedman and Sanders, 1978) prefer the term graded layer. Most graded units show a gradual trend from relatively coarse at the base to finer at the top (normal grading), although the opposite situation (inverse grading) may occur. Normal grading is commonly used to indicate stratigraphic top and bottom (Shrock, 1948).

The particle distribution may display two distinct styles (Middleton, 1967): one in which there are no fines in the lower part (distribution grading) and one in which fines are distributed throughout the unit with grading restricted to the coarse grains (coarse-tail grading) (Figure 3–19). Distribution grading is caused by successively

Figure 3–16 Photograph showing lenticular bedding. Scale is in centimeters. (Courtesy of H.-E. Reineck.)

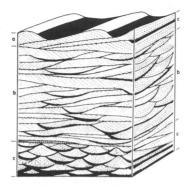

(a) Flaser bedding

(b) Wavy bedding

(c) Lenticular bedding

Figure 3-17 Block diagram showing (a) flaser, (b) wavy, and (c) lenticular bedding types. (After Reineck and Wunderlich, 1968a; Blatt et al., 1980, Figure 5-22.)

smaller particles being made available, whereas in coarse-tail grading all size grades are simultaneously available and settle, so that all sizes appear at the base. The range in particle sizes within a graded unit may be great but is typically within the sand and silt range. There is generally a direct relationship between the range in particle size and the thickness of the graded unit. Each graded unit may be from less than a cen-

Figure 3–18 Photograph of graded bedding showing very coarse units in the Ocoee Formation Precambrian of eastern Tennessee. (Courtesy of S. B. Upchurch.)

timeter to more than a meter in thickness; very thick units commonly have gravel-size particles at their bases.

Turbidity current deposition is a common origin for graded bedding. This was first demonstrated by the classic work of Kuenen and Migliorini (1950) and has since been confirmed by numerous workers. There are many other depositional environments where grading may develop; some occur in shallow water and some are intertidal. Almost any situation where at least a modest spectrum of particle sizes is in motion with a sudden decrease in competence will cause grading as particles settle through the fluid column. Flooding in streams or on river deltas may give rise to thin graded units. Waning currents or changes in current direction, such as on intertidal flats or the foreshore beach, may cause grading. In the latter situation the grading may be reversed (Clifton, 1969). Graded units have also been produced by extreme events such as hurricanes (Hayes, 1967a). It is also possible for organisms to produce grading through bioturbation (Warme, 1967). Inverse grading is commonly produced by grain flow. An excellent summary of the various ways by which graded bedding can be produced has been compiled by Klein (1965).

Imbrication and internal fabric

Although fabric, like graded bedding, is truly a textural property, it is commonly considered together with sedimentary structures. Fabric at the macroscale may be interfaced with data from sedimentary structures in order to interpret ancient depositional conditions, especially direction of sediment transport. Sediment par-

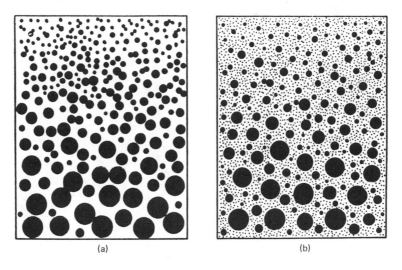

Figure 3-19 Sketch showing the two basic types of graded bedding: (a) grading with fines only near the top; (b) grading with fines distributed throughout. (Modified from Pettijohn, 1975.)

ticles with at least one short axis (i.e., nonspherical) may be useful in such interpretations, whereas nearly spherical particles can have no preferential orientation that can be observed by the unaided eye.

Some sediments or rocks lack orientation regardless of particle shape and are said to be isotropic, whereas those particles that show preferred orientation are anisotropic. This preferred orientation may result from the earth's gravitational field in combination with current flow such as in air, water, or ice (Potter and Pettijohn, 1977). Particles that may be preferentially aligned and observed in the field include gravel-size clasts and fossils. The orientation of such particles may be parallel with the primary bedding surface and show only direction (strike), or it may be inclined and therefore have dip as well.

In analyzing anisotropic fabric of sediment particles one must orient the particles in space by taking strike and dip measurements in the same fashion as is done for bedding plane surfaces and relating these measurements to the bedding planes.

Stream gravels afford an excellent opportunity to observe and analyze fabric and have been recognized for their preferential orientation since the work of Jamieson (1860). Many gravels contain flat or disk-shaped clasts which display an overlapping arrangement called **imbrication** (Figure 3-20). Imbricated particles typically dip upcurrent and therefore they provide an excellent criterion for direction of transport.

Elongate pebbles are more difficult to interpret because they may be arranged with long axes parallel to flow (Krumbein, 1940) or the long axes may be normal to flow orientation (Rust, 1972a; Boothroyd and Hubbard, 1975). The latter occurs as rolling of the clasts takes place in bed load transport. Prominent stratification results (Walker, 1975).

Glacial till may display fabric in the form of orientation of the gravel-size par-

Figure 3-20 Photograph showing imbricated pebbles in Pleistocene stream gravel near Lyons, Colorado.

ticles within the clayey matrix. Measurement of numerous clast orientations has shown that the long axes parallel the direction of ice flow.

The combination of particle orientation and very low relief ridges on bedding surfaces produces **parting lineations** in rocks. These linear features are prevalent in many thinly stratified sandstones and result in a flaggy condition. Flow direction is parallel to the lineation (Figure 3-21). Initial interpretations were that parting lineations indicated shallow sheet flow conditions; however, they also occur in turbidites.

Orientation of fossils. Numerous varieties and shapes of biogenic remains respond to current flow. In addition to obvious elongate particles such as high-spired gastropods, coral sticks, or large fusilinids, the concavo-convex valves of pelecypods may show preferred orientation. In both cases there may be some difficulty in making proper interpretations. For example, high-spired gastropods are typically oriented with the long axis parallel to flow, but they may have the apex pointing upstream or downstream (Kelling and Williams, 1967; Brenchley and Newall, 1970). The nature of the flow, the grain size of the bed materials, and the shell shape are important in determining the orientation of such shells.

Bivalve shells tend to display a fair variety of orientations, but some useful generalizations are possible. There is a strong tendency for the long axis to be aligned normal to the current direction (Brenchley and Newall, 1970). In addition, the most stable position under current flow is that of convex upward. Clifton (1971) has reported that widespread occurrence of concave-upward bivalves is related to bioturbation activity in quiet water and not to current transport of the shells.

Figure 3–21 Photograph of bedding surface showing parting lineation. Hammer handle is parallel to lineation and therefore current orientation. (Courtesy of E. F. McBride).

The discussion above considered preferential orientation due to current flow; however, waves may also cause similar orientation of shells. The convex-upward position is most stable, with elongation tending to be parallel to the wave crests.

Bedding Plane Structures

The first portion of this chapter dealt with structures that involve bedding styles and fabric; they therefore have some thickness. This section is concerned with sedimentary structures that develop on the interface between successive beds or sedimentation units; they are commonly called bedding plane structures. This large and diverse group of structures can conveniently be divided into two groups; bottom marks, which develop on the base of a bed as it is deposited, and surface marks, which form on the exposed surface of sediment after accumulation. Both types are abundant in the rock record, but the preservation potential of bottom marks is greater than that of surface marks because they are buried as they form or only shortly thereafter.

Bottom marks

This group of structures is also called **sole marks.** They may be produced either by currents or by gravity. Some of the structures are actually generated by particles which are propelled by currents or by gravity (Potter and Pettijohn, 1977). **Flutes** or **flute casts** (Crowell, 1955) are probably the most common structures of this type that are generated by currents. They are somewhat bulbous welts which have a rather steep upstream end and which flair out in a downcurrent direction. Typically, many occur adjacent to one another (Figure 3–22). The flute mark is the excavation on the existing sediment surface, and the flute cast is formed by overlying sediment filling in this excavation immediately after its formation. Although flutes and other bottom

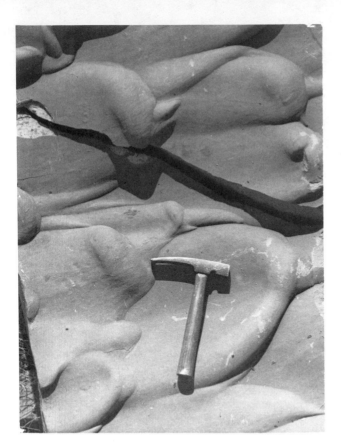

Figure 3-22 Photograph of flute casts with paleocurrent direction from right to left; Smithwick Formation (Pennsylvanian), central Texas. (Courtesy of E. F. McBride.)

marks are commonly associated with turbidite accumulations, they may occur in other depositional environments, especially those where sediment accumulation is rapid and associated with highly turbulent flow.

Flutes are excellent indicators of current direction and usually also indicate the paleoslope direction. As a result they have been widely used in paleocurrent studies (e.g., Yeakel, 1959; Hsu, 1960; McBride, 1962; Hubert, 1966; Sestini, 1970).

In cross section, flutes may be difficult to recognize. Small-scale cross-stratification may be visible, with the steepest inclinations upcurrent and near the blunt end (Potter and Pettijohn, 1977).

A variety of current-generated bottom marks exists that appears somewhat like flutes. Included are transverse scour marks, flute rill marks, and longitudinal furrows and ridges. **Transverse scour marks** (Figure 3-23) have a longitudinal shape similar to flutes, but they are only a few centimeters long and in plan view are ridgelike transverse to the paleocurrent direction (Dzulynski and Walton, 1965). Their origin is attributed to the shear produced by a current moving over a muddy bottom.

Flute rill marks (Figure 3-24) are long, narrow scour features which anastomose, but which have a distinct orientation parallel to current direction. Some authors call these features rill marks (Dzulynski and Walton, 1965), but that terminology is reserved for features associated with tidal flats (Reineck and Singh, 1980).

Figure 3-23 Photograph of transverse scour marks. (Courtesy of S. Dzulynski.)

Longitudinal furrows and ridges constitute another structure type, which is really a special kind of flute. They are nearly straight, long, closely spaced ridges and furrows (Dzulynski and Walton, 1965). Scattered flutes may be present and provide paleocurrent direction. The longitudinal ridges are oriented parallel to the current but do not generally show direction.

Various types of **tool marks** may be produced by the downcurrent transport of objects such as pebbles, shells, twigs or branches, and fish bones. These objects act as the tools that make contact with soft sediment when moved by currents and thereby

Figure 3-24 Photograph of flute rill; arrow indicates paleocurrent direction. (Courtesy of S. Dzulynski.)

create characteristic marks which may be preserved (Figure 3–25). As is the case for all bottom marks, burial is penecontemporaneous with formation; therefore, preservation of apparently ephemeral features is common.

Groove casts or striation casts are linear, whereas flutes commonly anastomose. Grooves indicate current orientation but have no linear asymmetry that would indicate direction. They are formed by currents moving particles which excavate the narrow features (Figure 3–26). Although not always associated with flutes, these features may be combined to give good indications of paleocurrent directions.

As a pebble is transported over a soft mud or fine sand surface, it may bounce, skip, or simply drag along this surface, creating characteristic marks (Figure 3–25). Similar effects are created as twigs, small branches, or skeletal pieces are moved over the substrate. This general type of phenomenon gives rise to **brush marks, skip casts, chevron molds, prod marks,** and **bounce marks.** Careful analysis of each of these types as well as others has been made to distinguish the various modes of origin. The reader is directed to the excellent comprehensive treatment of all bottom mark structures by Dzulynski and Walton (1965) for details and for numerous outstanding illustrations.

Surface marks

Numerous phenomena such as currents and climatic effects cause various types of structures to develop on the upper surface of sediment after it has accumulated but while it is still unconsolidated. Many of these structures are of value in reconstructing ancient depositional environments, but some are present in many environments, thereby lessening their value. Most of these surface structures involve only a few millimeters to a centimeter or so of relief and there is great variation in their preservation potential. Some surface structures indicate paleocurrent directions, but many are not current related.

Current-generated features. Current crescents, sand shadows, rill marks, and current lineations are the most common current-generated surface marks. When

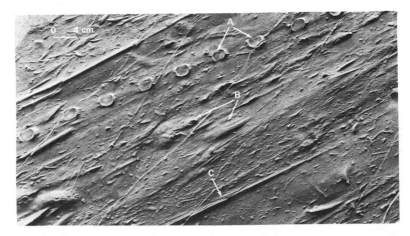

Figure 3–25 Photographs of various tools marks, including (a) skip casts, (b) brush marks, and (c) drag marks. (From Dzulynski and Sanders, 1962.)

Figure 3-26 Photograph showing groove casts which indicate direction only, Haymond Formation (Pennsylvanian), Marathon Basin, Texas. (Courtesy of E. F. McBride.)

large particles such as pebbles or shells are scattered on a sand or mud surface, they present an obstruction. Currents moving water past these obstructions cause crescentic scars around the upcurrent direction which open and lose definition downstream (Figure 3-27). These **current crescents** or obstacle marks have been described in detail by Sengupta (1966). They are good paleocurrent indicators but are not restricted to particular environments. Current crescents abound on tidal flats, foreshore beaches, and streams.

Sand shadows are very much like current crescents except that they are typically formed in aeolian environments. The pebble or shell causes the accumulation of a small, straight ridge of sand downcurrent of the obstacle. Although sand shadows are good paleocurrent indicators, they have a low preservation potential.

The small, bifurcating dendritic erosional features which are common on tidal flats are called **rill marks.** Numerous specific types are known, each with its own name (Shrock, 1948; Reineck and Singh, 1980). The most prevalent variety is that shown in Figure 3-28. Rill marks are good paleocurrent indicators; the bifurcation is always in the downstream direction. Unfortunately, they are uncommon in the rock record.

Climate-related features. Some surface marks, such as raindrop impressions, mudcracks, and frost cracks, give some information on paleoclimate. Cohesive sediments such as mud or muddy sand may develop small impact craters a few millimeters deep and up to a centimeter or so in diameter (Figure 3-29). These are **raindrop impressions.** Some have been observed to be asymmetric with elliptical surface expression (Shrock, 1948), indicating that the rain was falling oblique to the sediment surface, and some have a morphology like small ripples (Clifton, 1977). Raindrop impressions are preserved in the rock record most commonly in continental deposits of arid to semiarid climates.

Mudcracks are somewhat polygonal shrinkage cracks in muddy sediments (Figure 3-30). They are indicative of wet conditions followed by a dry period such as occurs in the supratidal zone, floodplains, playa lakes, or sabkhas. Mud cracks range

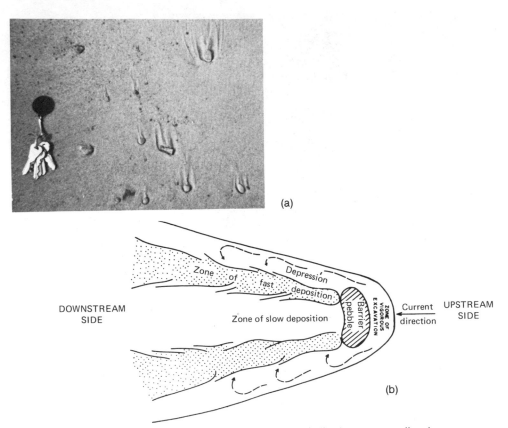

DOWNSTREAM
SIDE

Zone of fast deposition

Depression

Zone of slow deposition

Barrier pebble

ZONE OF VIGOROUS EXCAVATION

Current direction

UPSTREAM
SIDE

(b)

Figure 3-27 Current crescents with the arrow indicating current direction: (a) photograph from tidal flat; (b) diagram showing components of the feature. (From Sengupta, 1966.)

widely in size, both from the standpoint of the cracks themselves and the polygonal shapes that they generate in the mud.

Extreme cold and the resulting frost cracks and ice wedging may cause V-shaped cracks up to a few meters deep. These can then be infilled with sediment and may be preserved.

Miscellaneous surface features. The beach exhibits arcuate marks left by the uprush and backwash of the water on the foreshore. These are **swash marks.** They are extensive, concave seaward, and consist of small ridges only a few sand grains in diameter (Figure 3–31). Swash marks serve as diagnostic indicators of the shoreline when found in the stratigraphic record. Their spacing is inversely related to beach slope (Emery and Gale, 1951). Although they have been found in the rock record, swash marks are rarely preserved.

A special type of ripple may be generated when wind blows sand over a wet substrate. These **adhesion ripples** (Glennie, 1970) develop as sand accumulates in crests which migrate against the wind. The surface of the crests has a pock-marked appearance, which results in a crenulated ripple cross-laminated when viewed in cross

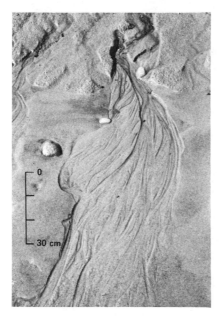

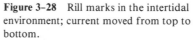

Figure 3-28 Rill marks in the intertidal environment; current moved from top to bottom.

section. Adhension ripples are present in the rock record and are good evidence for eolian deposition.

Vertical movement of fluids up through soft sediment may cause the formation of **pit and mound structures** (Shrock, 1948). These tiny volcano-shaped or pitlike features develop as the gas or water escapes the sediment.

Extreme conditions of salinity or temperature may cause crystals to form in the muddy sediments of tidal flats, sabkhas, or lagoons. These crystals of ice, halite, gypsum, or other evaporites are quite soluble. As a result, **crystal impressions** may form by solution, leaving a mold of the crystal in the sediment. Careful examination of the

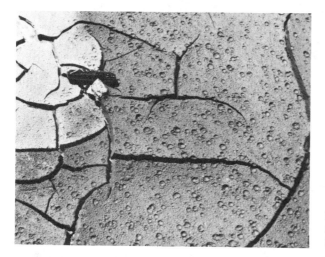

Figure 3-29 Raindrop imprints on mud surface in ephemeral stream bed. Pocket knife is 10 cm long. (From Huber, 1975.)

Chapter 3   Sedimentary Structures

(a)

Figure 3–30 Mudcracks (a) on the floodplain of a stream in the semiarid southwestern United States and (b) in Ordovician strata of western Maryland.

0   30 cm

(b)

shape of these crystals may enable recognition of the mineral species, which in turn can be indicative of some aspects of the depositional environment.

Penecontemporaneous Deformation Structures

Many sedimentary environments are subjected to rapid influx and accumulation of sediment. In some, this sediment is deposited on a nonhorizontal surface. Either or

Figure 3-31 Swash marks indicating maximum uprush of a given wave; they are always concave toward the water.

both of these factors may give rise to deformation of the sediment layers as the result of gravity. This may occur by downslope movement or simply by loading of large quantities of sediment upon unstable surfaces. Commonly, these phenomena produce well-defined and diagnostic sedimentary structures. These structures may be restricted to the bed surface between sedimentation units or may include the sediments within the unit.

The time of occurrence of these structures varies somewhat, but in all cases it is before consolidation of the sediment. It may be coincident with sediment accumulation but is typically shortly thereafter.

Load structures

Deformation of the interface between sedimentation units or of the underlying unit as the result of the mass of the overlying unit creates load structures (Kelling and Walton, 1957). These are the earliest of all deformation structures in that they form as the sediment accumulates. One of the most common types is caused by coarse sediment, sand, or gravel accumulating on soft mud. This gives rise to deformation of the interface between the contrasting textures resulting in **load casts** (Kuenen, 1953). These features are bulbous shapes on the bottom of the coarse unit (Figure 3-32). Although load casts may have some similarities to flute casts, they have no distinct linear alignment nor do they display a regular asymmetry.

A special type of load structure produces tonguelike protuberances of mud intruding the overlying coarse unit. These **flame structures** are good indicators of top and bottom. Some flame structures show paleocurrent direction as the result of transport during the loading (Figure 3-33). A simple type of load structure is formed when a large clast such as a rafted cobble or boulder falls onto a soft, well-bedded substrate. The large clast causes compaction and displacement of the underlying fine sediment. This deformation is rather regular and easily related to its origin.

Figure 3-32 Photograph of load structure caused by coarse, dense accumulation on fine mud.

Ball and pillow structures

Another penecontemporaneous structure developed in alternating layers of sand and mud is the ball and pillow structure, also known as pseudo-nodules. They are elliptical or pillow-shaped masses of sandstone typically discontinuous but numerous along a given stratigraphic horizon. The pillow may have internal bedding which is distorted. Ball and pillow is thought by some to be caused by slumping or

Figure 3-33 Photograph of flame structure that shows paleocurrent direction from left to right; Ochoee Formation, Tennessee. (Courtesy of S. B. Upchurch.)

Part One   Principles

downslope sliding, but there is not wide agreement on their origin (Potter and Pettijohn, 1977). They have been generated experimentally by shock, which causes a liquefaction of the mud and some mobilization of the sediments (Kuenen, 1958). Pillows range in diameter from a few centimeters to about a meter.

Dish structures

Recent interest in **dish structures** has resulted in a better understanding of their origin and significance (Lowe and LoPiccolo, 1974; Lowe, 1975). These structures are discontinuous, concave-upward laminations in sandstone (Figure 3–34). Compared to the surrounding sandstone, they are muddy, finer-grained sand, less sorted, or some combination. Formation is primarily by dewatering of sediment at, and immediately after, deposition. The dishes are separated by columns or pillars of massive sand; thus the term dish and pillar is commonly applied to these features. They are thought to be formed by upward flow of water through the unconsolidated sediment during compaction (Selley, 1964; Lowe and LoPiccolo, 1974).

Convolute bedding

Deformation of well-bedded sediments, typically mudstones or sandstones, produces crumpling or extremely complex folding called **convolute bedding** (Figure 3–35). This feature is generally formed in muds or fine sand and is characterized by remarkable continuity of the individual beds in the convoluted unit (Figure 3–35). Convolute bedding is commonly and incorrectly called a slump structure. Gravity-generated **slump structures** are due to gravity only and are associated with soft-sediment faulting.

Convolute bedding may form as the result of various phenomena, although

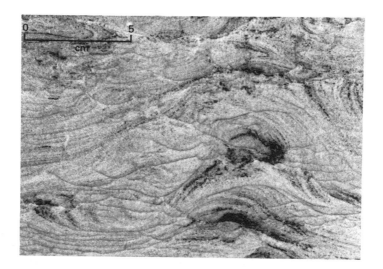

Figure 3–34 Photograph of dish and pillar structure; Mississippian of eastern Oklahoma.

Figure 3-35 Photograph of convolute bedding; Dimple Formation (Pennsylvanian), Marathon Basin, Texas. (Courtesy of E. F. McBride.)

some liquefaction appears to be necessary in all cases (Potter and Pettijohn, 1977) with some type of loading phenomenon being involved. Although convolute bedding is typically associated with turbidite sequences (Sanders, 1960), it occurs in other environments as well (Dott and Howard, 1962).

Slump structures

All sedimentary bedding features which are the result of penecontemporaneous deformation due to movement of sediment primarily by gravity are called slump structures. As is the case for many of the structures in this category, they are associated with rapid sediment accumulation and slope instability. In a general way slump features resemble complicated, "toothpaste"-type folding with overturned folds and faults. Kuenen (1965) has reproduced these structures in the laboratory. Slump structures are known from glacial deposits, point bars, dunes, deltas, and tidal flats as well as turbidite-type sequences. Paleoslope indications are possible by measuring the orientation of fold axes within the slump folds (Potter and Pettijohn, 1977).

Injection structures

The forceful injection of sediment, typically sand, through existing layers gives rise to sandstone sills, which are concordant, and sandstone dikes, which are discordant. These features have been recognized since Darwin's trip around the world (Potter and Pettijohn, 1977). They range widely in size from a few centimeters to several meters in thickness. These tabular sand bodies are injected as the result of a combination of liquefaction of the sand and loading. Although most sandstone dikes rise from a source below, some can be demonstrated as being injected from above. The source bed for the injection is traceable in most examples. Excellent summaries of these features are contained in discussions by Shrock (1948) and Potter and Pettijohn (1977).

BIOGENIC STRUCTURES

Geologists have long recognized that numerous structures preserved in the stratigraphic record are the result of various activities of organisms, although no actual part of the organism is included in these structures. These structures are generated by the interaction of living organisms with the sediment or rock substrate. The study of such structures is generally called **ichnology** and the structures themselves are referred to as **trace fossils** or **lebensspurren**. The term **bioturbation** (Richter, 1936) is typically applied to the reworking of sediment by organisms.

The study of trace fossils has taken on increased interest and sophistication beginning in the 1950s with the detailed work of Seilacher (1953, 1954, 1964), Schäfer (1956, 1972) and Reineck (1955, 1970). More recently, excellent books providing comprehensive treatment of the study and significance of trace fossils have been completed (Crimes and Harper, 1970; Frey, 1975). The interested reader is referred to these and other specific literature for details of the taxonomy, morphology, and significance of trace fossils. The following discussion will be limited to a general discussion of the important types of trace fossils and their significance in reconstructing sedimentary environments of deposition.

Borings and Burrows

One of the frequent improper uses of terminology that is prevalent among most geologists is the application of the terms "boring" and "burrow." Although used interchangeably, even unfortunately in the *Glossary of Geology* (Gary et al., 1972), there is a distinct and important difference. Both are verbs as well as nouns and relate to biologic activities of a broad spectrum of organisms. **Boring** is to be restricted to that penetration of a hard substrate such as rock, shell, or even wood. In contrast, a **burrow** is the result of that activity in *unconsolidated* sediment.

Borings

Organisms may bore into hard surfaces for protection or in search of food. Boring organisms include fungi, algae, sponges, worms, gastropods, bivalves, and echinoids (Warme, 1975). Sometimes the individual expires in the boring and enables one to ascertain the boring organism. Borings may occur either on shells, generally due to the search for food, and on rock substrates, primarily for protection. It is the latter type that is of greatest importance to the reconstruction of ancient sedimentary environments.

Borings in rock surfaces may indicate ancient shorelines, such as in beach rock; they may indicate unconformable surfaces where lithified sediment is exposed, as a marine substrate; or they may occur in the framework of reefs. Rock borings are particularly common in carbonate rocks because of the relatively short time necessary for lithification and the fact that it may occur at the surface.

Bored surfaces commonly display an iron oxide stain. Typically, the borings are filled with sediment of the overlying bed (Figure 3–36) and in some situations the borings will be filled with a sediment different from either the composition of the bored rock or the overlying unit. In such cases it is apparent that the unit deposited subse-

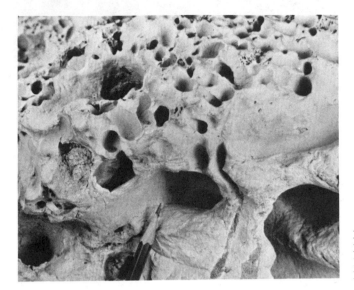

Figure 3-36 Bored surface on Miocene Pungo River Formation, North Carolina. Note pencil for scale (Courtesy of S. R. Riggs).

quent to the boring was also removed but that sediment in the borings remained and was covered by a succeeding sediment. An excellent summary on borings and boring organisms is presented in Warme (1975).

Burrows

A great variety of organisms makes holes in sediment. These burrows take on a broad spectrum of sizes, shapes, and orientations which may be related to specific environments. The more common groups of burrowers are some of the worms, arthropods, mollusks, and echinoderms.

Most burrows serve a combination purpose as both shelter and as a locus for feeding. Various suspension feeders live in burrows and circulate the water and its contained particulate matter through their burrow. The simplest burrow is a single vertical tube. Some organisms construct U-shaped burrows; some have very complex systems with many branching burrows (Figure 3-37). Burrows may be lined with mucus or detrital material to give them an aspect of permanence and also to keep them clean (Reineck and Singh, 1980).

It is possible to categorize burrows or bioturbation structures on the basis of the behavioral activities by which they are formed. Some are for feeding, which are temporary. Those for dwelling are at least semipermanent. Escape structures indicate the need for the organism to move relatively rapidly and are commonly transitional with feeding or resting structures.

Each burrowing species creates its characteristic burrow. This then enables us to determine the presence of various taxa or at least major groups of organisms within a sedimentary sequence even if no actual remains of the organism are present. As a result, these data add to the spectrum of information used in reconstructing ancient sedimentary environments.

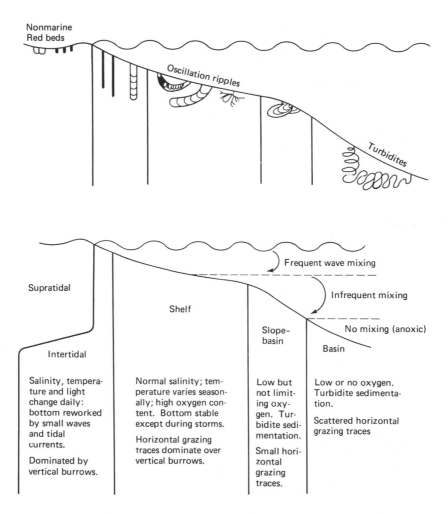

Figure 3–37 Diagram shows some of the typical burrow shapes in relation to the sediment-water interface. (After Rhoads, 1975, in Frey, 1975.)

Burrows may be of value in determining tops and bottoms, especially those that are U-shaped. They may also tell us something about the nature of the sediment substrate, that is, its degree of firmness or softness.

Tracks, Trails, and Resting Traces

Epifaunal organisms may interact with the soft sediment substrate and produce characteristic markings which, in turn, may be preserved in the stratigraphic record. There are essentially two broad types of these biogenic structures: those caused by organisms moving over the sediment surface and those generated by the organism coming to rest on the soft substrate.

Tracks

Virtually everyone who has walked over a sandy beach, tidal flat, or other exposed sediment surface has seen the tracks of animals. Most such tracks are generated by vertebrates, although arthropods also may create them. In fact, the initial studies of trace fossils involved vertebrate tracks (Sarjeant, 1975). Bird, amphibian, reptile, and mammal tracts have been preserved in the rock record, with some spectacular ones being the dinosaur tracks of the Triassic in Massachusetts (Figure 3-38) and the Glen Rose Formation (Cretaceous) in central Texas.

The value of tracks in interpreting ancient environments is primarily in providing information on the fauna present and by indicating subaerial exposure.

Trails

Many vagrant benthonic animals leave trails or traces as they slowly move from place to place. These trails may be crawling traces where the organism is simply moving from one location to another or they may be browsing traces made during feeding (Seilacher, 1953). Most trails are elongate, narrow, shallow depressions created as the organism moved across the soft sediment (Figure 3-39). The trails may be branching or anastomosing.

Most vagrant benthonic invertebrates except some arthropods make trails. Gastropods and bivalves probably account for most trails. Typically, trails are found in rather low energy environments, especially out of the zone of wave action. Tidal flats, lakes, lagoons, the shallow shelf, and the deep ocean are prominent sites for trails.

Some trails which are feeding traces are developed on hard rocky surfaces. These are similar in pattern to those on soft sediment but are made by organisms that

Figure 3-38 Photograph of dinosaur track from the Triassic near Holyoke, Massachusetts.

Figure 3-39 Trail of *Polinices duplicata* on intertidal flat.

scrape the surface of the rock for its contained organic material, such as algae, lichens, or fungi. Some gastropods (limpets), amphineurans, and echinoids make such trails.

Resting traces

Many mobile organisms come to rest on the unconsolidated bottom, causing crude impressions of part or all of their body to form on the substrate. These **resting traces** are somewhat like molds but generally lack the detail found in a true mold. The trace may be due simply to the mass of the organism or be the result of the organism partially digging into the sediment (Reineck and Singh, 1980). Some resting traces, such as those made by starfish, have shapes reminiscent of the organisms that made them, whereas fish that come to rest generally do not create shapes that are fish shaped.

Fecal Pellets

Numerous organisms ingest large quantities of sediment in their search for food. As a result of such feeding habits great quantities of this undigestible sediment are excreted and again become part of the substrate. These fecal pellets commonly may be

Figure 3-40 Photograph of fecal pellets made by the ghost shrimp *Callianasa.*

the only biogenic remains preserved. The size and shape of fecal pellets can often be related to the species of organism that produced them.

Some gastropods, bivalves, echinoids, holothurians, polychaetes, arthropods, and fish are among the prominent producers of fecal pellets. Pellets range from about fine sand to pebble size in diameter, but most are sand size and are ovoid in shape (Figure 3-40). They can occur in both terrigenous and carbonate rocks.

Certain lagoonal, estuarine, lacustrine, tidal flat, and shallow marine environments may be dominated by fecal pellets. One such example is the shallow Bahama Banks area west of Andros Island (Illing, 1954; Bathurst, 1975). Additionally, fecal pellets may form flaser or lenticular bedding. Diagenesis, especially compaction and recrystallization, may destroy their definition. As a result, one should be careful to look for them using thin sections or x-radiography.

In addition to coastal and shallow environments mentioned above, fecal material is an important constituent of the deep-sea environment. Worms and holothurians produce pellets and long stringlike fecal material (Holister et al., 1975). These materials are quite delicate but because of the general lack of physical energy in the deep sea, they may persist.

LATE DIAGENETIC OR CHEMICAL STRUCTURES

Features or structures that develop after lithification of the sediment have no significant bearing on our interpretation of depositional environments. It is important, however, that the geologist be able to recognize this type of structure for what it is and thereby not confuse it with primary structures. For this reason a brief description of the common late diagenetic structures is included here.

Mineral Segregations

Some constituents which may be present in quite low concentrations throughout the rock may be concentrated due to migration of ions and subsequent precipitation. Such mineral concentrations may give rise to concretions, nodules, septaria, geodes, and sand crystals.

Concretions or **nodules** are simply amorphous masses of various mineral compositions. They may be any size up to several meters in diameter and some have internal concentric layering. Nodules or concretions are commonly concentrated along a particular stratigraphic horizon. Composition may be chert, iron oxide, pyrite, or calcite, as well as other minerals. Some concretions have fossils in the center, such as insects, plants, or fish. It is assumed that the expired organism and its associated microgeochemical environment resulting from tissue decay caused formation of the concretions.

Prominent exceptions to the diagenetic nature of nodules are the manganese and phosphate nodules that are forming on the modern ocean floor.

Septarian structure is also related to concretionary forms. A developing concretion may have a shrinking mass in its interior. The cracks developed from this shrinkage fill with a mineral precipitate giving a boxwork-like vein structure (Figure 3–41). Calcite is the most common infilling mineral.

Solution may cause pockets or cracks, especially in carbonate rocks. Ground-

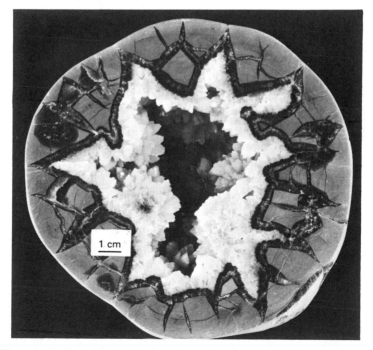

Figure 3–41 Septaria structure in a concretion; the septaria is formed by calcite precipitation in the cracks.

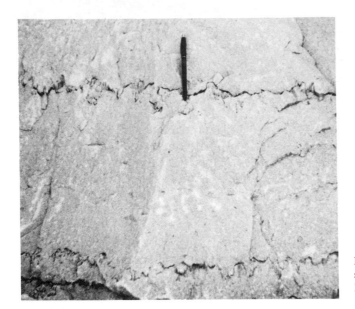

Figure 3-42 Stylolites in a carbonate section in the Mississippian of southern Indiana.

water provides ions in solution which eventually precipitate in these vacuities. Typically, calcite aragonite and quartz are the minerals. The subspherical infilled structures are called geodes and may display extremely beautiful crystal development.

Solution Structures

Pressure and the related solution give rise to distinctive structures. Although such features are most abundant in carbonates, they may also occur in quartz arenites and evaporites.

The irregular and interpenetrating nature of rocks on either side of a thin dark seam of clay characterizes the typical **stylolite** (Figure 3-42). Pressure causes solution, with insolubles concentrated along the stylolitic surface. These features are common in a variety of sedimentary rocks and also in marbles and quartzites (Pettijohn, 1975).

ADDITIONAL READING

COLLINSON, J. D. AND THOMPSON, D. B., 1982. *Sedimentary Structures*. George Allen & Unwin, London, 194p. A nice, succient treatment of the subject with good illustrations.

CONYBEARE, C. E. B., AND CROOK, K. A. W., 1968. *Manual of Sedimentary Structures*. Bur. Min. Resources, Geol. and Geophys., Canberra A.C.T., Australia, Bull. No. 102, 327 p. Little text information on structures and their formation but an excellent collection of photographs of structures from the rock record.

Middleton, G. V. (ed.), 1965. *Primary Sedimentary Structures and Their Hydrodynamic Interpretation—A Symposium.* Soc. Econ. Paleontologists and Mineralogists, Spec. Publ. No. 12, Tulsa, Okla., 265 p. A collection of papers, most of which are classics, which served as the springboard for most of our present concepts on bedform generation, migration, and quantification of bedform studies.

Pettijohn, F. J., and Potter, P. E., 1964. *Atlas and Glossary of Primary Sedimentary Structures.* Springer-Verlag, New York, 370 p. The best picture book of sedimentary structures. It is comprehensive and gives good coverage of both modern environments and the rock record.

Potter, P. E., and Pettijohn, F. J., 1977. *Paleocurrents and Basin Analysis,* 2nd ed. Springer-Verlag, New York, 425 p. and plates. This is an excellent book showing the value of sedimentary structures in interpretations of paleogeography; a good practical text for all geologists who deal with basin analysis.

Reineck, H. E., and Singh, I. B., 1980. *Depositional Sedimentary Environments.* 2nd ed. Springer-Verlag, New York, 549 p. (Part I). The best summary available on the broad spectrum of primary sedimentary structures, their recognition, and environments of formation.

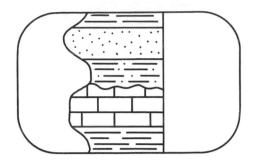

4 Principles of Stratigraphy and Basin Analysis

The title of this chapter is somewhat general. Usually, this topic is treated in an entire volume. However, a book on depositional systems must include at least some of the fundamental concepts of stratigraphy to provide a background that will enable proper understanding of the discussions on modern and ancient depositional systems in succeeding chapters of the book.

The discipline of stratigraphy has increased considerably in scope since the late 1950s. Much of the thrust of stratigraphy prior to that time had been classical from the perspective of today. Nomenclature was of great concern. Geologists were spending much effort on lithostratigraphic, biostratigraphic, and time-stratigraphic successions in a given area, and on the correlations among areas. They were generally concentrating on the recognition, definition, and distribution of various stratigraphic units. Although these pursuits are important, the combination of our relatively recent emphasis on modern depositional environments and the new concept of global tectonics has greatly stimulated the study of stratigraphy.

An understanding of depositional systems from the standpoint of modern sedimentary environments and from the rock record requires expertise in both sedimentology and in stratigraphy. In a general way the study of the modern environment requires more sedimentology than stratigraphy, whereas in the rock record the two disciplines come nearer being equal in emphasis. One of the primary tasks for the successful understanding of depositional systems is the ability to visualize modern environments as they would appear in the rock record to interpret the paleoenvironments represented by the rocks preserved in the stratigraphic record. The modern "soft-rock" geologist must be able to do this continually and must be equally competent both in modern sedimentary environments and in analyzing their resultant features in the rock record.

CONCEPT OF FACIES

The first use of the term **facies** is generally attributed to Amanz Gressly (1838, in Steinker and Steinker, 1972), a Swiss geologist who applied it to lateral changes he observed within time-stratigraphic units of the Mesozoic. "Facies" is now somewhat broadened to include the total of both lithologic and biologic characteristics of a stratigraphic unit. It is also used in metamorphic rocks and even into biological sciences (Krumbein and Sloss, 1963).

Beginning in the 1930s the term "facies" was used widely in classical stratigraphic studies (Caster, 1934; Stockdale, 1939). Applications to stratigraphy and to the analysis of the rock record were typically from either the lithologic or the paleontologic point of view (Longwell, 1949). The former are called **lithofacies** and the latter are **biofacies**. Although such usage is still proper and is still applied, there has been a more environmental and genetic use of the term in recent years. We now typically speak of "reef facies," "delta-front facies," or "tidal flat facies." In other words, each environment, no matter how broad or restricted in its definition, is characterized by its own facies. For example, "fluvial facies" covers the broad spectrum of stream deposits, whereas "point-bar facies" is restricted to a particular sedimentary environment within the fluvial system. The broad fluvial facies is defined by the total characteristics that identify it as stream deposited or at least as stream related. Similarly, the point-bar facies is defined by a more restricted set of criteria.

The operational use of the term "facies" is therefore likely to be based on somewhat different characteristics in the modern environment than it might be in the rock record. Characteristics of, and boundaries for, a modern sedimentary environment can be rather easily established using sediment parameters, biologic attributes if any, and sedimentary structures. Also, topography, water depth, and physical processes can be used in describing a modern facies. By contrast, the resulting rocks have only their preservable attributes, such as petrology, paleontology, sedimentary structures, and geometry, which geologists can use for the facies definition. It is not uncommon for sediments of adjacent and similar modern environments to exhibit the same or very similar appearance in the rock record. For example, it is often difficult to separate the beach and adjacent nearshore sediments in the rock records, as it may be to distinguish the fluvial from the deltaic facies. These potential problems provide yet another demonstration of the importance of a knowledge of modern sedimentary environments to the geologist working with the rock record, either on the surface or in the subsurface.

Walther's Law

Probably the most important single concept that must be thoroughly understood and applied by the sedimentologist and stratigrapher was formulated by Johannes Walther in 1894. He stated that "only those facies and facies-areas can be superimposed primarily which can be observed beside each other at the present time" (Walther, 1894, in Middleton, 1973, p. 979). Also called the Law of the Correlation (or Succession) of Facies, this fundamental principle of geology is used universally but without knowledge of its origin (Middleton, 1973) and is frequently misstated.

The practical use that is made of Walther's concept is the relationship between the lateral distribution of modern sedimentary facies and the vertical succession of facies in the rock record. It is common for the geologist who is examining the ancient record to experience some difficulty in interpreting depositional environments, due perhaps to lack of experience, absence of diagnostic criteria, or a multitude of other reasons. Perhaps one or two paleoenvironments can be recognized with some degree of certainty, but overlying and underlying ones cannot. Use of Walther's law enables one to make a logical interpretation of the stratigraphic succession. Typically, lagoons separate barriers from the coastal plain, and the inner shelf or shoreface is seaward of a barrier. It is axiomatic, therefore, that the vertical succession of environments or facies must reflect these relationships (Figure 4-1); if it does not, the succession may contain interruptions such as the erosion of some facies.

Detailed studies of the broad spectrum of modern environments and the stratigraphic record have shown that there is a limited number of associations of lithology, structures, fossils, and so on (Middleton, 1973). As a result, a number of sedimentary depositional models exists which characterizes various sedimentary en-

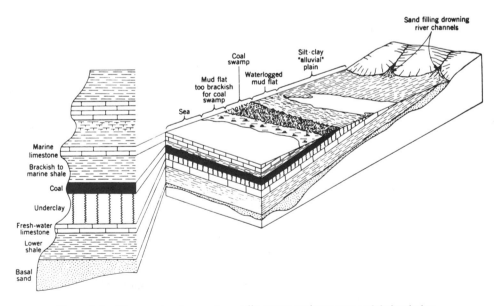

Figure 4-1 Diagram showing modern sedimentary environments and their relationship to stratigraphic succession in the rock record. (From Shaw, 1964).

vironments. Models of modern environments may be characterized by their lateral surficial associations, whereas emphasis on the ancient record is in the vertical sequence. Walther's Law is applied in order to properly understand the relationships between adjacent or related modern depositional systems and their counterparts preserved in the rock record.

A good summary of the use of Walther's Law in the interpretation of ancient sedimentary environments has been compiled by Visher (1965). Before discussing some principles and examples of these vertical sequences it is necessary to understand how any vertical sequence of sedimentary facies might come into being.

In order to accumulate a vertical sequence of changing rock types, changes must have taken place which caused the spatial shifting of the sedimentary environments that produced these rocks. The shift may be accomplished, for example, by tectonic activity, which in turn may cause sea level to change, or which may cause the relief to change; it may be caused by glacial activity, which can also change sea level and change climate, or by any other phenomena which will cause sedimentary environments to be displaced in space during time.

An association of environments which is useful as an illustration is found along the marine coast and related shelf. Commonly, a coastal plain is bounded by a lagoon or estuary, barrier island, nearshore, and shelf environments as one proceeds seaward. A major change in sea level will cause these environments to shift in response to the moving shoreline. Two basic situations exist: **progradation (regression)** as the shoreline moves seaward or **retrogradation (transgression)** as the shoreline moves landward. Most discussions on this topic center around transgression and regression, with sea level implied as rising in the former and being lowered in the latter. Another equally important consideration in the movement of the shoreline is the amount of sediment being transported to the coast. It is not uncommon for coastal deposits to build seaward or prograde during rising sea level, such as in a delta or in some barrier islands. As a result, "transgression" will be used in this text to refer to the situation of landward movement of the shoreline and "progradation" will be used in reference to the seaward migration of the shoreline. In either case the vertical succession of environments or facies shows those environments which are geographically adjacent to be stratigraphically adjacent. The sequence will be reversed but the relationships are the same (Figure 4–2). In a progradational situation the shallow water or the landward environments are on the top of the sequence, and in a retrogradational or transgressive sequence the deeper-water environments are on top at a given locality.

Depositional systems that are not coastal or marine undergo similar shifts, which result in similar sequences. For the most part the commonly preserved sequence is a progradational situation where one environment migrates over another. For example, in an arid region of at least moderate relief, one might find playa lake deposits over which dunes have migrated and an alluvial fan might cover the dunes. Again, environments which are geographically juxtaposed become vertically arranged in the stratigraphic record at a given locality.

Visher (1965) provided some rudimentary characteristics of various vertical sequences in the rock record which show how Walther's Law is applied. A prograding marine model contains all those environments mentioned above and more. In addi-

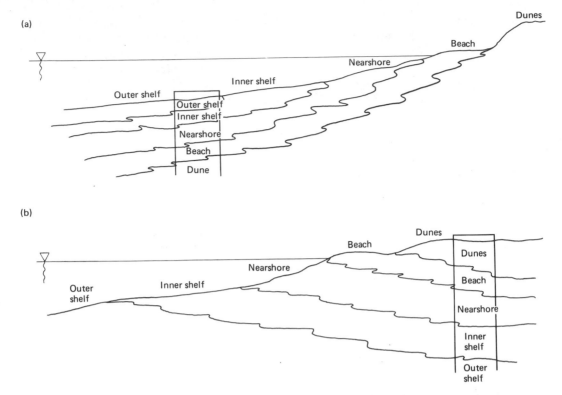

(a)

Dunes
Beach
Nearshore
Inner shelf
Outer shelf
Outer shelf
Inner shelf
Nearshore
Beach
Dune

(b)

Dunes
Beach
Nearshore
Inner shelf
Outer shelf
Dunes
Beach
Nearshore
Inner shelf
Outer shelf

Figure 4–2 Generalized diagram depicting (a) a transgressive sequence and (b) a progradational sequence. A typical stratigraphic section is designated for each.

tion to the textural and mineralogical parameters of the sediment, one must consider sedimentary structures, sediment body geometry, and biogenic constituents (Figure 4–3). This example represents the general types of successions that may be encountered in the rock record and also the criteria that are used to characterize them. It should be observed that these examples are general ones; many others have been formulated, several of which are discussed in succeeding chapters. In examining stratigraphic sequences for features that will be of value in environmental reconstruction, the geologist should utilize all the data available.

PRINCIPLES OF STRATIGRAPHY

To deal with sedimentary rock bodies, either on the earth's surface or from subsurface data, it is necessary to obtain a working knowledge of the terminology and concepts of stratigraphy. The discussion that follows is designed only to serve as a basic introduction. Detailed treatment of the subject may be found in the various references listed at the end of the chapter.

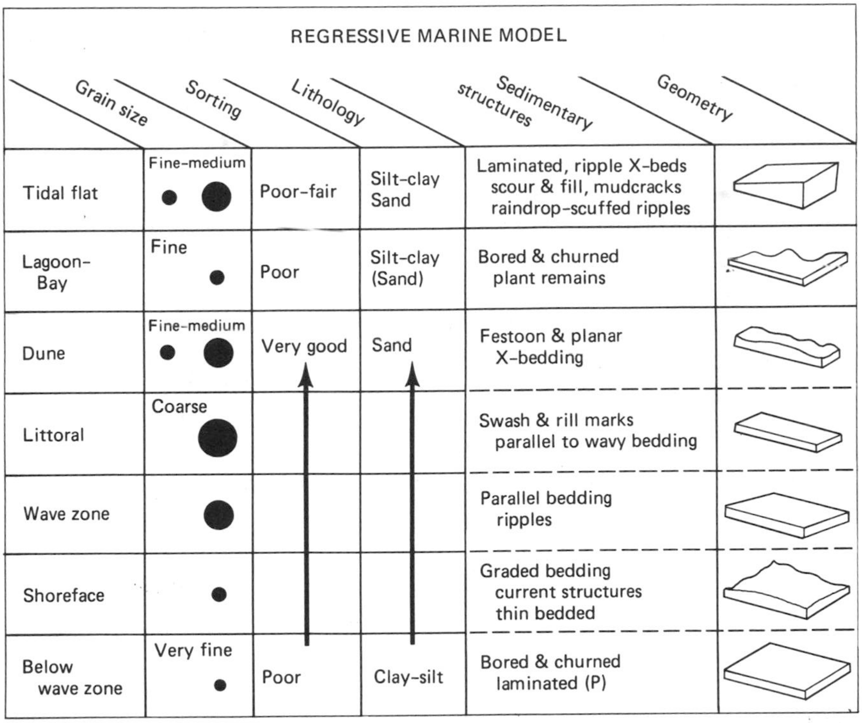

| REGRESSIVE MARINE MODEL | | | | |
Grain size	Sorting	Lithology	Sedimentary structures	Geometry
Tidal flat — Fine-medium ● ●	Poor-fair	Silt-clay Sand	Laminated, ripple X-beds scour & fill, mudcracks raindrop-scuffed ripples	
Lagoon–Bay — Fine ●	Poor	Silt-clay (Sand)	Bored & churned plant remains	
Dune — Fine-medium ● ●	Very good	Sand	Festoon & planar X-bedding	
Littoral — Coarse ●			Swash & rill marks parallel to wavy bedding	
Wave zone ●			Parallel bedding ripples	
Shoreface ●			Graded bedding current structures thin bedded	
Below wave zone — Very fine ●	Poor	Clay-silt	Bored & churned laminated (P)	

Figure 4–3 Various characteristics of a coastal and open marine depositional sequences. Line weight is suggestive of abruptness between successive units. (After Visher, 1965, p. 45.)

Stratigraphic Nomenclature

This section of the chapter may read like a glossary. Stratigraphic nomenclature is extensive and is governed by formal codes (e.g., American Commission on Stratigraphic Nomenclature, 1961). This code has been formulated to standardize terminology and practices used in stratigraphy; in a broad sense its intent is similar to that used in the life sciences, law, and other disciplines.

Only brief discussions of terms useful to a proper understanding of the primary subject of this book are considered here. The interested reader should consult either the North American code itself (American Commission on Stratigraphic Nomenclature, 1961) or the comprehensive volume on the international stratigraphic guide (Hedberg, 1976). The following discussion is based primarily on these two sources.

Categories of classification

Stratified rocks are classified according to their properties, including lithology, fossils, geophysical properties, chemistry, and others. They may also be classified according to depositional environment or time of accumulation (Hedberg, 1976). Three major categories of classification are available, plus a fourth which is based on indirect, geophysical data (Figure 4–4).

Lithostratigraphic units are those which are based on the lithology of the strata in the broad sense. Included are not only the mineral composition of the units but also the textures, sedimentary structures, and other nonpaleontologic characteristics. It is this category which may be based on, or at least related to, the environment of deposition that is emphasized in this book.

Stratigraphic units that are defined on the basis of fossil content are called **biostratigraphic units** (Figure 4–4). In such units no portion of their definition is dependent on the lithology, so that a biostratigraphic unit may contain many rock types or it may be uniform.

If only the age of the unit is considered, units are called **chronostratigraphic** (time-stratigraphic) (Figure 4–4). Their definition may involve use of physical, paleontologic, or radiometric dating, but all of these must be related to time.

Stratigraphic units based on mass geophysical properties of strata are now widely used. These **geophysical units** are especially important in petroleum exploration and in the study of the ocean basins. They may be based on electrical properties such as resistivity, on seismic velocities, or on magnetic characteristics.

Lithostratigraphic units. The fundamental type of stratigraphic unit used by the geologist is the lithostratigraphic unit. The definition of these units is based on

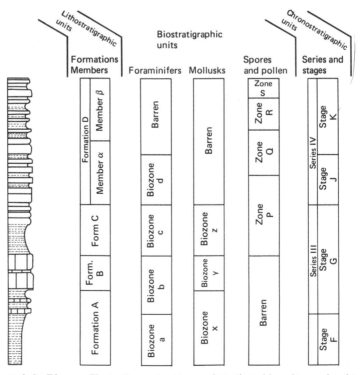

Figure 4–4 Diagram illustrating various types of stratigraphic units as related to a hypothetical section and to one another. (Modified from Hedberg, 1976, p. 8.)

observable criteria, and the extent of lithostratigraphic units is determined solely by the extent of their diagnostic features (Hedberg, 1976).

There is a hierarchy of formal lithostratigraphic units which includes group, formation, member, and bed. A **group** has no definition other than that it is comprised of two or more formations. The basic lithostratigraphic unit is the **formation,** which is a mappable lithologic entity characterized by a particular rock type or types. Formations range widely in both thickness and geographic extent. **Members** are subdivisions of formations displaying lithologic characteristics which distinguish them from adjacent portions of the formation. All, part, or none of a particular formation may be designated by members (Hedberg, 1976). In some lithostratigraphic units a particular bed or beds may be lithologically distinct. Such beds might include a coal bed or a bed of fossils, which would receive a specific designation because of its unique lithologic character.

A lithostratigraphic unit receives its name from the area where it is well exposed or thoroughly described. Geographic names are given to all levels of lithostratigraphic categories. These names may represent towns, rivers, physiographic features, or other recognized geographic places. For example, the St. Peter Sandstone is named after the town of St. Peter, Minnesota, and the Green River Shale is named after the Green River in southwestern Wyoming. There are some units which were named prior to this general policy, such as the Old Red Sandstone in England. These names have not been changed to conform to present policy because of the long history of the name.

All formal designations of stratigraphic units are proper nouns, including not only the geographic name but also the rock type or stratigraphic unit designation associated with it: for example, the Platteville Dolomite or the Knox Group. In some situations both the lithic type and stratigraphic unit are part of the formal designation (e.g., Dakota Sandstone Formation).

Biostratigraphic units. The basic unit in biostratigraphic classification is called a **zone.** The fossils that are used to characterize and define these units may represent a single taxon, a phylum, or the entire fauna. It is very important that such fossils be recognized as indigenous to the deposit in which they occur rather than as being reworked from previous deposits. The use of the term "zone" should not be restricted by scale; it may represent a single bed or the deposits of an entire geologic period.

Although biostratigraphic units are fundamentally different from lithostratigraphic units, their extent may actually coincide. By contrast, the boundaries may be quite different. Coincident boundaries could occur if a particular depositional environment contains an ecologically diagnostic benthonic organism. The opposite situation would occur if a zone were defined on the basis of floating organisms which could come to rest on many sediment types in a variety of depositional environments.

Chronostratigraphic units. This category of formally designated stratigraphic units includes the subdivisions of rocks which are based on particular time intervals. The boundaries may coincide with litho- or biostratigraphic units or they may not. Definition of chronostratigraphic units includes physical, paleontologic, and radiometric data. The system is the major category and is the basic unit for worldwide correlation. Each system, such as the Cambrian, includes all rocks that were deposited during Cambrian time (i.e., the Cambrian Period). The categories of series and stage are in decreasing rank below system.

Relationship between Depositional Environments and Stratigraphic Units

The genesis of lithostratigraphic units and their various components is of primary interest to the geologist who is concerned with the interpretation of ancient sedimentary environments. There is an obvious relationship between the characteristics of a given sedimentary environment and the materials that accumulate there. Some adjacent environments show quite contrasting conditions and sediments; therefore, the rocks produced by such environments will be strikingly different. These differences may be sufficient to warrant different designations as lithostratigraphic units. In other words, a quartz arenite accumulated as a barrier island may be adjacent to an estuarine mudstone (Figure 4-5). Each of these rock types may be designated as a formation. Thus adjacent lithostratigraphic units typically represent adjacent depositional environments.

If one looks closely at a quartz arenite that accumulated as a barrier island, it might be possible to recognize different facies within the lithostratigraphic unit. A well-sorted, well-rounded (supermature) quartz arenite with large-scale cross-stratification could be one lithofacies and represent dunes. Another portion of the quartz arenite units may display features characteristic of the beach (Figure 4-5), such as shells and thin beds.

Another example is that of a reef system of carbonate rocks surrounded by deeper water with low-energy conditions and the accumulation of fine terrigenous muds, producing a stratigraphic sequence of mudstone units and a limestone unit; let us designate them by letters (Figure 4-6). Unit B is the reef system and although it is all carbonate, there is a variety of rock types and specific environments represented. The reef core or framework may be one rock type, perhaps member 1. In the same fashion the fore-reef debris could be a separate unit, as could a back-reef unit (Figure 4-6).

Actual examples afford the best way to show how these relationships may be

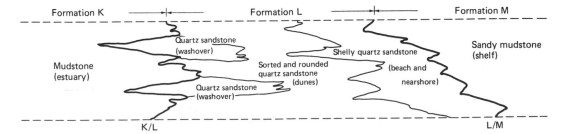

Figure 4-5 Diagram showing mudstone (estuary and shelf) with a quartz sandstone unit between (barrier). The barrier shows various facies designated, although it is dominantly quartz sandstone throughout. Approximate vertical exaggeration is 100×.

Part One   Principles

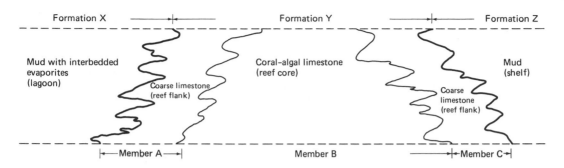

Figure 4-6 Diagram of reef carbonates surrounded by terrigenous mudstones of various lithostratigraphic units. Approximate vertical exaggeration is 100×.

developed and displayed in the rock record. Such an example is found in the Buelandet-Vaerlandet Formation (Devonian) of western Norway (Nilsen, 1969). A maximum of over 3000 m of breccia, conglomerate, and sandstone was accumulated unconformably on a fault scarp over green schists and other basement rock types.

This depositional system includes a half-graben setting, upon which an extensive alluvial fan complex developed (Nilsen, 1969). The total terrigenous sediment package is the Buelandet-Vaerlandet Formation, but there are three distinct lithic units within this formation: Melvaer Breccia, Vaeroy Conglomerate, and Sorlandet Sandstone. Each of these members is characterized by its own lithic attributes and each represents a different depositional environment within the alluvial fan complex (Figure 4-7). The Melvaer Breccia Member is thick and generally lacks stratification. It is comprised of a wide range of particle sizes and rests on the irregular green schist surface. In general, particles are angular and randomly oriented. The Vaeroy Conglomerate Member is also coarse but displays some rounding and a more closed fabric than the Melvaer Breccia. Imbrication is common and long axis orientation of clasts is widespread (Figure 4-7). The Sorlandet Sandstone Member is well stratified, with scattered pebbles and conglomeratic lenses. Both planar and trough cross-stratification are present.

Nilsen's (1969) interpretation shows each of the members as representing a somewhat distinct depositional environment (Figure 4-7). The reader should be cautioned to realize that there is much intertonguing and interfingering of each lithic type. It would be impossible to designate boundaries that in reality separate the lithic units. Pockets of Vaeroy Conglomerate lithology undoubtedly are present in the Sorlandet Sandstone; and so on.

Another similar example is provided by the Niger Delta area during the Tertiary (Burke, 1972). This depositional system prograded over Cretaceous rise deposits and Cretaceous–Tertiary pelagic deep-sea deposits (Figure 4-8). The prograding system consists of a thick basal unit of sandy deep-sea fan deposits which are unnamed by Burke (1972). Overlying this unit is the Akata Formation, which is a shale unit that represents deep, deltaic muds (prodelta). The overlying Agbade Formation, which is relatively thin, is a combination of mud and sand and represents the shallow portion of the delta. The Benin Formation caps the sequence and represents the upper delta sandy environments, which are intertidal to subaerial in depositional environment

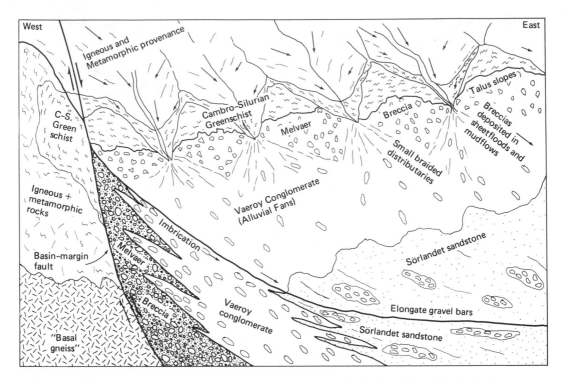

Figure 4-7 Diagram showing reconstruction of conditions during deposition of the Buelandt-Vaerlandt Formation (Devonian) in Western Norway. (From, Nilson, 1969).

(Figure 4-8). The result is a package of lithic units, each of which represents a depositional environment. Lines of isochroneity would be sigmoid in shape and pass through each of the lithic units.

In summary, therefore, it can be stated that lithostratigraphic units typically include paleoenvironments of deposition. Under some circumstances the rocks representing each separate environment deserve a separate lithostratigraphic designation, but in situations where only subtle differences represent different environments, such a designation may not be warranted or justified.

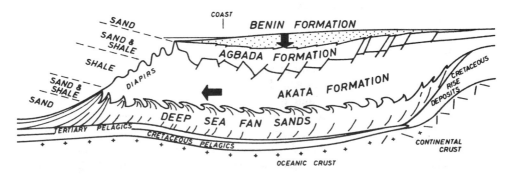

Figure 4-8 Diagram showing relationships between major tertiary lithic units of the Niger Delta (From Burke, 1972).

Breaks in the Record

The strata actually preserved in the rock record represent only a small portion of the material which has been deposited in sedimentary environments that have persisted on the earth for eons. Quite obviously, some strata are eroded and serve as the source for other sediment accumulations, which may themselves become part of the rock record.

Accumulation of sediment, regardless of the particular environment, is rarely continuous. As a result, layers are created due to changes in the sedimentation processes and the initiation and cessation of sediment accumulation. The individual bedding planes therefore represent minor breaks in the stratigraphic record, commonly called diastems.

The above-mentioned breaks are ubiquitous and represent only short periods of nondeposition, or **hiatuses.** There are also surfaces in the rock record that may represent major hiatuses of millions or even hundreds of millions of years. Such breaks are **unconformities,** buried surfaces of erosion. Some types are easily recognized, some are not. An unconformity implies that removal of previously existing strata was followed by accumulation of sediments and subsequent lithification. The amount of time represented by an unconformity, the hiatus, is not specified but is generally of great duration. The reader should observe that it is common for authors to write of ''missing time'' or ''time gaps'' when discussing unconformities (e.g., Matthews, 1974). These statements are erroneous; time is never missing nor are there any gaps in time. What may be missing are strata that represent a particular segment of geologic time.

Types of unconformities

There are three accepted types of unconformities plus another variety that has been described by some authors (e.g., Dunbar and Rodgers, 1957). An **angular unconformity** displays discordance between the underlying and overlying rock units and as a result is easily recognized (Figure 4-9). Also easily identifiable is the **nonconformity,** in which the strata overly eroded surfaces on igneous or metamorphic rocks (Figure 4-10). Both of the above are striking in appearance and require no detailed examination to establish the presence of a buried erosional surface. Implicit in their existence is tectonic activity and erosion (Dott and Batten, 1981).

The **disconformity** is perhaps the most common of the types of unconformities. The term is applied to those unconformities where strata above and below the erosion surface are parallel. Generally, there is little relief on the erosion surface itself (Figure 4-11). Various criteria are available for recognition of the erosion surface and are discussed below.

The **paraconformity** is a type of unconformity for which no physical evidence of erosion is present and the contact of overlying and underlying strata is a bedding plane (Dunbar and Rodgers, 1957). Primary evidence for the existence of a paraconformity is the absence of faunal zones or an abrupt faunal change. The paraconformity is not generally accepted as a valid type of unconformity. In fact, the type example chosen by Dunbar and Rodgers (1957) has been shown actually to be a disconformity (Freeman, 1968b). It is likely that physical and chemical evidence for erosion can be found at most heretofore designated paraconformities.

Figure 4-9 Photograph of angular unconformity with bedded Precambrian quartzite overlain by Cambrian quartz arenite; near Baraboo, Wisconsin.

Recognition of disconformities

When strata bounding an unconformity are indeed parallel, it is commonly difficult to demonstrate the presence of an erosional surface. Actually, many so-called "local" disconformities are in reality regional angular unconformities where the discordance between overlying and underlying strata is subtle or perhaps not demonstrable across a single exposure. Regardless of the geometric relationships of the strata, there are numerous physical criteria for recognition of unconformities. The following list includes most of the common types of evidence for erosion but is not intended to be complete.

Figure 4-10 Photograph of nonconformity with igneous complex overlain by sedimentary rocks, central Wyoming.

Part One   Principles

Figure 4-11 Disconformity with shale unit disconformably overlain by sandstone in the Pottsville Group (Pennsylvanian) near Crossville, Tennessee. Total exposure shown is about 4 m. (Courtesy of S. B. Upchurch.)

1. *Basal conglomerate.* The presence of a conglomerate at the base of the strata overlying a suspected unconformity may indicate erosion. Of particular significance is the presence of clasts in the basal conglomerate that display the same lithology as the underlying unit and were apparently derived from it.

2. *Weathering.* Buried and preserved soil horizons are excellent evidence for the presence of an unconformity. The existence of a soil indicates subaerial weathering. A related weathering phenomenon, the oxidation of iron sulfides, produces reddish brown coloration on the erosion surface. In some unconformities the soil profile or other unconsolidated weathering product is stripped away during transgression and deposition of overlying strata. It may be that the iron-stained surface is the only weathering phenomena preserved.

3. *Truncation.* Angular unconformities display obvious and marked truncation of strata. There are also small-scale varieties of truncation that can demonstrate erosion and be of great value in recognizing disconformities. Large particles such as pebbles or cobbles and fossils incorporated in a rock unit may be truncated at the erosion surface. This phenomenon is best displayed in carbonate rocks, where erosion is at least partially of a chemical nature; in terrigenous rocks the large particles would tend to be plucked from the eroding surface rather than truncated.

Truncation may occur at a scale not discernible with the naked eye. Using the petrographic microscope or the electron microprobe it is possible to identify microscale erosion (Freeman, 1968a, 1969). Both carbonate and terrigenous rocks may display such small-scale truncation.

4. *Borings.* The presence of a bored surface (see Chapter 3) may be indicative of an unconformity. The fact that borings must have been developed on a lithified surface implies that the surface was eroded. The period of time represented by such an unconformity, the hiatus, may be relatively brief. Bored surfaces are common on carbonate rocks. Cementation in carbonates typically takes a brief time and may even take place in the marine environment. As a consequence a carbonate sequence may contain numerous bored surfaces, but the total may represent only a brief time and little removal of strata (i.e., a diastem).

5. *Relief.* Erosion of any lithified surface creates some relief on that surface even if it is at the microscopic scale. Actually, the relief on a buried erosion surface may range up to tens of meters with underlying and overlying strata still being parallel. Obviously, large-scale relief on such a surface is good evidence for erosion and the more subtle types of evidence described above need not be considered (Figure 4–11). Small-scale relief, on the other hand, may be difficult to recognize and may be the only demonstrable evidence for erosion. The amount of relief is not necessarily proportional to the length of time represented by the uncomformity.

Recognition of disconformities and determination of some scale of the time represented by the disconformity are of great importance in the reconstruction of ancient sedimentary environments from the stratigraphic record. During the process of constructing a sequence of environments of deposition from the rock record, one must remember that according to Walther's Law, adjacent sedimentary environments must be juxtaposed in the vertical stratigraphic section. If it is determined that the reconstruction of environments for a particular vertical section is not a logical succession, there are two possible explanations: the depositional environment(s) has (have) been misinterpreted or an unconformity may be present.

TECTONICS AND SEDIMENTATION

Tectonics are the large-scale dynamics of the crust and related portions of the earth. This topic typically involves the outer few hundred kilometers and is concerned with features hundreds to thousands of kilometers across. Movements on the order of hundreds to thousands of kilometers horizontally and up to a few kilometers vertically are not uncommon. In summary, this is the really big picture of geologic features on the earth!

New concepts of tectonics that had their earliest roots in the early twentieth century did not really become well formulated until the 1960s. Now the new global tectonics, plate tectonics as it is commonly called, has caused a considerable change in how geologists view the dynamics of the earth's crust and underlying upper mantle. The basic fundamentals of these concepts are assumed knowledge in the following discussion. The interested reader is referred to excellent articles by Isacks et al. (1968), Dewey and Bird (1970), and Dickinson (1970, 1974) for general discussions of the plate tectonics concepts.

Basin Development

As crustal plates shift their position, major tectonic elements develop and depositional basins evolve. The various interactions and settings of the plates and their arrangement in space and during time determine the nature of the basins that develop. The major events that control basin development are the opening of ocean basins with associated rifting and the closing of ocean basins accompanied by the collision of continental blocks (Dickinson, 1974a).

Dickinson (1974a) has recently summarized basin evolution with a scheme that includes five primary types: oceanic basins, rifted continental margins, arc-trench systems, suture belts, and intracontinental basins.

Oceanic basins

There are basically three major elements of the typical ocean basin and they develop in all the various oceanic basin tectonic settings. These elements are the rise crest, the rise flank, and the deep basin. Rise crests are dominated by **ophiolites,** which are the rocks of the primary ocean floor produced along spreading centers (Vine and Moores, 1972). The rise flanks contain the cooling oceanic material which moves away from its place of origin. Pelagic, deep-sea sediments dominate sedimentation at these sites.

Turbidite sequences are typically dominant in the deep basins and may be interbedded with pelagic sediments. Turbidites may cover the abyssal plains and reflect their origin; adjacent to volcanic islands or carbonate platforms the turbidites will be volcanoclastic and carbonate, respectively (Dickinson, 1974a).

Rifted continental margins

Rifted continental margins are formed as continental plate separation occurs. These systems and their associated basins may be the most important tectonic elements in the rock world and certainly they are among the most complex sedimentary basins. Together with separation there is rapid sedimentation on the margin forming the typical two-part sediment prism. The nearshore and shelf systems have been termed the miogeocline (Dietz and Holden, 1966) and the deep-water prism of turbidites and related sequences at the base of the continental slopes is called the eugeocline. These terms are replacements for "miogeosyncline" and "eugeosyncline" as discussed in the classic work of Kay (1951).

Dickinson (1974a) has described a fivefold sequence in the evolution of rifted continental margins: prerift arch, rift valley, proto-oceanic gulf, narrow ocean, and open ocean. The prerift arch stage consists of broad domes with alkaline volcanics; these are the "hot spots" of some authors (Burke and Dewey, 1973). Such a tectonic setting gives rise to erosion of the domal areas and basinal accumulation of the terrigenous sediments that are produced.

The next step is the formation of rift valleys characterized by grabens and half-grabens. Crustal tension and extension causes high-angle block faulting and related volcanism, resulting in high relief and rapid erosion. The Triassic basins of the northeastern United States are excellent examples. As continued separation occurs, subsidence permits flooding in isolated fault-block basins, with thick evaporite accumulations in the proto-oceanic gulf (Hutchinson and Engels, 1972; Dickinson, 1974a, 1974b). The Red Sea is an excellent example of such a tectonic system.

Dickinson (1974a) has chosen to separate the true open marine ocean into two stages. The narrow ocean may experience sediment transport in such a fashion that the block on one side of the ocean may receive sediment from the other side. Once the two sides of the ocean cease to influence one another and the spreading center forms a midoceanic rise, the open-ocean situation exists. After separation and the onset of marine conditions, the continental margin develops with a basal terrigenous or clastic phase which develops as the result of initial high relief (Figure 4–12). This is followed by a carbonate-shale phase and the accompanying rise deposits, which develops a margin similar to that of the present east coast of the United States. Viewed areally, the breakup and separation typically yield irregular continental block margins due to rifts over hot spots (Dewey and Burke, 1974). Failures in the arms commonly give rise to river systems with prograding deltas in the reentrants of the continental block (Figure 4–13).

Arc-trench systems

When plates collide the resulting basinal development is in marked contrast to the rifted example discussed above. Plate collisions resulting in arc-trench systems typically occur in two major types: in one, oceanic plates collide; in the other, an oceanic plate collides with a continental plate. The former is typically of the arc-trench systems of the southwestern Pacific Ocean and the latter is typical of the western margins of Central and South America.

The arc-trench system is characterized by a trench, a subduction zone, and four distinct basin types which develop on the overriding plate (Dickinson, 1974a). Trench sediments are typically turbidites and are deformed or consumed during subduction. Landward of the trench-slope break is the fore-arc basin (Figure 4–14). These basins receive sediment from the adjacent arc structures where volcanic, plutonic, and metamorphic complexes are source rocks. The fore-arc basin may include shelf, deltaic, turbidite, and terrestrial facies. Intra-arc basins may develop within the island arc system and are dominated by volcanoclastic sediments. The back-arc area lies away from the island arc system and contains a variety of basins. Turbidite sequences are abundant, with some miogeoclinal strata in those basins which tend to receive their sediments from oceanic crust (Figure 4–14). Fluvial, deltaic, and miogeoclinal strata accumulate in basins developed on continental crust adjacent to the cratonic margin (Dickinson, 1974a).

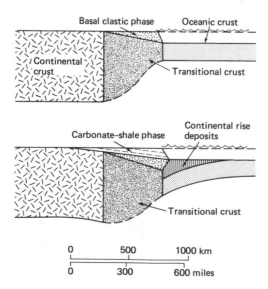

Figure 4–12 Generalized diagram depicting the evolution of rifted-margin prism along continent-ocean interface. (After Dickinson, 1974a.)

Part One   Principles

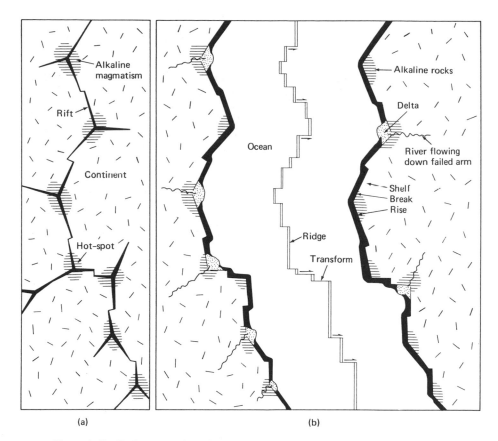

Figure 4–13 Early stages of continental plate separation, showing (a) crust starting to rupture over hot spots and (b) advanced stage of ocean development with irregular margin and river deltas developed at reentrants of failed arms. (From Dewey and Burke, 1974, p. 58.)

Suture belts

Collision between crustal elements of any combination of types causes a crustal suture adjacent to which basins may develop. These basins have been designated by Dickinson as peripheral basins (Figure 4–15). They are characterized by fluvial and deltaic sediments as well as thick turbidite sequences (Graham et al., 1975). The Himalayan suture belt that developed when the Indian plate collided with the Eurasian plate during the early Cenozoic is a good example of this system.

Intracontinental basins

Basins that are surrounded by anorogenic continental crust cannot be explained in terms of plate tectonics, at least on the basis of data currently available. Crustal warping provides sites of accumulation, and subsidence may accompany sediment deposition. This type of basin development is probably due to the movement of a continental block over irregularities in the underlying asthenosphere (Dickinson, 1974a).

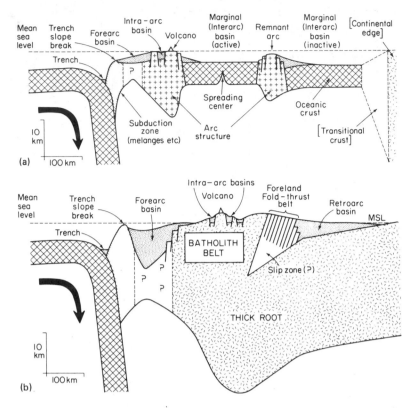

Figure 4-14 Generalized diagrams of sedimentary basins associated with (a) trench island-arc systems and (b) trench-mountain arc plate margins. (After Dickinson and Yarborough, 1976.)

Tectonics and Coastal Environments

Essentially all of the depositional and tectonic elements discussed in the preceding section are what can be called first-order features; that is, they are quite large. Large in this situation means about 1000 km along the plate, at least 100 km wide, and up to 10 km thick (Inman and Nordstrom, 1971). Most coastal depositional systems and shallow margin environments are smaller, except perhaps for the shelf system. Second-order systems are essentially an order of magnitude smaller and include large deltas, estuaries, barrier systems, and dune fields.

These features, both first- and second-order ones and smaller ones as well, form a diverse group of depositional systems which comprise a significant amount of the rock record. Inman and Nordstrom (1971) have applied the new plate tectonics model to a coastal classification scheme. They utilize a scheme of three major categories for the large-scale features. Coasts are of the collision or trailing-edge variety, with marginal seacoasts as a less abundant third type. Collision coasts may involve any combination of colliding plate types. They are characterized by a narrow continental margin and generally little or no coastal plain. Trailing-edge coasts may be young and

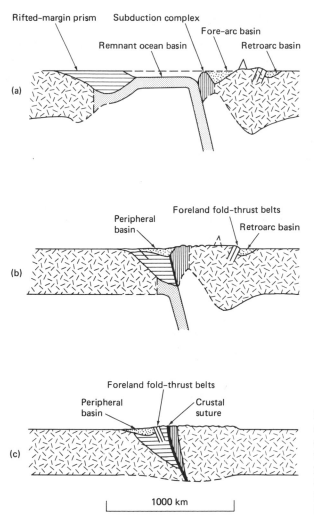

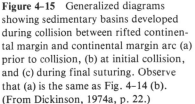

Figure 4–15 Generalized diagrams showing sedimentary basins developed during collision between rifted continental margin and continental margin arc (a) prior to collision, (b) at initial collision, and (c) during final suturing. Observe that (a) is the same as Fig. 4–14 (b). (From Dickinson, 1974a, p. 22.)

therefore near spreading centers (Red Sea, Gulf of California); they may be of the Afro type, where accumulation is low because both sides of the continent are trailing edges; or they may be of the Amero type, with extensive coastal plain and margin development (east coast of the United States). Marginal seacoasts are adjacent to water bodies that are protected by island arcs (Vietnam, China, and Korea) (Inman and Nordstrom, 1971). These are the interarc basins of Dickinson (1974b).

Collision coasts are characterized by narrow margins, generally with tectonic dams (Emery, 1969), and by abundant submarine canyons which head near the shore. Absence of a broad coastal plain and high relief does not permit well-developed drainage systems and large deltas. Trailing-edge coasts have prograding shelves or those dammed by reefs. Extensive coastal plains and well-developed drainage basins give rise to large river deltas (Inman and Nordstrom, 1971).

Role of Tectonics in Sediment Texture and Composition

Many of our present ideas concerning the role of tectonics in sedimentation are attributed to the late P. D. Krynine (1948, 1951) and are well summarized by Folk (1974). In its simplest form the earth's crust contains three layers: a veneer of sediments overlying metamorphic complexes, with plutonic igneous rocks at depth. As tectonic activity is increased, deep rocks are brought to the surface, where they can serve as source rocks for sediments.

Stable crustal plates with basins of accumulation within or a trailing-edge margin will be characterized by low relief and widespread, blanket-type accumulations of sediment (Figure 4–16) such as are present along the southeastern Atlantic Coastal Plain and adjacent continental margin today. Similar conditions were present in the midwestern United States during early Paleozoic times. These accumulations are dominated by sand-size quartz and carbonates with some muds. The terrigenous sediment is supplied by previously existing sediments and sedimentary rocks. Long-term quiescence may permit erosion to continue down through the metamorphic layer and into the plutonic rocks. For the most part this will not greatly affect the composition of sediments accumulating on the stable shelf because of prolonged weathering and transport. Some potassium feldspar may persist, however, as has been observed in Cambro-Ordovician rocks of the upper Mississippi valley (Odum, 1975). Stable tectonic conditions ideally produce sediments which are texturally and compositionally mature or supermature.

Modest tectonic activity such as would be brought about by nonviolent plate collision produces folding, thrusting, and metamorphism as well as at least a modest amount of relief. The relatively rapid erosion of both sedimentary and metamorphic

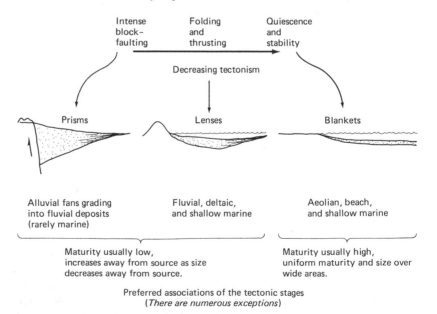

Preferred associations of the tectonic stages
(*There are numerous exceptions*)

Figure 4–16 General relationships between tectonic activity and sedimentation as developed by P. D. Krynine and presented in Folk (1974, p. 109).

Part One Principles

source rocks produces sediments which are less mature both physically and chemically than under quiescent conditions. These fluvial, deltaic, and shelf sediments will contain a significant amount of rock fragments and some mud. The geometry of these deposits is somewhat lenticular (Figure 4–16).

Violent plate collisions are associated with intense deformation, some volcanic activity, and subduction. Source rocks are diverse, relief is high, and the resulting sediments are chemically and physically immature. Various types of rock fragments are abundant. Rather thick, wedge-shaped accumulations of sediment result. An excellent summary of the relationships between sandstone composition and plate tectonics is presented by Dickinson and Suczek (1979).

High-angle block faulting is commonly associated with plate separation. The high relief causes rapid downcutting into the plutonic layer, where granitic source rocks provide sediments rich in potassium feldspar. These sediments accumulate in thick, wedge-shaped deposits which are not widespread (alluvial fans and fluvial deposits) (Figure 4–16). Most sediments are chemically and texturally immature to submature. Some volcanic activity is typically present, resulting in interbedded volcanic flows within the sediment accumulation. A classic example is the Triassic of the northeastern United States (Krynine, 1950).

Although there are some broad generalizations that can be made about the relationship between textural maturity and tectonics, many exceptions exist. The reader should be reminded that textural maturity is a rapid response to environmental conditions. All of the various maturity stages may be present in any of the tectonic settings described above (Folk, 1974).

BASIN ANALYSIS

In some respects the title of this section is redundant with the title of the book. However, it is necessary to consider some basic principles and techniques before plunging into a consideration of individual depositional systems. A sedimentary basin is a place where sediments accumulate. Basin analysis is simply the application of stratigraphy and sedimentology to the geologic history of a basin. It includes all aspects of both topics and some paleontology or paleoecology as well.

In this section consideration is given to various types of data that are useful in basin analysis, the synthesis of such data, and techniques for displaying the data.

Sediment Body Geometry

The size and shape of sediment bodies is probably the most fundamental type of data utilized in basin analysis; however, it is commonly overlooked. There is value in determining the overall dimensions of the entire basin as well as the size and shape of each of the lithostratigraphic units it contains. Such data are obtained through field mapping and measurement of stratigraphic sections, logging of cores or other subsurface samples, or from geophysical data.

It is not uncommon for the stratigraphic sequence in a given basin to be incomplete. That is, unconformities may be present, signifying the removal of a certain part of the rock record. These situations may cause the geologists's job of determining the true geometry of sediment bodies to be quite difficult.

Large-scale sedimentary sequences are grossly wedge-shaped (Potter and Pettijohn, 1977) regardless of composition or location. Within this overall wedge-shaped prism of sedimentary rocks there may be several lithostratigraphic entities each of which may have its own characteristic shape. Let us consider some general size and shape aspects of sediment bodies from various environments. At this stage in the analysis only the gross lithology and the geometry will be considered.

Probably the best example to use is that of quartz arenite with at least moderate sorting. At least four or five major depositional environments may fall within this general lithology. A tabular or blanket sand body could indicate the shoreface or an inland dune field. Linear sand bodies may represent barriers if they are straight or gently curved. Anastomosing patterns suggest the fluvial system. Branching or bifurcating linear sand bodies could develop as a tributary or distributary system; paleocurrent direction would determine which one was present.

This is but one example of how sediment body geometry may be useful. In most cases additional data beyond size and shape are necessary to accurately establish the particular depositional system of the sediment body in question. Certainly, however, a knowledge of the size and shape of the unit will permit the geologist to exclude certain depositional systems from consideration and thereby enable better utilization of time in analyzing the sediment body.

Provenance

Basin analysis includes more than interpreting the environment of deposition of various facies. To gain a comprehensive understanding of the basin it is also necessary to know where the sediments came from, that is, their source rock type and location. The subject of this section, the **provenance** of the basin sediments, covers just those points. Because the basic thrust of this book is at the megascopic scale, emphasis will be on those provenance characteristics that can be seen in the field with the unaided eye or the hand lens. The reader should know that detailed provenance studies require at least petrographic studies, and in some cases detailed mineralogy or trace element composition is required in order to trace the origin of some sediments.

For proper recognition of sediment particles that might be indicative of provenance, at least coarse sand is necessary. Typically, pebbles and cobbles form the basis for most provenance information in the field or at the hand lens scale. Because great detail, such as quartz types or specific mineral composition, cannot be determined at this level of observation, one must look for readily identifiable and traceable detrital particles. Examples are various rock fragment types, chert varieties, particles containing recognizable fossils, and any metallic ores in large clasts. Any of these particle types is exotic enough to be traceable to a source rock.

One of the primary techniques that has been utilized in determining direction of ice movement by Pleistocene glaciers is through the presence of exotic rock particles which can be tied to a known source. Large clasts of basaltic rock with native copper have been found in glacial drift in southern Illinois. The only source for this copper is the Keewenaw Peninsula of Michigan. This information not only establishes the source rock type but also provides general paleocurrent direction from the source to its present location.

Coarse conglomerates characterize the Cambrian section in the vicinity of the Baraboo syncline in south central Wisconsin. The major rock unit that forms the syncline is a red to maroon Precambrian quartzite. The clasts which are included in the basal Cambrian strata are clearly derived from this Baraboo Quartzite. These distinctive quartzite pebbles have been found in Cambro-Ordovician strata 85 km to the south, near Madison.

Trends in Texture of Sediment Particles

The distribution of particle size and shape through time and space can provide substantial information that is useful in basin analysis. In general, the attempt is made to establish the existence of certain trends in the texture of particles, both geographically and stratigraphically. It is not textural properties of the entire rock on which attention is focused but attributes of the individual particles, such as roundness and size.

The geographic variation in grain size has received important consideration by geologists since the 1920s (Potter and Pettijohn, 1977). Much effort has been directed toward grain size patterns in glacial drift and alluvial fans (Lustig, 1965). Students of F. J. Pettijohn at Johns Hopkins University have done comprehensive basin analyses in the Paleozoic strata of the Central Appalachians (e.g., Pelletier, 1958; Meckel, 1967).

Typical efforts involve the measurement and mapping of maximum particle size and roundness (Laming, 1966), although mean pebble size may also be utilized (Lustig, 1965). Increase in roundness and decrease in particle size indicates increasing distance from the source. Such trends also provide paleocurrent direction (Figure 4–17).

Stratigraphic trends in similar parameters are also useful in basin analysis. A decrease upward in maximum clast size implies that relief in the source area is decreasing, thereby decreasing the erosive power and competence of streams. On the other hand, an increase in particle size, particularly one that is abrupt stratigraphically, suggests that the source area has been rejuvenated; it is tectonically active.

Paleocurrents

A large number of sedimentary structures has directional, or at least orientational properties. Descriptions of many, as well as illustrations, were presented in Chapter 3. These structures serve as the fundamental basis for paleocurrent determinations. Actually, any feature that indicates direction of transport is a paleocurrent indicator.

Features such as cross-stratification and bottom marks serve as the best and most abundant structures for paleocurrent determinations. Regardless of the particular directional feature(s) being used, it is necessary for the investigator to measure the spatial orientations of many examples. It is not uncommon for paleocurrent studies to include at least several hundred directional measurements. The field geologist must remember to adjust any directional measurements to accommodate any

(a)

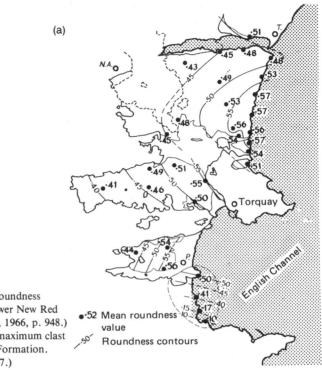

Figure 4-17 (a) Map of roundness variation in pebbles of Lower New Red Sandstone. (After Laming, 1966, p. 948.) (b) Map showing average maximum clast diameter in the Pottsville Formation. (After Meckel, 1967, p. 237.)

•·52 Mean roundness value

⌒·50 Roundness contours

structural movement of the strata involved; this is particularly critical in foldbelt areas such as the Appalachians.

Typically, paleocurrent data are portrayed graphically, generally in the form of a rose diagram. This type of diagram is simply a histogram plotted in such a way that direction is shown (Figure 4-18). Rather than plot the actual number of observations, it is preferable to plot percentages. This allows for comparisons and does not cause problems of large diagrams for some sites and small ones for others; however, one must record wide ranges in the number of observations from locality to locality. The modal class or classes can easily be recognized in a rose diagram or current rose, as it is frequently called (Potter and Pettijohn, 1977). A mean direction may also be calculated and shown on the diagram (Figure 4-18). It is appropriate to categorize directional data for purposes of rose diagrams by stratigraphic level as well as by geographic location. Paleocurrent directions may vary with time and therefore stratigraphic horizon, as well as with space.

Some studies include numerous geographic localities with a spread of directional data throughout the area. An arbitrary and suitable grid is used to partition the

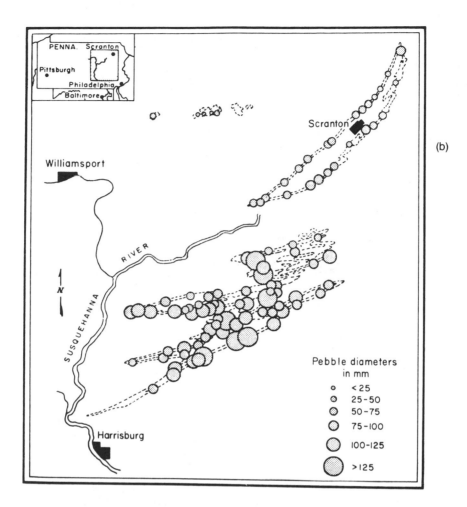

(b)

Pebble diameters in mm

○	< 25
◎	25-50
◉	50-75
◯	75-100
●	100-125
●	>125

data and moving averages are calculated (Potter, 1955; Pelletier, 1958). This approach gives a better visual depiction of directional trends.

Paleocurrent data may display a variety of patterns and may be rather complicated (Figure 4-19). Unimodal data may represent a stream, bipolar modes may be the result of ebb and flood tidal currents, and perpendicular modes could be due to combined influence of wave and longshore generated cross-strata. Polymodal data could be produced on the shelf by varying wind and wave orientations. It is also important to realize that the directional patterns of some paleocurrent data may be misleading. As a result, it is extremely important for the geologist to be able to recognize the feature producing the directional data and to understand its origin. Some examples of these misleading patterns pointed out by Selley (1968) show paleocurrent directions which may be perpendicular to the true direction of current movement (Figure 4-20).

Excellent summaries and pertinent generalizations on paleocurrent investigations are presented by Klein (1967), Selley (1968), and Potter and Pettijohn (1977). These authors point out the necessity of making numerous and careful observations

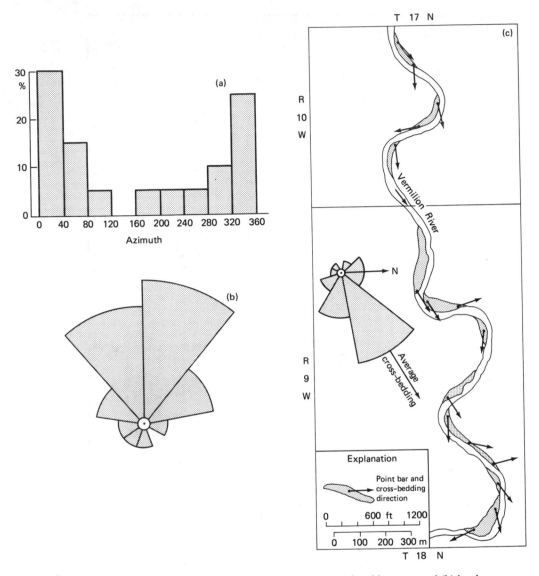

Figure 4-18 Cross-stratification orientation data as shown (a) in a histogram and (b) by the more commonly used rose diagram. An actual situation on the Vermilion River, Indiana, (c) shows cross-stratification data by location and as summarized. (Modified from Potter and Pettijohn, 1977.)

and especially the importance of understanding modern sediment dispersal schemes prior to making interpretations of paleocurrents in ancient depositional environments.

In most modern environments the cross-stratification or other paleocurrent data is more complicated than generally expected. For example, the beach environment may display at least two or three orientations of cross-strata because of dif-

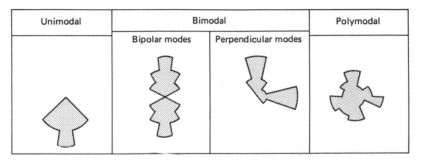

Unimodal	Bimodal		Polymodal
	Bipolar modes	Perpendicular modes	

Figure 4-19 Diagrams of various directional patterns of paleocurrents. (From Selley, 1968.)

ferent transport phenomena within a few meters of one another. Included would be (1) foreshore, seaward-dipping strata; (2) backshore, landward-dipping strata (Klein, 1967), and (3) shore-parallel strata in both directions from longshore currents. All of these can be preserved in very close proximity in the rock record both vertically and horizontally. Unless one has a sound understanding of the beach environment, proper interpretation of the record is impossible.

A similar situation also occurs on a much larger scale. Consider the continental margin in the area of the outer shelf and the slope. Here currents move along the shelf and slope paralleling the contours, and at the same time sediment gravity currents in the form of turbidity currents carry sediment downslope essentially perpendicular to the contour currents (Figure 4-21). The stratigraphic record from such a system would show dramatic differences in paleocurrent direction; however, a proper understanding of the situation is readily provided by the geologist who is familiar with the environments in terms of processes, sediments, organisms, and their total attributes.

Some general models have been proposed for paleocurrents by Selley (1968). Fluvial systems may show converging directions as basin fill takes place by sediment

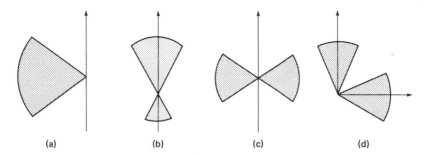

(a)	(b)	(c)	(d)

Figure 4-20 Misleading paleocurrent directional data with arrows showing true current movement: (a) mode perpendicular to current due to laterally inclined point bar cross-strata; (b) bipolar directions due to incorporation of antidunes and their contained cross-strata; (c) cross-strata perpendicular to current such as in seif dunes; (d) perpendicular modes in turbidites due to different directions at base of unit compared to the top. (From Selley, 1968.)

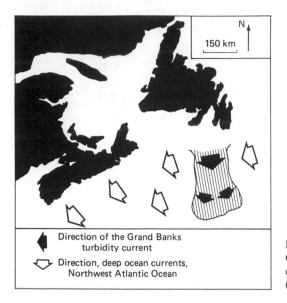

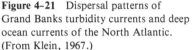

Direction of the Grand Banks
turbidity current

Direction, deep ocean currents,
Northwest Atlantic Ocean

Figure 4-21 Dispersal patterns of
Grand Banks turbidity currents and deep
ocean currents of the North Atlantic.
(From Klein, 1967.)

transport down the paleoslope or on a smaller scale by radiating paleocurrents such
as those which occur on alluvial fans. In the coastal regime a variety of possibilities
exists. When rivers flow toward the coast, paleocurrents may show a seaward direction in combination with longshore currents (perpendicular) or onshore currents
(bimodal in opposite directions). In the absence of riverine influence, the longshore
or onshore directions would prevail (Selley, 1968).

Regardless of the specific situation, the geologist who is interpreting paleocurrent data must be aware of the various current systems that operate in the broad spectrum of depositional environments and their potential for preservation in the rock
record.

An excellent summary of the information that should be included in a paleocurrent investigation is provided by Potter and Pettijohn (1977). They stress the need for
comprehensive data, not only the obvious directional features but also the description of the rock units. All measurements should be taken in a systematic fashion and
recorded in tabular form. It is important that there be a logical progression from the
raw outcrop data through various methods of plotting and culminating with the interpretation of paleocurrent patterns for the area of study.

Summary

Basins can be simple, with only a single source and few depositional environments, or
they may be quite complex, with numerous sources, polycyclic sediments, and many
varied depositional environments. Regardless of the size, complexity, or tectonic setting of the basin under consideration, it is important to conduct its analysis in a
systematic fashion. Some general models for basin development and for basin
analysis have been developed by F. J. Pettijohn and P. E. Potter together with some

of their students and colleagues (e.g., Potter and Pettijohn, 1977; Pettijohn et al., 1972) and they are briefly summarized here.

Probably the simplest general basin model is one involving direct sediment contribution from distinct and geographically separated sources [Figure 4–22(a)]. Here sediment body geometry, mineralogy, and paleocurrent indicators are likely to be straightforward and permit reconstruction of the system without difficulty. Some perturbations on this general model are possible depending on the positions of the source rocks and the amount of mixing of sediments during transportation and deposition [Figure 4–22(b)]. For example, although separated geographically, the sediments produced from the source rocks may be thoroughly mixed in the depositional basin or they may remain somewhat separated (Figure 4–23). Another variation on this theme is caused by the different source rocks being layered at one location. If they are horizontal, erosion should proceed from the topmost unit downward and resulting deposition in the basin should reverse the stratigraphic order of the source rocks (Figure 4–24). Strongly tilted source rocks could provide all three source types to the basin simultaneously and produce a mixed population of sediment.

Many basins receive sediments that are derived from other sedimentary rocks (Figure 4–22). This recycling of grains is common, especially in quartz, and it may be difficult to recognize. In most basin analysis investigations, the objective is to determine the immediate source rocks, and the paleocurrent patterns that produced the strata in a basin are a combination of recycled particles and direct contributions from source rocks [Figure 4–22(b)]. Paleocurrent studies form an even more important part of the analysis of this type of basin than those mentioned above.

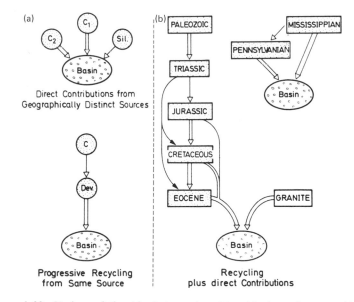

Figure 4-22 Various relationships between depositional basins and sources. (After Pettijohn et al., 1972, p. 314.)

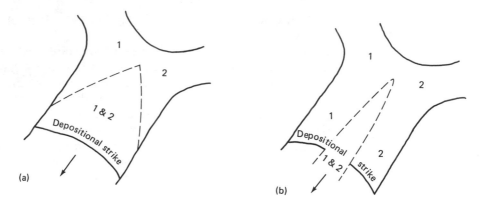

Figure 4–23 Comparison of basins having two sources but with different mixing; (a) nearly straight depositional strike implied by good mixing; (b) strongly curved depositional strike associated with little mixing. (After Potter and Pettijohn, 1977, p. 278.)

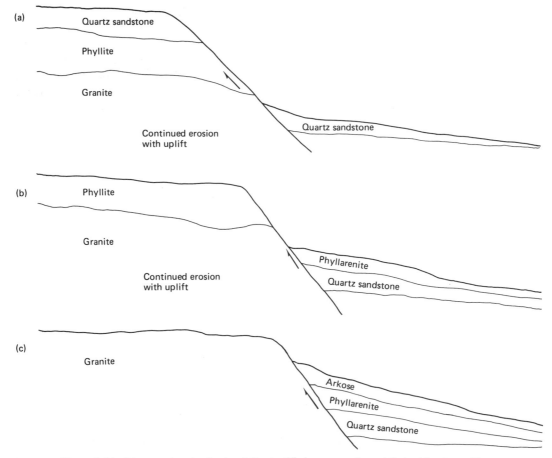

Figure 4–24 Diagram showing horizontally stratified source rocks and their ultimate position in the adjacent depositional basin.

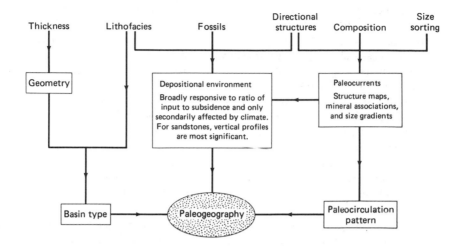

Figure 4–25 Flow diagram for basin analysis culminating with paleogeography. (After Potter and Pettijohn, 1977, p. 342.)

The last half of this chapter had dealt with some general principles and types of data that are necessary to understand the process-response phenomena that bring about the generally thick accumulations of sediment collectively referred to as basins. Regardless of the specific type of basin involved, certain types of data must be utilized in such a synthesis. In essence, these general types of data and their synthesis are what the remainder of this book is all about, but with the approach of considering specific depositional systems. In each of these systems one must determine at least certain standard information: thickness, sediment composition, textures, sedimentary structures. These data are then interpreted and an overall geometry, paleocurrent pattern, and general depositional environment are inferred (Figure 4–25). Further refinement and synthesis of the data allow one to reconstruct the depositional environment in some detail.

ADDITIONAL READING

AGER, D. V., 1973. *The Nature of the Stratigraphic Record.* John Wiley, New York, 114 p. Excellent book on stratigraphic principles that is also enjoyable reading.

AMERICAN COMMISSION ON STRATIGRAPHIC NOMENCLATURE, 1961. *Code of Stratigraphic Nomenclature.* Amer. Assoc. Petroleum Geologists Bull., v. 45, p. 645–660. The rules that North American geologists must follow in stratigraphic nomenclature as well as explanations and examples.

COMPTON, R. R., 1962. *Manual of Field Geology.* John Wiley, New York, 378 p. Good and comprehensive text on a broad spectrum of field procedures in geology. Chapter 12 is devoted to specific procedures involving sedimentary rocks.

DICKINSON, W. R. (ED.), 1974. *Tectonics and Sedimentation.* Soc. Econ. Paleontologists and Mineralogists, Spec. Publ. 22, Tulsa, Okla., 204 p. Excellent group of papers on relationships between tectonics and sedimentation in specific area or basins. Introductory paper by Dickinson is fine summary on the subject.

DUNBAR, C. O., AND RODGERS, J., 1957. *Principles of Stratigraphy.* John Wiley, New York, 356 p. Classical stratigraphy text with all the basic principles of descriptive stratigraphy. The book is dated and does not include much about depositional systems, but it is one all geologists should know.

HEDBERG, H. D. (ED.), 1976. *International Stratigraphic Guide: A Guide to Stratigraphic Classification, Terminology and Procedure.* John Wiley, New York, 200 p. Complete discussions of all aspects of the stratigraphic code; everything you wanted to know about stratigraphic nomenclature and more!

POTTER, P. E., AND PETTIJOHN, F. J., 1977. *Paleocurrents and Basins Analysis,* 2nd ed. Springer-Verlag, New York, 423 p. and plates. An excellent book dealing with all physical aspects of basin analysis. Numerous examples are used for all points of discussion; references are numerous and up to date.

SCHOLLE, P. A., AND SPEARING, D., 1982. *Sandstone Depositional Environments.* Memoir No. 31, American Assoc. of Petroleum Geologists, Tulsa, Oklahoma, 410p. Comprehensive and well-illustrated volume treating depositional environments and basin analysis from petroleum exploration approach.

Part Two

TERRESTRIAL ENVIRONMENTS

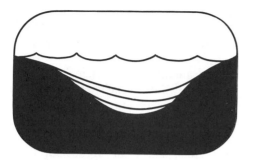

5 The Lacustrine System

Lakes are landlocked standing-water bodies that presently cover about 1% of the earth's surface. About 60% are fresh water (Reeves, 1968). In various portions of the geologic past, they have been much more widespread than at present. A reasonable distinction between lakes and ponds is that lakes are deep enough to prohibit vegetation other than subaqueous types across the water expanse (Reeves, 1968). Size is also a common distinction, although no specifics are used.

Although lakes occur throughout most climatic belts of the world, receive large volumes of various types of sediment, and in general represent a diverse depositional system, they are commonly neglected in considerations of sedimentary environments. Most lake studies emphasize the biological, chemical, and physical aspects of the environment but give little consideration to their geologic aspects. A good case in point is to glance over the table of contents for current texts on **limnology** (e.g., Cole, 1975); geology and sediments are not included. The geologic and related aspects of lakes are well covered in recent compilations by Lerman (1978) and by Matter and Tucker (1978).

General Characteristics and Distribution

Lakes are typically small in comparison to marine water bodies, although some lakes are larger than the smaller marine bodies (e.g., the Caspian Sea is larger than the Persian Gulf). In fact, many lakes are quite large and may also be hundreds of meters deep. Large lakes are much like the oceans in terms of physical processes and sedimentation. The most fundamental differences are the absence of significant tides and the chemistry of the water. The latter is quite variable and numerous lakes are saline.

Lakes tend to develop in regions of internal drainage wherever there is a basin in which water may accumulate and wherever water is available. The latter factor tends to be a limiting one, thus the strong relationship between climate and lakes. Extreme high latitudes do not allow lakes to develop due to nearly constant subfreezing temperatures. Such is the case in Antarctica. Desert environments typically lack moisture in the form of either rainfall or runoff that is necessary to develop lakes, although some do occur. There are ephemeral lakes or playas that form in arid climates; they are treated in Chapter 6. Some arid conditions give rise to saline lakes.

The basin in which a lake develops may have a broad spectrum of origins. Many of the large lakes have resulted from tectonic activity. Tertiary deformation created the basins now occupied by the Caspian Sea and Lake Victoria. Grabens host Lakes Baikal (USSR), Tanganyika (Africa), and Thoe (United States) as well as the Dead Sea (Reeves, 1968; Sly, 1978). Lakes developed in volcanic settings may be very deep but not extensive. Some are formed in calderas (e.g., Crater Lake in Oregon) and others result from lava damming (e.g., Sea of Galilee) (Sly, 1978). Mass-movement phenomena are also responsible for the creation of lake basins. Gros Ventre Lake in Wyoming and Earthquake Lake in Montana are examples of lakes that formed as the result of landslides damming valley drainage.

During the Pleistocene Epoch, and extending to the present, there has been extensive lake development associated with glaciation. Lakes may form due to ice damming, morainal damming, ice scour, kettle development, cirque formation, and other phenomena of glaciation. Excavations by continental ice sheets led to the Great Lakes of North America (Hough, 1958b). Channel shifting in meandering streams produces oxbow lakes.

Man has recently become directly responsible for the development of several large lakes by building dams across drainage systems. One of the most famous is Lake Mead in Arizona.

Types of Lakes

The classification of lakes has been discussed by Twenhofel (1950), Smith (1968), and Reeves (1968), all of whom propose genetic classifications. These categories are primarily the ones mentioned in the preceding section and each is not distinguishable by a particular type or suite of sediments. It is well known that climate is a great influence on lakes. This in turn affects the type of sediment accumulation that is found in lakes. As a result, it appears that a classification based on some combination of sediment type and climate may be a reasonable one for categorizing lakes. Two primary types of lakes are considered: those dominated by terrigenous sediment accumulation and those dominated by chemical precipitates. Superimposed on these two categories are climatic effects such as temperature, rainfall, and latitude. Depth of water and bottom configuration are also factors that affect sediment accumulation.

Related Environments

Other nonmarine standing water bodies may also have geologic significance. The difference between a lake and a pond has been stated. There is a tendency to use the terms "swamp," "marsh," and "bog" in a generally synonymous fashion when referring to poorly drained areas. A **swamp** is a saturated environment, generally with standing water and supporting a stand of trees. Some are very extensive, such as the Big Cypress of southern Florida and the Okefenokee of southern Georgia, but others cover only a few hundred square meters. The term **marsh** is applied to a saturated environment that supports grasses but lacks trees. All saturated environments that contain accumulations of peat are called **bogs,** so that bogs may occur within swamps or marshes.

Although this discussion is concerned with the nonmarine environment, both swamps and marshes develop marine counterparts, which are discussed elsewhere. Marine swamps are characterized by mangroves and are primarily intertidal. Salt marshes are widespread along the intertidal margins of estuaries and lagoons.

Nonmarine swamps and marshes may be of real geologic significance, due largely to their tendency to accumulate large thicknesses of organic debris. This becomes peat and eventually coal. Extensive modern swamps and marshes provide an excellent opportunity for studying the depositional environment of ancient coals (Cohen, 1973).

CHEMISTRY OF LAKES

Lakes are typically considered as being fresh water in comparison with the oceans. In an arid or semiarid climatic region this may not be a valid comparison, because many lakes in such a climate may be more saline than the ocean. That is to say that lakes show an extremely broad range in their water composition not only from one lake to another but in some cases there is great temporal variation within a given lake. A dissolved solids concentration of 5‰ is used to separate freshwater lakes from saline lakes. The chemistry of lakes is important, either directly or indirectly, to the nature and amount of sediment that accumulates within the basin.

The composition of lake water is relatively complex compared to seawater, but concentrations are quite low. The typical freshwater lake in a temperate climate contains a salinity or total dissolved solids concentration of about 120 parts per million (ppm). This compares with a concentration of 35,000 ppm for average seawater concentration. Calcium (Ca^{2+}) and carbonate ions (CO_3^{2-}) are the most abundant dissolved constituents in lakes, with Mg^{2+}, Na^+, K^+, SO_4^{2-}, Cl^-, and dissolved SiO_2 also present (Cole, 1975). A broad spectrum of other ions may be present in individual lakes depending on the composition of the drainage basin that serves as the source of water. Extreme concentrations of ions exist in saline lakes; their chemistry is discussed under "Chemical Sediments" later in the chapter.

One of the most critical factors in the natural history of a lake is the abundance and availability of oxygen in the water. Lakes that contain few nutrients and low productivity are called **oligotrophic;** they contain well-oxygenated waters. By contrast, lakes with high nutrient levels and high productivity tend to be oxygen deficient in the hypolimnion and are **eutrophic.** The lack of oxygen is caused by the excess of organic debris, which requires oxygen in the decay process. Production of great quantities of plants and therefore eutrophication may lead to peat accumulation and the development of bog conditions.

The pH of lakes may also be of importance as far as sediment accumulation is concerned. High pH levels permit precipitation of $CaCO_3$. However, volcanic lakes are typically low in pH and may give rise to solution of certain minerals.

Temporal changes in lake water chemistry are largely seasonal in nature and are strongly related to climate through precipitation, either directly or indirectly. Runoff into the lake changes in volume as rainfall and spring melting take place. Increased runoff causes great addition of dissolved ions, including nutrients. Typically, this should cause an increase in the total dissolved solids, but in shallow lakes the increase in volume caused by extreme periods of runoff may decrease the concentration of dissolved solids. The result is that each lake must be considered separately. It is more difficult to generalize about lake water chemistry than it is about the oceans, which are connected and relatively uniform in their chemistry.

PHYSICAL CHARACTERISTICS OF LAKES

There are numerous similarities yet some marked differences between the physical processes operating in the oceans and in lakes. Virtually all of these processes have some bearing on the distribution and accumulation of sediments. Coastal processes in lakes are the same as along the coasts of oceans except for tidal-related processes. The nearshore, beach, dunes, and related environments are well developed along many lake margins, especially the larger ones. Some lakes also contain spits, tombolos, and other geomorphic features typically associated with marine coasts. The discussion of coastal processes in the chapter on barrier islands (Chapter 12) is applicable to lakes as well.

Temperature

Although temperature distribution is itself not a process, it does give rise to important physical processes within lakes. Generally, water temperature varies greatly from lake to lake and seasonally within a given lake. In all but quite shallow lakes a distinct and fairly persistent stratification develops due to temperature differentials; warm, less dense water overlies colder, more dense water. In such stratified lakes, there is a narrow zone between these layered water masses where the temperature changes rapidly with depth; this is the **thermocline** region, also called the **metalimnion.** The warm surface water is called the **epilimnion** and the cold water mass below the thermocline is the **hypolimnion** (Figure 5–1). The position of the thermocline varies from lake to lake but is generally less than 30m below the surface.

Temperature distribution within the epilimnion may vary both with time due to air temperatures and with depth depending on circulation. Below the thermocline water is typically isothermal, and in deep temperate lakes the temperature is about

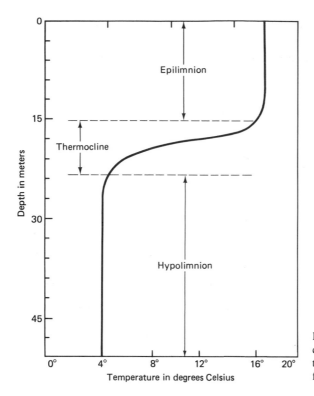

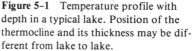

Figure 5-1 Temperature profile with depth in a typical lake. Position of the thermocline and its thickness may be different from lake to lake.

4°C, which is the temperature of maximum density. The temperature profile of a lake is actually a controlling force on the circulation of the lake.

Circulation

Large-scale water motion in lakes may be either wind driven or the result of density gradients. Sediment distribution may be dominated by either or both types of water motion and therefore dispersal mechanisms. Before considering different circulation types it is appropriate to consider the role of climate on lake circulation.

Annual circulation patterns

When the wind blows over a stratified lake, circulation is generated within the epilimnion, but there is little effect on the hypolimnion. If a lake lacks stratification, one might expect that the entire lake could be circulated or mixed by wind-generated motion. Herein then lies the critical characteristics that control lake circulation and therefore affect sediment distribution.

Some water bodies do not undergo mixing at all. They are continually ice covered and therefore are removed from the circulating effects of the wind. Such lakes are **amictic** and are restricted to very high latitudes or elevations. Many lakes are subjected to complete mixing due to wind circulation. This is commonly called overturn and occurs for short periods, usually once or twice during the year. **Monomictic** lakes are characterized by one period of overturn, and **dimictic** lakes show two periods per year. A single period of overturn may occur in low latitudes, where lakes are stratified in summer but are isothermal and therefore circulate in winter. The opposite situation exists in the high latitudes, where lakes are frozen over in winter but open and nearly isothermal for mixing in summer (Cole, 1975).

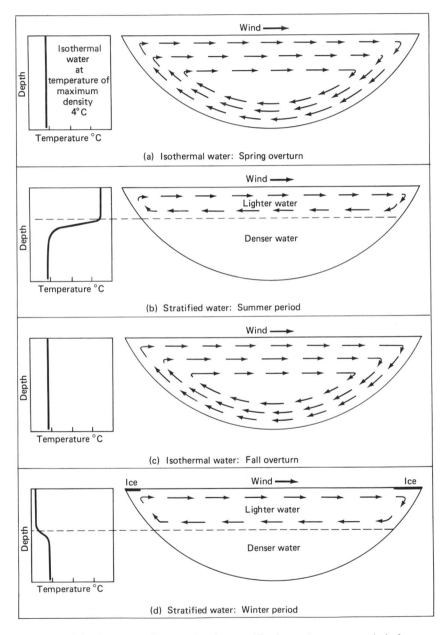

Figure 5-2 Four-stage diagram showing stratification and overturn periods for a dimictic lake. (After Hough, 1958b, Figure 22, p. 51.)

Temperate lakes are typically dimictic; that is, there are two periods of overturn annually. In the summer a well-developed thermocline restricts wind-driven circulation to the epilimnion. As air temperature cools in fall, the epilimnion becomes cooler and eventually isothermal conditions exist within the lake. This permits overturn and total mixing (Figure 5–2). During winter a less pronounced and negative thermocline develops, with the epilimnion actually colder but less dense than the hypolimnion. Spring warming again produces isothermal water and overturn prior to the development of a summer thermocline (Figure 5–2).

One might question the significance of such lake cycles in the accumulation of sediments within the lake basin. In fact, there is a distinct parallel between these lake cycles and the development of such features as varves (see Chapter 3). During summer there is a relatively high amount of oxygen in the hypolimnion due to runoff and plant productivity, whereas in the winter there is little runoff; instead of productivity, organic matter is decaying and consuming oxygen and ice cover may prohibit any wind-generated circulation.

Density-driven circulation

In some lakes there are currents, generally at or near the bottom, which owe their origin to density gradients. Any mechanism that produces a density gradient will cause circulation in an effort to eliminate this unstable gradient. Most common causes for such gradients are temperature differentials and suspended sediment. Influx of cold water or of sediment-laden water will cause density gradients which result in circulation. This circulation may be local, such as near the mouth of a river entering the lake, or it may be more widespread.

Upwelling and sinking

Surface circulation may have important effects on deep water. The results are anomalous circulation patterns and temperature distribution. When winds blow across the epilimnion, the warm surface water is pushed toward the downwind end of the lake basin. This causes some of the deeper, cold water from the hypolimnion to rise to the surface at the upwind margin of the lake basin (Figure 5–3). The result is that the warm water is pushed to the downwind side and is forced downward (sinking), whereas the upwind side experiences upwelling and anomalously cold water. An excellent example is Lake Michigan, which is within the prevailing westerly winds.

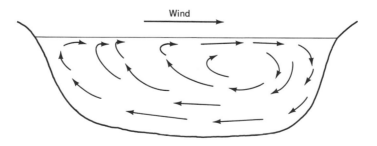

Figure 5–3 Profile of wind-generated circulation in a lake, showing upwelling on the upwind side and downwelling on the downwind side.

The west side of the lake is typically 8 to 12°C colder in summer than the downwind east side.

MODERN LAKE SEDIMENTS

Sediments that are presently accumulating in lake basins cover a broad spectrum of types if we consider particle size, mineralogy, source materials, or any other potential variables. There are, however, two general categories of lake sediments which lend themselves to separation for discussion purposes. These are terrigenous sediments and chemical sediments. Some lakes are wholly one or the other, but many accumulate a combination of the two types. Examples will be considered from lakes dominated by terrigenous sediments and by chemical sediments as well as those with nearly equal amounts of each.

Terrigenous Sediments

Lakes may receive terrigenous sediment particles through two primary mechanisms. Some sediment particles are derived from the action of waves and currents on the basin margins, although this is generally a minor source. Waves account for only a small portion of the dynamic energy of the lake system except in large lakes such as the Great Lakes or the Caspian Sea (Sly, 1978). Most terrigenous sediment that accumulates in lakes is derived from river input. The configuration of the basin, its size, and the circulation system are all prime factors in the distribution of sediment particles once they are debouched into the lake basin.

General distribution patterns

If a given lake is served primarily by a single point source of terrigenous sediment, the distribution of sediment in the lake will be markedly different than if several streams carry sediment to the lake. A generalized model for sediment distribution in a lake with several sediment sources would show a somewhat concentric pattern of facies (Twenhofel, 1932), with coarsest particles around the margin and finest particles, with perhaps some chemical precipitates, in the center (Figure 5–4).

A single, large source of sediment may develop a delta and produce a different overall pattern of sediment distribution than that described above. Examples of such well-developed deltas include the St. Clair Delta in the Great Lakes (Mandelbaum, 1966) and the Rhone Delta, which is in Lake Geneva (Houbolt and Jonker, 1968) (Figure 5–5).

Using Walther's Law it is rather easy to determine the expected stratigraphic relationships of the textural facies accumulating in terrigenous dominated lakes. The general model shown in Figure 5–4 will produce a coarsening-upward stratigraphic sequence. As the lake fills in, the coarser marginal sediments with continue to encroach upon the finer sediments near the center of the basin. A point source of sediment with a well-developed delta would also accumulate a coarsening-upward sequence but would display a marked asymmetry in its cross-section, with grain size decreasing away from the source.

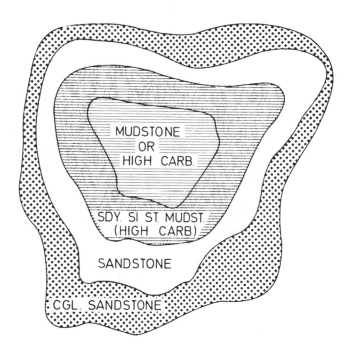

Figure 5-4 Schematic areal distribution of terrigenous sediments in a lake which receives sediment from several locations around its perimeter. (From Reineck and Singh, 1980, p. 242; after Twenhofel, 1932).

Pelagic sedimentation in lakes

The calm and deep portions of lakes are generally beyond the influence of waves and are characterized by fine-grained sediment particles; clay and silt (Figure 5-4). This sediment accumulates from suspension and does so at very slow rates. Measurements ranging from about 100 to 400 g/m^2 per year have been reported (Sly, 1978). Much particulate organic matter is frequently combined with these fine muds.

The suspended sediment is commonly distributed over broad areas through a combination of wind-driven circulation and phenomena related to the stratification of lake water. River water and its contained sediment may be lighter than the lake water or it may be more dense than the epilimnion but less dense than the hypolimnion. In the first situation, overflows or surface currents carry sediment to various parts of the lake and it then settles to the bottom, accumulating as pelagic mud. The latter situation is called interflow and sediment is more or less injected along the thermocline (Figure 5-6).

Although some lake muds are structureless, stratification may occur. In addition, some biogenic features may be widespread. The most common and significant form of stratification in terrigenous lake sediments is varves. These thin, seasonal layers are characteristic of midlatitude lakes throughout the world and are also associated with proglacial lakes. The thick, relatively coarse, light-colored layer which represents the spring and summer may show grading, whereas the thin organic rich layer is generally uniform in grain size. Although varves are indicative of the lacustrine environment, one should not expect that all lakes accumulate varved sequences.

Chapter 5   The Lacustrine System

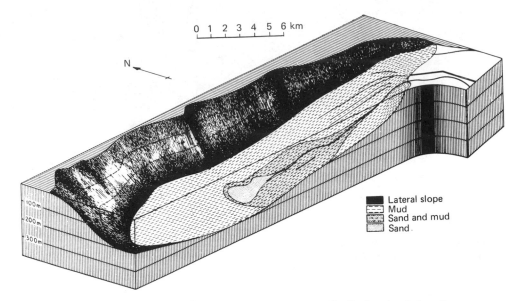

Figure 5–5 Diagrammatic representation of sediment distribution in Lake Geneva, Switzerland, where turbidites are abundant and where a point source of sediment supply is dominant. (After Houbolt and Jonker, 1968.)

The worm *Tubifex* may develop a dense network of burrows in lacustrine sequences. These somewhat circular burrows are about a millimeter in diameter and form a network that looks like the small root system of a plant (Figure 5–7).

Slumping may cause convolutions or other distortions of layering in lacustrine pelagic sediments, but diverse sedimentary structures are not common. Only the varves and *Tubifex* burrows are both diagnostic and widespread.

Turbidites in lakes

Some combination of cold runoff and considerable sediment in suspension will provide a sediment and water mixture that is more dense than ambient lake water. This situation results in underflows or turbidity currents. In fact, the first recognition and study of turbidity currents in nature took place in the deep lakes of the Swiss Alps (Forel, 1885). A number of studies has demonstrated the important role of turbidity currents and turbidite deposits in lacustrine sedimentation (e.g., Houbolt and Jonker, 1968; Sturm and Matter, 1978).

Turbidity currents are particularly common in deep lakes, where large pulses of runoff bring in much sediment. Mountainous lakes such as those in the Alps offer good examples, as do some proglacial lakes. In both types, cold, sediment-laden runoff enters the basins during spring thaw and summer melting seasons. In addition to the underflows bringing in sediment it is also possible that cold, clear water could generate underflows and cause reworking of already deposited sediment, with turbidity currents resulting (Collinson, 1978b).

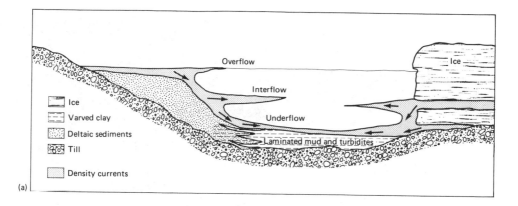

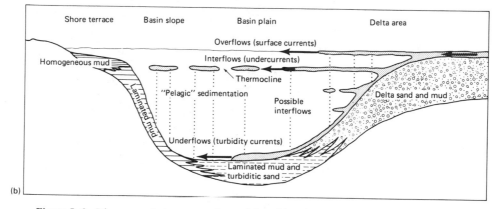

Figure 5-6 Diagrammatic representations of depositional models for terrigenous lakes with well-developed stratification. Both models show development of a delta, but (a) is proglacial and (b) is not. (After Sturm and Matter, 1978, Figure 10, p. 162; Gustavson, 1975a, Figure 14, p. 262.)

It is common for turbidity currents to develop subaqueous fans with channels, natural levees, and other features reminiscent of analogous features in the marine environment (Houbolt and Jonker, 1968; Sturm and Matter, 1978). Individual turbidites have been traced throughout their extent and studied in detail. In Alpine lakes there are thick and coarse-graded units which appear to be the result of catastrophic events that occur sporadically or perhaps only a few times each century. Thin, fine-grained graded units are numerous and varvelike. They represent seasonal melting (Sturm and Matter, 1978).

Somewhat similar turbidite accumulations have been studied from Malaspina Lake, a proglacial lake along the southern coast of Alaska (Gustavson, 1975a, 1975b). Lake sediments are provided by overflow, interflow, and underflow [Figure 5-6(a)]; however, turbidity currents appear to be the primary mechanism for sedimentation. In Malaspina Lake some of the thick varve layers are rippled as well as being graded. Gustavson (1975a) concluded that the thick unit which displays grading and ripples represents turbidity current deposition (underflow) and the thin clay units represent fallout from suspension (overflow or interflow) (Figure 5-6).

Very similar and excellent models for sedimentation in lakes dominated by ter-

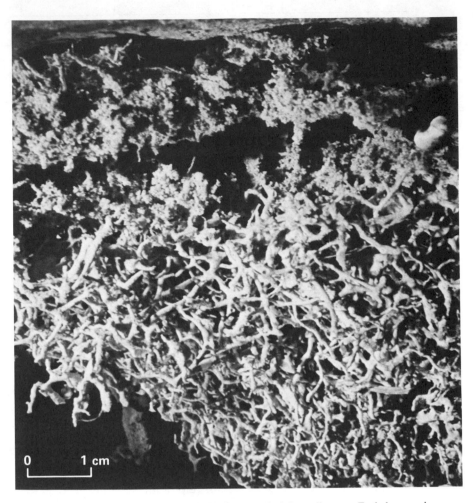

Figure 5-7 Resin casts of *Turbifex* burrows in lake sediments. Each burrow is nearly a millimeter in diameter. (Courtesy of H.-E. Reineck.)

rigenous sediment are presented for Alpine lakes (Sturm and Matter, 1978; Gustavson, 1975). The chronology and references cited in both papers indicate that the models were developed independently; however, they show marked agreement (Figure 5-6). Both show that deltas are significant in lacustrine systems, with laminated mud and sandy turbidites as the dominant basinal sediments. Mechanisms for sediment influx and distribution are the same. Gustavson's model is bounded on one side by the melting glacier [Figure 5-6(b)]. Data presented by Houbolt and Jonker (1968) for Lake Geneva suggest that it would follow the model closely.

Chemical Sediments

Although lakes may contain a very complex chemistry, there are basically two distinct types of inorganic chemical precipitates that are important as sediment accumulations in the lacustrine system. These are **saline lakes,** which produce evaporite sequences with some associated carbonates, and **carbonate lakes,** which accumulate various carbonate minerals as their only significant precipitates. Biogenic carbonate and silica materials such as that produced by diatoms, algae, or other organisms will not be considered; only physicochemical precipitates are emphasized.

Saline lakes

All lakes containing more than 5000 ppm (5‰) dissolved solids are considered as being saline (Beadle, 1974). To achieve such salinities, evaporation must exceed inflow and the lake basin should be closed or exhibit only restricted outflow (Hardie et al., 1978). Such conditions are associated only with arid or semiarid climates. This discussion will be restricted to perennial lakes and will not consider playa or ephemeral lakes, which are discussed in Chapter 6.

A dearth of rainfall and runoff may be achieved in various latitudes from the tropics to the arctic. Closed drainage systems may also be developed through various means, including tectonics, volcanic activity, and damming by glaciers, lava flows, or landslides. Some perennial saline lakes receive their inflow from a stream. An example is the Dead Sea, which is fed by the Jordan River. By contrast, Lake Eyre in Australia receives inflow only during flash floods (Hardie et al., 1978).

There is a wide range in compositions of saline lakes, with SiO_2, Ca^{2+}, Mg^{2+}, Na^+, K^+, HCO_3^-, CO_3^{2-}, SO_4^{2-}, and Cl^- being the most abundant. Sodium is by far the most concentrated cation, but anion abundance varies (Hardie et al., 1978). Many factors contribute to the composition of saline lake waters, the composition of the bedrock in the drainage basin being most important.

As evaporation in a perennial lake takes place, there is concentration of the surface brine with eventual nucleation of minerals. This concentration and precipitation causes sinking of the brine due to density increase, with less dense runoff floating on the surface. This low-concentration surface water will eventually increase its density and the process repeats itself. Although the chemistry of the brines and the species of minerals that eventually precipitate will vary greatly, the basic mechanism prevails (Hardie et al., 1978). Such is the origin of the stratified perennial saline lake. The relative rates of evaporation and inflow to a lake tend to determine the concentrations of the brines and therefore the suite of minerals that precipitates. As a consequence, one can interpret the geochemical history of a lake from its chemical sediments.

The sequence of minerals precipitated follows the general evaporite sequence first recognized by the Italian chemist Usiglio (Figure 5-8). As the brine becomes more and more concentrated, the first minerals to become supersaturated are carbonates. Continued concentration causes co-precipitation of carbonates and gypsum; this stage is represented by the present-day Dead Sea (Neev and Emery, 1967). After much more concentration of the brines, halite is precipitated (Figure 5-8). This would require enough evaporation so that the lake has less than 1/100 the volume that it had during precipitation of only carbonate (Hardie et al., 1978). The Great Salt Lake is at this stage of maturity.

At this point there are two courses that the lake may follow if concentration continues. In shallow lakes the brine commonly dries up and the lake becomes

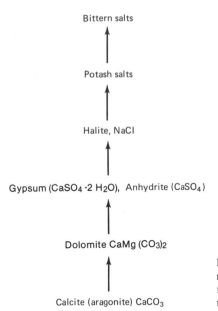

Bittern salts

Potash salts

Halite, NaCl

Gypsum (CaSO₄ ·2 H₂O), Anhydrite (CaSO₄)

Dolomite CaMg (CO₃)₂

Calcite (aragonite) CaCO₃

Figure 5-8 Succession of evaporite minerals after Usiglio's experiments. The first mineral to precipitate is on the bottom.

ephemeral or remains dry. The Great Salt Lake may be subjected to this fate in the not too distant future. Deep Springs Lake in California is now a playa but was deep (>20 m) as recently as several hundred years ago (Eugster and Hardie, 1978). The other course for a lake at the halite precipitation stage is to maintain its status as a perennial saline lake, which can occur only if it is hundreds of meters deep (Hardie et al., 1978). Eventually, of course, the lake will dry up or become ephemeral, but not until very thick accumulations of evaporites have precipitated. Other evaporites, such as the potash salts and bittern salts, are typically associated with playa lakes and are not common in a nonalkaline lacustrine sequence.

Hardie et al. (1978) have presented a simple model for the stratigraphic sequence that accumulates as a perennial saline lake matures. The initial basal portion is comprised only of carbonates. This is succeeded by a sequence of gypsum with some carbonates. The last or upper sequence is halite with some gypsum and a trace of carbonate. Thicknesses vary depending on several variables, such as lake volume, composition of inflow, rate of evaporation, and others.

Few sedimentary structures are developed in saline lakes. Stratification is common, with thin layers of fine terrigenous particles interbedded with evaporites. The couplets of evaporite and terrigenous layers have been interpreted as varves representing seasonal accumulations (Reeves, 1968); however, other investigators have found the layers to be irregular and to represent longer durations (Neev and Emery, 1967; Smith, 1979).

Examples of perennial saline lakes. Probably the most famous of all of the saline lakes in the United States is the Great Salt Lake. This lake is large but very shallow; its maximum depth is about 10 m. The Great Salt Lake was much larger during the Pleistocene than it is at present and it was not saline (Gilbert, 1890). The first detailed study of its chemical sediments was made by Eardley (1938), who reported a doubling of the salinity from 1877 to 1935. The lake is fed by three perennial streams: the Bear, Weber, and Jordan Rivers (Figure 5-9). It is surrounded by mountains with alluvial fan complexes and extensive old dried-up lake beds. Brines in Great Salt Lake have been monitored only since 1963. It is known, however, that halite precipitation has occurred only twice in historical time: in 1933–34 and 1959–63 (Eugster and Hardie, 1978). Sediments in the Great Salt Lake are primarily carbonates with some gypsum. Algal stromatolites and ooids are abundant and have been studied by various authors (Eardley, 1938; Carozzi, 1962; Kahle, 1974; Sandberg, 1975).

Another famous but scientifically less well known saline lake is the Dead Sea, which is located in an extremely arid region, fault controlled in its location, and served by the Jordan River. This lake is nearly 100 km² in area and reaches a maximum depth of 400 m (Eugster and Hardie, 1978). The Dead Sea is stratified into three water masses, with salinities ranging from 300 to 332‰. Present precipitation includes both carbonates and gypsum (Neev and Emery, 1967).

There are numerous other perennial saline lakes in the world. Typically, each is quite different from all the rest. For example, Lake Magadi in Kenya is not only accumulating typical evaporites but is also the site of bedded chert precipitation (Eugster, 1967, 1969). Lake Chad in Africa, Deep Springs Lake, California, and the

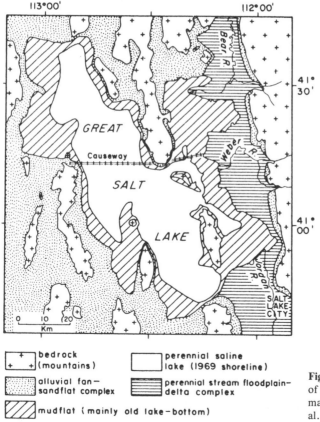

bedrock (mountains)

alluvial fan– sandflat complex

mudflat (mainly old lake-bottom)

perennial saline lake (1969 shoreline)

perennial stream floodplain– delta complex

Figure 5-9 General geologic map of Great Salt Lake and vicinity, showing major environments. (After Hardie et al., 1978, Figure 8, p. 36.)

Basque lakes of British Columbia are all unique perennial saline lakes. The Gulf of Karabugas in the Caspian Sea is an area of great evaporite deposits.

Carbonate lakes

Unlike saline lakes, the carbonate lakes are not restricted to dry climatic regions and therefore the carbonate precipitates are not caused by evaporation. In contrast with marine carbonate environments, lacustrine carbonates are primarily inorganic in nature. There are, however, important biological processes that affect the precipitation of carbonate minerals. In addition, calcite is the dominant mineral in freshwater sediments, whereas aragonite and high-magnesium calcite are most abundant in the marine environment. Freshwater **chalks** and **marls** have been recognized for nearly 150 years and many are in lakes which owe their origin to the Quaternary glaciation (Kelts and Hsu, 1978).

Calcite precipitation. Most freshwater bodies contain $CaCO_3$ as the most abundant dissolved compound; however, its solubility is low in pure water. When carbonic acid is present, as it is in natural waters, then $Ca(HCO_3)_2$, calcium bicarbonate, is present; $Ca(HCO_3)_2$ is quite soluble (Cole, 1975) and therefore dissociates readily. If one considers a simplified carbonate-water system (Figure 5–10), the various compounds and ions present can be seen as they occur in the atmosphere, the lake, and the sediment on the lake floor. Much carbon dioxide is present in the lake together with some carbonic acid and various anions and cations which are products of the dissociation of calcium bicarbonate.

Before considering the chemical reactions that might occur in this lacustrine model consideration must be given to the effects of pH level on the system. pH is a measure of the hydrogen ion activity and is expressed as

$$pH = -\log(H^+) \tag{5-1}$$

Most natural aquatic environments fall within a pH range of 3 to 12, with the acid end of the spectrum represented by some volcanic lakes and the basic end typically represented by evaporite lakes. pH has a marked effect on the presence and relative proportions of carbonate species in the aquatic system (Figure 5–11). Low pH values are dominated by CO_2 and H_2CO_3, whereas high pH is characterized by CO_3^{2-}. The middle range, about 7 to 8, has HCO_3^- as dominant. Most lacustrine environments have pH values in the middle range.

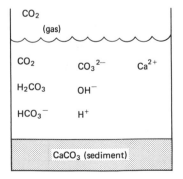

Figure 5–10 Simplified schematic of carbonate-water system, showing common ions and compounds.

To understand the basic chemistry of calcium carbonate formation, a few simple reactions must be considered. These same equations are applicable to carbonate precipitation in the marine environment (Chapter 14). If an acid is added to calcium carbonate,

$$CaCO_3 + 2H^+ \longrightarrow Ca^{2+} + H_2O + CO_2 \qquad (5\text{-}2)$$

or
$$CaCO_3 + H^+ \longrightarrow Ca^{2+} + HCO_3^- \qquad (5\text{-}3)$$

By contrast, if a base is added,

$$Ca^{2+} + HCO_3^- + OH \longrightarrow CaCO_3 + H_2O \qquad (5\text{-}4)$$

These reactions indicate the significance of pH in the stability and therefore the precipitation of calcium carbonate. A pH of about 8 or higher is required to precipitate calcium carbonate.

Factors affecting precipitation. It appears that lake waters are rarely in equilibrium but that their concentrations vary with the seasons or with respect to high-impact events (Kelts and Hsu, 1978). Saturation of lake water with respect to calcite can be accomplished in two basic ways. During photosynthesis, plants utilize much carbon dioxide, which raises the pH and facilitates calcite precipitation. Both diurnal and seasonal changes may occur, with the latter being most important. The seasonal temperature change of lake water in temperate climates is great. This affects the solubility of calcite and CO_2, which results in precipitation during warm temperatures when solubility of calcite is low. Minor effects on precipitation can be generated by evaporation and by waves driving off carbon dioxide.

Minerals precipitated. Calcite is by far the dominant carbonate mineral that precipitates in lakes, but other species are present as well. Rates of accumulation of calcite are about 1.5 to 2.0 g/m² per day (Brunskill, 1969; Otsuki and Wetzel, 1974) in small lakes in the United States and 0.5 to 0.7 per year in Lake Zug, Switzerland (Thompson and Kelts, 1974). In addition to calcite, high-magnesium calcite, dolomite, magnesite, and siderite, together with some uncommon exotic carbonates,

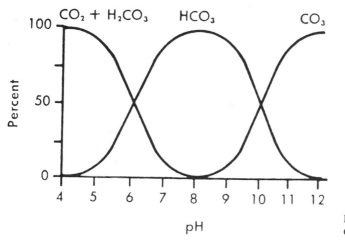

Figure 5-11 Relative proportions of CO_3 species in relation to pH.

are also known to precipitate in modern freshwater lakes (Kelts and Hsu, 1978).

The Mg/Ca ratio of the water has been shown to be the controlling factor in determining the species of carbonate mineral that precipitates. Muller et al. (1972) have analyzed data from 25 lakes in Europe, Asia, and Africa, a broad climatic belt. They found that three categories are convenient for generalizations. A Mg/Ca ratio of < 2 yields calcite, a ratio of 2 to 12 produces high-magnesium calcite, and a ratio of > 12 is characterized by aragonite. Extremely high Mg/Ca ratios cause precipitation of hydrous magnesium carbonates. Some dolomite may form as a primary mineral in Mg/Ca ratios between 7 and 12; however, the secondary carbonates are essentially all dolomite (Figure 5–12).

Stratigraphic characteristics. Quiescent conditions and accumulation of very fine grained sediment particles yield a sequence of thinly stratified mud with an abundance of varved couplets (Figure 3–1). Lake Zurich in Switzerland contains a high amount of carbonate sediment, as do many other deep lakes located in the mountainous terrain of Europe. The nonterrigenous varves appear much the same as those described in the early portion of this chapter. Light-colored layers represent the summer and autumn, and dark layers containing organic debris and diatoms represent the winter and spring. The deep carbonate lakes of Europe contain about 50% calcite but show significant range. Total carbonate content is closer to two-thirds of the bulk volume (Kelts and Hsu, 1978). An extreme variation is Lake Balaton in Hungary, which is dominated by high-magnesium calcite (Muller and Wagner, 1978).

Biogenic lake sediments

There are various organisms which make significant contributions to lake sediments in the form of both skeletal and nonskeletal particles. The most obvious are the contributions by rather large invertebrates such as pelecypods and gastropods, which provide large aragonitic material from their shells. Diatoms are probably the most widespread of all lacustrine biogenic contributors. There are hundreds of taxa, all siliceous and all small, typically less than 100 μm in maximum diameter.

A special form of stromatolite called **oncolites** is a common constituent of many carbonate and saline lakes. The subspherical to tabular bodies are formed by the encrustation of filamentous blue-green algae around some type of detrital

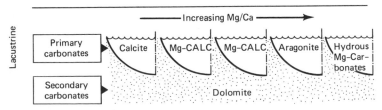

Figure 5–12 Changes in carbonate mineral precipitation as Mg/Ca ratio increases in perennial lakes with minor fluctuations in water level. (Modified from Muller, et al., 1972, Figure 4, p. 163.)

nucleus. Carbonate mud and coarser debris adhere to the sticky algal layers in irregular but concentric layers, forming oncolites which range up to a few tens of centimeters in diameter. An excellent study of freshwater oncolites has been published by Schafer and Stapf (1978). They found soft spongy varieties to be abundant in areas of strong current and smooth, hard oncolites to be located along the shorelines. Size tends to increase with current strength.

ANCIENT LAKE DEPOSITS

The marked similarities between lakes and shallow marine seas suggests that it is difficult to distinguish between the two depositional systems when analyzing the rock record. Although this may sometimes be the case, there are numerous distinct differences. Except for some difficulties in making the distinction at a given outcrop, there should be little reason to confuse these depositional systems when considering all criteria.

There are three separate but broad categories which tend to characterize lake deposits. These are the biological, chemical, and physical criteria. Certainly, these categories are not different from those utilized in interpreting most depositional environments. In the case of lacustrine rocks, chemical characteristics are the most important; in marine environments the other two are stressed.

Most lakes are quite small compared to even small marginal seas. This distinct size difference coupled with expected facies relationships permits recognition of lacustrine facies in the rock record. Excellent summaries of criteria for recognizing lake deposits are given in Reeves (1968) and in Picard and High (1972, 1981).

General Characteristics

Available data indicate that lacustrine deposits are prominent in the geologic record beginning in the Mesozoic. Greatest abundance is in Cenozoic strata, with the peak occurring during Pleistocene time and markedly related to the extensive changes in climate during that portion of geologic time. It is possible that Paleozoic lacustrine deposits are really more abundant than we believe but are well beneath the surface and consequently unavailable for study. Some extensive lacustrine deposits have been studied from the Precambrian (Belt Group).

Among the most fundamental characteristics of suspected lacustrine strata that one must consider are the stratigraphic relationships between the rocks under scrutiny and the adjacent strata plus the nature of the stratigraphic section within the sequence of interest. The typical shape of lake deposits is somewhat tabular, with a crudely equidimensional areal extent. Because lakes are landlocked it should be expected that adjacent strata be fluvial related in nature. The stratigraphic sequence is also typically predictable in terms of general textural trends. Lakes are ephemeral with respect to geologic time and are ultimately filled with sediment or destroyed tectonically. Consequently, the regressive or progradational sequence of an upward-coarsening trend is accumulated (Twenhofel, 1932; Visher, 1965; Picard and High,

1972, 1981). No particular group of sedimentary structures is diagnostic of the lacustrine environment.

In many respects the paleontology of lacustrine strata is dull. Generally, there is little fossil material preserved and the diversity is low compared with marine deposits. Although fossils may provide diagnostic criteria for recognition of the lacustrine environment (Feth, 1964), they may be difficult to distinguish from stream faunas and saline faunas may be similar in both lakes and marine waters (Picard and High, 1972). Some spores, pollen, diatoms, and algae are preserved and may be indicative of lacustrine conditions, but these organisms are very diverse and have adapted to a wide variety of environmental conditions. Some mollusks and ostracods are restricted to fresh water, but there is difficulty distinguishing between streams and lakes unless physical or chemical criteria are also utilized (Picard and High, 1972; 1981). Fish and insects are preserved in some lacustrine strata and certain taxa are diagnostic.

The mineralogy of lacustrine deposits offers one of the best types of information for recognizing this environment in the rock record. Evaporite sequences as well as various carbonates may provide diagnostic criteria for the lacustrine system. Also, siliceous banded iron deposits are thought to be lacustrine by some workers (Hough, 1958a; Govett, 1966). In general, the chemistry of lakes is more complicated and more variable than that of the marine environment. The result is precipitation of a great variety of minerals, many of which are diagnostic. More than 50 authigenic minerals have been identified from the lake deposits of the Green River Formation (Eocene) in the western United States (Picard and High, 1972).

Green River Formation (Eocene)

Beginning in the Late Cretaceous and extending through the middle of the Eocene, large lake basins were present in the area of Wyoming, Utah, and Colorado around the Uinta uplift (Figure 5–13). The Eocene Green River Formation, which is the largest and most studied lacustrine sequence in the world, accumulated in these basins (Eugster and Hardie, 1978). It extends over 100,000 km^2, with a maximum thickness of 3 km in Utah. Deposition covered a period of more than 10 million years. Although four depositional basins are recognized (Figure 5–13), two large lakes were present, separated by the Uinta Mountains: Lake Gosiute on the north and Lake Uinta on the south. Lake Uinta was the deeper of the two, whereas much of the history of Lake Gosiute may be characterized by playa-type deposition (Picard and High, 1972, 1981; Smoot, 1978). These lacustrine rocks are economically significant in the form of both petroleum and trona ($NaHCO_3 \cdot Na_2H_2O$); they have a diverse suite of evaporite minerals, and they contain an unusually well preserved fauna and flora.

The Green River Formation and its adjacent lithostratigraphic units comprise a complex continental depositional system not unlike a modern strand plain complex. In addition to perennial lakes, there were playas, fluvial systems, alluvial fans, and perhaps aeolian depositional environments present. Discussion here is centered on the perennial lakes, in keeping with the theme of this chapter.

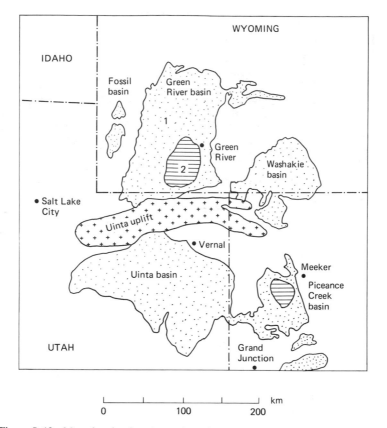

Figure 5-13 Map showing locations of Tertiary Basins around Uinta Uplift. Dot pattern (1) indicates Green River Formation; horizontal lines (2) indicate bedded salts. (After Eugster and Hardie, 1978, Figure 42, p. 283.)

General stratigraphy

The definition and extent of the Green River Formation differs from one author to another (Ryder et al., 1976) and it is not the purpose of this discussion to worry over stratigraphic nomenclature. As a consequence, consideration is given to the sequence of perennial lacustrine deposits which accumulated in Lake Uinta during the early Tertiary. Emphasis is on the open lacustrine facies, with marginal lacustrine facies considered only as they relate to deeper-water sediments. The open lacustrine facies is characterized by gray to brown micrite and calcareous mudstone with a maximum thickness near 900 m (Ryder et al., 1976). This facies is bounded laterally as well as above and below by calcareous mudstone and sandstone of the marginal lacustrine facies (Figure 5-14). Local carbonate beds of the marginal facies are also present. Evaporites include some gypsum in the marginal facies and in a saline facies in the upper part of the Green River Formation (Eugster and Hardie, 1978).

Most of this facies is characterized by well-developed horizontal laminations. Some convolutions in bedding and microfaults are present, together with scattered

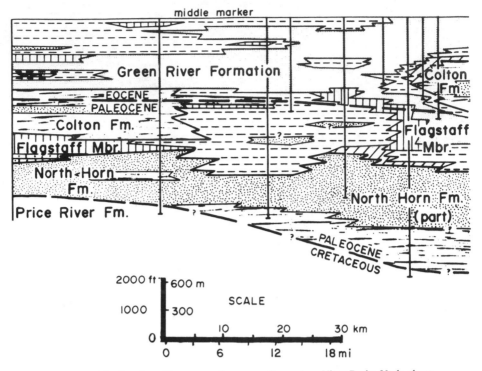

Figure 5–14 Stratigraphic cross section across the western Uinta Basin, Utah, showing general relationships of various lithic units. (After Ryder et al., 1976, Figure 4, p. 501.)

mud cracks near the margins of the facies (Ryder et al., 1976). The basal part of this facies (Parachute Creek Member) contains what have been interpreted as deep-water cycles of sedimentation. Each cycle consists of a relatively thick terrigenous mudstone overlaid by a thinner carbonate unit which typically contains oolites and may also include algal structures or micrite (Figure 5–15). The thickness of each cycle ranges from about 1 m to 3 m, with the carbonate portion accounting for only 15 to 25 cm (Picard and High, 1972, 1981).

Local sandstone units contain small-scale cross-stratification, linear bottom marks, and burrows. Fossils in the open lacustrine facies include ostracods, gastropods, pelecypods, fish, and various algae (Ryder et al., 1976).

Depositional environments

Cyclic patterns appear to be important (Picard and High, 1981) in at least part of the depositional environment, which has been proposed for the open lacustrine facies of the Green River Formation (Ryder et al., 1976). The general model for the entire system includes not only the open lacustrine environment but adjacent environments going shoreward to the north toward the Uinta Mountains and toward a deltaic plain in a southerly direction (Figure 5–16). To the north are the marginal lacustrine environments, containing carbonates, sandstones, and mudstones. These merge landward with fluvial and alluvial fan environments. To the south the

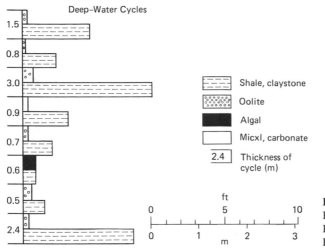

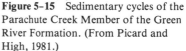

Figure 5-15 Sedimentary cycles of the Parachute Creek Member of the Green River Formation. (From Picard and High, 1981.)

marginal lacustrine environment is more extensive but similar in character to the opposite side. The south is bordered by a low-lying delta plain complex (Figure 5–16).

Some sediments contain abundant kerogen. This is generally associated with calcite layers and is believed to have accumulated in conjunction with algal blooms. Quiet water conditions prevailed and depths ranged from 5 to 30 m (Ryder et al., 1976).

All data indicate that the open lacustrine facies of Lake Uinta represents a fresh to slightly brackish perennial lake. No indications of playa conditions are present, in the form of either evaporite minerals or sedimentary structures. Only anhydrite is present and that is restricted to the landward edge of the marginal facies (Ryder et al., 1976).

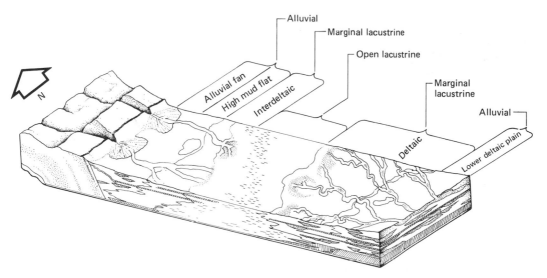

Figure 5-16 Depositional model for lacustrine and related environments of the Uinta Basin during Green River time. (After Ryder et al., 1976, Figure 6, p. 503.)

Triassic Lakes

Lakes were widespread in North America during the Triassic, which was the first period of extensive lake sedimentation on this continent with the possible exception of the Precambrian (Picard and High, 1972). In the western United States extremely large lakes and related environments accumulated the Chugwater Group in Wyoming, Idaho, Utah, and Colorado. Linear grabens associated with plate rifting gave rise to large lakes along the northeastern United States and into the maritime provinces of Canada. Because of their markedly different geologic settings but similar depositional environments, both areas are discussed here.

Western United States (Chugwater Group)

The Chugwater Group extends over 100,000 km^2 and covers more than half of the state of Wyoming. This sequence may be up to 150 m thick and contains marine facies overlaid by fluvial and lacustrine units (Figure 5–17). The lacustrine facies is best developed within the Popo Agie Formation (High and Picard, 1969). Four distinct lithic units are present in the Popo Agie; two of them are interpreted as representing the lacustrine environment. The basal unit is a carbonate conglomerate which represents a fluvial situation. It is overlaid by a purple mudstone containing montmorillonite (smectite) and representing the floodplain environment. The next unit is an ocher mudstone with montmorillonite and abundant analcimolite representing a persistent lake. Lacustrine conditions continued but sediment composition shifted to a carbonate-rich situation (Picard and High, 1972, 1981).

The ocher unit is up to 20 m thick, massive, and contains abundant burrows

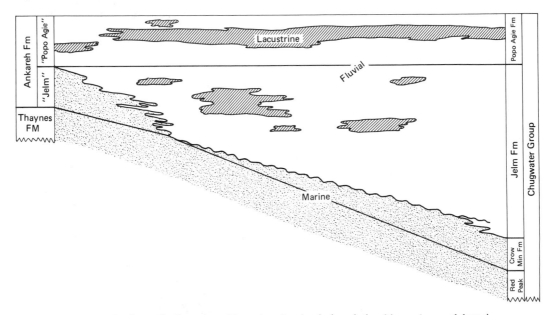

Figure 5–17 Generalized stratigraphic section showing facies relationships and general depositional environments of the Chugwater Group in Wyoming. (From High and Picard, 1969, Figure 12, p. 1102.)

(Keller, 1952). There are two general lithologies in this unit: those with abundant montmorillonite (smectite) and little analcime and those with the opposite ratio (High and Picard, 1965). The overlying carbonate unit in the lacustrine facies is less than 4 m thick and is a quartzitic dolomite with some beds of gray mudstone. Analcime and montmorillonite are absent; the clay mineral present is illite. A fair diversity of reptile types has been found in the Popo Agie together with freshwater plants (*Equisetum*) and a freshwater clam (*Unio*) (High and Picard, 1965).

Environment of deposition. All data gathered on the Popo Agie mudstone points to a lacustrine depositional environment, although one must at least consider the possibility of a fluvial environment. The reports by High and Picard (1965) and Picard and High (1972, 1981) systematically consider the various possibilities. The presence of large predaceous reptiles favors a permanent lacustrine situation rather than a stream. The acidity of streams does not favor the precipitation of carbonates or analcime; both are abundant in the Popo Agie Formation. The bedding characteristics and general absence of significant current-generated features also support a lacustrine environment rather than deposition in a stream.

Eastern United States (Newark Group)

The Newark Basin in New Jersey and Pennsylvania and the Hartford Basin in Connecticut and Massachusetts are two of several graben structures which received considerable sediment influx during the Late Triassic. Classic fault-block-related environments are represented in the sequence of sediments that accumulated in these basins. Alluvial fan, fluvial, and lacustrine facies dominate the sequences, although volcanics are also interbedded with the sediments. The Newark Basin contains the Stockton, Lockatong, and Brunswick Formations, with the Lockatong being lacustrine and the others some complex of alluvial systems (Van Houten, 1964; Picard and High, 1972, 1981). A similar sequence is present in the Hartford Basin, with the East Berlin Formation, which may be Early Jurassic, being the unit ascribed to a lacustrine environment (Sanders, 1968; Hubert et al., 1976).

Lockatong Formation. The Lockatong Formation reaches 1200 m in thickness and lithologically is comprised of various mudstones, with carbonates, analcime, and pyrite also present. The most comprehensive account of the stratigraphy and depositional history of this unit is that of Van Houten (1964, 1965), which will serve as the primary basis for this discussion.

The Lockatong is characterized by sedimentary cycles of two varieties: terrigenous cycles which are 4 to 6 m thick and chemical cycles which are 2.5 to 4.0 m thick. The terrigenous cycles consist largely of dark mudstones which may be calcareous. These cycles coarsen upward and may be capped by cross-stratified fine sandstone with convolutions (Figure 5–18). The basal mudstone is pyritic and is overlaid by mudcracked and bioturbated calcareous mudstone. Each of these terrigenous cycles indicates decreasing depth and increasing energy. The presence of a thermocline is suggested, at least for the lower deep portion of the cycle, by the presence of pyrite and reducing conditions (Figure 5–18).

The chemical cycles are more abundant than the terrigenous ones. They consist of thinly bedded mudstone throughout but with much range in mineralogy. Pyrite is abundant at the base, carbonates are abundant throughout, and analcime becomes

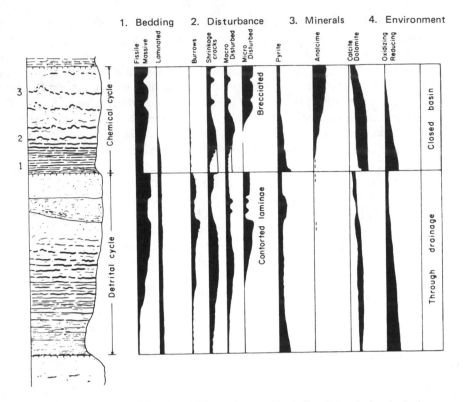

Figure 5-18 Depositional model for terrigenous (detrital) and chemical cycles in the Lockatong Formation, New Jersey. (After Van Houten, 1964.)

quite abundant, up to 40%, at the top of a cycle (Figure 5-18). Mudcracks and small-scale brecciation related to syneresis are prevalent.

The Lockatong Formation is interpreted as representing deposition in a perennial lake which was thermally stratified through much of its history. The climate was warm, with cyclic rainfall that led to changing water levels and the cyclic sedimentation (Van Houten, 1964). Terrigenous deposition characterized periods of relatively high rainfall and through drainage, whereas chemical cycles indicate a closed saline lacustrine environment. The related Stockton Formation represents fluvial and alluvial fan deposition, whereas the overlying Brunswick Formation is interpreted as a mudflat environment developed as lacustrine deposition terminated with the infilling of the lake (Van Houten, 1965).

East Berlin Formation. The East Berlin Formation is a terrigenous sequence between the Holyoke and Hampden formations, both lava flows. This sequence has been placed in the Meriden Group but is essentially the time equivalent of the Newark Group farther to the southwest (Hubert et al., 1976). The East Berlin ranges in thickness from 150 to nearly 300 m, thickening toward the north in the Connecticut Valley. The unit is quite diverse in its composition and in the depositional environments in which the sediments accumulated. Arkosic conglomerates, sandstones, and mudstones are all present.

Two apparent lacustrine facies occur in the East Berlin: an interbedded mudstone and siltstone–fine sandstone, and dark pyritic mudstone. The coarser, red beds contain ripple cross-stratification, lineations, and some grooves (Hubert et al., 1976). This facies in interpreted as being a lacustrine deposit (Krynine, 1950; Sanders, 1968); however, it is probably a marginal facies with some exposure. The environment was well oxidized and part of the section suggests a floodplain or mud flat type of environment (Hubert et al., 1976). This is comparable to the marginal lacustrine environment in the model discussed for the Green River Formation (Figure 5–16).

Dark gray and black mudstone in the East Berlin Formation is a few meters thick, contains pyrite, and is interbedded with the red, marginal lacustrine facies. Some carbonate is present, together with mudcracks, ripples, cross-stratification, and dinosaur tracks. The depositional environment for this facies is interpreted as that of a perennial lake, shallow but with some thermal stratification. Fine cyclic layers in the form of alternating calcareous and terrigenous sediment suggest changes in climatic conditions such as rainfall but are probably not varves in the usual sense (Hubert et al., 1976).

The overall depositional system for both the East Berlin and Lockatong Formations is quite similar and could be fit well into the model suggested by Ryder et al. (1976). Probably the north side of the model (Figure 5–16) better approximates the Triassic situation, due to the geologic setting in a graben.

SUMMARY OF CHARACTERISTICS OF SEDIMENT/SEDIMENTARY ROCK BODIES REPRESENTING THE LACUSTRINE DEPOSITIONAL SYSTEM

Tectonic setting. Virtually all tectonic conditions provide possible settings for the lacustrine system. They are typically not common on coastal plains of the trailing-edge portion of a plate, due to the lack of relief to provide a basin in which the lake might develop.

Shape. Lacustrine sediment systems are generally somewhat circular in plan view and lenticular in cross-section; however, in tectonic basins wedge-shaped lake basins may form.

Size. A great range in sizes is known, with some of the largest exceeding 100,000 km². Most lacustrine sequences are thin (< 200 m), but some exceed 1000 m.

Textural trends. Although virtually any grain size may be present, there is a tendency for most lakes to display a fining-outward trend into the basin and a coarsening-upward stratigraphic trend as the basin fills. This pattern is common to lakes with numerous sediment sources around the perimeter. Lakes with a point source of sediment show a tendency to fine away from the source. They may or may not display a coarsening-upward trend.

Lithology. Much variety of particle composition is present. Chemical precipitates may yield important information in interpretations. Marl is typical of freshwater lakes, and saline lakes may contain evaporite minerals. Some zeolite minerals, especially analcime, are common lake deposits.

Bedding and sedimentary structures. Lacustrine strata are typically thin and continuous relative to other continental depositional environments. Varves suggest lakes if present. Small ripples and ripple cross-strata are common. Some desiccation features may develop along lake margins. No diagnostic structures are present.

Paleontology. Fossils present may be indicative of the lacustrine system; however, freshwater lakes and saline lakes display a different fauna and flora. Various algae, pelecypods, gastropods, ostracods, and vertebrates may be diagnostic. Some bioturbation may be present but is not of much value in recognizing the lacustrine system.

Associations. Facies of the lacustrine depositional system may be associated with any of the other continental systems or they may rest unconformably on older sequences. Fluvial facies and periglacial facies are the most commonly associated strata. Strata that accumulated in saline lakes would be expected to be associated with facies of the desert system.

ADDITIONAL READING

COLE, G. A., 1975. *Textbook of Limnology.* C. V. Mosby, St. Louis, Mo., 283 p. This is a concise, well written and current book covering most aspects of limnology. It is weak in geologic aspects of lakes.

ETHRIDGE, F. G., AND FLORES, R. M. (EDS.), 1981. *Recent and Ancient Nonmarine Depositional Environments: Models for Exploration.* Soc. Econ. Paleontologists and Mineralogists, Spec. Publ. No. 31, Tulsa, Okla., 349 p. Contains three papers on lacustrine deposits, one of which (Picard and High) is an excellent summary article.

HUTCHINSON, G. E., 1957. *A Treatise on Limnology,* v. 1: *Geography, Physics, and Chemistry.* Wiley, New York, 1015 p. This classic and comprehensive reference on limnology remains the standard in the field. Although some of the methodology is dated and the bibliography is not current, the book is the most widely cited volume in the field.

LERMAN, A. (ED.), 1978. *Lakes: Chemistry, Geology, Physics.* Springer-Verlag, New York, 363 p. Excellent collection of current summary papers on lacustrine research. Geology of modern lakes is well covered.

MATTER, A., AND TUCKER, M. E. (EDS.), 1978. *Modern and Ancient Lake Sediments.* Internat. Assoc. Sedimentologists, Spec. Publ. No. 2. Blackwell, London, 290 p. Good collection of papers on lacustrine sedimentation with both terrigenous and chemical sediments included. Good companion volume to Lerman (1978) above.

PICARD, M. D., AND HIGH, L. R., 1972. Criteria for recognizing lacustrine rocks, *in* Rigby, J. K., and Hamblin, W. K. (eds.), *Recognition of Ancient Sedimentary Environments.* Soc. Econ. Paleontologists and Mineralogists, Spec. Publ. No. 16, Tulsa, Okla., 108–145. This is the best single summary article on recognition of lacustrine environments in the rock record. Its only serious drawback is lack of coverage of examples beyond North America.

REEVES, C. C., 1968. *Introduction to Paleolimnology.* Elsevier, New York, 228 p. This is the only fairly comprehensive treatment of the subject. Emphasis is on evaporite lakes. The book contains many excellent photos.

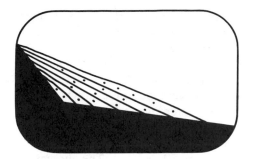

6 The Desert System

Environments of deposition which are within the landmasses of the world are markedly affected by climatic conditions, whereas deep marine environments typically are not. This chapter and the following one are distinct in their climatic restrictions. The concern here is primarily with depositional systems as opposed to individual environments. Such being the case, it is most appropriate to consider the desert system as a whole and to integrate the various depositional environments with one another.

Modern deserts are characterized by absence of extensive vegetation due to a general dearth of precipitation; in fact, some definitions are based on the absence of vegetation. This type of definition would be inappropriate for usage here because concern is with both modern and ancient deserts, some of which existed prior to the evolution of land plants. For purposes of this discussion a desert is considered as an environment with little precipitation in which evaporation rates are considerably in excess of precipitation.

General Characteristics and Distribution

To most individuals the term **desert** carries a connotation of a hot, dry place with abundant sand dunes. In fact, there are also deserts in cold climates, although they are not presently important depositional environments. According to Holmes (1965) only about 20% of the world's deserts are dune covered. On a worldwide basis most deserts are concentrated in a low-latitude belt where high pressure prevails and where the westerly and trade wind systems diverge (Figure 6–1).

Rainfall in deserts is less than about 25 cm per year and often this amount comes in a few cloudbursts or at least in a short period of time. Consequently, there may be only a few days each year when precipitation occurs, but these days may each experience such high rates of rainfall that flash flooding and much sediment transport occurs. Most of the time the atmosphere is dry, with evaporation rates high. The consequence of this condition is rapid loss of water to the atmosphere, causing precipitation of evaporites from inland seas or saline lakes. The absence of moisture and therefore vegetation permits much eolian transport of sediment. There are, therefore, three major sedimentologic processes operating at various times in the desert: fluvial processes, during and shortly after rainfall periods; eolian transport; and chemical precipitation, the latter two operating for long periods of time between the few days associated with rainfall events.

As a result of these conditions, the desert shows a rather diverse suite of depositional sedimentary environments. Fluvial processes may produce arid-type alluvial fans and they may also produce **wadis,** which are channels of intermittent streams. Evaporite conditions cause development of **playa lakes** or inland **sabkhas.** These terms are commonly used synonymously for intermittent lakes that occur in arid and semiarid climates. Dunes are also common and are accumulations of mobile eolian sand.

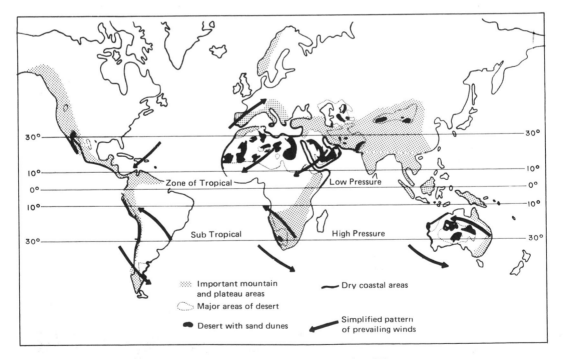

Figure 6–1 World distribution of deserts. (After Glennie, 1970.)

Geologic Setting

For the most part, deserts are not the direct result of tectonics or of the overall geologic setting. There are, however, some indirect relationships. Fault-controlled basins may give rise to thick accumulations of sediment which accumulate in the desert environment, such as in the Basin and Range Province of the southwestern United States. Similar desert accumulations are also developed on the stable craton of the Sahara Desert (Collinson, 1978c). Data from the rock record indicate that similar desert environments occurred in the geologic past.

Those deserts that develop in fault-controlled basins where relief is high typically contain well-developed alluvial fan and wadi environments near the basin margin, and eolian, sabkha, and playa lake environments near the center (Figure 6–2). Coarse sediment particles found in the fans and wadis are the result of short transport from high-relief sources. Such tectonically controlled deserts may also display extensive exposed bedrock surfaces and **pediments** may be developed.

Deserts where high relief, and therefore coarse sediment, is absent are dominated by sand accumulations and evaporite deposits with some eolian silt accumulations. Such deserts with extensive eolian sand are called **ergs** (Wilson, 1973).

ARID ALLUVIAL FANS

Development of alluvial fans is largely the result of the proper geomorphic setting. The accumulation is a fan-shaped wedge of sediment which typically develops at the base of a rather steep mountain slope where there is an abrupt change to a flat or only slightly sloping area and where sediment looses its confinement from a high-relief valley. Alluvial fans are particularly numerous in tectonic regions where long fault blocks produce an extensive scarp. Adjacent and coalescing alluvial fans form a **bajada**.

General Morphology and Location

Mountainous areas near the fault scarp serve as the ultimate sediment source and are characterized by bare rock. The sediment is transported downslope and erodes the upper slopes, forming a pediment, which is an erosion surface that may have a veneer of sediment over it.

The fan sediment typically thins and spreads away from the source. Sediment is typically supplied via a single stream valley, but upon exiting the confinement of the valley, spreading predominates. Fans are generally subdivided into three zones (Blissenbach, 1954): the upper fan or fanhead, which is near the apex; the middle segment or midfan; and the lower fan or base. The latter zone is where most of the spreading and coalescing occur (Figure 6–3). The reader should pay careful attention to the fact that fan illustrations typically display much vertical exaggeration. Steepest slopes on arid fans are about 10°. Upper fan slopes are generally greater than 5°, midfan is 2 to 5°, and lower fan slopes are less than 2° (Blissenbach, 1954). The fan surface exhibits a slightly concave-upward profile (Figure 6–3). A given fan, espe-

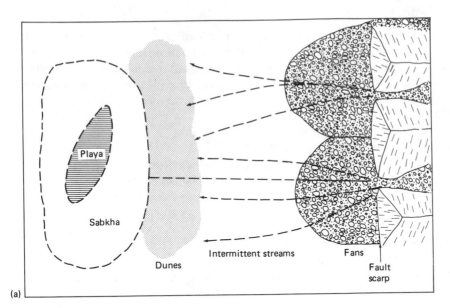

(a)

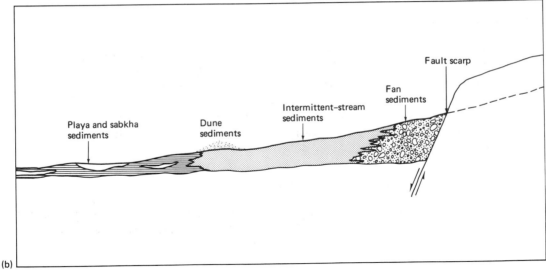

(b)

Figure 6-2 Schematic diagram showing major sedimentary environments of the desert depositional system: (a) plan view; (b) cross section. (After Friedman and Sanders, 1978, p. 203.)

cially if it is fairly large, will actually comprise several fans or lobes, due to shifting depositional areas of wash from the mountainous source (Figure 6-4). The result is a very complicated stratigraphy.

Arid fans show a wide range in size, although the overall morphology and operational processes are irrespective of size. A fan may be only a few hundred square meters or more than 1000 km²; most present-day arid fans are less than 100

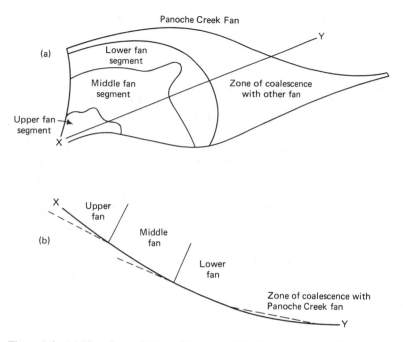

Figure 6-3 (a) Plan view and (b) profile section of the Panoche Creek alluvial fan in southern California. (After Bull, 1964.)

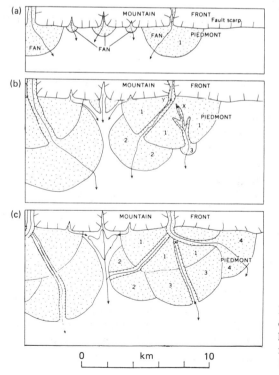

Figure 6-4 Map view of succession of fan lobes in the development of a hypothetical fan complex along a steep fault scarp. (After Denny, 1967, Figure 1, p. 86.)

km^2 in areal extent. The area covered by a fan tends to have a relationship to the drainage basin area such that

$$A_f = CA_d^n \qquad\qquad (6\text{-}1)$$

where A_f is the fan area; A_d is the drainage basin area; n is an exponent, empirically derived from log plots of known fans; and C is an empirically derived coefficient which has great variation and is controlled by several variables, including lithology, climate, and tectonics (Bull, 1968). It has been shown, for example, that in a single basin, the San Joaquin Valley of California, fans from mudstone sources cover nearly twice the area as those derived from sandstone (Bull, 1964).

The fan itself may display up to hundreds of meters of relief between the elevation of the apex and the valley floor (Figure 6–5). This indicates great thickness of sediment in the proximal area, thinning to nothing in the distal portions.

Relief on the fan surface is confined to the channels cut during discharge from the adjacent mountains and levees which form on channel margins. Stream incision may be severe, up to several meters, and is typically most pronounced near the apex, where discharge is maximal and both competence and capacity are high.

Processes

Fluvial processes dominate sedimentation on arid fans. The general conditions that cause sediment to be deposited are mostly due to the abrupt change in morphology of the channel through which the discharge flows; deposition is not due to an abrupt slope change (Bull, 1968). As the discharge exits the confinement of its channel and spreads laterally, there is an increase in flow width and a decrease in velocity and depth which causes deposition of sediment. Another important factor is the general permeability of the fan sediment. Much water percolates down into the fan, thereby reducing discharge and facilitating sediment deposition (Bull, 1968).

Most authors (e.g., Blissenbach, 1954; Bull, 1964, 1972; Hooke, 1967) recognize four common sedimentation processes on arid fans: debris flow, sheetflood, channelized flow, and sieve deposition. Only debris flow is characterized by high viscosity (see Chapter 2); the others are low-viscosity fluid flow.

Although some authors might wish to separate the terms, this discussion treats debris flow and mudflow as synonymous. This fluid-sediment mixture can support large boulders and tends to flow in well-defined masses (Collinson, 1978c). Sheetflooding is most prevalent at the downstream ends of channels where confinement ends. This is most commonly in the distal portions of the fan or lower midfan areas. Similar low-viscosity flow of sediment-laden water can occur as overbanking during extreme periods of discharge where existing channels cannot handle the volume. Stream channel deposition is fairly well restricted to the upper fan or midfan and is similar to typical braided stream deposition.

Sieve deposition (Hooke, 1967) is apparently unique to fans. Percolation of water and fine sediment takes place in relatively coarse, permeable sediments. The result is that sediments become low in mud content and are clast supported. This percolation retards flow and enhances deposition, which takes place in the form of a

(a)

(b)

Figure 6-5 (a) Aerial photograph and (b) ground-level photo of arid alluvial fan. [(b) from Malde, 1966.]

sieve lobe with concentrations of coarse material at the front (Figure 6–6). Because of the dependency on coarse sediments beneath and within the sieve deposit, this process is predominantly on the upper fan (Figure 6–7) and on some fans that lack such fine sediment.

Sediments and Sedimentary Structures

Arid fan sediments are identified as such primarily from their physical characteristics, with only minor influences from chemical features. Oxidation is widespread and such minerals as calcite and gypsum may be present as precipitates, especially in distal portions of the fan (Bull, 1972). Fossils are generally absent or if present would not be distinguishable from those of other continental environments.

As a general rule, fan sediments are coarse, poorly sorted, and commonly display cross-stratification. There is a general decrease in grain size from the apex to distal portions of the fan and a subtle overall decrease in grain size from the base of the fan to the active surface. These generalizations are broad in nature. The wide range in energy conditions and transport mechanisms on a fan gives rise to great local variations. Many individual layers are well sorted. It should be noted that during much of the time the arid fan is static in terms of sediment transport. Only during runoff events is sufficient water available for sediment transport, and then the discharge varies tremendously. The result is that the stratigraphy of a fan is very complex in detail, but there are some broad generalizations that are valid.

Debris flow deposits

Because debris flows show high strength and contain much sediment, they act more like a plastic mass than a fluid (Bull, 1972). The result is very poor sorting; there is no mechanism to preferentially remove certain size particles. The debris flow sediment body is lobate and tabular in shape (Figure 6–7), with little or no internal organization of particles. This sediment facies of the fan is most common in the upper fan or proximal area (Hooke, 1967).

The amount of water in a flow will be a determining factor in the internal organization. Relatively low strength flows contain smaller maximum clast size, may have some graded bedding, and may show imbrication compared to high-strength flows which carry large boulders (Bull, 1972). Some desiccation cracks may develop on the surface of a flow. A crude stratification may be recognized in situations where succeeding flows have accumulated. Typically, flows show a uniform thickness throughout their extent.

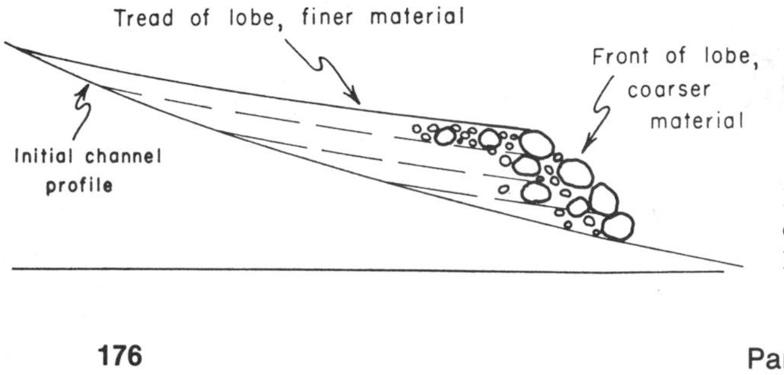

Tread of lobe, finer material

Front of lobe, coarser material

Initial channel profile

Figure 6–6 Diagrammatic cross section of a sieve lobe. (From Hooke, 1967, Figure 9, p. 454.)

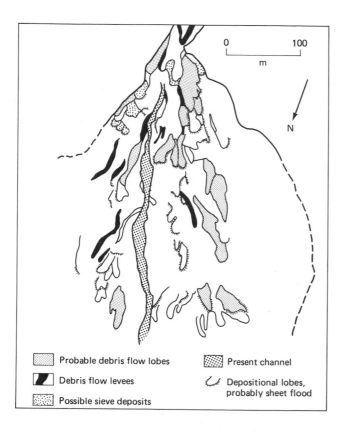

Figure 6-7 Diagrammatic view of debris lobes, channels, sieve deposits, levees, and so on, on the Trollheim Fan, Death Valley, California. (From Hooke, 1967.)

Legend:
- Probable debris flow lobes
- Debris flow levees
- Possible sieve deposits
- Present channel
- Depositional lobes, probably sheet flood

Sheetflood deposits

Sediments that accumulate due to spreading of sediment-laden water as it exits a stream channel are quite similar to braided stream deposits. Streams carrying sand and pebble-size gravel dump their load rather quickly upon leaving the confinement of a well-defined channel. Anastomosing distributary channels are branching and interlacing, producing a netlike or braided pattern. They fill and shift location as a thin but rather widespread sheet of sand and fine gravel is deposited (Bull, 1972).

The sediments are predominantly sand, with some silt and gravel. Small-scale cross-stratification is common in the sand and some imbrication may be present in the gravel. Plane beds are the most common bedding type, indicating that upper-flow-regime conditions prevailed. Sheetflood lobes are similar in size and shape to those of debris flows. Sheetflood deposits are somewhat more common in the midfan than in other areas.

Stream channel deposits

Channel deposits develop as a result of cut-and-fill accumulation primarily in the upper half or so of the fan. These channel deposits are much like any other, but they are not large or thick, due to the nature of channels on fan surfaces. Sediments are sand and gravel with some grading; gravel layers are typically concentrated at the base of the channels (Figure 6–8), and cross-stratification of sand layers is common (Bull, 1972).

Sieve deposits

Coarse gravel deposits that contain little or no material of sand size or finer have been called sieve deposits by Hooke (1967), who studied them extensively. The coarse lobes are most common near the fan apex, where slopes are steep. The gravel

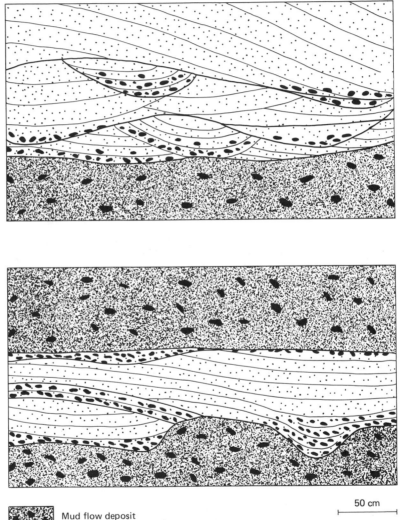

50 cm

 Mud flow deposit

Stream channel deposits

Figure 6–8 Bedding types from alluvial fan deposits, showing the interlayering of mud flow deposits with stream channel deposits. (From Blissenbach, 1954.)

acts as a sieve through which the finer sediment percolates, thus essentially eliminating any runoff. Sieve deposits are the least abundant of the four types of fan accumulations, but they are the most diagnostic.

Summary

Fan sediments are a series of rather small overlapping lobes and channels which display a complex sediment facies pattern (Figure 6–8). Stratigraphically, there are trends in the distribution of these lobes and their contained sediment textures. Well-developed channels, sieve deposits, and coarse debris flows are most common in the upper fan, whereas sheetfloods and graded debris flow deposits occur in more distal parts of the fan.

The fan surface at any given time may also acquire special characteristics, some of which may be recognized stratigraphically. Oxidation of surficial particles such as mafic and feldspathic rock fragments may be recognizable. Deflation surfaces with desert pavements are also common, due to long periods when moisture is absent from sediment (Figure 6–9). It should be noted that there is great variety in the texture and mineralogy available in different fans.

WADIS

Intermittent streams of the desert are known as wadis or, in the southwestern United States, as arroyos. Like desert fans, wadis receive water sporadically and in large bursts. Flash floods are the only events of significant sediment mobilization on a wadi.

General Morphology and Processes

Wadis are located on the gently sloping, nearly flat portions of the desert; commonly, they begin on the distal portions of fans and extend to inland sabkhas or playas. In basin and range topography, wadis trend along basin axes at the terminus of fans. Permanent channels do not exist, due to the lack of continual water flow (Glennie, 1970). A wadi may be considered an end member in the spectrum of available sedi-

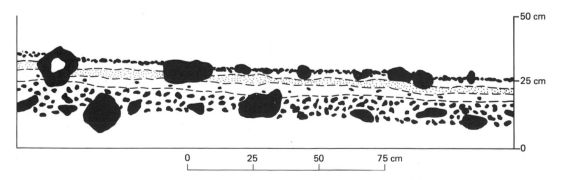

Figure 6–9 Cross section from trench in alluvial fan, showing desert pavement on the surface. (From Denny, 1967, p. 19.)

ment versus water discharge. Wadis display a braided pattern due to the extreme sediment volume and dearth of water in the stream bed. It is common for water to persist for only a few hours or at most a few days after rain. Typically, the water percolates down into the stream bed, resulting in abrupt loss of carrying power and rapid sediment accumulation in the channel. Wadis may develop dams of sediment which can cause changes in the stream path during succeeding flash floods (Glennie, 1970).

In addition to running water, the wind plays an important role in sediment transport in the wadi environment. The limited length of time during which sediment is underwater or is moist permits the wind to redistribute sand-size and finer particles such that a typical section of wadi-deposited sediment contains both wind-and water-laid deposits.

The most extensive wadi complexes are located in northern Africa (Glennie, 1970), the Middle East (Karcz, 1972), Australia (Williams, 1971), and the southwestern United States (Picard and High, 1973). Although relief and geologic setting vary from site to site, the general morphology and processes remain markedly similar throughout.

Sediments and Sedimentary Structures

Wadi sediments consist of terrigenous clasts that range through the entire grain size spectrum. In a typical sequence of wadi sediments one would find coarse gravel to mud, although the latter is not abundant. There is also a wide range of sedimentary structures incorporated within the wadi sequence. The result is that the typical wadi sediment sequence is rather diagnostic and easily recognized, especially when placed within the context of the desert environment.

Flash floods represent the most severe energy conditions on the wadi. They start suddenly and wane almost as suddenly. Consequently, there is a broad spectrum of energy conditions, which results in all bedform types being generated. However, as flow decreases, features such as antidunes are washed out, so that preserved bedforms include ripples, rarely megaripples, and most commonly plane beds (Glennie, 1970). Gravel commonly displays imbrication. Mud drapes resulting from the settling of suspensed fine sediments commonly show desiccation cracks (Figure 6–10).

Wind activity also generates its own set of sedimentary structures. Large-scale cross-stratification showing discontinuities and wedge-shaped cross-sets is the result of small migrating dunes. Deflation surfaces with concentrations of pebbles also may be present. Some of these surfaces may show pebbles incorporated in the upper portion of cross-sets as the result of wind scour and sand drift.

The typical wadi sequence consists of alternating units of wind- and water-deposited sediment; each unit ranges from about 10 to 30 cm (Glennie, 1970). The water-laid deposits commonly show a crude fining-upward grain size, culminating with a mud-cracked surface (Figure 6–11). Sand layers within the unit are generally cross-stratified. Gravels may display imbrication, crude grading, or they may be structureless; they are all water laid, but care should be exercised to be aware of deflation gravels. Wind deposits consist largely of cross-stratified sand overlying mud-cracked water-laid deposits (Figure 6–11). Some horizontal stratified **loess** may also be present.

Figure 6-10 Dessicated muds in wadi sediments. (From Glennie, 1970.)

ERGS

Extensive areas of wind-transported sand accumulation are called ergs. These seas of sand have until recently simply been called deserts and the assumption has been made that they are covered with wind-blown sand (i.e., dunes). This is an oversimplification: first, in that much of the desert, or in fact all of it, may not contain dunes, and second, because eolian sand accumulates in other forms in addition to dunes. Ergs may cover as much as thousands of square kilometers or they may be localized eolian sand accumulations. They are covered by a complex assemblage of ripples, dunes, and **draas.** The latter term is used for extremely large bedforms with wavelengths of about a kilometer or more (Wilson, 1972).

General Morphology and Processes

Basic mechanics of sediment transport by wind are similar to those by water, although there is an extreme difference in density and viscosity of the fluid. The reader is directed to the classic comprehensive work of Bagnold (1941) for details of eolian transport. As pointed out by Sharp (1963), the most important difference in sediment transport between the two media is that in the atmosphere, intergranular collision is paramount, whereas in subaqueous conditions, fluid turbulence is most important in mobilizing and supporting grains.

Sediment of ergs is generally made available from adjacent arid fans or wadis, but in some cases continual reworking of sediment by wind has resulted in extensive

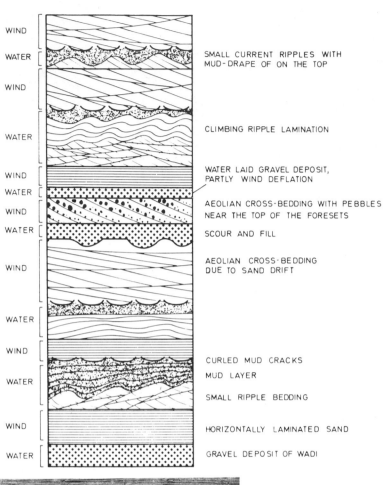

WIND

WATER — SMALL CURRENT RIPPLES WITH MUD-DRAPE OF ON THE TOP

WIND

WATER — CLIMBING RIPPLE LAMINATION

WIND — WATER LAID GRAVEL DEPOSIT, PARTLY WIND DEFLATION

WATER

WIND — AEOLIAN CROSS-BEDDING WITH PEBBLES NEAR THE TOP OF THE FORESETS

WATER — SCOUR AND FILL

WIND — AEOLIAN CROSS-BEDDING DUE TO SAND DRIFT

WATER

WIND

WATER — CURLED MUD CRACKS
MUD LAYER
SMALL RIPPLE BEDDING

WIND — HORIZONTALLY LAMINATED SAND

WATER — GRAVEL DEPOSIT OF WADI

ca 20 cm

(a)

column for wadi

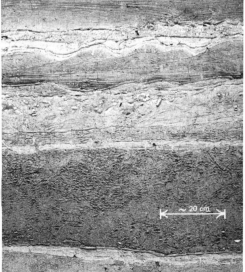

~ 20 cm

(b)

Figure 6–11 (a) Schematic sequence and (b) corresponding photograph of wadi deposits, showing alternating wind- and water-laid sediments. (From Reineck and Singh, 1980, Figure 317; Glennie, 1970, Figure 43.)

sand blankets with no apparent source at the surface. Bedforms that are generated fall in distinctly separate classes of wave length (Wilson, 1972). Although grain size does show a relationship to wave length, the ripples, dunes, and draas are easily subdivided (Figure 6–12). Most ergs display the complete spectrum of eolian bedforms except in ergs where sediment supply is limited (Wilson, 1973). Australian ergs lack draas, whereas the extensive ergs of the Sahara, Middle East, and central Asia display all eolian bedform varieties.

The actual shapes taken by the bedforms, especially dunes, are almost infinitely variable. Many authors devote much discussion to a categorization of these dune shapes (e.g., Glennie, 1970; Cooke and Warren, 1973). Such efforts are valuable and necessary for the geomorphologist; however, the ultimate primary purpose of this book is to provide the reader with the tools necessary to interpret ancient depositional environments from the stratigraphic record. Detailed dune morphology is therefore not considered in this chapter.

Sediments and Sedimentary Structures

The general conditions of the eolian environment plus the relative homogeneity of eolian sediments in comparison to those of most environments results in a relatively

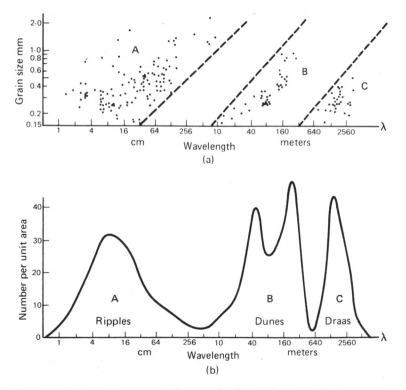

Figure 6–12 Plots of the coarse 20th percentile of (a) grain size and (b) frequency of wave lengths for eolian bedforms. The fields correspond to ripples, dunes, and draas, respectively. (After Wilson, 1972, Figures 2 and 3, p. 193.)

narrow suite of sedimentary structures compared to other continental sedimentary environments. Various obstructions that interfere with or interrupt the transport of sediment cause particles to come to rest on their lee side. Such sand shadows or sand drifts (Bagnold, 1954) form in association with pebbles, vegetation, exposed rock, or any other pertubation of the otherwise smooth eolian surface. Deflation pavements commonly produce **dreikanters,** which are three-sided pebbles or cobbles shaped by wind-blown sand. They contain a dull side which rested on the pavement and generally two polished sides formed by abrasion.

Bedforms are the dominant feature formed in eolian sediments. Dunes are typically from tens to hundreds of meters in wave length (Wilson, 1972). The cross-stratification in these sand dunes is persistent and is commonly preserved in the stratigraphic record, where it may be diagnostic of the desert environment. Other bedforms are also present. Ripples with wave lengths of a few to about 30 cm are present nearly everywhere on the dune surface. Some eolian environments accumulate sand in horizontal layers representing the equivalent of plane bed conditions. According to Bagnold (1954) and Glennie (1970), an abundance of well-sorted sand with strong winds causes these sheet sands to develop. The term accretion deposits has been used for sheet sands, in contrast to encroachment deposits for cross-stratified sediment (Bagnold, 1954). This is not to imply that these sheet sands represent plane bed deposition; climbing wind ripples are ubiquitous (R. Hunter, personal communication).

Internal structures in dunes are not well known because of the general difficulty of observing the bowels of a dune. A few deep and large trenches have been cut using large heavy earthmovers (e.g., McKee, 1966), but most studies rely on small, shallow trenches or natural scour by the wind. Probably the most fundamental approach to studying dune cross-stratification is measurement of the direction of dip, which provides information on wind direction. The magnitude of dip ranges with location on the slip face; typically, it is highest near the dune crest and becomes nearly horizontal at the base.

Cross-stratification takes on various geometries and orientations with respect to dune morphology (Andrews, 1981). This dune stratification can be divided into two major categories (Bagnold, 1954): accretion strata and encroachment strata. The accretion strata accumulate on the gently sloping windward side of the dune and comprise only a small portion of eolian cross-strata. Encroachment deposits develop as particles move down the slip face of the dune as they do in subaqueous bedform migration.

Some dunes display a single mode of cross-strata orientation, as in barchan dunes, whereas the longitudinal or sief dunes show a bimodal orientation of cross-strata due to the slip face being nearly normal to the wind direction (Glennie, 1970). It is common to observe small dunes and their resulting cross-strata moving over the surface of larger dunes of a different shape.

Bounding surfaces that separate cross-sets are nearly horizontal. Stokes (1968) showed how groundwater may control bounding surfaces by anchoring dunes while their crests are blown off (Figure 6–13). Brookfield (1977) has shown that bounding surfaces may also be produced by migration of eolian bedforms, with the inclination inversely related to bedform size.

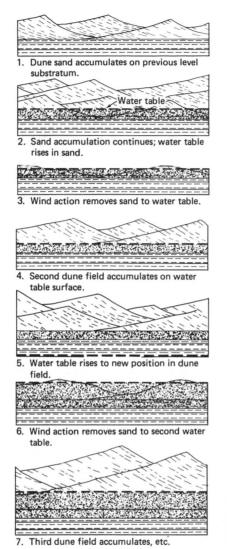

1. Dune sand accumulates on previous level substratum.

2. Sand accumulation continues; water table rises in sand.

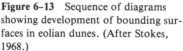

3. Wind action removes sand to water table.

4. Second dune field accumulates on water table surface.

5. Water table rises to new position in dune field.

6. Wind action removes sand to second water table.

7. Third dune field accumulates, etc.

Figure 6–13 Sequence of diagrams showing development of bounding surfaces in eolian dunes. (After Stokes, 1968.)

Detailed study of dune cross-strata has revealed six major types (Hunter, 1977b). Four of these involve traction deposition on rippled surfaces and plane beds. Grainfall deposition occurs on smooth surfaces in the zones of flow separation leeward of dune crests, and avalanche or grain flow deposition takes place on the slip face of the dune (Figure 6–14). Some distortion of cross-strata may be associated with steep, unstable slip face dune surfaces.

Most dunes of the world are terrigenous in composition, with quartz the dominant mineral. Carbonate dune sands may be found on deserts but are typically coastal in their location. The White Sands area of New Mexico has unusual dunes in that they are composed of gypsum sand (McKee and Moiola, 1975). Most dunes are fine to medium sand in grain size; they are well sorted, well rounded, and have

Figure 6-14 Diagram of a transverse dune, showing location of various bedding types. (From Hunter, 1977a, Figure 1, p. 365.)

slightly positive skewness. Frosting of the surface of sand grains is a common phenomenon in dunes. Some authors attribute this to impact of one grain against another (Glennie, 1972) but others believe that the pitting may be chemical in origin (Kuenen and Perdok, 1962). SEM studies have demonstrated that both origins of this surface texture are common.

INLAND SABKHAS AND PLAYAS

The internal nature of the drainage system in a desert causes formation of broad, shallow lakes in central depressions caused by deflation or by tectonic setting. Most of these lakes are dry most of the year, although some are semipermanent (Reineck and Singh, 1980). The term **playa** has been applied to these intermittent lakes in North America; however, usage of the term **sabkha,** taken from Arabian desert areas, is becoming prevalent on a worldwide basis.

These environments form where the water table is very near the ground surface, although it may be above or below at a given time (Figure 6–15). They are supplied with water by wadis and form the final resting place for runoff from adjacent high elevations. The size of sabkhas ranges from a few hundred square meters to Lake Eyre in Australia, which covers 8000 km² (Twidale, 1972). Some authors (e.g., Hardie et al., 1978) choose to distinguish between the sabkha environment and the ephemeral saline lake environment. However, they are combined in the following discussion. These evaporite basins produce a unique suite of minerals and sedimentary structures which make the sabkha sequence rather diagnostic in the stratigraphic record.

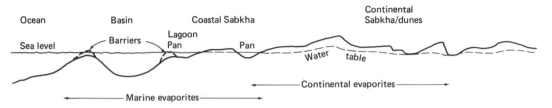

Figure 6–15 Profile sketch of a sabkha, showing position of the water table and location of evaporite environments. (From Kinsman, 1969.)

Processes

The primary events that affect the inland sabkha environment are the catastrophic deluges of water associated with rain showers and flash flooding. The actual physical movement of sediment by such events is restricted largely to adjacent alluvial fans and wadis; only the outer margins of the sabkha are commonly subjected to sediment entrainment. The most significant phenomena that affect the sabkhas are those involving evaporation and related chemical processes.

The sabkha environment may change its size and character significantly over just a few months or it may remain somewhat constant in its nature. Where flash flooding is prominent, such as in the desert southwest of the United States, sabkhas (playas) commonly receive large quantities of water which may persist for months. As the playa lake dries up, the marginal mud flats dry and evaporite precipitation takes place, giving rise to somewhat concentric patterns in the sediment (Hardie et al., 1978). A classic example of this situation took place in Lake Eyre, South Australia, after flooding in 1950.

Physical processes during flooding consist of some sheetwash along marginal mud flats, producing plane bed deposition, possibly turbidity current deposition, and some settling of suspended fines. Typically, only mud and fine sand are carried to the playa basin.

In northern Africa and Arabia the sabkhas are seldom subjected to even an occasional flash flood. As a result, standing water is almost never present. Although minor amounts of terrigenous sediment is carried to the sabkha via wadis, most terrigenous sediment is carried to the sabkha by wind. The moistness of the sabkha surface, caused by capillary rise of water from the water table, permits eolian sediment to adhere to the sabkha surface, thus causing terrigenous sediment accumulation (Glennie, 1970).

In summary, the North American sabkha (playa) is dominated by runoff as a sediment supply, with eolian processes being less important. This is due to a combination of a high-relief sediment source and generally more rainfall than in northern Africa and Arabia, where eolian sediment supplies dominate.

Actually, much of the physical and mineralogical character of the inland sabkha is due to chemical processes. Not only do they produce complex assemblages of minerals, but during precipitation of these minerals there is physical rearrangement and disruption of bedding features (Reeves, 1968; Glennie, 1970).

Sediments

Terrigenous sediments on sabkhas or playas are rarely coarser than mud and accumulated in thin, planar layers only a few millimeters thick. Such sediments are not abundant, due to a general absence of weathering that produces such particles in the arid environment (Hardie et al., 1978). Chemical precipitates form a primary component of the mineral matter that accumulates in this environment. Salinities reach extreme concentration only weeks after influx of runoff. Gypsum and halite are widespread and abundant. Other evaporite minerals may also be precipitated. The area over which salts precipitate may, however, be only a small portion of the playa environment. Typically, there is extreme evaporation and therefore reduction in area of the ponded water to achieve the salinities necessary to precipitate salts (Eugster and Hardie, 1978).

The combination of terrigenous mud layers and precipitation of evaporites produces a couplet which may be repeated to accumulate tens or hundreds of meters of sediment (Hardie et al., 1978). The mud is subjected to bacterial reduction of SO_4, producing hydrogen sulfide (Reeves, 1968). This dark mud layer may contain some salt crystals and is in turn overlain by a thicker layer of salts.

Sedimentary Structures

Desiccation features dominate the sabkha sequences with wrinkled or mud-cracked algal mats and muds. The presence of large salt polygons characterizes the sabkha surface (Figure 6–16). Typically, this surface forms a hard but thin crust over the wet sediment underneath (Glennie, 1970).

Adhesion ripples are common on the sabkha surface. As deflation takes place on adjacent dry areas, small quantities of eolian sediment accumulate on the wet surface and are molded into small bedforms by the wind. These ripples may be the cause of the characteristic wavy to lenticular bedding that is found in sabkhas (Figure 6–17). Such bedding may also be formed by the physical deformation of bedding by growth and dissolution of gypsum crystals, bubble growth, and collapse or growth of algal mats (Glennie, 1970).

STRATIGRAPHY OF DESERT DEPOSITS

Individually, the various environments of deposition that characterize the desert system may be difficult to recognize. The alluvial fan deposits, for example, strongly resemble glacial outwash or braided stream deposits. When the entire spectrum of desert facies is viewed as each relates to the other and in their proper geometric form, one should have little difficulty in recognizing the system.

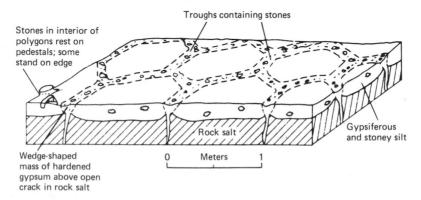

Figure 6–16 Polygons formed by cracks in rock salt and infilling by sediment. (After Hunt and Washburn, 1960.)

Figure 6-17 Trench in sabkha sediments, showing wavy bedding. (From Glennie, 1970.)

Commonly, the desert margin or margins are bounded by high-relief sediment sources which provide the ultimate source for the desert system. Alluvial fans form at the base of the slope and grade into a sandy, flat area. Both the fans and adjacent desert flats may contain extensive wadis and the sand flats may develop dune fields (Figure 6-2). Playas and inland sabkhas may be found in the central portion of the internally drained region.

Excellent examples of this complete arid depositional system have been summarized by Hardie, et al. (1978). Saline Valley, California, serves as a good model for such a system. The valley is only about 12 km wide but has all of the typical elements. Rather high relief bedrock complexes serve as the physiographic margins and as the primary sediment source for alluvial fans (Figure 6-18). These coarse fan deposits intertongue with sand flat and mud flat deposits, with the center of the valley being occupied by evaporite deposits from an ephemeral lake.

Extensive desert areas that have no high-relief boundaries or sediment sources would be similar but without the coarse alluvial fan facies. We might expect a stratigraphic section representing such an arid complex to have sequences of wadi, dune, and sabkha deposits interlayered.

ANCIENT DESERT DEPOSITS

Recognition of the desert depositional system in the rock record would be rather simple if the entire system were both preserved and exposed, as shown in Figure 6-18. As is typically the case, the geologist must rely on but a few pieces of the puzzle, with many missing. Nevertheless, perserverance, common sense, and some understanding of the systems in question can provide a proper interpretation.

The following discussion considers various criteria which may be helpful in

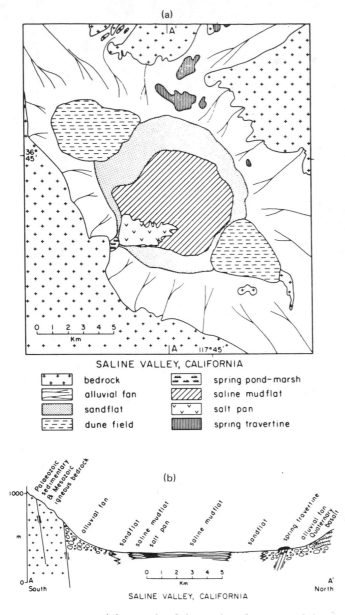

(a)

SALINE VALLEY, CALIFORNIA

bedrock		spring pond-marsh	
alluvial fan		saline mudflat	
sandflat		salt pan	
dune field		spring travertine	

(b)

SALINE VALLEY, CALIFORNIA

Figure 6–18 Generalized map and cross section of Saline Valley, California, showing typical desert sedimentary environments. (From Hardie et al., 1978, Figure 5, p. 33.)

recognizing each of the major elements of the desert system. Finally, well-known examples of arid depositional systems from the rock record are considered.

Alluvial Fans

Sediments and sequences from alluvial fans may be quite similar to glacial outwash of glaciofluviatile sediments. They may also be like braided stream deposits. Consequently, one must closely scrutinize suspected alluvial fan deposits before reaching any conclusions about their depositional setting. Once the deposits in question have been identified as alluvial fans, it is necessary to determine if they are arid fans or humid fans.

Bull (1972) has given what is probably the best summary of criteria for recognizing ancient alluvial fan deposits in the rock record. Fans are typically thick deposits that are lenticular or wedge shaped. Oxidation rate is high; therefore, organic matter is rarely preserved. Sediments are deposited by both water and debris flows, generally in sheets with channels as minor constituents. Alluvial fans contain a wide range of textures. Grain size ranges from mud to large boulders, and sorting is varied. Individual sedimentation units show great range in hydraulic conditions, reflecting the sporadic nature and intensity of sediment transport. There is an overall decrease in grain size down the paleoslope and paleocurrent direction. Paleocurrent data may reflect a radial transport pattern, due to the point-source nature of the sediment supply from the adjacent high-relief areas.

Distinction between arid and humid fans may be difficult. Typically, arid fans contain little mud compared with that in humid fans; however, accumulation of mud on the fan itself may not reflect this difference. Mud on arid fans is typically in thin layers or drapes above sand and gravel layers, indicating deposition by waning current. Humid fans may have little mud or may have a lot. Mud is readily available on humid fans because of the weathering processes enhanced by abundant moisture. Because of the relatively constant discharge of humid fans compared to arid fans, the fan may be nearly devoid of mud in the case of a rather high discharge fan, or may have considerable mud if currents are slow. Arid fans may be expected to contain or be adjacent to evaporite deposits and eolian accumulations, whereas these would not be expected as part of humid fans. Some reports have stressed that "red beds" are indicative of arid conditions (Walker, 1967), whereas others have suggested that they signify humid conditions (Krynine, 1949; Van Houten, 1964; Walker, 1974). This controversy continues and consensus has not yet been reached.

Ancient examples

Precambrian fans have been described from the Torridon Group of Scotland. Although there is no doubt about the fact that these arkosic conglomerates and sandstones are fans, there is not agreement as to the climate that prevailed during deposition. Williams (1966) suggested that the data point toward a humid fan, but a later paper (Williams, 1969) suggests that a semiarid climate is indicated by these sediments. A similar conclusion was reached by Maycock (1962, in Bull, 1972), based on the presence of shaly playa deposits in the distal facies of these Torridonian strata.

Tertiary intermontane deposits of the Ridge Basin area of southern California have been interpreted as arid fan-lacustrine sequences (Crowell, 1954; Link and Osborne, 1978). The fan deposits are thousands of meters thick and contain abundant debris flow deposits. As in the Torridon deposits, the distal, thin mud facies are interpreted as representing the playa environment.

Wadis

Both wind- and water-laid deposits are involved in wadi deposition, making it necessary to be able to recognize both accumulation types. Frequently, this must be accomplished from small-scale data such as cores or cuttings from the subsurface. Glennie (1970) has provided a good synthesis of criteria for recognition of these sediments (Table 6–1). The various criteria listed in Table 6–1 are for recognition of wind- or water-laid deposits in general, so that when using this list for recognition of suspected wadi deposits, allowances must be made. For example, the wind deposits in a wadi sequence are generally from a few to a few tens of centimeters in thickness and any cross-stratification that may be preserved would be rather low-angle and small-scale ripple cross-stratification.

Glennie (1970) has developed a reconstruction of a wadi sequence which shows that wind and water paleocurrent directions may be in opposition to each other and that these directions may differ from that indicated by the overlying sand sheet.

TABLE 6-1 RECOGNITION OF DESERT SEDIMENTARY ROCKS

Wind-deposited sands

1. Sequences of sands that may vary in thickness from a few centimeters to several hundred meters and whose laminae dip at angles from horizontal to 34°.
2. Laminae commonly planar, but ripples occasionally seen on steeply dipping foresets.
3. Individual laminae well sorted, especially in finer grain sizes: sharp differences in maximum grain size between laminae common.
4. Larger sand grains tend to be well rounded.
5. Percentage of silt and clay generally well below 5% or even absent.
6. Clay drapes very rare, and when present should be accompanied by evidence that they were water-laid.
7. Quartz sands at shallow depths commonly friable or lightly cemented with hematite.
8. Presence of adhesion ripples with associated increase in clay content and common presence of gypsum or anhydrite cement.

Water-laid sands

1. Most sedimentary features similar to those of water-laid sediments from non-desert continental environments but modified by the presence of one or more of the following:
   a. Commonly calcite-cemented, or locally cemented by gypsum or anhydrite.
   b. Many grains coated with hematite.
   c. Conglomerates may be common, and sometimes with several cycles of deposition that lack a sand-size fraction at the top of the cycle (deflation of the sand and silt).
   d. Presence of mud-flow conglomerates.
   e. Sharp upward decrease in grain size (especially from sand to clay), indicating a rapid fall in water velocity.
   f. Common presence of clay pebbles and curled clay flakes.
   g. Presence of mud cracks with sandy infill.

SOURCE: After Glennie, 1970.

Ancient example

The Permian Lower New Red Sandstone of England contains a sequence of sandstone beds in the Exe Group which may be interpreted as wadi deposits. These strata occupy the upper several hundred meters of nearly 3000 m of red, terrigenous rocks in southwestern England. The primary features of these strata are the interbedded sandstones of fluvial and eolian origin (Laming, 1966). Paleocurrent directions indicate that the fluvial sediments show similar patterns to the coarser-grained facies, whereas eolian sandstones have paleocurrent directions nearly at right angles to the fluvial deposits. The paleogeographic setting envisioned by Laming (1966, Figure 8) is quite similar to the present-day depositional environment of much of the Middle East.

Dunes

Recognition of ancient dune deposits in the rock record has long been taken for granted but may actually be rather difficult. Such criteria as fine to medium sand, well sorted, well rounded, spherical, and lacking in mud or gravel are typically used to summarize the characteristics of dune sediment (Table 6–1). It is true that these, plus additional features such as large-scale cross-stratification, characterize most eolian sand (Hunter, 1981). It is also true that the same features may characterize some fluvial or marine deposits. In fact, some heretofore accepted "eolian" deposits have recently been interpreted as being marine (e.g., Pryor and Amaral, 1971). Overall geometry of the sandstone body may also be important; eolian sand bodies are typically tabular.

Ancient examples

Numerous well-sorted sand units with large-scale cross-stratification have been ascribed to an eolian origin. Two well-documented examples are the Lyons Sandstone (Permian) in Colorado and the Entrada Formation (Jurassic) of the Four Corners Region in the southwestern United States.

Walker and Harms (1972) have carefully examined various aspects of the Lyons Sandstone from the type area near Boulder, Colorado, and have concluded that it is a classic eolian blanket sand deposited over alluvial sediments. This is contrary to the work of Thompson (1949), who concluded that the Lyons is a beach-nearshore deposit. The criteria presented by Walker and Harms are compelling and leave little doubt about the eolian origin of the Lyons. This unit is well sorted, with cross-stratification sets ranging from 3 to 13 m in thickness. Dip angles on the cross-strata are commonly 25 to 28°. Other eolian-type structures include ripples with high ripple indices. These ripples have been found on the lee slope of steeply inclined cross-strata with crests parallel to dip (Figure 6–19). Raindrop impressions, avalanche bedding, tracks of reptiles, and coarse lag deposits are also present, although they are rare (Walker and Harms, 1972).

In a study of the Entrada Formation and related units, Tanner (1965) reached the conclusion that the Four Corners Region was occupied by an extensive dune field

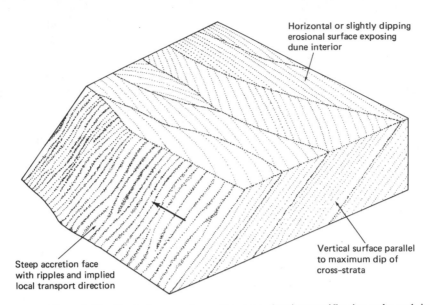

Horizontal or slightly dipping
erosional surface exposing
dune interior

Steep accretion face
with ripples and implied
local transport direction

Vertical surface parallel
to maximum dip of
cross-strata

Figure 6–19 Block diagram of an eolian dune, showing stratification styles and rippled prograding face. (After Walker and Harms, 1972, Figure 7, p. 285.)

with a general wind direction toward the south-southwest. The Entrada is overlain by the Todilto Formation, a complex of carbonates and evaporites, many of which are also interpreted as eolian (Tanner, 1965). The large-scale cross-strata, ripples, and stratigraphic relationships with adjacent units provide ample evidence for the eolian origin of the Entrada.

Inland Sabkhas and Playas

The primary feature associated with these environments is the presence of evaporites interbedded with generally fine grained terrigenous sediments. Various desiccation features, such as mud cracks and other shrinkage features, sedimentary dikes, and disturbance of bedding by crystal growth are also widespread. Although evaporite precipitation is common and generally dominated by gypsum and halite, a great variety of less common evaporite species may be present. The variety and abundance of these is generally unique to a given playa deposit.

Wilkens Peak Member, Green River Formation (Eocene)

Eocene strata in southwestern Wyoming and adjacent Utah and Colorado have been studied extensively for nearly a century. The Green River Formation consists of more than 1000 m of terrigenous and chemical sediments which comprise three members: the Laney, Wilkens Peak, and Tipton. The Wilkens Peak Member displays a broad spectrum of features which typify the desert depositional system and are remarkably similar to the present-day desert complex (see Figure 6–18).

The Green River Basin is bounded on the south by the uplifted Uinta Mountains (Figure 5–13). Adjacent to this high-relief front are alluvial fan deposits characterized by poorly sorted conglomerates, cross-stratified sandstones and gravels, and lenses of quartz arenites with planar horizontal to gently inclined strata (Smoot, 1978). These coarse, fan deposits are typically included in the Wasatch Formation (Eugster and Hardie, 1975), which is characterized by fluvial deposition. The braided alluvial deposits represent upper and midfan deposits. The large clasts are dominantly carbonates and quartz arenites which were derived from the Upper Paleozoic sequences in the Uinta Mountains source area.

Distal fan deposits are characterized by sandstones composed of both quartz and dolomite grains. The presence of terrigenous carbonate particles testifies to the arid conditions of the environment; carbonates are extremely soluble compared to silicate minerals. Some muds, generally detrital dolomites, are present as drapes over the sandstone strata (Smoot, 1978). Most of the coarse terrigenous sediments are restricted to what were the margins of this depositional basin. There are, however, several sandstone tongues which extend northward into the central basin. These are interpreted as fluvial deposits (wadis) which were related to tectonic activity in the Uintas and reworking of the alluvial fan sediments (Eugster and Hardie, 1975).

The central portion of the basin is characterized by thin-bedded but thick and extensive deposits of terrigenous and carbonate mud as well as evaporites. This rock unit contains some of the most extensive trona ($NaHCO_3 \cdot NaCO_3 \cdot 2H_2O$) deposits in the world (Eugster and Hardie, 1978). Fine-grained laminated dolomite mudstones display various desiccation features. On a microscopic scale grading can be recognized in these deposits, which Smoot (1978) interpreted as sheetflood deposits. He also investigated the origin of this fine-grained dolomite and concluded that it was derived from local sources such as travertine in the alluvial fans, tufa, and caliche. It was not derived from Paleozoic strata in the Uinta Mountains, as were the coarse carbonate clasts. These extensive desiccated mudstones are interpreted as mudflats along the playa margins (Eugster and Hardie, 1975; Smoot, 1978).

The central portion of the basin, north of the Uintas, is dominated by various evaporite minerals with some interbedded detrital mudstones and intraclastic carbonates. The evaporites include halite as well as trona. Beds of the latter, which number 42, may reach 11 m in thickness (Eugster and Hardie, 1978). The central basin strata are characterized by cyclic accumulations. Increased runoff expands the lake, producing a reworking of the mudflat deposits during transgression and yielding the intraclastic carbonates. Sheetflood deposits typically overlie these "basal conglomerates" and are in turn overlain by evaporite accumulation as the playa dries up (Smoot, 1978).

The Wilkins Peak Member of the Green River Formation and the laterally adjacent portion of the Wasatch Formation provide an excellent example of the desert depositional system. Uplifted mountains are bounded by coarse arid fans which laterally grade through wadis and mud flats, culminating in evaporites which accumulated in a large playa (Figure 6–20).

Rotliegendes (Permian) of Northwest Europe

Extensive red beds, primarily of Permian age, are known collectively throughout northwestern Europe, including the British Isles, as the Rotliegendes.

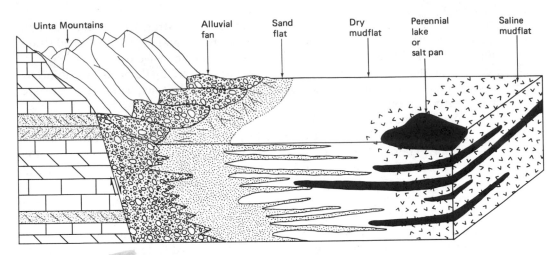

Figure 6-20 Reconstruction of desert sedimentary environments in which the Wilkins Peak Member of the Green River Formation accumulated. (From Smoot, 1978, Figure 2, p. 113.)

These deposits unconformably overlie various Paleozoic strata and are overlain by the Zechstein evaporite deposits. This thick red bed sequence is of great economic significance because it is an important gas reservoir rock.

The Rotliegendes sequence has been investigated in part by numerous authors and has recently been summarized by Glennie (1972) who interpreted the sequence to be an ancient desert depositional system. The following discussion of the various Rotliegendes facies is taken largely from his work.

The Rotliegendes is typically divided into the Lower, which contains volcanics throughout, and the Upper, which does not. Emphasis here is on the Upper Rotliegendes, which consists of the Slochteren and Ten Boer Members (Figure 6-21). The total thickness is near 200 m in the south and southwest but expands to more than 1000 m in the central part of the Rotliegendes basin (Glennie, 1972).

Although neither widespread nor covered in Glennie's synthesis, there are Permian alluvial fan deposits in southwestern England (Devonshire) (Laming, 1966). Coarse arid fan deposits of the New Red Sandstone are at least in part equivalent to the Rotliegendes. Most of these coarse fan deposits may be in the Lower Rotliegendes, because they contain volcanics. In any case they serve as a source for reworking of sediment and as the coarse, marginal deposits of the typical desert basin.

Throughout much of the Rotliegendes in the southern part of the basin, the basal part of the section is comprised of interbedded sandstones and conglomerates with some calcite cement. In addition, thin, discontinuous mud layers are present and show desiccation features (Glennie, 1972). Bedding may be horizontal or cross-stratified. Some grading is detectable in these units. Sediments in and associated with the conglomerates are moderate to poorly sorted. Interbedded with these strata are well-sorted sands with frosted grains. This facies is interpreted as representative of the wadi environment with alternating fluvial and eolian deposition (Figure 6-21).

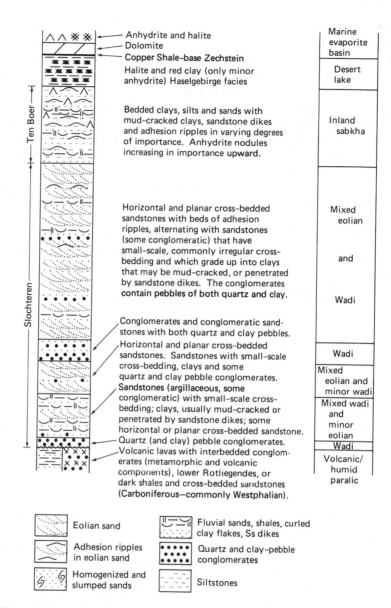

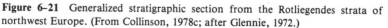

Figure 6–21 Generalized stratigraphic section from the Rotliegendes strata of northwest Europe. (From Collinson, 1978c; after Glennie, 1972.)

The desiccated mudstones represent drapes on the upper surface of the fluvial portion of the sequences.

Overlying the wadi sequence there is typically well-sorted and well-stratified sandstone. The sandstone units are separated by surfaces of discontinuity. Stratification is nearly horizontal at the base of each unit, increasing to 20 to 27° at the top, where truncation occurs (Figure 6–21). Cross-stratification directions show a distinct mode to the southwest throughout the Rotliegendes (Glennie, 1972). These well-sorted and locally frosted sandstones represent eolian dunes, probably of the transverse type. The surfaces of truncation are bounding surfaces common to this general environment (see Chapter 3).

Interbedded with these cross-stratified dune deposits are lenses and thin tabular sand bodies characterized by what may be adhesion ripples. Cementing may be by anhydrite and locally by halite. Anhydrite nodules are also present. Glennie (1972) interprets this facies as representative of rather local, interdune sabkhas with moisture supplied by groundwater.

The foregoing three desert environments are characteristics of the Slochteren Member of the Upper Rotliegendes. To the north and laterally equivalent or overlying the Slochteren is the Ten Boer Member, which is characterized by mudstones and evaporites with some thin sandstones (Glennie, 1972). The mudstones are thinly stratified with mudcracks and other evidence of desiccation. Anhydrite nodules are also present and increase in abundance upward (Figure 6–21). Adhesion ripples are present in the sandstone beds. Halite is also present, accounting for up to 30% of the total section above the Ten Boer Member. This is commonly referred to as the Haselgebirge facies and represents the thickest accumulation of Rotliegendes sediments (Figure 6–22). The Ten Boer and overlying evaporites represent a large and persistent sabkha-playa complex (Glennie, 1972).

The entire Rotliegendes sequence is representative of a classical desert depositional system, with a terrigenous source to the southwest in the Variscan Highlands of France and Holland (Figure 6–22). The fans grade through wadi and sabkha environments into extensive playas in an area now covered by the North Sea and Denmark. This Permian sequence compares well with the Eocene of the western United States and the present environments of the southern California desert (see Figures 6–18, 6–20, and 6–22).

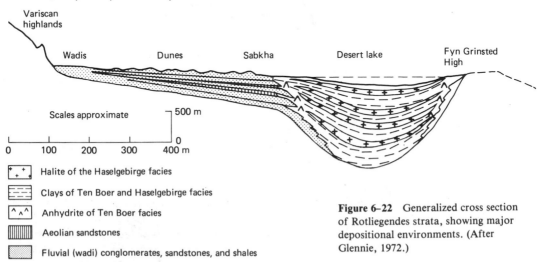

Figure 6–22 Generalized cross section of Rotliegendes strata, showing major depositional environments. (After Glennie, 1972.)

SUMMARY OF CHARACTERISTICS
OF SEDIMENT/SEDIMENTARY ROCK BODIES
REPRESENTING THE DESERT DEPOSITIONAL SYSTEM

The desert system is represented by three broad and distinct facies, each representing one of the major depositional environments in the system. These are the alluvial fans, dunes, and playas. In order to be recognized these facies must be characterized separately.

Tectonic setting. Deserts may form in any tectonic setting; however, a large amount of relief must be present in order to develop fans.

Shape. Fans are wedge shaped, dunes tend to accumulate tabular bodies, and playa deposits are thin and tabular to lenticular.

Size. Both fans and playas individually tend to be local, although numerous fans may coalesce along an extended high-relief face. Dune facies may extend for many thousands of square kilometers.

Textural trends. Fans contain a wide range of particle sizes but show a distinct fining away from the source. Theoretically, there should be a tendency for fining upward, but this is commonly masked by the varied particle sizes throughout the fan. Sediments tend to be submature to mature for a given sedimentation unit. Dunes are well sorted and rounded, generally supermature. Grain size is typically fine to medium sand and most dune accumulations are positively skewed. Playas are typically mud sized but may have some sand at the margins or from wind-blown sources. The latter would yield a textural inversion.

Lithologic trends. Fans display a variety of particle compositions, reflecting the variety in the nearby source rocks. Labile minerals decrease in abundance away from the source or may show a wide range in weathering. Generally, rapid deposition and close proximity to the source permit preservation of unstable minerals and rock fragments, including terrigenous carbonate fragments. Theoretically, the dunes reflect the composition of the fans but tend to be dominated by quartz and feldspar simply because of their relative abundance. Playa lakes contain evaporite minerals interbedded with terrigenous muds.

Bedding and sedimentary structures. Fans are characterized by discontinuous strata, lenticular bodies of sand, and gravel with some mud drapes. Grading is common and loading structures may be preserved. Various sizes of cross-strata are common in the fine gravel and the sand. Paleocurrent data show a spread from the source with an overall fan-shaped pattern.

Large-scale cross-strata with dips commonly between 20 and 30° characterize desert dunes. Bounding surfaces separate cross-sets. Climbing ripple cross-strata are generally present within the large-scale cross-strata.

Playas commonly exhibit extensive desiccation features, raindrop impressions, and deformation of thin strata by growth of evaporite mineral crystals. Some small ripple cross-strata may be present.

Paleontology. There is an overall absence of fossils preserved in strata of the desert system. The combination of sparsely populated environments, rapid oxidation, and slow burial is not conducive to preservation. Pollen and spores plus some algae might become incorporated in playa deposits.

Associations. Strata within the desert system may display a variety of stratigraphic associations. Fluvial or lacustrine facies could be found above or below the desert strata in conformable relationship. Typically, a desert system produced by tectonic events would result in desert strata resting unconformably over older strata.

The sequence accumulated in a desert developed in a basin such as in the southwestern United States would be one of a coarsening upward trend with playa facies at the base, dunes in the middle, and an alluvial fan at the top. This is a generalization and many possibilities exist.

ADDITIONAL READING

COOKE, R. U., AND WARREN, A., 1973. *Geomorphology in Deserts.* University of California Press, Berkeley, Calif., 374 p. Outstanding book on desert landforms which includes processes as well as morphology.

ETHRIDGE, F. G., AND FLORES, R. M., (EDS.), 1981. *Recent and Ancient Nonmarine Depositional Environments: Models for Exploration.* Soc. Econ. Paleontologists and Mineralogists, Spec. Publ. No. 31, Tulsa, Okla., 349 p. Contains four fine papers on eolian deposits; two on modern environments, and two from the stratigraphic record.

GLENNIE, K. W., 1970. *Desert Sedimentary Environments.* Developments in Sedimentology, v. 14. Elsevier, Amsterdam, 222 p. This is the best single book on the sedimentology of the desert. It is well illustrated and well written. Its only drawback is overemphasis on African and Arabian deserts.

McKEE, E. D. (ED.), 1979. *A Study of Global Sand Seas.* U.S. Geol. Survey, Prof. Paper 1052, 429 p. Recent compilation of several papers on various aspects of desert sedimentation, with emphasis on eolian dunes. Ancient examples are included and illustrations are both numerous and excellent in quality.

RIGBY, J. K., AND HAMBLIN, W. K. (EDS.), 1972. *Recognition of Ancient Sedimentary Environments.* Soc. Econ. Paleontologists and Mineralogists, Spec. Publ. No. 16, Tulsa, Okla., 340 p. Papers on alluvial fans (W. B. Bull) and on dunes (J. J. Bigarella) are excellent and give good criteria for recognition of these environments in the rock record.

VAN HOUTEN, F. B. (ED.), 1977. *Ancient Continental Deposits.* Benchmark Papers in Geology, v. 43, Dowden, Hutchinson & Ross, Stroudsburg, Pa., 367 p. Although this volume goes well beyond the desert, several papers deal with the desert or adjacent environments.

7 The Glacial System

A glacier is a large mass of ice formed by compaction and recrystallization of snow that is at least partly on land and that flows due to gravity. Glaciers may flow downslope or upslope due to their own mass. Important sedimentary environments are presently associated with glaciers and have been at intervals throughout geologic time. Because of a dependence on net accumulation of snow and ice for their existence, glaciers are restricted by climate. This restriction may be due to latitude, elevation, or both.

Glaciers are typically classified according to their morphology, which is in part determined by their geomorphic setting. The smallest glacier variety is the cirque glacier, which is limited in extent to a cirque (Flint, 1971). Valley glaciers (also called mountain or alpine glaciers) are confined to rock-bounded valleys and flow downslope. These valley glaciers may be fed by ice caps or ice fields and are typically quite elongate in proportion to their width. The outward spreading and coalescing of distal portions of multiple valley glaciers produces a piedmont glacier. In some respects it is morphologically similar to a bajada, where several alluvial fans coalesce. The largest of all glacial types is the continental glacier (ice sheet), which may cover millions of square kilometers. These glaciers may be thousands of meters thick and are not restricted in their extent by topography.

This dicussion will not consider how the ice accumulates and how the glacier itself is formed. Such information can be obtained from references by Flint (1971),

Paterson (1969), Sharp (1960), and Sugden and John (1976). Primary emphasis in this chapter is on how glaciers erode, how sediment is transported and eventually deposited, and the nature of the materials that accumulate as the result of these processes.

Although some valley glaciers are found at high altitudes on a nearly worldwide basis, there are only two major continental or ice sheet glaciers in existence today, in the Antarctic and Greenland. Smaller ones are present on Baffin Island, Iceland, and Spitsbergen. Eighteen thousand years ago more than half of North America, half of Europe, and parts of northern Asia were covered by ice sheets. Many geologists and climatologists believe that we are currently in an interglacial stage, with perhaps much more ice to come in the future. Best available data show that glaciers cover about 15 million square kilometers of the earth's surface today, whereas during maximum extent of glaciation in the Pleistocene it was nearly three times that amount (Flint, 1971).

In studying the glacial depositional system, it is not only important to be able to recognize glacial sediments as such, but also to pay strict attention to the size and geometry of the deposits as well as to the terrain on which they accumulate. This will permit determination of the type of glacier involved, it will indicate something about the geology and geomorphology of the setting, and in general it will allow for a thorough analysis of the depositional environment.

GLACIAL PROCESSES

With the exception of the glacial environment, all other sedimentary environments involve only fluids as the dominant mechanism for sediment entrainment, transport, and deposition. Whereas ice is a weak crystalline solid that flows, it does represent a unique transporting medium for sediment.

Ice Dynamics

Glacial ice is quite brittle at or near the surface, where large cracks or crevasses are developed. In contrast, glacial ice behaves like a plastic at depth, especially near its contact with the material, generally bedrock, over which it moves. This is the result of the pressures created by the weight of the ice. Partially as a consequence of its plastic behavior, the entire ice mass may slide over the underlying bedrock (Paterson, 1969).

Velocity of ice flow is controlled primarily by the thickness of ice, the slope of the ice surface, water content, and by temperature. Rates are typically on the order of a few to 100 or so meters per year, obviously slow compared to fluid motion in other sedimentary environments. Maximum velocity is at or near the ice surface, with slowest movement at the contact between the ice and the bedrock over which it moves (Figure 7–1). Friction is the primary factor in reducing the velocity.

In glaciers the ice mass may slide over the bedrock surface as well as the ice itself flowing. Wet-base glaciers are characterized by basal temperatures at the melting point of ice and slide over the rock surface. Cold glaciers may be frozen fast to rock

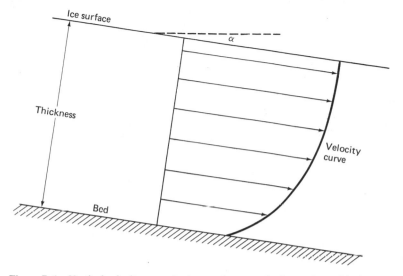

Figure 7-1 Vertical velocity curve for ice moving over sloping surface with slope of ice surface. (Simplified from Flint, 1971, p. 40.)

below. Their movement is within the ice mass. Thus there are two important components to the net movement. Not many measurements of sliding velocity have been made, due primarily to the difficulty in drilling through the ice mass to bedrock. Data indicate that sliding accounts for about half the total movement of the glacier (Paterson, 1969). Such sliding of ice over the bedrock is quite important in the erosional processes and sediment entrainment by the glacier.

Glacial Erosion

The ice mass moves slowly and with great friction over the earth's surface. By so doing, ice passes over the bedrock surface and incorporates sediment particles by enveloping them during its motion and by the thawing and refreezing process at the ice-bedrock interface. Some particles are probably made available for incorporation into the ice through frost wedging as well. The result is that at and near the lower or lateral margins of the ice mass, large quantities of sediment are incorporated into the ice and move over the bedrock surface, resembling a piece of sandpaper moving over a wood surface. The sediment particles, especially those of gravel size, act as tools shaping the bedrock surface and provide more sediment for entrainment.

Subglacial erosion may take place as the result of meltwater streams which flow under the ice and over the bedrock surface.

Glacial striations and **glacial grooves** are linear excavations (generally small) that are formed by the sediment-laden ice as it passes over the bedrock (Figure 7-2). The large grooves may be several meters deep in soft carbonates or mudrocks (Flint, 1971). Such lineations are valuable in paleoenvironment interpretations because they provide an orientation for ice movement; only rarely do they exhibit directional properties. The latter is possible when the striations or grooves show an asymmetry with the blunt end implying the upstream direction.

Another related erosional feature on bedrock surfaces is the **crescentic gouge.** These are features, usually several centimeters from tip to tip, which are thought to

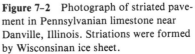

Figure 7–2 Photograph of striated pavement in Pennsylvanian limestone near Danville, Illinois. Striations were formed by Wisconsinan ice sheet.

be formed by large particles moving over the bedrock under great pressure (Figure 7–3). They are concave upstream.

Numerous large ice-flow features may also be formed on the bedrock surface. Some examples are roche moutonnées, whalebacks, and glacial troughs (Sugden and John, 1976). Because such features would be rare or difficult to recognize in the rock

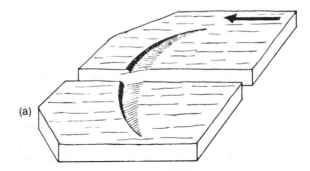

(a)

Figure 7–3 (a) Schematic drawing of crescentic gouge and (b) photograph of same feature formed on bedrock surface during Permian glaciation, Hallet Cove, South Australia. Ice movement is right to left in both illustrations.

(b)

record, they are omitted in this discussion. The interested reader should consult the suggested references on glacial geomorphology at the end of the chapter.

In addition to the above-mentioned features, glaciers commonly impart a polish or smoothed surface to the bedrock surface. The presence of such a surface and the degree to which it is developed is due primarily to the nature of the bedrock and the abundance of fine-sediment particles (sand or smaller) at the base of the ice mass. Typically, those surfaces which display glacial polish also contain striations, grooves, or other similar features.

Sediment Transport by Ice

Sediment particles transported by a glacier are referred to as the load of the glacier (Flint, 1971). The nature and distribution of sediment particles within the ice mass show some variation, but most is concentrated at the base or margins of the ice. Some sediment may also be concentrated at the upper surface when the glacier is experiencing ablation or melting.

Unlike a stream or other transporting mechanisms for sediment particles, glaciers exert considerable pressure on the sediment particles which are between the ice and the bedrock. This pressure, or the combination of pressure and movement, can crush or facet particles. In addition to causing the erosional features mentioned above and the crushing of particles, this pressure also results in melting at the base of the ice and formation of a thin layer of water between the bedrock and the glacier (Boulton, 1975).

Some sediment particles are carried in suspension by the glacier within the ice and some in a traction mode by the thin layer of water. Sediment particles transported by glaciers come from two sources: most are from the subglacial bed (especially in an ice sheet) and some are derived from valley walls and are on or near the upper surface (Boulton, 1975). In valley glaciers sediment is concentrated in elongate accumulations on the surface. These accumulations form **lateral moraines** when they occur at the contact between the glacier and the valley wall or medial moraines where a glacier has been fed by tributaries. Basal sediment in a glacier may be transported toward the surface of the ice mass near its terminus due to shearing of the ice. The result is a ridgelike accumulation of sediment and may occur in both valley glaciers and ice sheets (Figure 7–4).

Glacial Sedimentation

Deposition of sediment particles transported by a glacier takes place in four general areas:

1. Subglacial sedimentation occurs at the base of the ice mass and just above the bedrock.
2. Englacial sedimentation takes place within the ice mass.
3. Supraglacial sedimentation involves materials on the upper surface.
4. Proglacial sedimentation is in front of the glacial terminus and is dominated by discharging meltwater.

Figure 7–4 Photograph of terminal moraine at Logan Pass, Glacier National Park, Montana. Observe the large piles of unsorted glacial drift. (Courtesy of L. J. Schmaltz.)

In general, most deposition occurs first at the ice-rock interface and thereafter at the ice–glacial drift interface. Glaciers are quite dynamic and much of what is deposited in the glacial environment is only temporary. Englacial or supraglacial sediment is finally deposited only when complete melting occurs. As melting proceeds, mass movement of supraglacial material is very important.

The broad spectrum of glacial-deposited sediment is referred to as **drift** and is generally subdivided into two major categories. Glacial **till** is unsorted and unstratified glacial drift; it typically contains a significant amount of fine matrix and a wide range in particle sizes. Sorted and stratified drift is dominated by sand and gravel-size particles in **outwash** deposits, and by silt and clay particles in glacial lake deposits.

The general mode of deposition from the ice or meltwater determines which of these types will accumulate. As a generalization, the subglacial and superglacial sediment accumulates as till and the englacial and proglacial material accumulates as outwash or stratified drift. Till results from concentration of the entrained sediment by melting of the glacier with little or no winnowing or sorting of materials. It is essentially dumped by the ice. Till may be strongly modified by seawater, rainfall, or meltwater. It is common to observe sorted lenses of sand within till bodies. This sorting may take place in the thin layer of water at the sole of wet-base glaciers. Flowing meltwater washes away the silt and clay and locally transports the coarser particles, resulting in stratified drift.

The statements above are, with exceptions, rather broad generalizations; however, for purposes of recognizing and interpreting glacial environments from the rock record, they are useful. The reader should consult references such as Embleton and King (1968), Flint (1971), Wright and Moseley (1975), or Sugden and John (1976) for details of sediment transport and accumulation in glaciers.

GLACIAL SEDIMENTS

Sediments that result from glacial and related processes are perhaps the most heterogeneous of any depositional system that will be considered in this book. Certainly, that is the case within the spectrum of terrestrial environments, their only rival being turbidites in the marine realm (see Chapter 15). There is essentially no restriction with respect to particle size; some boulders bigger than a house have been transported and deposited by ice. Because glaciers may traverse an infinite variety of rock types, they incorporate and transport an equally diverse suite of particle compositions.

Till

The unique character of glacial till makes it the best single criterion for recognition of past glaciation. It is the dominant drift type in many moraines. In addition to the petrologic and textural variation, there are some other more specific features that one should look for in recognizing a till.

It would be expected that at least some of the identifiable sediment particles have their source at considerable distances from their site of deposition in the till. Actually, there have been contradictory results concerning this aspect. For example, one study showed bedrock over 300 km away contributed as much debris as that located only 3 km away, whereas another study in a different area demonstrated that 90% of the particles came from within 80 km of the site of deposition (Goldthwait, 1971).

A rather diagnostic characteristic of till is the presence of striated gravel-size particles. As a particle moves over the bedrock or another entrained particle, these striae are produced. It is not uncommon for the striated particles to contain a faceted surface or two where the particle was held by the ice and dragged over the subglacial bedrock under high pressure. This is most common in relatively soft rocks such as carbonates or mudstones.

Good and succinct discussions of till and its characteristics can be found in Goldthwait (1971) and Dreimanis (1976). The following paragraphs are largely excerpted from these references.

Stratigraphic studies of till must rely on considerable laboratory analysis of the sediment—both texturally and mineralogically. To make correlations and to reconstruct the history of events leading to the accumulation of a till sequence, it is necessary to have some basis for definition of the sediment bodies involved. In contrast to most lithologic units, this sometimes cannot be done at all in the field for tills. There are some situations where abrupt contacts between tills are delineated due to obvious textural changes or perhaps distinct color changes which reflect mineral composition.

Most of the time it is necessary to conduct either particle size analysis, mineralogic analysis, or perhaps both, in order to permit definition of a till unit or to correlate between localities. Such differences in composition of the till may reflect different sources of the particles, differing conditions of transport and deposition, weathering changes, or some combination of these. Perhaps the simplest and most common approach is to determine the sand-silt-clay ratios of the till [Figure 7-5(a)]. This characteristic is one that may show local variation even in a single till unit, so that it may not provide the desired results. Compositional analysis seem to be of more

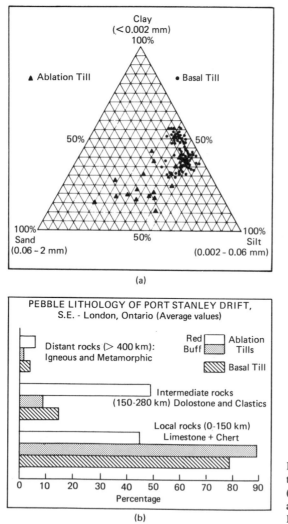

(a)

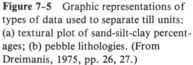

(b)

Figure 7-5 Graphic representations of types of data used to separate till units: (a) textural plot of sand-silt-clay percentages; (b) pebble lithologies. (From Dreimanis, 1975, pp. 26, 27.)

value in characterizing tills. These include clay mineralogy, pebble lithology [Figure 7-5(b)], and heavy mineral assemblages (Goldthwait, 1971).

Composition of till

Although till represents what is considered a texturally immature type of deposit, it may have some internal arrangement of particles. The most common form of preferred arrangement is the alignment of long axes of large sediment particles with the direction of ice movement (Figure 7-6). Imbrication may also be present, with the long axes dipping in the direction from which the ice advanced (upstream) (Lineback, 1971). Because of the general lack of stratification of till units, it may be possible to distinguish between till units, and therefore ice mass advances, by using the orientation of gravel particles in the till (Figure 7-6).

Stratified Drift and Outwash

Unstratified glacial drift (till) represents deposition directly from the ice with little or no involvement of fluids. Stratified drift, on the other hand, is the result of deposition by glacial meltwater and may take place upon (supraglacial), within (englacial), or underneath (subglacial) the ice mass (Allen, 1970a), all of which are considered

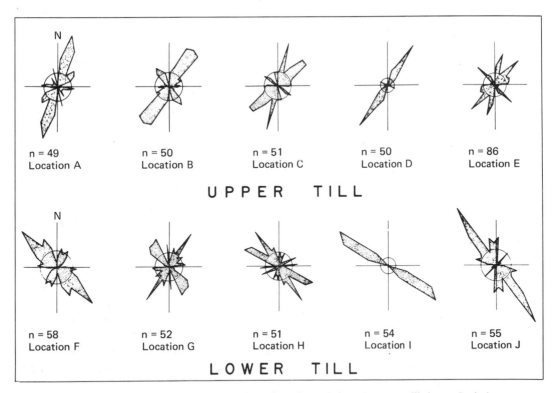

Figure 7-6 Rose diagrams showing orientations of clasts from two till sheets. Such data are useful in defining till units. (From Ramsden and Westgate, 1971, p. 337.)

ice-contact depositional sites. In this discussion proglacial stratified deposits which eminate from glacial melting are also included. The term "outwash" is typically applied to such deposits and it is quite appropriate because they do indeed come from a washing out of sediment from the ice mass. In actual fact, the total amount of outwash which accumulates in a given glacial system typically is much greater in volume than the nonstratified drift produced by the same system.

Stratified glacial drift (also called ice-contact deposits) comprises the landforms known as **kames, kame terraces,** and **eskers** (Figure 7-7). Kames are local mound-shaped accumulations of stratified sand and gravel. They originate from pockets of sediments in crevasses or other openings in the ice mass which are deposited as the ice melts. Kame terraces are composed of similar sediments and are developed along the lateral margins of valley glaciers by streams. Eskers are stream deposits that are let down on the subglacial surface after melting; they are common in all types of glaciers, as are kames.

The ice contact deposits described above are rather easy to recognize because in addition to being sorted and stratified, there commonly are slump or collapse features preserved in them. Included are contorted bedding and small gravity faults.

The most complex and widespread type of glacially derived sediment is outwash. This sediment is dominated by moderate to well-sorted sand and gravel which accumulates immediately in front of the ice mass. As sediment-laden streams debouch from the glacier they become unconfined, resulting in a decrease in velocity and a loss in their competence and capacity. The result is that coarse material (sand and gravel) accumulates with the fines (mud) being carried farther downstream or settling out in small impoundments (proglacial lakes) which are common in outwash plains or sandurs (Figure 7-8). Some fines may be carried out to the open ocean.

Because outwash deposits are ultimately transported as coarse bed load in proglacial streams, they are not only stratified but are moderate to well sorted and they contain a modest spectrum of cross-stratification. The depositional environment is analogous to the braided stream environment (Chapter 8) with an outwash fan being a complex of numerous coalescing braided streams and the width sometimes exceeds the length (Figure 7-8).

Texturally, outwash material is sorted; probably as a rule it is better sorted than ice-contact drift. The grain size ranges widely, with sand, pebble, and cobble size particles dominating. It should be noted that whereas individual beds or sedimentation units are well sorted, there is commonly great stratigraphic variety in mean grain size; the general rule is interbedded sand and gravel (Figure 7-9).

Outwash represents an excellent example of the necessity for sampling a single sedimentation unit in order to characterize the texture, especially sorting, of a sediment. If one wants to relate sorting (or textural maturity) to energy and a depositional environment, it is critical to sample only a single sedimentation unit and therefore one set of environmental parameters. A "grab" sample from the outwash illustrated in Figure 7-9 would yield a broad spectrum of grain sizes and therefore produce a low sorting value, lower than would accurately reflect conditions in the environment of deposition.

Pebbles and cobbles in outwash sediments commonly display well-developed

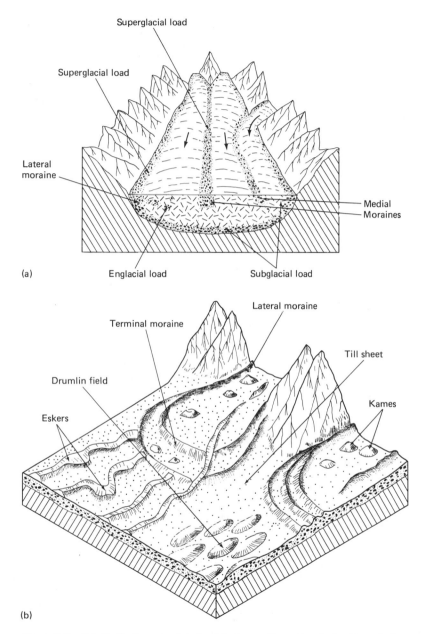

Figure 7-7 Schematic diagrams showing (a) where sediment is entrained in a glacier and (b) the location and geometry of sediment bodies. (After Allen, 1970a, pp. 228-229.)

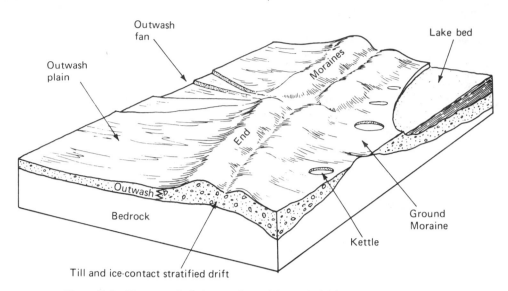

Figure 7-8 Diagram of glacier terminus with proglacial lake, end moraines, and outwash sediment accumulations. (After Bloom, 1978.)

Figure 7-9 Outwash sediments showing interbedded gravels and cross-bedded sands typical of this depositional environment; Kalamazoo, Michigan.

imbrication with the dip oriented upstream as it is in typical braided streams (Schlee, 1957; Doeglas, 1962). In addition, there may be a preferential arrangement of the long axes of pebbles and cobbles such that particles are elongate transverse to flow direction (McDonald and Banerjee, 1971; Boothroyd, 1972).

Sedimentary structures

Sediment transport involved in the deposition of stratified drift and outwash material is strongly dominated by unidirectional flow. As a result of this and the abundance of sand and gravel size particles, there is the generation of a wide variety of bedforms on the depositional surface and the resultant cross-bedding from bed-form migration. For purposes of discussion we will begin at the high-energy end of the spectrum and proceed downward in energy level.

Upper-flow-regime conditions are common where large quantities of water eminate rapidly from the ice mass during melting. It is common for antidunes and standing waves to be generated on the water surface with flow velocities up to 350 cm/sec (Boothroyd, 1972). These conditions of high velocity and supercritical flow are most common on the upper or proximal part of the outwash fan. Here grain size is very coarse, with pebble and cobble gravels dominant. A structure produced under these conditions is the concentration of coarse particles in transverse ribs (McDonald and Banerjee, 1971) or clast stripes (Boothroyd, 1972). The ribs of coarse particles rest on sand-size sediment and represent antidunes (Figure 7–10). To date, these features have not been recognized in the rock record, although they have been described from Holocene deposits (Rust and Gostin, 1981).

Plane beds are developed on the surface of some sediment bars within the braided system and show current lineation with trains of small pebbles generally parallel to the channel orientation.

Lower-flow-regime conditions generate widespread development of ripples and some megaripples. Locally in low-energy pools, mud drapes occur over the bed-forms. Stratigraphically these bedforms produce a broad spectrum of cross-bedding both in terms of the size and style (Shaw, 1972). Sediment transport in eskers typically produces similar structures to those above but with somewhat more sand and lower-energy conditions. Virtually all features present on outwash complexes are also present in eskers except perhaps the transverse gravel ribs. Miscellaneous surface structures such as current crescents are widespread in stratified drift and outwash as they are in any environment where gravel particles occur scattered over sand and where currents dominate.

Detailed studies of **glaciofluvial** environments and their contained sedimentary structures are presented in Jopling and McDonald (1975; also, see the references at the end of this chapter).

Periglacial Deposits

There are three rather significant environments associated with glacial systems that each accumulate their own unique type of sedimentary deposit. There are glacial lakes, eolian environments, and the glacial marine environment.

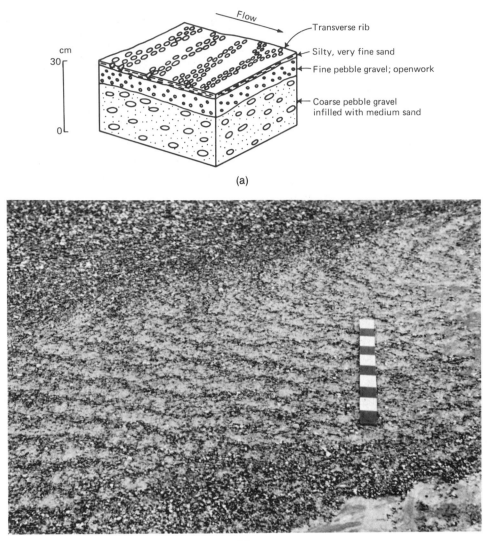

Figure 7-10 (a) Diagram of transverse ribs, an upper-flow-regime structure in proglacial sediments (After McDonald and Banerjee, 1971, p. 1290.) (b) Small transverse ribs with 15 cm spacing and clast size of 2 cm on longitudinal bar on Yana fan, Alaska. (Courtesy of J. C. Boothroyd.)

Glacial lake deposits

Although glacial lakes may form in a variety of ways, only one widespread type will be treated in this discussion; these are the ice-marginal lakes which form as the result of ponding meltwater typically bounded by moraines. Such lakes are quite short-lived geologically but accumulate significant and diagnostic sediments.

Most lakes of this type include typical shoreline sediments and morphology comparable to marine coasts except for size (see Chapter 12). Small deltas are produced where sediment-laden meltwater enters the lake (see Chapter 9). In many cases both of the related environments described above produce sediment accumulations which are nearly indistinguishable from the adjacent outwash deposits.

The fine-grained muds that accumulate in regular thin layers in glacial lakes represent a unique sediment accumulation in the glacial depositional system. Because of seasonal variation in meltwater supply and therefore sediment supply, varved sediments are characteristic of glacial lakes. The general absence of organic material typically results in varves which are primarily defined by textural means: thicker and coarser layers in spring/summer with thinner and finer-grained layers in fall/winter. Floating ice blocks containing sediment may contribute various sizes of ice-rafted particles to the fine varved sediment.

Excellent references on glaciolacustrine sedimentation are Embleton and King (1975) and Jopling and McDonald (1975).

Loess

Extensive till and alluvial plains are widespread during glaciation. Deflation of these unvegetated and fine-grained sediment bodies by strong winds causes transport and eventual accumulation of widespread deposits of silt-size material called loess. These deposits are rather unusual in that they represent one of the few sedimentary environments where the dominant grain size is in the silt range. Loess is typically sorted but unstratified (Figure 7–11). Lack of stratification is due primarily to the overall homogeneity of the sediment within a loess body. Most of the grains are quartz, although some clays and carbonate may also be present.

Although loess is generally associated with glacial activity and sediments, it can be derived from other sediment sources by similar means (Flint, 1971). Loess is very porous and permeable and may be found interbedded with glacial and proglacial

Figure 7–11 Photograph of loess bluff from the Pleistocene of the Mississippi River valley. (Courtesy of D. R. Crandell, U.S. Geological Survey.)

sediments of all types. Deposits more than 50 m thick are known and some loess sheets cover many thousands of square kilometers.

Texturally, loess generally has a mean grain size in the coarse silt range (4 to 5ϕ) and is positively skewed. Sorting is between 1 and 3ϕ, which is good sorting for fine-grained sediments. Further information on loess can be obtained from comprehensive references by Schultz and Frye (1968) and Smalley (1975).

Glacial marine sediments

When ice masses reach the marine coast they commonly extend out over marine waters due to the buoyancy of the ice, which is less dense than seawater even though it contains much sediment. A combination of processes eventually leads to the melting of the ice, which then frees its contained sediment to fall through the water column and accumulate as **glacial marine sediment** on the ocean floor. This sediment is not truly glacial drift and is therefore considered as a special category of glacially derived sediment.

Such sediment is quite similar to till in texture and in its heterogeneous mineralogy, but it also contains some characteristics which permit its proper recognition. Probably the most important is the incorporation of marine fossils. Also, there is commonly some stratification with thin discontinuous layers of sand, probably the result of reworking on the ocean floor by waves and current (Flint, 1971). Depending on the configuration of the sea floor, glacial marine sediments may display features of slumping and sliding caused by turbidity currents generated during periods of extreme sediment supply on a sloping surface. Evidence of ice rafting might also be present, such as large pebbles or cobbles which were dropped onto a soft mud surface.

GLACIAL SEDIMENT BODIES AND DEPOSITIONAL SYSTEMS

The glacial depositional system is quite diverse and complicated, with numerous processes, sediment types, and sediment body shapes resulting. Although there are several components of the system which are common to all types of glaciers, there are also some aspects which are unique or which collectively permit recognition of not only the presence of glaciation, but also the specific glacial type responsible. In characterizing glacial environments and their specific components, attention must be given to the sediments, sedimentary structures, geometry, and size of the sediment bodies or facies.

Consider a glaciated region after the ice mass has melted and characterize what is present. Regardless of the type of glacier that covered the area there will be evidence of erosion on the bedrock surface immediately below the glacial drift. Small-scale features such as grooves, striations, and crescentic gouges, as well as the smoothing of the bedrock surface, can be expected. Areas of high relief display the typical U-shaped valley.

Sediment accumulations that may be preserved in the rock record are widespread and diverse in many respects. Unsorted and unstratified drift (till) is confined to ground moraine or till sheets with tabular shapes and to linear lateral or end moraines (Figure 7–6). In general, there is no difference in sediment texture and composition between morainal types. Detailed study should show some particle orientation in ground moraines and probably less mud content in end or lateral moraines. The latter is due to slight washing of the sediment during ablation.

Local accumulations of ice-contact, stratified drift in the form of kames, eskers, and so on, are distributed over the till surface. These sediment bodies are commonly only a few meters thick.

Beyond the actual extent of the glacier or at least beyond the terminal moraine are located outwash fans and plains (sandurs) which may contain lacustrine deposits and/or eolian deposits (loess). The topographic profile of outwash fans is similar to that of alluvial fans (see Chapter 6), with a steep proximal area grading to a gently sloping distal fan and eventually to the outwash plain. There are some distinct textural and stratigraphic trends in proglacial sediment which are valuable aids in interpreting analogous ancient deposits. The maximum clast size decreases from the proximal to the distal fan and is directly related to the slope of the fan (Boothroyd and Ashley, 1975) (Figure 7–12). Pebble and cobble particles of the upper fan display well-developed imbrication with dips upstream. In general, grain size in the outwash complex fines outward from the proximal fan and also upward stratigraphically in a receding glacier.

Sediment facies that comprise the outwash fan-plain complex include not only the active areas of braided stream deposition but contain inactive areas as well. These

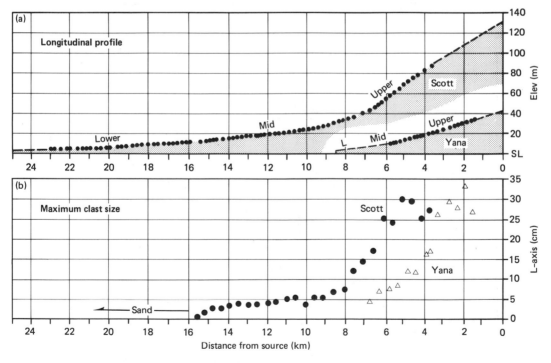

Figure 7–12 Longitudinal profiles of (a) the Scott and Yana outwash fans and (b) the maximum clast size relative to position on the fan. (After Boothroyd and Ashley, 1975, p. 200.)

may be vegetated (Figure 7–13) and contain deposits similar to those of the currently active stream complex. Fine sediments may form marshes or mudflats, and depressions accumulate lacustrine deposits (Boothroyd and Ashley, 1975).

In addition to the textural trends noted above, there are widespread sedimentary structures, dominated by cross-stratification, which are important in characterizing the outwash facies. Flatbed and large-scale trough cross-strata representing megaripple migration characterize the upper fan area. As the distal portion of the fan is reached, cross-stratification becomes smaller, with much ripple drift and planar-type features. This may be interbedded with muddy overbank deposits and grade upward into marsh and swamp muds (Boothroyd and Ashley, 1975) (Figure 7–14). Cross-stratification orientations are strongly unimodal and dip down-fan away from the source (ablating ice mass).

Generalizations and Summary

The characteristics of the glacial depositional system described above apply to all glacier types. In distinguishing the three primary varieties, greatest attention must be given to the size and geometry of the deposits. Valley glaciers produce local and elongate tills with very arcuate end moraines compared with ice sheets and piedmont glaciers, which may produce similar but more widespread deposits.

Outwash fans and plains may enable accurate recognition of glacier type by

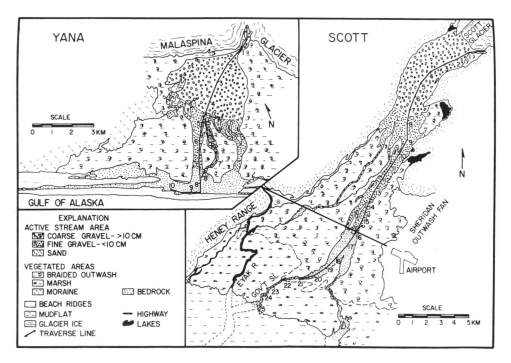

Figure 7–13 Surface sediment facies distribution on the Yana and Scott outwash fans in Alaska. Numbers on the traverse line indicate distance from source in kilometers. (From Boothroyd and Ashley, 1975, p. 195.)

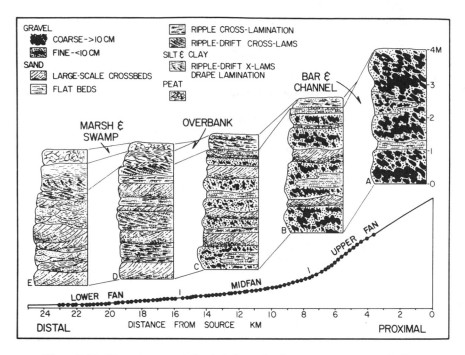

Figure 7-14 Downstream variation in facies and sedimentary structures on the Scott outwash fan. (From Boothroyd and Ashley, 1975, p. 218.)

using similar guidelines. In addition, outwash deposits show marked thinning away from the terminus of the facies. Small-scale features such as kames or eskers do not provide good data for such determinations.

There is little with which glacial till can be confused in the rock record. Some turbidite, mud flow or gravity slump, and slide accumulations might bear a resemblance, but they can be distinguished by compositional or stratigraphic means. Turbidites generally display stratification and contain sedimentary structures such as graded bedding or small-scale cross-strata. In addition, marine fossils might be present. Whereas gravity slump or slide sediments may be unsorted and unstratified, they typically do not contain the lithologic variety contained in most tills; they also have a very limited extent and different geometry.

ANCIENT GLACIAL DEPOSITS

In many respects ancient glacial deposits are relatively easy to recognize when compared to sedimentary sequences from other depositional environments. Even though this is the case and it is possible to reach agreement on interpretation from a great majority of geologists, there are typically some dissenters who remain unconvinced. The object of the following discussion will be to present general characteristics for recognition of ancient glacial activity and then give some examples of sequences that have been interpreted as being glacial in origin. For purposes of discussion the Pleistocene glacial deposits will not be considered; criteria for recognition and examples will be for pre-Quaternary glaciation. Excellent general references on ancient periods of glaciation and on paleoclimatology are those by Nairn (1961, 1964), Schwarzbach (1963), and Wright and Moseley (1975).

Recognition of Ancient Glaciations

Geologists are presently living in an excellent period of time for studying ancient glaciations. The presence of modern glaciers and their proximal Pleistocene counterparts permit some direct applications of the Law of Uniformitarianism. The presence of widespread and easily accessible Pleistocene and Recent glacial deposits provides a complete framework of data with which to compare materials from probable ancient analogs.

Initially, one might begin searching for ancient glacial deposits in the high latitudes and altitudes where the appropriate climatic conditions are present. This, of course, is nonsense because of tectonic activity, continental drift, and other factors that have caused extreme changes both in climate and elevation throughout geologic time.

Probably the best single evidence for a past glaciation is the presence of **tillite**, the lithified version of glacial till (Figure 7–15). It is commonly underlain by a glaciated bedrock surface containing striations, grooves, or other erosional features of ice movement. The establishment of a **polymictic conglomerate** or **diamictite** as a till is not always easy. Numerous "tillites" have been found to be deposited by forces other than glaciers (Crowell, 1957; Dott, 1961). Some general criteria for tillite recognition include grain size, sorting, particle shapes, fabric, lithologic composition, and size and geometry of the unit (Flint, 1961).

By contrast, glacial outwash is difficult to recognize as such unless it can be stratigraphically related to known tillites; it strongly resembles alluvial deposits. Thinly laminated, rhythmically deposited mudstones which resemble modern and Pleistocene glacial lake deposits are also inconclusive indicators of glacial activity unless they can be related to more conclusive evidence. In summary, then, conclusions on ancient glaciation should rely on unequivocal identification of tillite; however, glacial marine deposits are becoming widely used to identify glacial episodes.

Periods of ancient glaciation

Except for some evidence of local glaciation, such as valley glaciers in the Tertiary, extensive pre-Quaternary glaciation was confined to Precambrian and Paleozoic times. North America was subjected to glaciation in Middle and Late Precambrian. Late Precambrian glaciation is also well documented in northern Europe and in continents of the present southern hemisphere. The next widespread glaciation occurred in Ordovician time in northern Africa, Australia, and Scotland. Perhaps the most extensive and certainly the best documented pre-Quaternary glacial period was during late Paleozoic times. Carboniferous and Permian glacial deposits are known from Africa, India, Australia, Antarctica, and South America. Other glaciations have been postulated but deposits are either quite local in extent or not well documented.

Figure 7-15 Permian tillite near Korkuperimal Creek, Bacchus Marsh Area, Victoria, Australia.

Late Paleozoic Glaciation in the Southern Hemisphere

Widespread and well-known pre-Quaternary glacial deposits occur throughout the Gondwanaland landmasses and range from Carboniferous through Permian in age. All the landmasses in the present southern hemisphere plus India contain numerous exposures of these sequences. Although there is agreement throughout the geologic community as to the presence of widespread glaciation during this period, some of the rock sequences that resulted have been interpreted differently by various authors. The situation is somewhat like that described for Precambrian glacial deposits, in that some polymictic conglomerates are interpreted as tillites, as glacial marine, or as gravity slump deposits, depending on the particular author one chooses to read. Two classic areas and stratigraphic sections will be considered, the Dwyka Formation (Upper Carboniferous) from South Africa and the Lower Permian from Australia.

Dwyka Formation

Dwyka glacial deposits and thin stratigraphic equivalents occur throughout the southern Africa continent including Madagascar (Frakes and Crowell, 1970; Adie, 1975) but are best developed in the Karoo Basin in South Africa (Crowell and Frakes, 1972). The Dwyka is in the Upper Carboniferous or Lower Permian and consists of about 1300 m of section with the middle 800+ m comprising the tillite units.

The Dwyka tillites rest on widespread striated pavements and contain striated and faceted clasts in the lower portion of the section (Adie, 1975). Well-preserved glacial valleys are also exposed throughout much of the area. Dwyka rocks contain the typical characteristics of till, including clasts orientations. The tillites contain both locally and remotely derived gravel particles.

Detailed studies of the Dwyka from the Karoo Basin by Crowell and Frakes (1972) provide a good description of this classic tillite sequence. The typical lithology is a gray diamictite dominated by mud with scattered gravel clasts. Detailed compositional studies has shown that matrix particles are similar in composition to the larger clasts (Hamilton and Krinsley, 1967). Stratification is absent except near the top of the tillites, where it is poorly developed (Crowell and Frakes, 1972). Large, apparently ice rafted boulders are present at this horizon, suggesting rafting and deposition in a quiet water body, probably near the margin of the glacier. These poorly stratified beds grade upward into the overlying shales.

Reconstruction of this area during Carboniferous glacial times is made possible by the numerous glacial pavements with striae and the preservation of the glacial valleys and other geomorphic features. Systematic mapping of this area has enabled Mathews (1970, in Crowell and Frakes, 1972) to interpret the paleogeography as shown in Figure 7–16. This represents one of the best such studies for any of the pre-Quaternary glaciations.

Permian of Australia

Widespread glaciation took place throughout Australia during the Late Paleozoic with glacial deposits preserved in every state in the country. Valley glaciers dominated during the Late Carboniferous, with continental ice sheets reaching their maximum extent during Permian time (Crowell and Frakes, 1971). These deposits have been recognized and investigated for more than a century. Among the best known are the areas at Hallet Cove, south of Adelaide, and at Bacchus Marsh, west of Melbourne.

The entire spectrum of glacial and periglacial sediments is represented in these areas, including outwash, glaciolacustrine, and glacial marine accumulations. Glacial pavements containing striations, grooves, and crescentic gouges [Figure 7–3(b)] indicate that the primary direction of ice movement was from south to the north, the source region being the then adjacent land mass of Antarctica (Crowell and Frakes, 1971).

At Hallet Cove and nearby Fleurieu Peninsula there are clasts of various plutonic and metamorphic sources as well as nearby Precambrian tillites incorporated in the diamictites (Hamilton and Krinsley, 1967). In addition to the structureless tillites there are interbedded sandstones and mudstones, with the sandstones displaying small-scale cross-strata indicating a northerly transport direction (Crowell

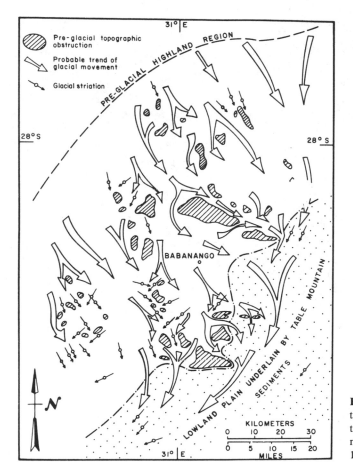

Figure 7–16 Paleogeographic map of the Karoo Basin, showing preglacial topographic highs and directions of ice movement. (From Crowell and Frakes, 1972, p. 2903.)

and Frakes, 1971). Ice rafting was also a contributor, as evidenced by large dropstones scattered throughout the laminated sediments (Figure 7–17). The presence of some marine foraminifera denotes a marine origin of the rafted materials and indicates that glaciers are adjacent to the marine shelf.

Several hundred meters of glacial and periglacial sediments are preserved near Bacchus Marsh in Victoria, Australia. These exposures also contain well-preserved bedrock surfaces which exhibit polishing and tool marks as well as ancient roche moutonnées. The Bacchus Marsh sections exhibit a wide variety of paleoenvironments. Much of the diamictite displays a crude and irregular stratification. Crowell and Frakes (1971) interpret this as being the result of remobilization of the till into mudflows and debris flows, probably in glacial lakes. Abundant, well-bedded mudstones and fine sandstones are present and probably represent glaciolacustrine and distal outwash environments.

Here also, marine influence is recorded; there are *Notoconularia* preserved in coarse gravels associated with the periglacial sequences. These gravels may represent strandline reworking of diamictites by wave action (Davis and Mallet, 1981).

It is apparent from the details of the stratigraphy that there were multiple

Figure 7-17 Photograph showing rafted clast in bedded periglacial Permian sediments, Hallet Cove, South Australia.

glacial advances in southeastern Australia during Permian time. The number has not been accurately determined; however, one author has suggested that ice sheets passed over the Bacchus Marsh area dozens of times (Bowen, 1959).

Late Precambrian Glacial Systems

Tillites are well documented from a large number of locations throughout the world and have been dated at 650 to 700 million years B.P. (Pringle, 1973). This places them just below the Cambrian Period and much of the literature on this glacial episode uses the term "Infra-Cambrian" as the time period for this glaciation (Harland, 1964a, 1964b). Tillites from this glacial period are known from every major landmass in the world except for Antarctica.

Northern Europe

Probably the best known and certainly the first tillites described from this period are those from northern Europe. The Varangian Ice Age and resulting deposits takes its name from the type locality along the northern coast of Norway. Glacial deposits from these glaciers are found throughout northern Europe, the British Isles, and eastern Greenland (Figure 7-18).

There is abundant evidence to support a glacial origin for these polymictic conglomerates or **pebbly mudstones,** which have been interpreted as tillites. These rocks contain a diverse grain size and composition with little or no stratification. Many

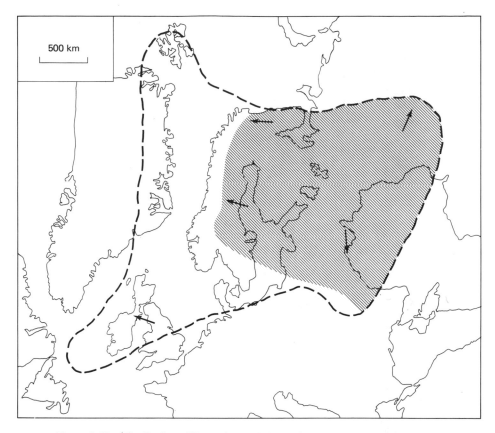

Figure 7-18 Distribution of Varangian glacial deposits in Europe and the north Atlantic region. (After Spencer, 1975, p. 234.)

clasts have been collected with well preserved striations (Bjørlykke, 1974). Some evidence for slumping and downslope transport of sediment has been found at many localities, causing debate over the details of the mechanism of deposition of these glacial sediments. Excellent discussions of these sequences are presented by Reading and Walker (1966) and Spencer (1975). The following paragraphs are summarized largely from their work.

These tillites and other rock units with which they are interbedded range in thickness from a few hundred meters to several kilometers. Tillite thickness reaches 750 m in Scotland. At only a few localities do the tillites rest directly on striated pavements. Large clasts in the tillites show a wide range in composition but generally reflect that of the immediately underlying rock units (Spencer, 1975). In addition to the tillites there are other lithic units which are of glacial origin: varved siltstones and ice-rafted clasts. The Port Askaig Tillite found in Scotland and Ireland represents an excellent example of the complex stratigraphy and glacial history of tillites and related periglacial units (Figure 7–19). The sequence figured is over 700 m thick and contains glacial marine, glaciofluvial, and nonglacial sediments as well as tillites. Spencer (1971, 1975) interprets more than half of the sequence as being the result of

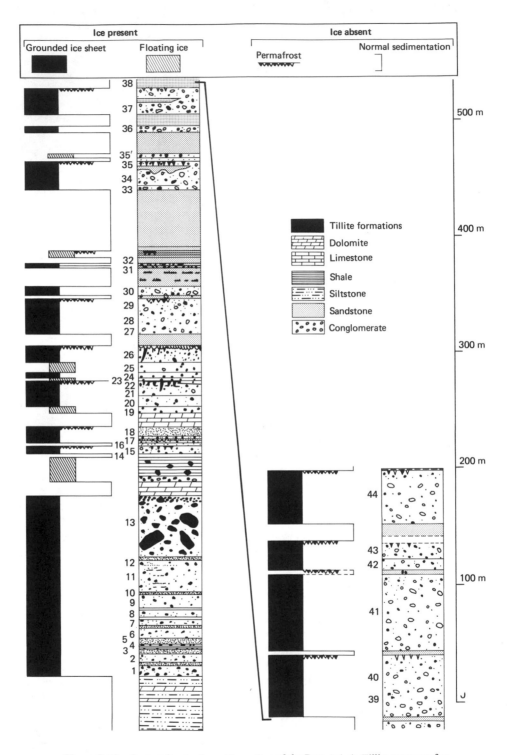

Figure 7-19 Composite stratigraphic section of the Port Askaig Tillite sequence for Scotland and Greenland. (After Spencer, 1975, p. 226.)

grounded ice sheets, with the remainder divided between sediments that accumulated from floating ice and those accumulating from normal marine sedimentation. He cited the sharp contact at the base of the tillites as evidence for their origin from grounded ice, whereas floating ice would produce gradational contacts as normal sedimentation was overcome by ice-rafted sedimentation. Additional evidence for grounded ice is provided by lenticular beds of sandstone, shale, and conglomerate which are interpreted as subglacial or englacial stream deposits. As a result of these and additional lines of evidence, Spencer (1975) has interpreted the tillites as representing ground moraines.

By way of contrast with Spencer's interpretation of the Scotland glacial section, Reading and Walker (1966) have interpreted most of the Lower and Upper Tillite Formations of northern Norway as being glacial marine in origin. These investigators recognize several facies: tillites, turbidites, mudstones, and sandstones, with each having multiple subfacies. In this area no striated clasts or striated pavements are present. Tillite units contain poorly sorted sediments, with the majority of clasts being angular; horizontal stratification is common, and the tillite units are thin (< 50 m) and widespread (500 km²) (Reading and Walker, 1966). These characteristics, plus the association of the tillites with turbidites, mudstones, and sandstones, have caused the authors to propose that the tillites are actually a result of glacial marine sedimentation.

Two models are proposed for the accumulation of the Precambrial glacial sediments in northern Norway: one involves the grounded portion of the ice sheet and the other includes the floating shelf ice and its related iceberg zone (Figure 7–20). Grounded ice may be dry based, where the ice temperature is below the melting temperature, or wet based, where the opposite situation exists and meltwaters deposit sediments. The model for shelf ice and iceberg sedimentation from shelf ice yields no separation of grain size or stratification, whereas intermittent sedimentation permits stratification due to different settling rates of various particles sizes.

Reading and Walker (1966) have interpreted the sequence of events leading to the composition section for northern Norway as shown in Figure 7–21. Their conclusion of a glacial marine origin for the tillites (unit A) is based on the proposed models shown in Figure 7–20.

Gowganda Formation, Ontario, Canada

The Gowganda Formation of the Huronian sequence (Precambrian) crops out in central Ontario, Canada, from the north shore of Lake Superior eastward to Quebec. The age has been placed at about 1300 million years B.P. and its origin has been debated in the literature for more than 75 years (Lindsey, 1969). The Gowganda sequence is nearly 1000 m thick and is comprised of polymictic conglomerates, mudstones, and sandstones. The polymictic conglomerates have been interpreted as tillites and are the subject of discussion here. Much of the following is excerpted from the excellent summary by Lindsey (1969).

The supposed tillites of the Gowganda contain the typical characteristics of a glacial till: poor sorting, lack of apparent stratification, variety of particle shapes, and a wide spectrum of particle lithologies. There are sandstone lenses and some dropstones, suggesting ice rafting. Lindsey (1969) measured thousands of pebble

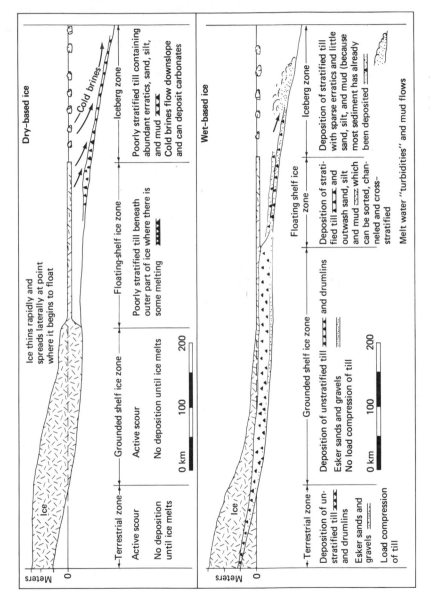

Figure 7-20 Diagrammatic cross sections showing models for deposition from grounded and floating ice in both dry-based and wet-based conditions. (From Reading and Walker, 1966, p. 201.)

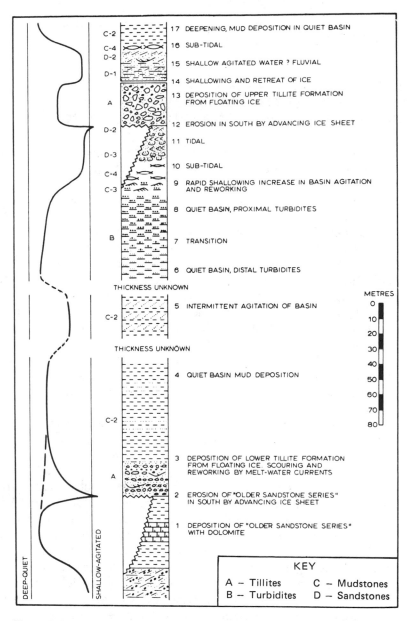

Figure 7-21 Generalized composite stratigraphic column for northern Norway, showing depositional environments. (After Reading and Walker, 1966, p. 198.)

orientations at various horizons throughout the polymictic conglomerates and found distinct preferential orientation which is compatible with grooves and striations at the base of the Gowganda. Cross-strata and ripple mark orientations from the sandstone lenses also coincide with pebble orientations.

The conclusion reached is that the polymictic conglomerates of the Gowganda are indeed tillites (Lindsey, 1969). One of the compelling arguments is the long-axis orientation of pebbles; glacial marine deposition (see Reading and Walker, 1966) would not give rise to such orientations.

SUMMARY OF CHARACTERISTICS
OF SEDIMENT/SEDIMENTARY ROCK BODIES
REPRESENTING THE GLACIAL DEPOSITIONAL SYSTEM

The glacial system is characterized by three major facies which are closely related both genetically and stratigraphically; they are glacial till, glaciofluvial, and glacial marine.

Tectonic setting. There is no association between glaciers and any particular tectonic conditions.

Shape. Glacial system deposits are typically tabular for ice sheet systems but linear for valley glacier systems. Within the ice sheet type of glacial system there may be a variety of sediment body geometries. End moraines are rather linear and glaciofluvial deposits may be tabular, linear, or somewhat wedge shaped.

Size. Valley glaciers deposit small sediment bodies as compared to those produced by ice sheets. Till sheets may cover many thousands of square kilometers, whereas stratified drift and outwash tend to be more local in extent. Glacial marine units may extend for thousands of kilometers.

Textural trends. No generalities are appropriate for glacial deposits in either the stratigraphic or the geographic context. Glaciofluvial deposits would be expected to exhibit a decrease in grain size away from the glacial terminus or source of sediment.

Lithology. Because glacial deposits are generally heterogeneous in composition it is not possible to generalize about trends. Commonly, the composition of glacial sediments reflects the composition of the material underlying the ice sheet.

Sedimentary structures. No structures are indicative of the glacial system. Unstratified drift is structureless and stratified drift contains cross-stratification and other structures common to numerous environments of deposition. Varves are widespread in glacial lake deposits. Large dropstones which are the result of ice rafting and occur in stratified glacial marine deposits are indicative of that environment.

Paleontology. Extensive work has been done on pollen and spores associated with Pleistocene glaciers, but these are the only widespread organic remains in glacial sediments and do not extend throughout the geologic record. Glacial marine sediments incorporate marine organisms, especially planktonic varieties.

Associations. The glacial system itself may be associated with nearly any of the continental depositional systems or with coastal and marine systems. Typically, the terrestrial glacial system would be related to other continental systems and glacial marine deposits to other marine facies. Within the terrestrial glacial system, some facies associations should be expected; extensive till accumulation could be overlain by varved glaciolacustrine facies, outwash sands and gravels, or loess deposits in various stratigraphic arrangements. At or near the terminal moraine complex glaciofluvial facies and loess may be dominant. The glacial marine system may also have an extensive basal till with thinly stratified glaciomarine muds containing dropstones overlying or laterally adjacent. Sandy turbidite units may be interbedded with these layered muds.

ADDITIONAL READING

EMBLETON, C., AND KING, C. A. M., 1975. *Periglacial Geomorphology.* 2nd edition, Halsted Press, New York, 203 p. Thorough coverage of the physiographic settings and related features for glacial activity and the resulting landscapes. Also contains some aspects of glacial sedimentology.

FLINT, R. F., 1971. *Glacial and Quaternary Geology.* John Wiley, New York, 892 p. This is probably the best single book on the subject of glacial geology. Suffers somewhat from the sedimentologists' point of view in that processes are not detailed.

GOLDTHWAIT, R. P. (ED.), 1971. *Till: A Symposium.* Ohio State University Press, Columbus, Ohio, 402 p. Excellent collection of papers dealing with nearly all aspects of Pleistocene glacial till.

GOLDTHWAIT, R. P. (ED.), 1975. *Glacial Deposits.* Dowden, Hutchinson & Ross, Stroudsburg, Pa., 464 p. Most of the classic papers on glacial and periglacial deposits are included in this reprint volume. Few of the papers deal with the processes of formation of the deposits in question.

JOPLING, A. V., AND MCDONALD, B. C. (EDS.), 1975. *Glaciofluvial and Glaciolacustrine Sedimentation.* Soc. Econ. Paleontologists and Mineralogists, Spec. Publ. No. 23, Tulsa, Okla., 320 p. The only comprehensive volume of process-oriented papers dealing with periglacial sedimentation. Outstanding illustrations make it unique among this group of references.

LEGGET, R. F. (ED.), 1976. *Glacial Till: An Inter-disciplinary Study.* Roy. Soc. Canada, Spec. Publ. No. 12, 412 p. Excellent collection of papers that somewhat parallels those in the Goldthwait (1971) volume; however, Legget includes much on the applications of till in soil science, mineral exploration, and the geotechnical aspects of till.

SUGDEN, D. E., AND JOHN, B. S., 1976. *Glaciers and Landscape.* Edward Arnold, London, 376 p. Excellent general reference on glacial dynamics, landforms, and sediments. It is well written and well illustrated.

WRIGHT, A. F., AND MOSELEY, F. (EDS.), 1975. *Ice Ages: Ancient and Modern.* Seel House Press, Liverpool, England, 320 p. A unique volume in that it is not restricted to the Quaternary Period. Excellent papers on general aspects of glacial geology as well as summaries of some of the classic ice ages of the Precambrian and Paleozoic.

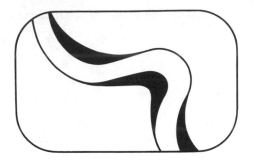

8 The Fluvial System

The river has long been recognized as the primary agent for removal and transportation of sediment from the continents to the oceans. In this chapter the entire fluvial system is considered except for the delta, which is discussed in Chapter 9.

Most early research on streams was concentrated on their geomorphology and hydrology until the classic works of H. N. Fisk on the Mississippi River (Fisk, 1944, 1947). Subsequent work on riverine processes followed (e.g., Leopold and Wolman, 1957). Perhaps the most significant contributions to both modern and ancient aspects of fluvial sedimentology were the concept of flow regime (Simons and Richardson, 1961; Simons, et al., 1965) together with related papers in a symposium volume (Middleton, 1965) and the application of facies models to the fluvial system (Allen, 1964b, 1965a, 1965b). These classic papers have served as the bases for numerous research efforts, many of which were summarized in a "state of the art" symposium volume on fluvial sedimentology (Miall, 1978).

Fluvial systems vary greatly in both space and time. Climate, regional geology, and geomorphology are significant contributors to the overall characteristics of fluvial environments. Relief, discharge, substrate composition, and other factors determine the process-response systems that operate in the fluvial system.

In a general way, one can characterize a riverine system in a hypothetical drainage basin as consisting of the following major elements, going from the most distant tributaries toward the delta:

1. *Humid fan environment*. Alluvial fan is a high-relief environment, somewhat like arid fans but with a continuous water supply.
2. *Braided stream environment*. Broad, generally shallow streams with more sediment available than can be transported by the stream. Typically, the discharge is "flashy" or sporadic.
3. *Meandering stream environment*. Generally develops on areas of low relief such as a coastal plain. Discharge is uniform compared with braided streams.
4. *Delta environment*. When the stream quickly loses its capacity on reaching a standing body of water, much sediment is deposited, forming a delta, generally covered with distributary channels.

Although this sequence of fluvial environments can occur, it is not common that all four are present in a given drainage system. Most often missing would be the humid fans, which commonly are associated with braided streams only. Each of the first three fluvial environments listed is discussed in this chapter, with both modern and ancient examples considered.

HUMID ALLUVIAL FANS

Alluvial fans that develop in humid climates show many similarities with their counterparts in arid environments. All fans are in similar geologic and geomorphic settings; that is, they require an area of high relief and an adjacent low-lying area for sediment accumulation. Alluvial fans in general are related as least indirectly to tectonic activity because of the need for high relief. Humid fans require high rainfall in addition, which tends to place them on the oceanic side of mountain ranges, generally not far distant from the moisture source, the ocean. The western margin of North and South America provides such a setting. Many humid fans also develop as outwash fans in front of glaciers, such as in Alaska.

A variety of humid fans has been recognized from the rock record (McGowen and Groat, 1971; Vos, 1975). Unfortunately, the spectrum of modern counterparts is not great; probably the best examples are the proglacial outwash fans from Alaska and Iceland (Boothroyd and Nummedal, 1978) and along the southern flank of the Himalayas, although one might argue that fluvial deposits on the western coast of Guatemala also qualify (Kuenzi et al., 1979). These authors have proposed a humid alluvial-fan model which appears to be applicable to the broad spectrum of humid fans, not just those of glacial origin.

General Morphology

In most respects the shape of humid fans is comparable to that of arid fans. Humid fans may become large, with areas of a few hundred kilometers in Iceland (Boothroyd and Nummedal, 1978) and up to 10,000 km² in India (N. Smith, personal communication). The geologic setting of humid fans is typically much more varied than for arid fans, which commonly front fault scarps. Relief is certainly important to the overall development of humid fans and it may confine the fan to a valley or permit development of broad coalescing fans (Figure 8–1). Some humid fans are longer than they are wide.

One of the striking contrasts between humid and arid fans is the longitudinal profile and slope (Ryler, 1972). Both types may conveniently be subdivided into proximal, middle, and distal fan segments. Humid fans display long, gently sloping profiles in comparison to the steep arid fans (Figure 8–2). Humid fan gradients rarely exceed 50 m/km, whereas arid fans are generally at least twice that value.

Channel patterns in Alaskan humid fans are characterized by from one to three deeply incised channels in the proximal zone. The midfan zone contains numerous braided channels with numerous longitudinal bars separating the channels (Boothroyd and Nummedal, 1978). The Samala River of Guatemala is incised nearly 50 m into its fan near the apex, 36 km from the Pacific Ocean; 22 km downstream but still in the upper fan area, this incision decreases to less than 10 m (Kuenzi et al., 1979). Distal fan areas contain braided channel patterns similar to the middle fan but with smaller channels and less-dense braiding. During flooding conditions the channel

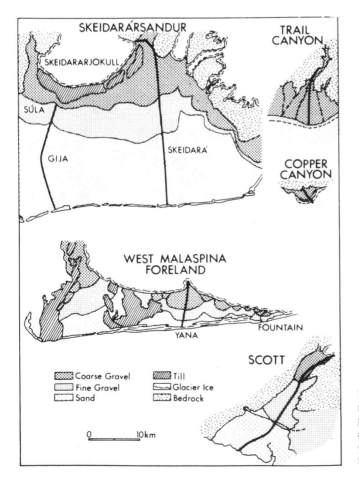

Figure 8-1 Surface sediment facies maps of selected humid fans with comparison to arid examples (Trail Canyon and Copper Canyon, Death Valley, California). (From Boothroyd and Nummedal, 1978, p. 647.)

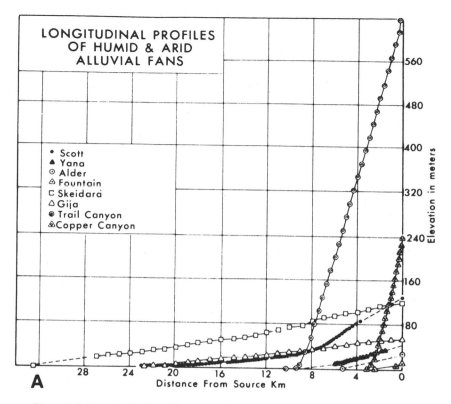

Figure 8–2 Longitudinal profiles comparing humid and arid fans. Observe that arid fans displays slopes that are 5 to 10 times greater than humid fans. (From Boothroyd and Nummedal, 1978, p. 649.)

bars in this region are inundated and the active channel complex contains a continuous sheet of water (Boothroyd and Nummedal, 1978).

Processes

Continuous discharge is typical of humid fans, although there may be strong seasonal variation. Those related to glaciers experience extreme discharge during spring and summer. Low-latitude areas may have strong seasonality due to precipitation patterns. Boothroyd and Ashley (1975) measured velocities and other flow parameters in the Scott (Figure 8–3) and Yana fans in Alaska. Upper fan velocities over 2m/sec were common, with midfan values about one-half that. Froude numbers at or in excess of critical flow (1.0) were common. All data point to a very high energy situation, at least on the upper fan.

Sediments and Sediment Body Morphology

The general characteristics of sediments on humid fans are much like those of arid fans; the upper fan is characterized by the coarsest particles, with a marked decrease toward distal portions of the fan. Humid fans seem to display relatively low strati-

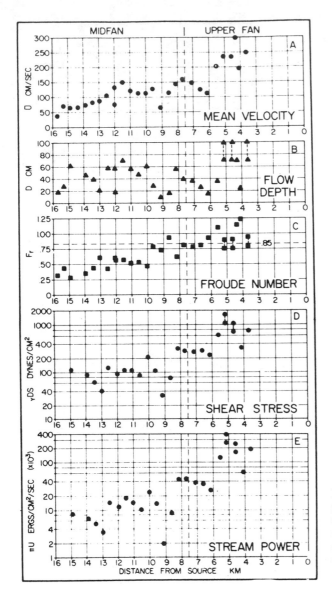

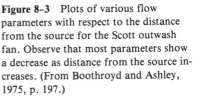

Figure 8-3 Plots of various flow parameters with respect to the distance from the source for the Scott outwash fan. Observe that most parameters show a decrease as distance from the source increases. (From Boothroyd and Ashley, 1975, p. 197.)

graphic variability in sediment texture. A hypothetical section near the apex would be expected to contain nearly all particles of pebble size or greater. This is in some contrast to arid fans, which tend to show more interbedding of fine and coarse strata. Such textural sequences merely reflect the relatively constant flow conditions of humid fans in contrast to extreme ranges of flood and static conditions on arid fans.

Maximum clast size shows a distinct relationship to humid fan gradient (Figure 8-4). There is also a pronounced increase in roundness in a down-fan direction (Gustavson, 1974). This is not necessarily the case for arid fans, which are typically steeper (Boothroyd and Nummedal, 1978). It seems logical to conclude that the con-

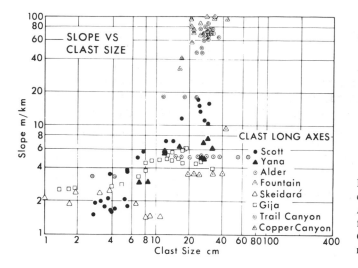

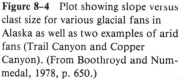

Figure 8-4 Plot showing slope versus clast size for various glacial fans in Alaska as well as two examples of arid fans (Trail Canyon and Copper Canyon). (From Boothroyd and Nummedal, 1978, p. 650.)

tinuous sorting and winnowing of humid fans produces at least near-equilibrium conditions between texture, morphology, and processes; that is not the case for arid fans, which are more like a "dumped" deposit.

Imbrication of large clasts is widespread, with the plane of the long and intermediate axes dipping upstream (Rust, 1972a; Boothroyd and Ashley, 1975). It should be observed, however, that the long axis is typically transverse to flow, although long axes parallel to flow have also been observed. Laboratory studies indicate that such orientation results from "contact" or rolling transport (Johannson, 1963).

Sediment bodies within the fan are typically longitudinal bars, but they display rather systematic changes across the fan profile (Boothroyd and Ashley, 1975). Upper fan bars are diamond shaped, are composed of coarse gravel, and occur on interstream areas between incised channels (Figure 8-5). They are rarely subjected to flooding. Midfan bars are typically composed of both gravel and sand, with prograding slipfaces developing on the finer-grained bars (Boothroyd and Ashley, 1975).

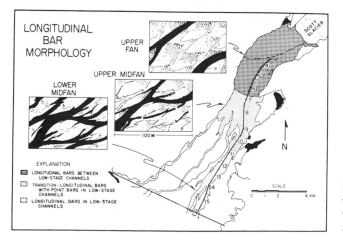

Figure 8-5 General surface facies and variation in bar type along the Scott outwash fan, Alaska. (From Boothroyd and Ashley, 1975, p. 206.)

Some bedforms may develop on the bars. Longitudinal bars in the midfan area are more elongate than upper fan bars and are activated during most flood conditions. Sandy bars of the lower fan display both the longitudinal and linguoid bar form (Boothroyd and Ashley, 1975, Figure 19) in the braided portions of the fan. Some of the distal areas contain meandering channels which produce point bars.

The general facies model for humid fans consists of three core facies which are associated with various lateral facies (Boothroyd and Nummedal, 1978). The core facies are coarse gravel, fine gravel, and sand, which correspond to upper, middle, and lower fan, respectively. The laterally adjacent facies depend upon many variables but would include such environments as swamp, marsh, eolian dune, and if near the coast, wind tidal flats (Figure 8-6). Into this basic framework one must incorporate various textures and sedimentary structures in order to produce a humid fan model.

BRAIDED STREAMS

In many respects it is at least difficult if not impossible to separate a discussion of humid fans from one covering braided streams. Many aspects of processes, sediments, and morphology are markedly similar between these two depositional en-

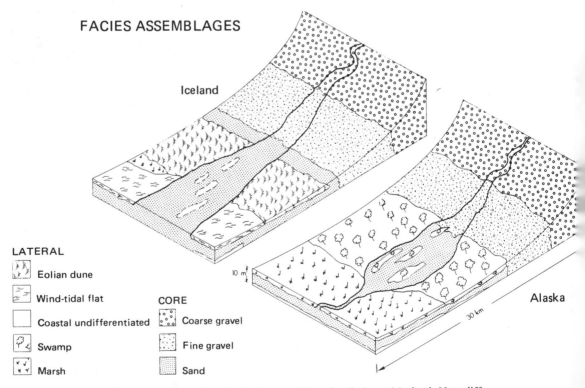

FACIES ASSEMBLAGES

Iceland

Alaska

LATERAL
- Eolian dune
- Wind-tidal flat
- Coastal undifferentiated
- Swamp
- Marsh

CORE
- Coarse gravel
- Fine gravel
- Sand

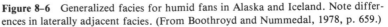

Figure 8-6 Generalized facies for humid fans in Alaska and Iceland. Note differences in laterally adjacent facies. (From Boothroyd and Nummedal, 1978, p. 659.)

vironments. Distinction between humid fan and braided stream deposits in the stratigraphic record is also problematical. Such similarities between these environments lend themselves to close inspection in order to ascertain criteria to assist geologists in the proper recognition of each.

Until the excellent study by Doeglas (1962) there was no systematic attempt to distinguish between meandering and braided streams as far as processes are concerned. Succeeding studies by Ore (1964) and Moody-Stuart (1966) dealt with criteria for the recognition of various stream types from sediment deposits. Miall (1977) has written an excellent and comprehensive summary of the braided stream environment. Braided rivers are characterized by numerous channels separated by bars and small islands. They show a low sinuosity and contain coarse sediment with both bed load and suspended load modes of sediment movement (Miall, 1977). Deposition is characterized by shifting of channels and bar aggradation. Braided streams have relatively steep slopes and width/depth ratios that may exceed 300. A combination of slope and bankfull discharge can be used to distinguish braided and meandering streams. Braided streams can develop on steeper slopes than meandering streams for a given discharge (Leopold and Wolman, 1957).

Shifting of channels and bars typifies braided stream sedimentation. Some bars are relatively stable and develop vegetation. They are however, inundated during floods. Many variables control the stream channel pattern, including discharge, sediment load, channel shape, velocity, and bed roughness (Leopold and Wolman, 1957). Climate is also important in that it controls the amount and variation in discharge.

Processes

It is difficult to discuss this aspect of a dynamic environment such as a braided stream without including the response portions of the system, such as sediments, structures, and sediment body morphology. Such separation is somewhat artificial but necessary for orderly organization. The processes that operate in the braided stream environment are markedly similar to those of alluvial fans, particularly humid fans and most particularly the lower or distal portions of the fan.

Braided streams have been called "overloaded" because of the abundance of sediment in the stream. A primary reason for such a condition is related to temporal variation in discharge. Braided streams are frequently characterized by short flooding periods and relatively extensive period of low discharge or perhaps none at all. According to Leopold and Wolman (1957), braiding results from the inability of the stream to move a coarse component of its load. This coarse portion is left behind, developing as channel bars. During flood stages all sizes of particles in the stream bed are moved and the limiting factor for particle movement is the availability of particle sizes, not the velocity of the stream (McKee et al., 1967).

In some respects stream flow acts like flow patterns in tidal estuaries except that flow reversals are not present. The change from low discharge or low water stage to that of flood conditions is similar to the rise from low to high tide. During low-water conditions flow is restricted to the channels and is diverted around the channel bars [Figure 8–7(b)]. As discharge increases the bars are eventually flooded, producing

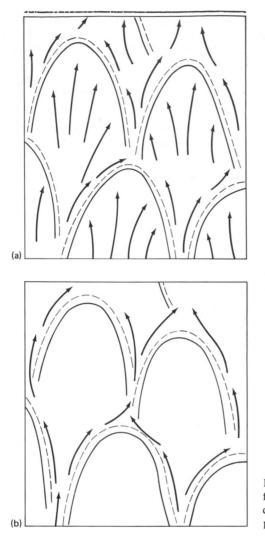

Figure 8-7 Areal diagram of bars with flow patterns at (a) high and (b) low discharge stages. (From Collinson, 1970, p. 42.)

complex flow patterns [Figure 8-7(a)]. Under these conditions flow separation and eddying develop on the lee of slip faces of the migrating bars, which develop mainly in sandy streams.

Sediments and Sedimentary Structures

Great variability in sediment textures exists in braided streams. An excellent summary of depositional facies in this environment is presented by Miall (1977), who recognizes 10 distinct lithofacies based on the combination of textures and sedimentary structures.

Gravels are widespread and may range to the boulder category, although pebbles are typical. Sorting of these gravels shows a wide range. Miall (1977) subdivides the gravels into three distinct facies based largely on bedding characteristics. These facies are planar cross-stratified, trough cross-stratified, and essentially massive (Table 8–1). Planar cross-stratification occurs in sets up to more than a meter in thickness, with reactivation surfaces being common. The trough cross-strata develop in shallow channels that cut into one another. Erosional bases with lag deposits are common and the dip of foresets may reach 30°.

The massive to crudely bedded gravel facies is most widespread in previously studied braided streams (Miall, 1977, Table III). Although most are clast supported, some are matrix supported, indicating the possibility of debris flow deposition (see Chapter 6). Small lenses of sand and mud may be present with ripples and cross-strata in the sand lenses. Imbrication of pebbles is also common.

Although most gravels are grain supported with pores filled with sand and mud, some have no fine material included in the gravel portion. These sedimentation units

TABLE 8-1 FACIES AND THEIR CHARACTERISTICS FOR BRAIDED STREAMS

Facies identifier	Lithofacies	Sedimentary structures	Interpretation
Gm	Gravel, massive or crudely bedded, minor sand, silt, or clay lenses	Ripple marks, cross-strata in sand units, gravel imbrication	Longitudinal bars, channel-lag deposits
Gt	Gravel, stratified	Broad, shallow trough cross-strata, imbrication	Minor channel fills
Gp	Gravel, stratified	Planar Cross-strata	Linguoid bars or deltaic growths from older bar remnants
St	Sand, medium to very coarse, may be pebbly	Solitary or grouped cross-strata	Dunes (lower flow regime)
Sp	Sand, medium to very coarse, may be pebbly	Solitary or grouped planar cross-strata	Linguoid bars, sand waves (upper and lower flow regimes)
Sr	Sand, very fine to coarse	Ripple marks of all types, including climbing ripples	Ripples (lower flow regime)
Sh	Sand, very fine to very coarse, may be pebbly	Horizontal lamination, parting or streaming lineation	Planar bed flow (lower and upper flow regimes)
Ss	Sand, fine to coarse, may be pebbly	Broad, shallow scours (including cross-stratification)	Minor channels or scour hollows
Fl	Sand (very fine), silt, mud, interbedded	Ripple marks, undulatory bedding, bioturbation, plant rootlets, caliche	Deposits of waning floods, overbank deposits
Fm	Mud, silt	Rootlets, desiccation cracks	Drape deposits formed in pools of standing water

SOURCE: Miall, 1977, p. 20.

are characterized by an open gravel network overlain by sand; such units may be repeated in cyclic fashion. Smith (1974) has interpreted the mixing of sand and gravel as representing relatively low discharge compared to the open, matrix-free gravels.

Miall (1977) described five sand facies but mentions that several others may also be present. The five facies are trough cross-stratified, planar cross-stratified, ripple cross-stratified, horizontally stratified, and scour and fill. The trough cross-stratified type is characterized by medium to coarse sand with scattered pebbles. Sets of cross-strata range up to near 0.5 m and are generally wide (Miall, 1977). The planar cross-stratified facies is markedly similar to the trough type except for the geometry of the cross-strata and the scale of each set. Planar sets may reach a few meters in thickness, although most are less than 1 m. General continuity and geometry of the cross-strata suggest that migration of lower-flow-regime bedforms such as megaripples is responsible for the cross-strata. Trough-shaped sets are due to sinuous or cuspate bedforms, and planar sets result from linear crested forms. Larger cross-sets may result from sand wave migration or migration of solitary bar forms.

Ripple cross-strata are commonly in medium sand, although a range in sand sizes may occur. Various styles of ripple cross-stratification (Allen, 1968) may be included in this facies. The horizontal stratification facies may be composed of the entire range of sand sizes and rarely pebbles. Thickness of the facies ranges widely and although typified by horizontal strata, nonstratified areas are present locally. The ripple cross-stratification represents lower-flow-regime conditions, whereas planar or horizontal strata indicate upper-flow-regime plane bed conditions (Table 8–1).

The scour and fill facies is characterized by coarse sand with some pebbles. Minor channels, tens of centimeters deep and a meter or so in width, become filled with sediment; some of it is massive and some shows cross-strata (Miall, 1977).

Muds are not an abundant constituent of braided streams because the flashy discharge tends to keep them in suspension. They occur primarily as local and thin layers on the top of bars or interbedded in sand sequences. Laminated muds comprise lenses in sand sequences of braided streams. They represent localized or temporal low-energy conditions and may include various small-scale physical or biogenic structures. Laminated muds are generally only a few centimeters thick. Mud drapes represent the other type of fine-grained facies in braided streams. They form thin, sometimes discontinuous layers over bedforms or sand bodies.

Bedforms represent the dominant group of sedimentary structures in braided streams. In addition to the traditional varieties, it is appropriate that transverse ribs of gravel which have been described previously in this text and are equivalent to antidunes, also be considered as a bedform type. Bedforms are most abundant in, but not restricted to, sands and cross-stratified gravels (Table 8–1). The entire spectrum of scales and geometries has been observed; however, straight crested varieties are not common, due to the complex channel and flow patterns. Perhaps the largest fluvial bedforms are those reported from the Brahmaputra River (Coleman, 1969), which may reach 15 m in height and more than a kilometer in wave length. Migration rates of bedforms in fluvial systems have been studied little except by extrapolation from flumes. Coleman's (1969) studies showed great variation, with ripples moving about 2 to 3 m per day and megaripples more than 100 m per day. Detailed studies of reactivation surfaces on migrating bedforms testify to the discontinuous nature of the migration process (Collinson, 1970).

Particle orientation is quite prevalent throughout braided stream deposits and represents one of the most preservable and potentially useful types of information for the geologist studying the ancient rock record. There are several types of spatial positioning of sedimentary particles. The most useful and obvious is imbrication of elongate or disk-shaped clasts showing dip in the upcurrent direction (Figure 3–20). Orientation of long axes parallel to flow is common but must be viewed with care because high discharge and therefore high Froude number conditions produce orientations with long axes transverse to flow (McDonald and Banerjee, 1971; Boothroyd and Ashley, 1975). Asymmetrical clasts with one end blunt in comparison to the other may show orientation with the blunt ends pointing upcurrent. A detailed study of clasts orientation by Doeglas (1962) shows distinctly imbricate patterns as well as some which include transverse orientations and some with both longitudinal and transverse orientations. Most such studies of particle orientation include only gravel-size particles; however, Doeglas (1962) also measured the long axes of sand grains. Such measurements are tedious and require properly oriented thin sections for accurate determinations.

Sediment Bodies

In braided streams it is necessary to separate sedimentary structures such as small-scale periodic bedforms from the larger-scale and aperiodic types commonly known as bars which are prevalent throughout braided streams. Terminology associated with these sediment bodies is complex and confusing; according to Smith (1978), it is in dire need of revision. Rather than become involved in a needless discussion of excessive terminology, the bases for discussion of these bars will be three general variables: composition, shape, and mobility.

Bar composition is essentially either sand, gravel, or a mixture of both. Shape presents a more complicated variable, although only two are very common. **Longitudinal bars** are longer than they are wide and display a somewhat diamond shape (Miall, 1977), whereas **linguoid bars** are rather lobate in areal configuration. Transverse bars are much like linguoid bars but are wider with straighter crests (Figure 8–8). Although typically associated with meandering streams, point bars and side bars may also occur in braided streams. They are attached to the banks and occur in zones of relatively low current energy. Some bars may become stabilized by vegetation, whereas others tend to migrate due to flow.

Longitudinal bars are most commonly formed of gravel or a mixture of sand and gravel (Rust, 1972b; Smith, 1974). This type represents the classic bar form of gravelly braided streams and commonly displays evidence of erosion (Figure 8–9). In actuality, longitudinal bars may be quite different than the original bars from which they developed. This is well shown in the flume studies of Leopold and Wolman (1957). Textural trends are apparent on the bar, with coarsest gravel being located in the central-axis and upstream end (Smith, 1974; Boothroyd and Ashley, 1975). Bed-

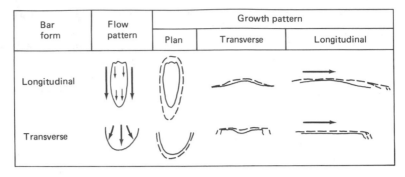

Bar form	Flow pattern	Growth pattern		
		Plan	Transverse	Longitudinal
Longitudinal				
Transverse				

Figure 8-8 Schematic views of longitudinal and transverse bars with related flow patterns and growth tendencies. (From Smith, 1974, p. 218.)

ding is crude to absent and upper-flow-regime conditions are indicated by transverse ribs of gravel. Imbrication is widespread.

Linguoid bars are most commonly developed in sandy braided streams (Miall, 1977), although both linguoid and longitudinal bars are present in braided fans of Alaska (Boothroyd and Ashley, 1975). The bar exhibits a longitudinal profile much like a migrating lower-flow-regime bedform with an avalanche slope in the downstream direction. They are typically covered by cyclic bedforms such as ripples

Figure 8-9 Longitudinal bar in braided stream with erosion along the sides; Little Thompson River south of Fort Collins, Colorado.

and megaripples. In fact, the linguoid bar itself developed as a large-scale bedform during flood conditions (Collinson, 1970).

A variety of situations may cause bars to become stabilized through vegetation. Regions such as the high latitudes may be characterized by one short but intense flood event per year, typically during the spring thaw when meltwater is abundant. In the interim the bars may be exposed and plant growth begins. If a similarly intense flood is not experienced within the next few years, the bar may develop a dense vegetation and maintain stability. Another situation is the diversion of channels such that a bar is removed from active flow, thereby allowing vegetation to become established. Some of these islands in the braided stream may become quite large (Boothroyd and Ashley, 1975).

Reference Sequences

Although the braided stream is complex in its sediments and morphology, it is possible to consider some models which characterize these environments. According to Miall (1977), there are four types of sedimentologic events that may take place in the braided stream environment: (1) flooding, where beds formed under decreasing velocity are superimposed; (2) lateral accretion as side or point bars develop; (3) channel aggradation as a channel fills due to waning energy; and (4) reoccupation of a channel, causing cut and fill. Any or all of these episodes may represent a given braided stream. They depend on the channel cross sections, bed load grain size, and discharge.

Using the various lithofacies typical of braided streams and the four cyclic sedimentologic units, Miall (1978) has developed six vertical profile sequences based on empirical data from six locations. Three of these, the Trollheim (California), Scott (Alaska), and Donjek (Yukon), contain gravel at least as a prominent constituent and represent an arid fan, humid fan, and perennial glacial stream, respectively (Figure 8-10). The other three, the South Saskatchewan (Canada), Platte (Nebraska), and Bijou Creek (Colorado), are dominated by sand (Figure 8-11). Bijou Creek is an ephemeral stream; the other two are perennial.

Both the Trollheim and Scott models (Figure 8-10) are representative of braided systems in fan environments. The Donjek model represents what might be considered as a perennial braided system. The model is based primarily on the data of Williams and Rust (1969) and Rust (1972b) from the Donjek River, a north-flowing tributary of the Yukon River, located in the western part of the Yukon Territory, Canada.

Principal morphologic features of the Donjek are channels, bars, and islands, with channel patterns developed in complex braided networks (Figure 8-12). Stream channels are coincident with levels 1, 2, 3, and 4. Each level is distinguished on the basis of elevation, discharge, vegetation, and relative age (Williams and Rust, 1969). The oldest and highest elevation is at level 4 (Figure 8-12) and progresses downward to the youngest, least stable level, 1. Longitudinal, linquoid, and point bars are present in the system. Cross sections of the valley-fill system display complex, lenticular accumulations of sand and gravel. At the time of study, the Donjek was experiencing general degradation, probably due in part to high discharge of meltwater.

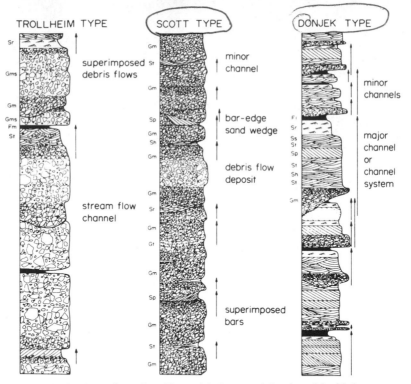

TROLLHEIM TYPE

Sr
superimposed
Gms   debris flows

Gm
Gms
Fm
St

stream flow
channel

SCOTT TYPE

Gm
St   minor
channel
Gm
Sp   bar-edge
Gm   sand wedge
Sh
Gm   debris flow
deposit
Gm
Sr
Gm
Gt
Gm
Sp   superimposed
bars
Gm
St
Gm

DONJEK TYPE

minor
channels

Fl
Sr
Ss   major
St   channel
Sp   or
St   channel
Sh   system
St
Gm

Figure 8-10 General stratigraphic models for gravel-dominated braided streams. Descriptions of individual facies are given in Table 8-1. (From Miall, 1978, p. 600.)

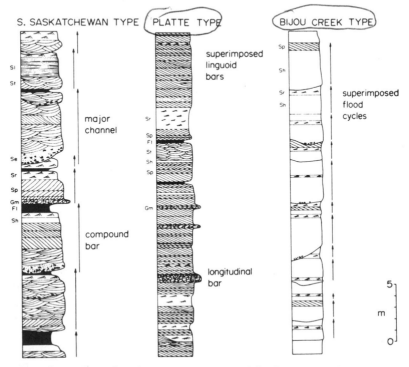

S. SASKATCHEWAN TYPE   PLATTE TYPE   BIJOU CREEK TYPE

Sl
St

major
channel

Se
Sr
Sp
Gm
Fl
Sh

compound
bar

Sp
superimposed
linguoid
bars

Sr
Sp
Fl
St
Sh
Sp

Gm

longitudinal
bar

Sp
Sh

Sr
Sh   superimposed
flood
cycles

5
m
0

Figure 8-11 General stratigraphic models for sand-dominated braided streams. Descriptions of individual facies are given in Table 8-1. (From Miall, 1978, p. 601.)

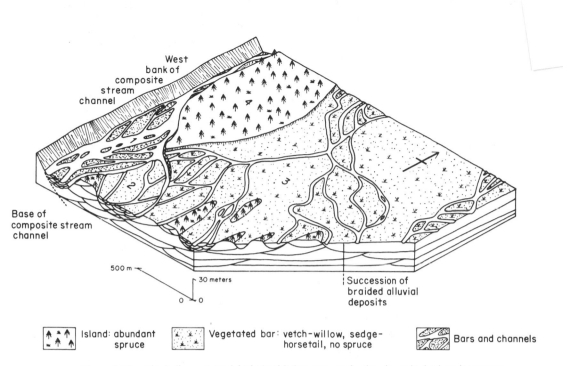

West
bank of
composite
stream
channel

Base of
composite stream
channel

500 m

30 meters

0

0

Succession of
braided alluvial
deposits

| Island: abundant spruce | Vegetated bar: vetch-willow, sedge-horsetail, no spruce | Bars and channels |

Figure 8-12 Block diagram model of a braided stream reach, showing principal environments. Numbers refer to the major channels or levels; 1 is the deepest and 4 is the highest level. (After Williams and Rust, 1969, p. 676.)

The generalized vertical section of the Donjek model as compiled by Miall (1977, 1978) shows a series of cycles, each of which fines upward and ranges from about 1 to 20 m. Each cycle has a conglomerate at the base and goes through a sand sequence of trough-shaped, megaripple cross-strata followed by ripple cross-stratification and generally capped by a thin mud drape (Figure 8-10). Miall (1978) suggested that the Donjek model be used for those braided systems with 10 to 90% gravel, whereas the Scott type (Figure 8-10) be used for those with greater than 90% gravel.

The generally sandy braided systems have also been characterized by Miall (1978) (Figure 8-11). Each of the three model types represents a markedly different system which is reflected in the vertical profile. The South Saskatchewan type is quite similar to the Donjek model but is representative of braided systems with less than 10% gravel. Cycles are based largely on the work of Cant (1978, 1980) and Cant and Walker (1978). The South Saskatchewan River is a perennial stream which experiences modest flooding in late spring, but its islands are inundated only every 2.2 years (Cant, 1978).

Bijou Creek, a north-flowing tributary of the South Platte River in Colorado, is an ephemeral stream which experiences extreme events in the form of flash floods. These dominate the sequence of sediments which accumulate during waning energy. Each flood event is represented by a distinct fining-upward cycle (Figure 8-11). Upper flat beds, ripples, and convolute bedding dominate each cycle and typically appear in this ascending order (McKee et al., 1967). Larger-scale cross-stratification is present locally. Each flood cycle is a meter or so in thickness.

The Platte River system flows west out of the Colorado Front Range, across Nebraska, where it eventually meets the Missouri River. Studies of sediments and morphology by Smith (1970, 1971) have enabled fairly detailed characterization of this braided system. The river typically experiences annual flooding. Braiding develops at intermediate and low discharge, whereas at high discharge the bed is covered and braiding is not present (Smith, 1971). The general morphology of the bars changes from longitudinal in the upper reaches, mixed longitudinal and linguoid in the middle, and primarily linguoid bars in the lower reaches of the river (Smith, 1970). Similar trends are present for grain size, which decreases, and sorting, which increases. Bedforms are dominated by megaripples and ripples. Bar migration produces large-scale relatively planar cross-stratification (Smith, 1972).

Miall's (1977, 1978) vertical profile model of the Platte-type braided system is dominated by sandy, linguoid bar deposits with only scattered gravel and longitudinal bar accumulations (Figure 8–11). Distinct cycles are not apparent, although some fining-upward sequences can be identified. Planar cross-stratification dominates, with scattered ripple cross-strata. Trough-shaped cross-strata from megaripples are scarce, probably because these forms are destroyed by slowly falling water levels over the bars. Miall (1977) attributes the Platte-type model to shallow rivers or those without topographic differentiation.

MEANDERING STREAMS

Most of the initial detailed sedimentological research on alluvial systems, beginning with the classic study of Fisk (1944), was carried out in meandering streams. These streams display relatively ordered conditions of riverine processes and sediment accumulation. Although there are numerous environments within the meandering stream complex, each with its own processes and characteristic sediments, the changes in both space and time are typically predictable based on well-documented process-response interactions in this relatively ordered system.

The geologic literature abounds with excellent articles on fluvial sedimentology, far more than can be mentioned here. There are also a few excellent summary references, the one by Allen (1965b) being the most comprehensive.

There are two types of meandering systems based on a combination of sediment particle size and discharge characteristics (Figure 8–13). The **coarse-grained meanderbelt** (McGowen and Garner, 1970) is actually somewhat of a transition between typical braided systems and the classical meandering stream. Sediments are sand and gravel with almost no mud accumulated. The stream pattern is only slightly to moderately sinuous (Figure 8–13). Typical meandering streams lack gravel, have at least a modest suspended load, and exhibit a broadly meandering pattern. Specific fluvial sedimentary deposits such as point bars, levees, crevasse splays, and floodplain sediments are common to both. For purposes of discussion these types will be combined, but distinctions will be made when appropriate.

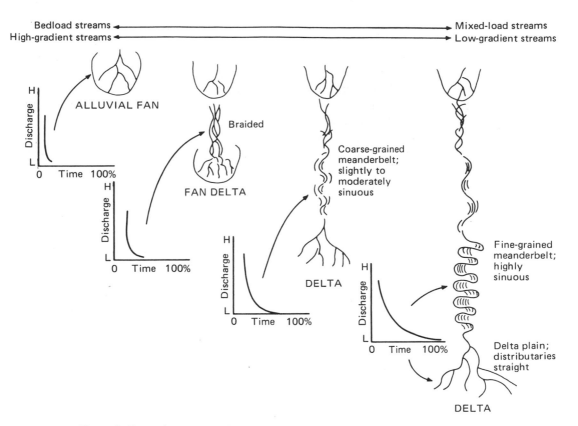

Figure 8–13 Various patterns in fluvial systems and their discharge characteristics. (From Brown et al., 1973, p. 13.)

General Morphology

Meandering streams typically occupy a position downstream from braided streams and upstream from deltas in the fluvial system (Figure 8–13). They occur in a variety of geologic settings but are most common on coastal plain regions, where they flow more or less perpendicular to the coast. Meandering streams are characterized by single channels in contrast to the multichannel braided streams. Discharge varies and generally includes a period of overbank flooding, which typically occurs less than once each year.

A ratio of channel length to valley length, called sinuousity (Leopold et al., 1964), has been used to describe the channel pattern of streams. Values of less than 1.5 are considered straight and those 1.5 and above are meandering. Streams are rarely straight for distances of more than 10 channel widths unless modified unnaturally or by unusual geologic control. Meanders develop as the result of perturbations in uniform flow across a channel. They may be caused by variation in sediment, slope or gradient, bed roughness, or other factors. Detailed discussion of channel shape and development may be found in Leopold and Wolman (1960), Leopold et al. (1964), or Schumm (1977).

The channel profile at the bends of a meandering stream display a characteristic asymmetry with one side quite steep in comparison to the other, which gently slopes upward from the stream bed. The steep side experiences lateral erosion, whereas the gently sloping side accumulates sediments in the form of point bars. This continual migration of the channel gives rise to various modes of shifting or even abandonment of the channel (Figure 8–14). Chute cutoff due to development of a new channel leaves much of the meander loop abandoned. The more common neck cutoff abandons the entire meander and develops an oxbow lake. Avulsion, due to downstream aggradation, causes abandonment of the channel and development of a completely new one [Figure 8–14(c)]. This is an important process in developing widespread fluvial deposits.

Processes

The primary processes in streams are based on the principles of fluid mechanics in open channel systems. There has been much research on this topic, including the theory, laboratory experiments, and field studies (see Chapter 2). Streams are characterized by turbulent flow. Velocity in streams varies both horizontally and vertically across the channel; it varies along the channel and also with time. Such changes in space and in time produce complex process-response patterns of sedimentation, but patterns that have enough regularity to be characterized.

Streams typically transport considerable sediment in both the bed material load and the wash load. Most of the wash load passes by the fluvial plain except for special circumstances such as flooding, when stream competence changes greatly. The wash load commonly is accumulated in the deltaic system, or much of it may be transported onto the shelf or beyond. As geologists we are interested primarily in that portion of the stream's load which accumulates in the meander belt; this is dominantly bed load in channel deposits and wash load in overbank deposits.

By contrast to braided streams, the meandering stream provides a regular pattern of flow in its channel. The consensus among fluvial sedimentologists is that helical flow is a major characteristic of flow in meander bends (Reineck and Singh, 1980; Hayes and Kana, 1976). This flow causes a superelevation of the water level on the outside of the meander. The helical flow produces a component of flow normal to the stream bank that is toward the eroding bank near the surface and toward the accreting side near the bottom. This has the net effect of producing a circulation cell which interacts with the bed to carry sediment upslope along the accretion surface

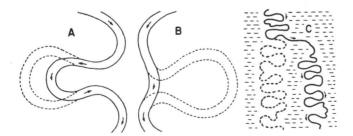

Figure 8–14 Various styles of channel shifting: (a) chute cutoff; (b) neck cutoff; (c) new meander belt after avulsion. (From Allen, 1965d, p. 119.)

(Figure 8–15). Lateral velocity components have been observed to be 10 to 20% of the downstream flow (Leopold and Wolman, 1960), with greatest values at the **thalweg** and general decrease up the accretion slope. Surface streamlines shift from the convex bank of one meander bend to the concave bank of the next meander downstream (Allen, 1965b). This causes sediment eroded from one bank to be deposited on the accretionary surface of the next downstream meander. The result of this process is downstream as well as lateral migration of meanders. In fact, most erosion takes place on banks which are nearly perpendicular to the trend of the meander belt (Allen, 1965b).

Early studies of meanders by Fisk (1947) described two primary types of bank erosion. Sloughing is a gradual and continuous removal of noncohesive sediment, whereas slumping takes place in cohesive materials where large blocks fail and give rise to scalloped banks (Turnbull et al., 1966).

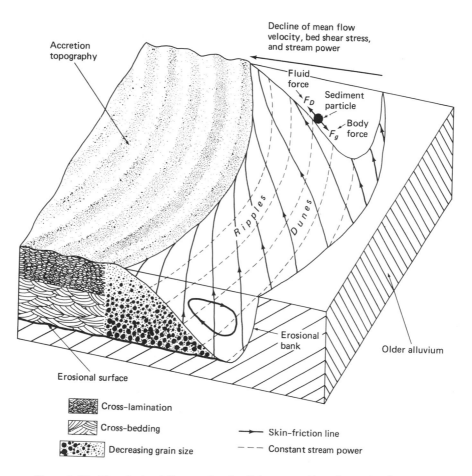

Figure 8–15 Meander bend diagram showing fining-upward lateral accretion deposits of the point-bar and fluid flow patterns. (From Allen, 1970a, Figure 4.5.)

Environments

Numerous local environments or subenvironments exist within the meandering stream system, each forming characteristic deposits (Figure 8–16). Each one produces its own sediment body morphology and accumulates a distinctive facies in the preserved fluvial record. The following discussion is based on the subdivisions of Reineck and Singh (1980), who recognized three primary categories: channel deposits, bank deposits, and flood basin deposits. Each has numerous subenvironments within it.

Channel deposits

Meandering streams accumulate two distinct sediment types within the channel itself: the channel lag deposits, which are located in the deep portion of the channel near the thalweg, and the point bars, which form on the inside of the meander loops.

Channel lag deposits. In many streams the load includes very coarse particles, at least in the pebble-size category. Meandering streams generally do not have enough competence to move such particles except during periods of high discharge. Finer particles are winnowed out and transported downstream, leaving the coarse particles as a lag deposit. The general mechanism of formation is similar to formation of a desert pavement. The nearly constant migration of the channel produces a continuing source of sediment for this gravelly lag as the channel banks erode and coarse particles fall to the channel floor.

Bedding is typically indistinct or lacking, but imbrication is common. The channel lag deposits typically accumulate in thin and discontinuous lenses. Lack of gravel in some meandering streams results in the absence of this facies or it may be represented by pieces of wood, skeletal material, or other debris (Reineck and Singh, 1980).

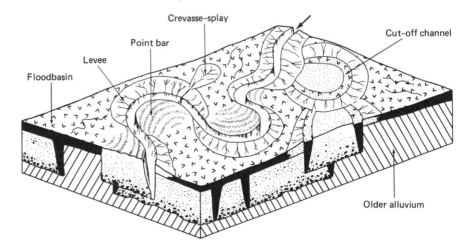

Figure 8–16 Model of meandering stream, showing major depositional environments. Solid pattern represents mud. (From Allen, 1970a, Figure 4.7.)

Figure 8-17 Aerial photograph showing point bars along reach of meandering stream; near Launceston, Tasmania.

Point-bar deposits. The greatest accumulation of sediments in the meandering system is in the point-bar environment. These conspicuous sediment accumulations develop on the convex side of the meander loop (Figures 8-15 and 8-16), typically somewhat downstream from the bisectrix of the loop (Figure 8-17). Point-bar deposits extend from the channel lag deposits usually upward to above mean water level. Therefore, their thickness is equivalent to the depth of the channel occupied.

Although point bars vary in size in proportion to the river which they occupy, their general morphology is similar. Large point bars exhibit surface topography consisting of **scroll bars,** which are positive ridge features, and **swales,** which are the alternating linear depressions (Fisk, 1947; Sundborg, 1956). These scroll bars and swales may have several meters of relief. They display prominent patterns and represent point-bar migration achieved by a flood (Figure 8-18).

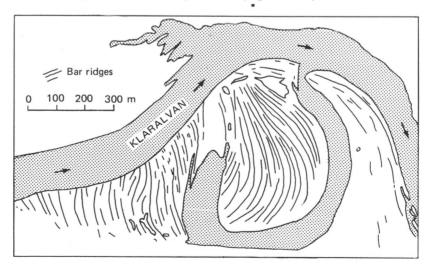

Figure 8-18 Map view of a point bar on the Klaralvan River (Norway), showing numerous accretion ridges or scroll bars. (After Sundborg, 1956.)

Chapter 8   The Fluvial System

One typically thinks of point-bar deposits as being rather well sorted sands, but a wide range of material may be present. Pebbles may be common (Wolman and Leopold, 1957; Folk and Ward, 1957) and silt may also be present. In fact, some entire point bars may be of coarse materials from medium sand to gravel with an absence of fine sand and mud (McGowen and Garner, 1970). Point-bar sediments may display a distinct trend in grain size from coarse at the base near channel lag deposits to fine on the top. Commonly, mud drapes form a thin veneer over the point bar or at least partially fill in the swales between scroll ridges. These trends in grain size depend on availability of a range in particle size. Some point bars may be dominantly fine grained (Jackson, 1978).

The surface of the point bars is commonly covered with bedforms. These range from ripples through megaripples and include plane beds. The uppermost portion of the bar commonly includes mudcracks developed in the thin mud drapes. The only organic remains generally found are leaves and plant debris, which is incorporated in point-bar sands. Skeletal animal material is relatively rare (Allen, 1965b).

Throughout much of the literature dealing with point-bar accumulations in meandering streams, there is no distinction made as to types of point-bar accumulations. Such a designation has been proposed by McGowen and Garner (1970) in an excellent paper on coarse-grained meander belts. Although classified as meandering systems, the rivers studied, Colorado in Texas and Amite in Louisiana, are actually somewhat transitional between braided and the typical broadly meandering stream. They are characterized by single channels, distinct but not pronounced meandering, and coarse, abundant sediment. During periods of extreme discharge these streams try to shorten their course by straightening the meanders (McGowen and Garner, 1970). This is accomplished by trough-shaped scour channels on the upper point-bar surface. They are called chutes and commonly terminate downstream in a coarse parabolic-shaped chute bar (Figure 8–19).

Channel sequences. Unlike their braided stream counterparts, the channel deposits of meandering streams produce a regular and cyclic sequence of sediment accumulation that is quite characteristic of this environment. Like most fluvial sedimentary sequences it has a fining-upward trend in grain size. The typical fine-grained point-bar sequence for meandering streams has the channel lag at its base, which is comprised of coarse sand and gravel. Above, and sometimes interbedded with the coarse material, is large-scale, trough cross-stratification [Figure 8–20(a)] produced by megaripples. Horizontal lamination representing upper-flow-regime flat bed conditions forms only a small portion of the section. Next are various types of ripple cross-strata representing low-energy conditions near the top of the bar. This fining-upward sequence is capped by a thin, laminated mud formed by lower flat bed conditions as suspended sediment settled [Figure 8–20(a)]. Actually, it is common that the upper portion of the sequence is removed by erosion prior to the development of the succeeding sequence.

The point bar sequence represents a classic example of application of Walther's Law. At any given time in the natural history of the stream, each of the environments in the point-bar sequence is present, from the channel lag up across the bar to the

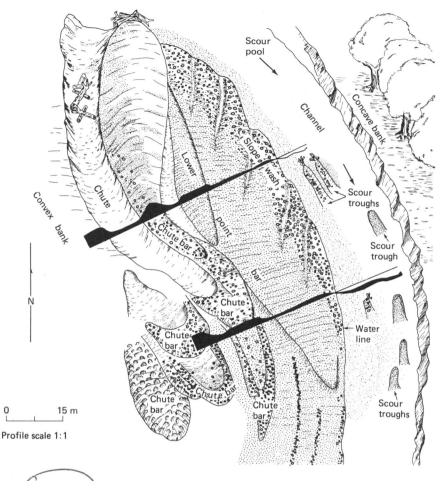

Figure 8-19 Coarse-grained point bar, showing sediment bodies and related features. (After McGowen and Garner, 1970, p. 31.)

mud drape on the upper surface. As the bar migrates laterally, the sequence accumulates [Figure 8-20(b)].

In the lower portion, the sequence for a coarse-grained point bar is somewhat similar to that described above, but it differs markedly in the upper part. Channel lag with large-scale cross-strata formed in scour troughs (Figure 8-19) comprises the basal unit [Figure 8-21(a)]. The overlying unit in the sequence consists of tabular and trough cross-strata formed by megaripples. The thick upper unit of the coarse point bar is comprised of chute bar deposits which show large cross-strata developed as these bars migrate downstream [Figure 8-21(a)].

A similar analogy holds true for fine-grained point bars; application of Walther's Law is readily apparent [Figure 8-21(b)]. Brown et al., (1973) list four

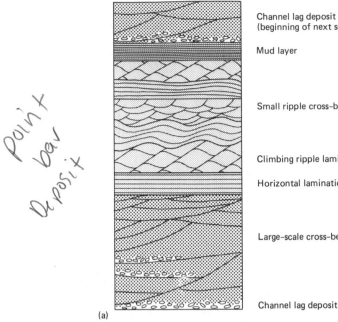

Channel lag deposit
(beginning of next sequence)

Mud layer

Small ripple cross-bedding

Climbing ripple lamination

Horizontal lamination

Large-scale cross-bedding

Channel lag deposit

(a)

Point bar Deposit

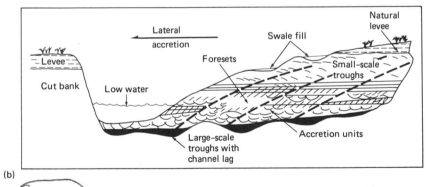

(b)

Figure 8–20 (a) Sequence of sediments and structures for the idealized, fining-upward point-bar deposit. (After Reineck and Singh, 1980, Figure 394, p. 277.) (b) Diagrammatic cross section of point bar and its contained strata with adjacent environments. (Modified after Bernard et al., 1963.)

primary criteria for recognition of the coarse-grained point-bar sequence: (1) coarse grain size and little mud, (2) incomplete sequence as compared to fine-grained system, (3) large wedge cross-strata in upper portion, and (4) laterally adjacent sand bodies. Care should be exercised to distinguish between the basal units and the chute-bar deposits.

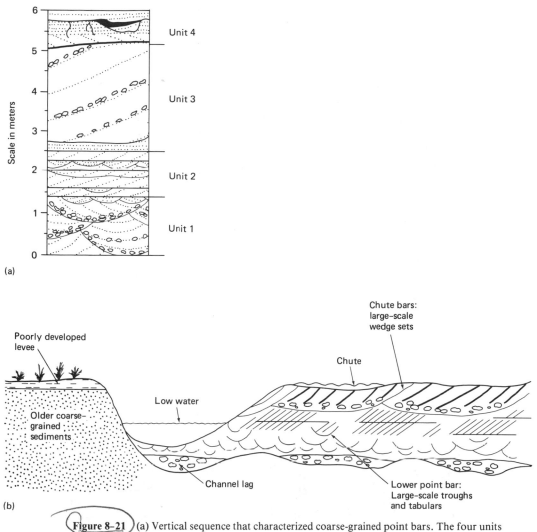

Figure 8-21 (a) Vertical sequence that characterized coarse-grained point bars. The four units represent scour pool deposit (1), lower point bar (2), chute bar (3), and floodplain (4). (Modified from McGowen and Garner, 1970.) (b) Generalized cross section shows a transverse profile across the coarse-grained point bar. (Modified from Brown et al., 1973.)

Overbank deposits

Extreme discharge events typically cause the stream to overtop its banks. As it does so, sediment is carried out of the channel and deposited in the form of overbank deposits. Such accumulations occur in various forms, each of which exhibits a rather distinctive vertical sequence.

Natural levees. Levees are the somewhat linear sediment accumulations situated along the stream bank immediately adjacent to the channel. These sediment bodies are thickest at the channel bank and thin toward the flood basin. They accumulate as the result of the sudden loss of competence by the stream as it overflows the confining bank. Natural levees are typically well developed on the outside or concave side of the channel where the bank is steep, whereas on the convex side they are subtle or not present. This is a reflection of the abrupt loss in competence on the steep-banked side of the stream, compared to a gradual loss on the point-bar side.

The thickness of natural levee deposits tends to vary with the size and capacity of the stream. They may be prominent in large streams with well-developed meanders (Figure 8–22), but in small streams they may be so thin as to be undetectable in their morphology. The crest of the natural levee is somewhat below the highest flood level. Sediments of natural levees accumulate in the typical fining-upward sequence of fluvial deposits (Allen, 1965c). Grain size also decreases away from the channel.

Although levee deposits are generally thinner and finer grained than point-bar deposits, they are similar to the upper portion of the point-bar sequence. Individual sequences are commonly a few decimeters thick and consist of rippled and horizontally stratified sand overlain by laminated mud. The horizontal stratification is considered to be the result of upper-flow-regime conditions (Singh, 1972). Some interlayering of the rippled and horizontally stratified sand occurs and probably represents pulsations in currents generated by flooding. Laminated mud is dominant away from the channel and represents accumulation from suspension. Desiccation features and raindrop impressions may develop and be preserved. Vegetation is abundant on these muds, which may cause bioturbation of laminations and incorporation of plant debris in the top of the natural levee sequence (Allen, 1965b; Coleman, 1969).

Figure 8–22 Oblique aerial photograph of a meander reach on the Arkansas River showing scroll bar accretion and swell and swale relief on the point bar. Shrubs along the cutbank denote natural levees. (Courtesy of S. A. Schumm.)

Crevasse-splay deposits.   Whereas the natural levee accumulations represent rather widespread and uniform overbank accumulations, crevasse-splay deposits are localized. Under flood conditions there may be situations that cause or permit the natural levees to become breached. As a consequence of this breakthrough, sediment-laden water debouches onto the flood basin, causing fan-shaped splays of sediment to accumulate. These splay deposits are most common on the concave banks of meanders and they take a fan-shaped aerial configuration (Figure 8–23). This sediment body commonly contains its own set of distributaries and it covers the distal portion of the natural levee (Allen, 1965b). Although splays are generally not large, some may extend over a kilometer from the channel, such as along the Rio Grande (Happ, 1940) and Brahmaputra Rivers (Coleman, 1969).

Crevasse-splay deposits are characteristically coarser than the related natural levee deposits. Although the internal organization is not as orderly as other fluvial accumulations, fining does occur upward and outward. Sands may display horizontal and ripple bedding with interbedded mud. Diverging dips on cross-strata reflect the splay or fan shape. These splay deposits are interbedded with natural levee and other overbank deposits. A given splay may be reactivated during multiple flood conditions or it may be a single-event accumulation.

Flood basin deposits.   Away from the channels and occupying the topographically lowest portion of the stream's floodplain is the flood basin environment. It is characterized by low relief, generally poor drainage, slow rates of accumulation, and fine, organic-rich sediment. A rather wide range of sedimentary environments may occupy this setting, with water supply and drainage being critical factors in their development. Swamps, lakes, and exposed plains may be present. Regardless of the specific environment, the sediment that is carried to this basin by floodwaters is

Figure 8–23 Aerial photograph of a crevasse-splay accumulation along a channel in the Mississippi Delta area. (Courtesy of J. Coleman.)

generally fine silt or clay and accumulates slowly from suspension. Individual flood events may cause a centimeter or so of sediment to accumulate. Broad flat plains give rise to thinner accumulations than in lakes or swamps, which trap the sediment-laden water. Lacustrine flood basin facies would be expected to show bioturbation, and swamps accumulate much plant debris in the bioturbated sediment (Figure 8–24).

Channel-fill deposits (oxbow lakes)

Allen (1965b) has designated channel-fill deposits as a combination of over-bank and channel deposits because of their unique circumstances. Once a cutoff of a meander occurs, that abandoned portion of the stream channel occupies a markedly different role in the fluvial depositional system. There is no longer a continuous motion of its bed and it becomes a very low energy sedimentary environment.

The cutoff portion of the channel is initially plugged by bed material load, but as the plugging of the channel proceeds the sedimentation rate declines. Eventually, washload is all that reaches the abandoned channel. Once the channel has been com-

Figure 8–24 Photographs of cores taken from selected flood basin environments of the Mississippi River. Both photos are from the floodplain of the Mississippi River near Baton Rouge, Louisiana: (a) oxidized, laminated muds; (b) mottled muds with roots.

(a)   (b)

pletely plugged, only flooding supplies sediment to the oxbow lake. As a result, infilling may take a long time, even centuries. These channel-fill deposits are well stratified and may include some skeletal material from ostracods or mollusks as well as plant debris. The sediments are commonly oxygen deficient, so that sulfides may precipitate.

ANCIENT FLUVIAL DEPOSITS

The various styles of fluvial deposition are well represented in the geologic record. The three primary types of fluvial environments described earlier—humid fan, braided stream, and meandering stream—are also treated in this section. Numerous studies indicate that there is a change in fluvial style through the duration of geologic time. Humid fans, although not common, have been recognized back to the Precambrian (Long, 1978). It has been speculated that evolution of land plants led to related changes in fluvial style (Schumm, 1968). A recent compilation of literature (Cotter, 1978) has shown that braided streams dominated until the mid-Paleozoic, when a mixture of braided and meandering became prevalent. This change is attributed to the development of land vegetation and its effects on retarding erosion rates.

Humid Fans

Whereas alluvial fan deposits may be distinctive from other fluvial facies, it is commonly difficult to separate fan deposits into their main depositional types: arid or humid. This discussion considers only humid alluvial fans; other types have been covered in Chapters 6 and 7.

Conglomerates are typical of alluvial fan deposits and may be of use in distinguishing between fan types in the rock record. Generally, conglomeratic textures can be considered as being of two styles: paraconglomerates and orthoconglomerates. **Paraconglomerates** are massive, lacking in organization, and contain a mixture of gravel, sand, and mud. They are interpreted as being the result of mass movement or highly viscous gravity flow (Bluck, 1967). By contrast, **orthoconglomerates** are stratified, lack mud, and typically display cross-stratification. These are framework gravels and are the result of deposition by water in a channel. As a general rule, paraconglomerates would be likely to result from sudden and extreme events, such as flash flooding on arid fans, whereas widespread accumulation of orthoconglomerates is typical of humid fans where discharge is more continuous. Exceptions have been recognized, however, with short-term fluctuations in rainfall a likely cause (Rust, 1979).

The best data for recognition of fan type would be to place the facies in question in the broad context of the depositional system and geologic setting. Arid fans show diagnostic adjacent facies (see Chapter 6), as do proglacial fans (Chapter 7).

An example of a well-documented ancient humid fan is that described by McGowen and Groat (1971) from the Precambrian(?) of west Texas. The Van Horn Sandstone consists primarily of conglomerates and cross-stratified sandstones, essentially all orthoconglomerates. Using the three standard subdivisions of fans, upper, middle, and lower, these authors have recognized three facies which grade laterally into one another.

Proximal fan facies is dominated by gravel, with individual clasts reaching a meter in diameter. Bedding is indistinct and imbrication is widespread. The mid-fan facies is characterized by interbedded units of gravel and cross-stratified, pebbly sandstone. Scouring is evident at the base of individual units. The distal facies is the most varied and is characterized by both tabular and trough cross-stratified sandstone.

McGowen and Groat (1971) have reconstructed the paleogeography of the Van Horn Sandstone as a series of coalescing fans each of which was probably 30 to 40 km at its widest part. The coarse proximal deposits were confined to valleys separated by divides (Figure 8–25). The midfan was characterized by coarse braided channels with

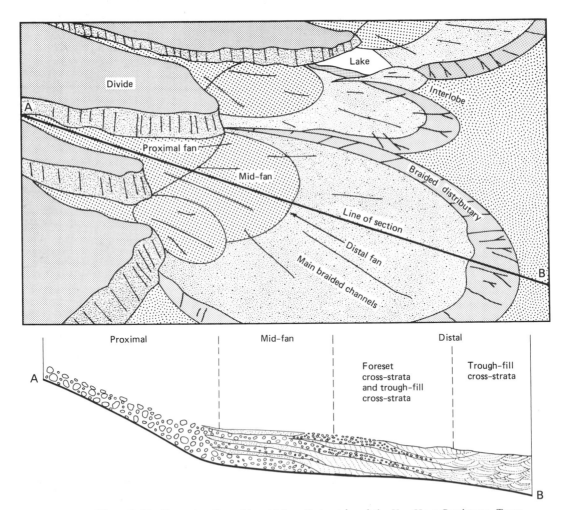

Figure 8–25 Reconstruction of humid fans that produced the Van Horn Sandstone, Texas. (After McGowen and Groat, 1971, pp. 38, 39.)

elongate bars. The distal environment was most widespread and was dominated by braided channels in which large bedforms and migrating bars produced planar and trough cross-stratification. The most distal portion was dominated by small distributaries with abundant bedforms. Topographic lows between fans may have formed lakes where some fines accumulated (Figure 8-25).

Braided Streams

Published studies indicate that the braided stream environment has been important throughout geologic time (Cotter, 1978), with the oldest rocks interpreted as representing this environment being 3.5 billion years old (Long, 1978). There are two relatively distinct facies in the rock record representing braided fluvial deposition: one contains abundant gravel and the other essentially lacks gravel. Actually, each of these two facies represents two subenvironments: the braided river, an elongate deposit, and the alluvial plain, which produces tabular and laterally extensive similar deposits (Rust, 1978).

Coarse braided facies
Braided gravels are typically bedded and have widespread imbrication of clasts. This is facies Gm (Table 8-1) of Miall (1977). These gravels are interbedded with and grade laterally into sandstones. The coarse lenses represent longitudinal bars and are thin, reflecting the low relief in these braided systems.

An excellent example of the coarse braided facies is the Devonian Malabie Formation, which is exposed on the Gaspé Peninsula of Canada (Rust, 1976, 1978). This unit consists of conglomerates and coarse sandstones (Figure 8-26) which reach a maximum thickness of 200 m. Clast-supported conglomerates are typically horizontally bedded, with maximum clast size 10 to 40 cm. Most clasts show imbrication. Paleocurrent directions range rather widely but display a similar mean vector (Rust, 1978).

Strata displaying similar characteristics but with more sand-size material were deposited during the Cretaceous along the southern coast of Victoria (Australia). These rocks probably represent a braided environment more distal from the source than that described above (G_{III} of Rust, 1978). Conglomerate is lenticular and may contain pieces of coally material, suggesting that large pieces of plant material were being incorporated into the conglomeratic lenses during high-discharge events when vegetated bars were eroded.

Sandy braided facies
One of the major problems in fluvial sedimentology at this time is the distinction of sandy braided sequences in the rock record from those deposited by meandering streams. At present there are no absolute criteria.

One of the best documented interpretations of a sandy braided sequence is the study of the Battery Point Formation (Devonian) in Quebec, Canada (Cant and Walker, 1976; Cant, 1978). The unit is about 110 m thick and consists largely of trough and planar cross-stratified medium to coarse sandstone. Several fining-upward sequences are present in the Battery Point. These characteristics, together

Figure 8–26 Photograph of Malabie Formation (Quebec), which is interpreted as a braided stream deposit. (Courtesy of B. R. Rust.)

with freshwater fish fossils and in situ plant remains, led to the interpretation of a fluvial origin for this unit (Cant and Walker, 1976).

The composite fining-upward sequences in the Battery Point contain up to eight different characteristic units or facies (Figure 8–27). The basal unit (SS) consists of coarse sandstone and mudstone clasts resting on an erosion surface. Units A and B are poorly defined and well-defined cross-stratified facies, respectively. They, together with the planar cross-bedded facies (C), may be repeated in a single sequence (Figure 8–27). Well-bedded, small-scale, planar cross-strata comprise facies D, which may be overlain by a scour facies (E). Interbedded rippled sandstone and mudstone (F) and low-angle to horizontally stratified sandstone (G) complete the sequence. Note that paleocurrent directions are consistent throughout the sequence except for the planar cross-stratified facies (C) (Figure 8–27).

The sequence described above is interpreted as representing a sandy braided system such as the South Saskatchewan River (Canada) (Cant and Walker, 1976). Unit SS represents channel scour. Infilling is by trough-shaped cross-stratified sandstones (A and B) generated by megaripples. Planar cross-strata with divergent directions (C) represent the migration of small bars across or diagonal to the channel. The small, low-angle cross-strata are thought to represent long-wave-length, low sand waves. The upper units (E to G) represent the highest elevations in the braided system, crests of bars or floodplains, and are vertical accretion deposits (Cant, 1978).

The question naturally arises as to how one distinguishes this type of sandy braided sequence from the typical meandering sequence described earlier in the

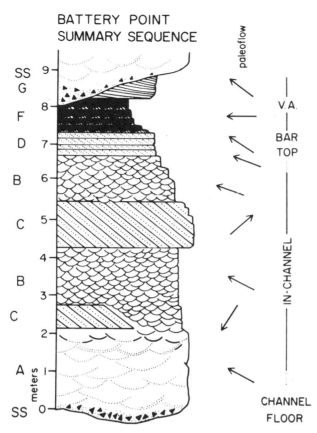

BATTERY POINT
SUMMARY SEQUENCE

paleoflow

SS G 9

F 8 — V.A.

D 7 — BAR TOP

B 6 —

C 5 —

4 —

B 3 — IN-CHANNEL

C 2 —

A 1 —

SS 0 —

meters

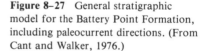

CHANNEL
FLOOR

Figure 8–27 General stratigraphic model for the Battery Point Formation, including paleocurrent directions. (From Cant and Walker, 1976.)

chapter. Walker and Cant (1979) have considered this problem and provide some suggestions. The nature of the braided system is such that bar migration is rather unpredictable and channels are ephemeral relative to meandering streams. As a consequence, repetition of the trough and planar cross-stratified facies is expected. In the meandering stream lateral accretion by point-bar migration is rather unidirectional and repetition is not expected within a single sequence. Also, vertical accretion deposits including flood basin deposits are better developed in a meandering system.

Meandering Streams

Strata attributed to the meandering stream environment have received considerable attention from geologists, especially since the early 1960s. During geologic history meandering streams did not become abundant until about Devonian time (Cotter, 1978).

Recognition of the meandering stream environment may be based on various characteristics, but most commonly it is based on the stratigraphic sequence of sediment textures and structures. Meandering fluvial systems result in a more regular cyclic pattern of fining-upward sequences than do braided systems. The following section considers a few examples of well-documented meandering systems from the rock record. The reader is again referred to the recent volume on fluvial sedimentology (Miall, 1978) for additional examples and an extensive bibliography.

Tertiary (Miocene) of Eastern Spain

Several sequences of fluvial deposits have been described from the Southern Pyrenees Mountains of Spain (Puigdefabregas and Van Vliet, 1978). Although these sequences range widely in total accumulation, one of the smallest represents an ideal example of an ancient analog for the meandering stream environment. Miocene strata from the Ebro Basin can be observed in three dimensions and also have the ancient point-bar surface preserved and exposed at the surface (Puigdefabregas, 1973).

Although the relict features of old point bars are easily seen on topographic maps, there are few situations where these features can be viewed in their preserved state in the rock record (Figure 8-28). They have been shown from the Carboniferous of Morocco (Padgett and Ehrlich, 1976) and the Jurassic of England (Nami, 1976), but the Miocene examples in Spain permit the best examination and reconstruction.

As the point bar migrates in the meandering system, a ridge and swale type of relief develops; this is sometimes referred to as accretion topography or scroll bars. Related to this is low-angle cross-stratification (called **epsilon cross-stratification**), which marks pulses in the point-bar accretion and is essentially normal to stream flow. One particular exposure, the Murillo Point Bar (Puigdefabregas, 1973), shows all these features well (Figure 8-29). The channels in which these sediments accumulated were small, 3 to 5 m wide, 1 to 2 m deep, with 25 cm of relief on the accretion units. The aerial view clearly shows the typical point bar–meandering stream configuration, with epsilon bedding normal to the ridges and trough cross-strata oriented with current direction.

The point bars are only 1 to 2 m thick and the entire fluvial sequence of 50 m is

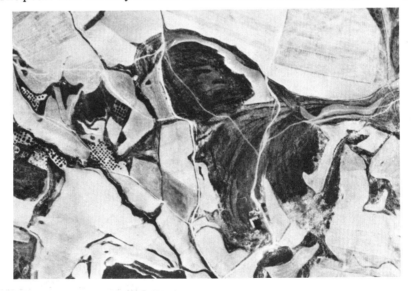

Figure 8-28 Aerial photograph of Miocene meander scars (Murillo point bar) from Spain. (Courtesy of C. Puigdefabregas.)

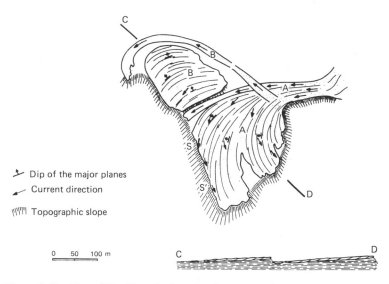

Figure 8-29 Map of Murillo point bar, showing preserved scroll bars. (From Puigdefabregas, 1973, p. 137.)

dominated by overbank deposits (Puigdefabregas, 1973). Three channel deposits are present, each having the typical fining-upward sequence with associated cross-stratification varieties (Figure 8-30). Most of the overbank deposits are mudstones with lenses of sandstone or interbedded sandstones. The sandstones show ripple and climbing ripple cross-stratification typical of natural levee and flood basin deposits. Some thin (5 to 10 cm) graded units are present which may represent splays or proximal overbank deposits. Desiccation features are common.

Jurassic of Yorkshire, England

Numerous examples of well-documented meandering systems have been found in Mesozoic strata and are listed by Cotter (1978). One that is particularly well exposed and shows interesting stratigraphic trends is along the North Sea coast of Yorkshire in the Scalby Formation of the Middle Jurassic. These strata crop out at the base of bluffs in the intertidal zone. They extend for a few kilometers along the coast with a total thickness near 50 m (Nami, 1976; Nami and Leeder, 1978). As in the case of the previously discussed Miocene example from Spain, meander loops with accretion topography are exposed.

The vertical sequence is comprised of cross-stratified sandstone which grades upward into siltstones and clayey siltstones. Several of these units are present, separated by epsilon cross-stratification. Some clay clasts, wood fragments, and convolute bedding may be present in the upper portion. Cross-stratification directions are normal to the trend of accretion topography for the epsilon sets and parallel to ridges for the contained trough cross-sets—basically as in the Miocene example. The epsilon units have an aggregate thickness of 3 to 4 m with each unit from 30 to 70 cm thick (Nami, 1976).

The cross-stratified sequences are interpreted as point-bar deposits that were

Chapter 8   The Fluvial System

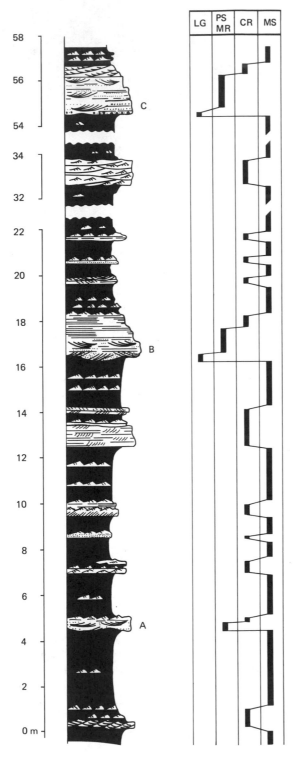

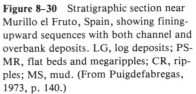

Figure 8–30 Stratigraphic section near Murillo el Fruto, Spain, showing fining-upward sequences with both channel and overbank deposits. LG, log deposits; PS-MR, flat beds and megaripples; CR, ripples; MS, mud. (From Puigdefabregas, 1973, p. 140.)

being cut into cohesive muds. Flood basin mudstones and sandy mudstones are widespread along the exposures and represent about one-half of the total section. A composite stratigraphic section shows an overall fining-upward trend within the Scalby Formation, which is interpreted by Nami and Leeder (1978) as resulting from a change in channel morphology in response to rising sea level. The basal sequence unconformably overlies a coastal marine barrier system and consists of coarse, lenticular, cross-stratified sandstone interpreted as representing a coarse meander system or possibly a braided system (Figure 8–31). Above this is a sequence of lateral-accretion sandstones containing epsilon cross-strata which the authors interpret as a sinuous river with little discharge.

The upper portion of the Scalby consists of alternating coarse and fine sequences; the relatively thin, coarse sequences are cross-stratified sandstones with epsilon bedding (lateral accretion surfaces), whereas the thicker mudstones have some sandstone lenses, mudcracks, and dinosaur footprints (Figure 8–31). The coarse sequence represents point-bar accumulation, whereas the fine sequences were deposited as overbank deposits. They accumulated on a coastal plain with broadly meandering streams and frequent meander cutoff or avulsion.

Lower Old Red Sandstone (Devonian), England

Among the earliest and best documented studies of ancient fluvial deposits are those by Allen (1964b, 1965c, 1974), who investigated the Lower Old Red Sandstone of Devonian age. His efforts were concentrated in southern Wales and southeastern England, where more than 1000 m of this unit are present. Allen examined numerous exposures of supposed fluvial strata and demonstrated not only their depositional framework but showed that the now-classic fining-upward fluvial sequence may be preserved in numerous variations (Allen, 1965a).

The following discussion is based largely on Allen's work and will include two of the six sequences that were described in detail (Allen, 1964b). Other sequences have also been described from the Lower Old Red Sandstone and show further variations (Allen, 1965b).

Abergavenny Sequence. A fining-upward sequence of strata about 7 m thick shows most of the common fluvial components. A scoured basal surface is overlain by three rather distinctive lithofacies. The basal meter or so consists of conglomerates and sandstones, with the latter displaying convolutions, flat beds, and some ripple cross-strata (Figure 8–32). This is interpreted as a combination of channel lag and channel-fill material. Some large plant debris is indicative of flood events (Allen, 1964b). Some aggradation of the channel took place.

The next unit is relatively homogeneous and consists primarily of ripple cross-stratified sandstones with some channeling (Figure 8–32). This represents lower-flow-regime conditions and lateral accretion deposition by point bars under low flow rates. More than half of the sequence is comprised of siltstone and mudstone, with some burrows and sand lenses which show ripple cross-stratification. Desiccation features are not evident. These fine-grained deposits represent overbank deposits and consist of natural levees as well as floodplain accumulations. The latter were probably dominated by swamps and lakes, thus the absence of mudcracks (Allen, 1964b).

This sequence probably represents a rather shallow, small meandering stream

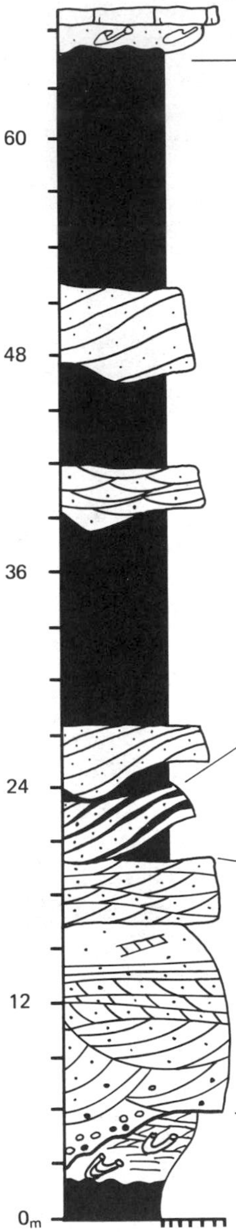

Description	Interpretation	Controls
Bioturbated sandstones and thin bioclastic limestones	Nearshore marine	Arrival of marine transgression
1. Fine members Gray kaolinitic mudstones with in situ Equisetites and rare thin coals. Thin interbedded sharp based sandstones are laterally continuous and show basal scours, infilled desiccation polygons and dinosaur footprints. Top surfaces with symmetrical ripples. **2. Coarse members** Fine to medium sharp based lenticular sand bodies 4–8 m thick and up to 600 m wide. Rarely multistorey, usually isolated. May be lenticular bedding but usually with lateral accretion surfaces.	Coastal plain of alluviation with extensive vegetated floodbasins undergoing very rapid vertical accretion. River channels dominantly of high sinuosity with frequent avulsions. Lateral density of active channels unknown.	Basin alluviation due to rising sea level
Upward fining fine sandstone to siltstone with crosscutting curvilinear lateral accretion units seen on surfaces parallel to bedding.	High sinuosity river with low relative discharge	
Regional sheet sandstone. Multistorey in part. Sharp erosive base. Medium to coarse grained with gross upward fining. Siltstone lenticles at top. Tabular cosets with rarer very large scale trough sets. Mature quartzarenite.	Low sinuousity, braided (?) river with high relative discharge	Degrading alluvial system; high slope; entrenchment
Upward coarsening from bioturbated mudstone to low angle cross-laminated, well sorted very fine sandstones.	Coastal marine beach/barrier system	Wave-dominated clastic shoreline

Figure 8–31 Composite section and interpretation of depositional environments for Scalby Formation fluvial deposits. (From Nami and Leeder, 1978, p. 438.)

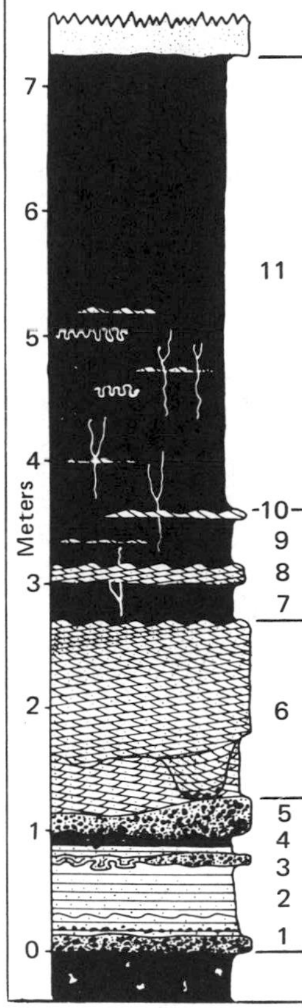

MAIN FACTS	INTERPRETATION
Red coarse siltstones with invertebrate burrows, ripple-bedded sandstone lenticles, and convolute laminations. No evidence of exposure.	Vertical accretion deposit from overbank floods. Backswamp area, probably a permanent lake.
Red coarse siltstones alternating with beds or "biscuits" of ripple-bedded, very fine sandstone. Invertebrate burrows. No proofs of exposure.	Vertical accretion deposit from overbank floods. Levee and backswamp deposits with area possibly a lake for long periods.
Red, flat-or ripple-bedded very fine to fine sandstone with a channeled scoured surface in lower part. Scattered siltstone clasts.	Probably mixed channel-fill and lateral accretion deposits. Deposition of suspended and bed loads on channel bars and sand flats. Deepening or wandering of channel at times.
Intraformational conglomerates on scoured surfaces alternating with green siltstones and very fine to fine sandstones, showing ripple-bedding, flat-bedding or convolute lamination. Concentrations of plant debris and ostracoderms, same of latter articulated.	Mixed channel-fill and channel lag deposits. Repeated migration and partial aggradation of channel. Flotsam of floodplain plants and riverine ostracoderms deposited in or near active channel.
Scoured surface of low relief cut on siltstones.	Erosion at floor of wandering river.

Figure 8-32 Generalized stratigraphic section at Abergavenny, showing channel and overbank deposits. (From Allen, 1964b, p. 184.)

which is subjected to intense flooding. Both the channel-fill deposits and the thick overbank sediments support this contention. During periods of low discharge the small point bars exhibited only small-scale bedforms. At the same time, floodwaters remained ponded on the poorly drained flood basin, with swamps and shallow lakes persisting.

Ludlow Sequence. This 6 to 7 m-thick sequence is nearly uniformly divided into a coarser sandstone lower portion and a massive red siltstone above (Allen, 1964b). The sequence rests on a scoured surface with about 15 cm of relief. Clasts of the underlying siltstone occur at the base and scattered throughout the sandstone unit (Figure 8–33). Trough cross-stratification is present throughout this unit and convolute bedding is presently locally. This portion of the section represents a classic example of lateral accretion deposition in the point-bar environment. Megaripples were the predominant bedform, as evidenced by the size and type of cross-stratification. Some upward fining is evident within the sandstone unit.

A thin, fine sandstone unit with ripple cross-stratification caps the basal sand-

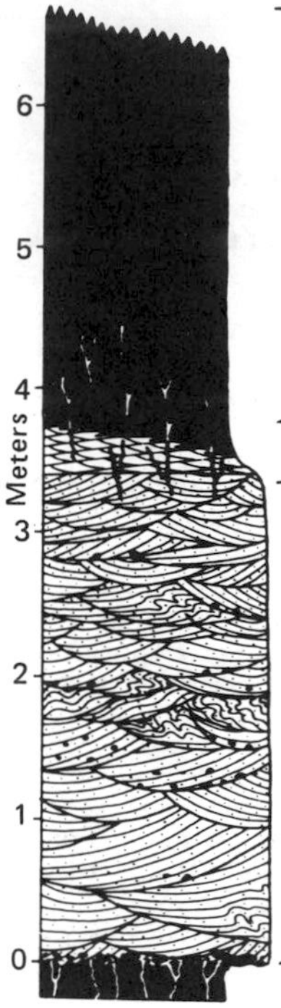

MAIN FACTS		INTERPRETATION
3	Red, coarse siltstone devoid of bedding. Sparse calcium carbonate concretions. Suncracks absent.	Vertical accretion deposit from over bank floods. Probably deposited in backswamp area, perhaps a more or less permanent lake.
2	Variable thickness of red, ripple-drift bedded, very fine sandstone. Grades up into siltstone. Invertebrate burrows.	Vertical accretion deposit from overbank floods. Possibly a levee deposit or a point-bar swale filling.
1	White to purple, fine to medium, well sorted sandstones. Siltstone clasts concentrated at base and scattered throughout. Trough cross-stratified, units 10-90 cm thick. Contorted cross-strata near base and middle.	Channel deposit probably formed by lateral accretion on a point-bar. Sand transported as bed-load over river bed formed into lunate "dunes". Strong, variable currents. Siltstone clasts form lag concentrate where channel was deepest.
	Cut on siltstone. Maximum relief 15 cm. Few directional scour structures.	Erosion at deepest part of wandering river channel.

Figure 8–33 Generalized stratigraphic section from Ludlow. (From Allen, 1964b, p. 171.)

stone (Figure 8–33). This unit is interpreted as an overbank deposit, either a natural levee or a crevasse splay. It is burrowed with burrows extending into the overlying siltstone. This massive unit is devoid of any diagnostic physical or biogenic structures. It does contain some carbonate concretions, which may represent carbonate soil (calcrete) horizons (Allen, 1974). The siltstone unit is an overbank sequence probably dominated by standing-water conditions such as a lake.

SUMMARY OF CHARACTERISTICS OF SEDIMENT/SEDIMENTARY ROCK BODIES REPRESENTING THE FLUVIAL SYSTEM

Within the fluvial system there are several extensive and recognizable components which may occur in various relationships with one another or which may be absent. Typically, there are three of these: humid fans, braided streams, and meandering streams. The latter two are well summarized in a diagram by Galloway (Figure 8–34). Any or all of these may be preserved as the fluvial depositional system in the rock record.

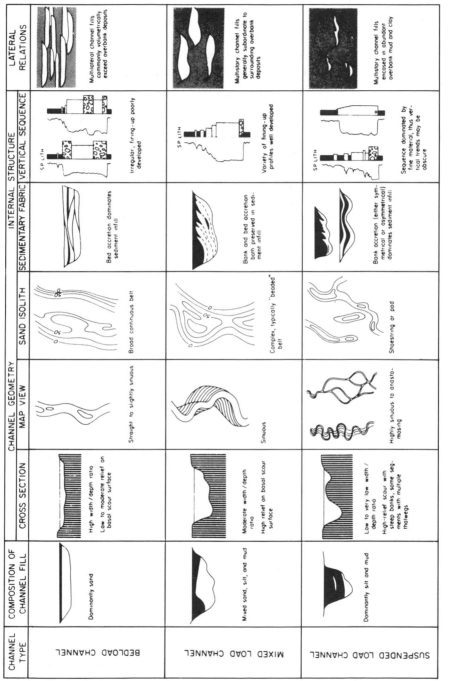

Figure 8–34 Geomorphic and sediment characteristics of stream channel segments of different types. (From Galloway, 1977.)

Tectonic setting. No particular setting is characteristic of the fluvial system in general, but there are some settings which tend to favor specific components. For example, humid fans need high relief as well as abundant moisture. Coastal mountainous areas are therefore favorable sites. By contrast, extensive meander belts are common on low-relief coastal plains.

Shape. A single meander belt would be expected to display a linear shape; however, it is common for numerous adjacent meander belts to form a fluvial plain. The fans would each have a wedge shape, but several adjacent ones may combine. Channel deposits tend to be linear but sinuous and overbank deposits are tabular.

Size. No generalities can be made about the size of fluvial deposits. They range widely in thickness and areal extent.

Textural trends. Streams and their rock record counterparts display a decrease in grain size away from the source (downstream). There is also an increase in rounding and sphericity. More important are the stratigraphic trends in grain size. In general, fluvial sequences display a pronounced fining-upward grain size trend. Typically, each environment and its accumulated sequences must be viewed individually to perceive this trend. Braided streams are an exception in that their setting and flashy discharge are not conducive to the establishment of regular grain size trends, although they may be present.

Lithology. There is nothing regarding composition that can be used to characterize the fluvial system.

Sedimentary structures. No specific structures are characteristic of fluvial sediments; however, some important trends are present. Just as grain size decreases upward, there is a tendency for the sedimentary structures—bedforms and their resulting cross-strata—to produce a stratigraphic trend. There is a decrease in flow regime upward in the fluvial sequences. An exception is that upper plane beds may accumulate near the top of sequences as water levels subside, giving high Froude numbers. Each environment displays a suite of features, with floodplain facies having some of the more diagnostic ones (e.g., raindrop impressions and mudcracks).

Paleontology. Fluvial sediments are typically unfossiliferous. Certain pollen, spores, bivalves, and crustaceans may be indicative of stream environments.

Associations. Almost an infinite variety of arrangements may exist between sediment bodies within the fluvial system and between fluvial strata and other continental deposits. The transition between humid fans, braided streams, and meandering streams makes it essentially impossible to draw real boundaries between the facies representing these environments. Fluvial facies are commonly related to, and difficult to distinguish from, glacial outwash and delta plain deposits.

ADDITIONAL READING

BERNARD, H. A., MAJOR, C. F., JR., PARROTT, B. S., AND LEBLANC, R. J., SR., 1970. *Recent Sediments of Southeast Texas: A Field Guide to the Brazos Alluvial and Deltaic Plains and the Galveston Barrier Island Complex.* University of Texas, Bur. Econ. Geol., Austin, Tex.,

Guidebook 11, 16 p. and appendices. Classic reference on modern fluvial sediments, structures, and sequences.

ETHRIDGE, F. G., AND FLORES, R. M. (EDS.), 1981. *Recent and Ancient Nonmarine Depositional Environments: Models for Exploration.* Soc. Econ. Paleontologists and Mineralogists, Spec. Publ. No. 31, Tulsa, Okla., 349 p. Contains numerous papers on both modern and ancient fluvial systems; some are local and some are regional in scope.

LEOPOLD, L. B., WOLMAN, M. G., AND MILLER, J. P., 1964, *Fluvial Processes in Geomorphology.* W. H. Freeman, San Francisco, 522 p. Although somewhat out of date from the sedimentology point of view, this text covers the basic principles of stream processes.

MIALL, A. D. (ED.), 1978. *Fluvial Sedimentology.* Can. Soc. Petroleum Geologists, Mem. No. 5, Calgary, Alberta, 589 p. This is the state of the art in the subject. Good selection of papers on both modern and ancient systems by the world's experts.

SCHUMM, S. A., 1977. *The Fluvial System.* John Wiley, New York, 338 p. A good basic text on the subject by one of the world's leading researchers.

VAN HOUTEN, F. B. (ED.), 1977. *Ancient Continental Deposits.* Benchmark Papers in Geology, v. 43, 367 p. Papers 4 through 10 are on fluvial deposition and include most of the classic works on the subject.

Part Three

TRANSITIONAL ENVIRONMENTS

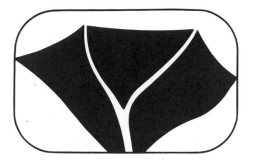

9 The Delta System

Deltas are accumulations of sediment at the end of a channel where it discharges into a standing body of water. The term **delta** is taken from the Greek capital letter delta (Δ) and was first used by ancient Greeks, who applied it to the accumulation of sediment at the mouth of the Nile River. Deltas represent ''gifts'' from the rivers to the sea (Wright, 1978). They are probably the most complex of all depositional systems, with more than a dozen distinct environments of deposition within the deltaic system. Deltas are also of great economic importance because ancient deltaic deposits represent one of the prime sources of fossil fuels, including coal, gas, and oil.

Although there has been a great deal of research directed toward the investigation of modern deltas, this effort has come about only in the last few decades. Prior to World War II only seven articles on modern North American deltas appeared in the literature: two on the Fraser Delta in Canada, three on the Mississippi, and two on the Colorado (LeBlanc, 1975). The explosion of research on both modern deltas and their ancient analogs in the rock record has led to a wealth of literature. Many symposia and short courses on deltas have been the bases for outstanding publications, including Shirley (1966), Fisher et al. (1969), Morgan and Shaver (1970), Broussard (1975), Weimer (1976), and Coleman (1976).

DISTRIBUTION

The primary factor in determining whether or not a delta will develop along a coast is the balance between sediment yield and the energy flux in the site of accumulation. As long as there is more sediment supplied at the river mouth than can be removed by the tides, waves, and longshore currents, a deltaic system will evolve. In addition to the proper imbalance between these processes, there must be a site upon which the delta can develop.

Although deltas occur nearly worldwide, there are some climatic, geomorphic, and tectonic constraints on their development. A significantly large and active drainage system must be present to provide proper discharge of water and sediment. In general, only those land masses presently covered by ice sheets (Antarctica and Greenland) lack major deltas (Figure 9–1).

A glance at the world distribution of major deltas shows that global tectonics is a significant factor; there are few large river deltas on collision coasts. This is primarily due to tectonic activity; drainage divides are close to the ocean and there is typically no wide and shallow shelf on which a delta can accumulate. Trailing-edge coasts and marginal seacoasts (Inman and Nordstrom, 1971) and their associated well-developed drainage basins contain numerous deltas. Of 58 major rivers, only five, the Columbia and Colorado (United States), Fraser (Canada), Ebro (Spain), and Po (Italy), debouch on collision coasts. The first three listed do not occur on the list of major deltas of the world (Wright, 1978).

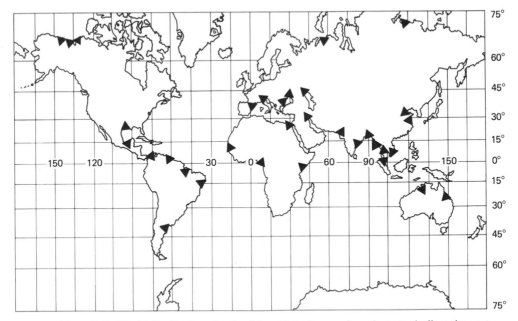

Figure 9–1 World distribution of major deltas. Note absence along the tectonically active west coasts of North and South America. (From Wright et al., 1974.)

CHARACTERISTICS OF THE DELTA SYSTEM

The delta is comprised of a diverse complex of depositional environments that span a wide range from terrestrial through coastal to purely marine. The processes and resulting morphology of a delta are the result of a great variety of interdependent factors. Although there is a considerable range in the specific characteristics among the world's deltas, they also share some general features and processes that are universal regardless of size or location of the deltaic system.

Deltas can typically be subdivided into three broad environments: the delta plain, which is largely subaerial but contains subaqueous portions, the delta front, and the prodelta. There is a general seaward fining of sediment particle size within the latter two regions of the delta (Wright, 1978). The distal portion of the delta (prodelta) is dominated by clays settling from suspension.

Whereas one commonly thinks of deltas in terms of large accumulations of sediment and as *constructional* systems, there is typically a *destructional* phase as well. Erosion may actually dominate at some areas or during certain periods of time when little sediment is being supplied. Waves and currents, both wave generated and tidal, are the primary agents of erosion on the delta. The destructive phase dominates on abandoned portions of the delta when sediment influx from the river has ceased. The active delta is dominated by the constructional phase. Beaches, beach ridges, and dunes may develop along the shore of the abandoned delta, whereas the active delta exhibits progradation of the deltaic plain, with greatest rates concentrated at the major **distributaries.** Emphasis in this chapter is on the active delta, where constructive processes dominate.

DELTAIC PROCESSES

Knowledge of sedimentary processes associated with the development of deltas has only recently approached the level of that for delta morphology and stratigraphy. Although G. K. Gilbert recognized the importance of processes in delta development in his classic work on Lake Bonneville (Gilbert, 1885, 1890) and Barrell (1912) utilized these concepts, there was no comprehensive effort to understand deltaic processes until the important paper on delta formation by Bates (1953). Beginning in the 1950s and continuing to the present, much effort has been expended in understanding the complexities of deltaic processes, which are complicated by the interaction of riverine processes with those of the marine environment. Foremost among these efforts have been those of the Coastal Studies Institute of Louisiana State University, whose staff has examined virtually all the major deltas of the world.

A number of processes shares control over delta development and maintenance. Some of these operate on, or adjacent to, the delta and others may be geographically remote from the delta itself. Among the latter are climate, relief in the drainage basin, sediment yield, and the water discharge regime (Coleman and Wright, 1975). In some respects climate may be the most influential of all the processes that affect deltas. It controls not only the amount and temporal distribution of water discharge but also the vegetation, weathering, and soil development, and to some extent, the relief in the drainage basin. Obviously, the latter factors strongly influence the nature and volume of sediment discharged through the river mouth.

Although the processes described above are important in deltaic sedimentation, most have been treated in Chapter 8. In this discussion, emphasis is placed on those processes that are operating on the delta itself: riverine processes and marine processes, including tides, waves, and coastal currents.

Riverine Processes

For initial purposes of discussion it is best to consider the relatively simple river mouth situation, where tides are negligible and wave power is minimal (Wright and Coleman, 1974). These circumstances result in a river-dominated situation such as occurs in lakes, estuaries, enclosed seas, or where there are broad, flat, offshore slopes (Wright, 1977). Under these conditions three primary forces dominate: inertia, bed friction, and buoyancy. Factors such as discharge, outflow velocity, water depth, size and amount of particles in the sediment load, and the density contrast between the river and basin waters determine which of the three forces controls the riverine processes on the delta.

Inertia-dominated effluent

High outflow velocities, deep water immediately seaward of the river mouth, and negligible density contrasts give rise to dominance by inertial forces, causing the effluent to spread and diffuse as a turbulent jet (Wright, 1977, 1978). This turbulent mass of sediment-laden water is termed **homopycnal**; that is, there is a uniformity in density between the effluent and the ambient water body. Such a situation represents the simplest conditions of river mouth sedimentation and was the basis for Gilbert's classic deltaic model with topset, foreset, and bottomset components (Gilbert, 1885). These conditions are generally restricted to high gradient streams entering a deep lake or where tidal activity homogenizes the water masses, thereby destroying any density gradients (Wright, 1978).

An idealized depositional pattern for this simple but rare type of river mouth accumulation is shown in Figure 9–2. The primary sediment body is a lunate bar, convex seaward, with the coarsest sediment particles located landward of the bar crest. Low spreading angles of the turbulent jet restrict the lateral dispersion of sediment particles (Wright, 1977). The relief on the bar is low and a relatively steep bar front is developed (Figure 9–3). Laboratory experiments by Jopling (1963) have substantiated this model from the two-dimensional point of view.

Bed friction-dominated outflow

In many river mouth environments, continual discharge of sediment causes substantial shoaling just beyond the mouth; in fact, it is common that depths in this area are a maximum of slightly less than in the outlet. The result is a lateral spreading of the effluent as a plane jet accompanied by shear between the outflow and the bot-

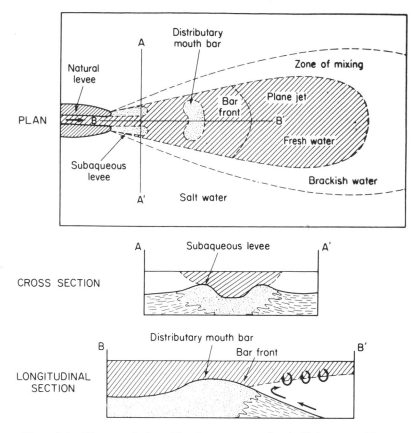

Figure 9-2 River mouth depositional patterns associated with buoyant effluents. (After Wright, 1977.)

tom with significant frictional effects (Wright, 1977, 1978). This is also a type of homopycnal flow, but requires only that density contrasts be eliminated by marine processes (e.g., tides, waves, currents).

As outflow occurs there is shoaling, but in this situation the lateral spreading is enhanced by the friction-induced deceleration. This in turn increases the shoaling rate and results in a divergence or bifurcation of the outflow, causing formation of a middle ground bar (Figure 9-4).

Buoyant outflow

Most of the major rivers flow into marine waters. The density of fresh water is essentially 1.00 g/cm^3, whereas that of seawater is typically 1.026 to 1.028 g/cm^3. Even though these rivers carry a significant sediment load in suspension, the density of the sediment-laden water is rarely at or above the density of seawater. The result is that the river outflow "floats" on the ambient seawater due to the density contrast. This is called **hypopycnal** flow. Strong tidal influence and mixing caused by wave or current action may decrease this buoyant effect so that it is dominated by inertial or

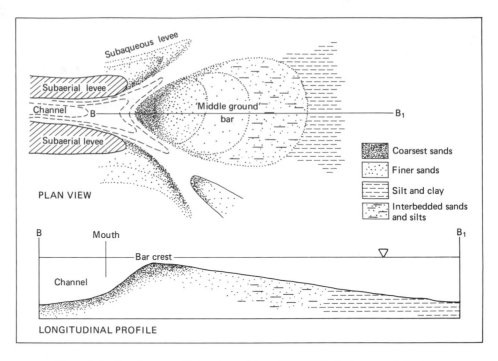

Figure 9-3 Development of plane jet and deposition of distributary-mouth bar in a delta. (After Morgan, 1970, and Wright, 1977; Figure 19–25 of Blatt, Middleton, and Murray, 1980.)

bottom friction flow (Wright, 1977). In rivers with high discharge rates or during flood stages of some rivers when outflow is strong, the buoyant effect is prominant. Rivers that experience small tides typically are characterized by a stratified circulation. Such is the case for the Mississippi, the Danube (Romania), and the Po (Italy) Rivers (Wright, 1978).

Convergence of flow near the bottom results in sediment accumulation in the form of straight subaqueous levees (Figure 9-2). A distributary-mouth bar accumulates about four to six channel widths to seaward. Due to flow convergence, the coarsest particles are restricted to a narrow region (Wright, 1977). As the distributary-mouth area prograzes, these bar sands develop into the bar-finger sands described by Fisk (1961). Particle size decreases seaward down the distal front of the bar (Figure 9-2).

Marine Processes

In virtually all river mouth environments there is a complicated interaction between the riverine processes described above and those of the basin into which the river debouches. Although here these are called marine processes, with the exception of tides they also apply to rivers entering nonmarine water bodies.

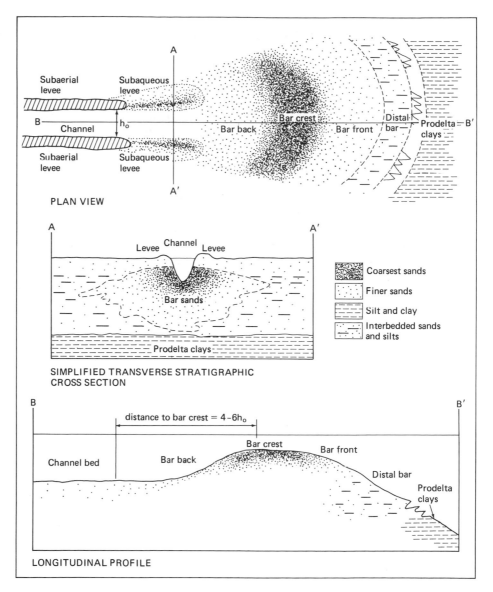

PLAN VIEW

SIMPLIFIED TRANSVERSE STRATIGRAPHIC
CROSS SECTION

LONGITUDINAL PROFILE

Figure 9-4 Channel bifurcation and middle-ground bar typical of friction dominated river mouth system. (After Wright, 1977.)

Tides

Several of the major deltas in the world are either strongly influenced or are dominated by tidal activity. Although only the Amazon falls into that category for the western hemisphere, the Ganges-Brahmaputra (Bangladesh) and the Ord (Australia) are but two of several deltas that experience severe modification by tides.

The role of tides in river mouth processes is essentially threefold:

1. Mixing destroys density gradients and negates effects of buoyancy.
2. During low discharge periods tides may be the dominant sedimentation processes.
3. The zone of marine-riverine interaction is extended in range, especially horizontally (Wright, 1977).

Tidal dominance is marked by a distinctly bidirectional nature to the processes, which is reflected in the sediment bodies accumulating seaward of the river mouth. The most common morphology is a system of linear shoals with their long axes parallel to flow (Figure 9–5). These ridges have been described by Off (1963) from several locations and by Wright et al. (1975) from the Ord Delta in Australia.

In tide-dominated rivers the channel widens markedly near the mouth, forming a bell-shaped configuration (Figure 9–5). Because of the relatively strong tidal currents, this zone is generally sand filled. Marked meandering with related point-bar deposits is common upstream from the bell-shaped channel near the mouth (Wright, 1977).

Waves

Typically, deltas are not well developed or are not present at all along coasts with a rigorous wave climate. There is a general relationship between broad, gently sloping shelves, low wave energy, and well-developed deltas. Some deltas do, however, develop on steep nearshore slopes under high-wave-energy conditions.

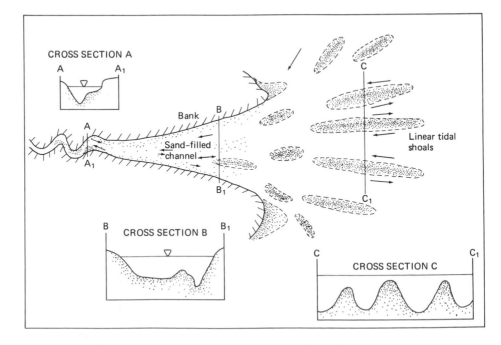

Figure 9–5 Bell shaped river mouth with associated linear tidal shoals characteristic of tide-dominated coasts. (After Wright, 1977.)

As waves approach and impact on the active delta they cause a spreading and deceleration of the effluent. Riverine outflow interacts with waves, causing them to break in abnormally deep water (Wright, 1977). This has the effect of abruptly decreasing the competence of the outflow and causes sediment to accumulate closer to the river mouth than under low-wave-energy conditions. This sediment accumulation is in the form of a crescentic river mouth bar and subaqueous levees upon which **swash bars** develop (Figure 9-6). This river mouth sediment body configuration is typical for waves that approach normal to the shore (Wright, 1977, 1978).

Coastal currents

Although shallow currents may be generated by wind, waves, tides, or deep ocean currents impinging on the continental margin, their net effect on deltaic sedimentation is generally to modify river mouth morphology in a shore-parallel direction (Coleman, 1976). These currents typically display an alongshore component, and it is this aspect that markedly affects river mouth sedimentation and will be considered in this discussion. Undoubtedly, the most prominent of these currents is the wave-generated longshore current which results from refraction in shallow water (see Chapter 2).

Even though there may be a wide range in the direction of these currents through time at a given location, there is typically a dominant direction. It is this current direction that is responsible for the river mouth morphology. The most common modification is the formation of a spit and accompanying channel mouth migration (Figure 9-7). In the example illustrated, waves have generated longshore currents that flow from the top to the bottom of the diagram, causing the river mouth and the sand spit to migrate toward the bottom of the diagram. Note that the basic sediment body elements are the same as in the wave-dominated river mouth, where no appreciable longshore currents exist (Figure 9-6).

DELTAIC SEDIMENTS

Deltas commonly exhibit numerous sedimentary environments within the depositional system. Each of these is characterized by its own assemblage of sediments and sedimentary structures. By far the most complex and diverse of the major portions of the delta is the upper surface or delta plain. Here there is a gradual transition from the continental fluvial environment (see Chapter 8) to the river mouth delta with its marine influences. As one proceeds seaward to the delta-front and prodelta portions of the system, there is a relatively simplistic and uniform pattern of deposition.

Delta Plain

The upper surface of the delta complex is characterized by numerous sedimentary environments; some are dominantly subaerial, some dominantly subaqueous, and some are at least partially intertidal. Included are (1) distributary channels and their associated natural levees, point bars, and crevasse splays, all of which mimic the fluvial system; and (2) interdistributary deposits (bays and marshes). The delta plain

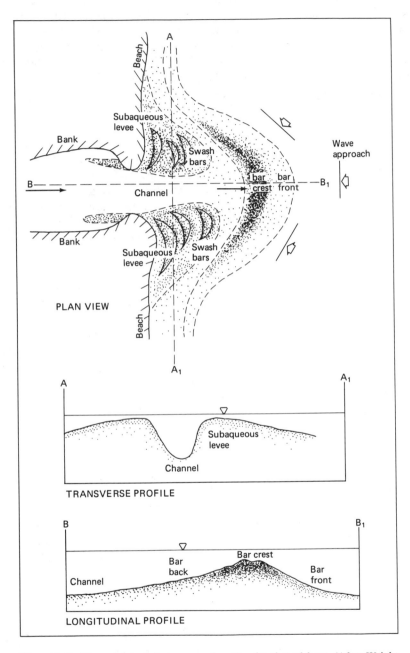

Figure 9–6 Wave-dominated river mouth with related sand bars. (After Wright, 1977.)

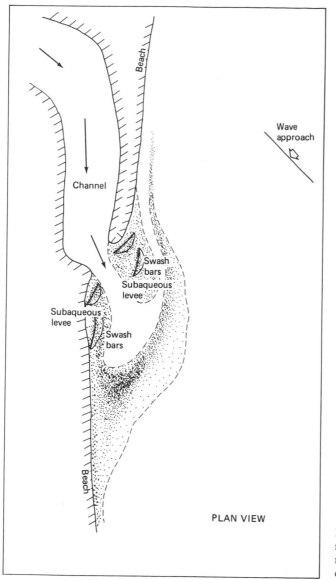

PLAN VIEW

Figure 9-7 Depositional patterns at river mouth dominated by longshore currents generated by oblique-wave incidence. (After Wright, 1977.)

is subjected to a wide range of energy conditions and physical processes, which vary through both space and time. Extremes are during river flood stages, when riverine processes dominate, and during periods of onshore-moving storms, when marine processes dominate.

Distributary channels

The various processes described in the preceding section give rise to the development of generally numerous distributaries on the delta plain. These channels serve as the primary paths for discharge of sediment-laden water into the basin. As such they experience similar processes and deposit similar sediments in accumulations much like that of their counterparts that do not occupy a river mouth area. Distributary channels are the site of deposition for most of the coarsest sediment that is carried to the delta. This discussion is abbreviated to avoid repeating that of the analogous environments considered in Chapter 8.

Although distributary channels typically show little tendency to migrate laterally and form point bars (Coleman, 1976), such situations do exist. The result is development of point-bar sands which show cross-bedding oriented downstream or generally toward the basin, except in tide-dominated situations, where bidirectional cross-stratification is common (Wright, 1978). These sands also display ripple cross-stratification, scour and fill structures, and may include thin, discontinuous clay lenses. The sand bodies are small and discontinuous throughout the delta plain. Distorted bedding due to slumping is also common (Coleman et al., 1974).

Abandoned channels fill slowly with a mixture of silt, clay, and organic material over a scoured sand base (Wright, 1978).

Distributary mouth bars are formed mostly of sand with some silt which is complexly cross-stratified in response to the many processes operating in this location on the delta plain. Some organic debris is generally trapped in the rapidly accumulating sediment but tends to oxidize, releasing gases and producing small gas-heave structures (Coleman et al., 1964).

Natural levees may be subaerial or subaqueous, with the latter associated with the mouth of distributaries (Figure 9–4). Subaerial levees extend throughout most or all of the delta plain and reflect varying river stages. Sediments that comprise subaqueous levees are fine sand with some clay or organic debris as laminae (Coleman, 1976). Cross-stratification is present and shows a wide range of orientations due to complex wave and current patterns (Reineck and Singh, 1980; Wright, 1978).

Although similar in general morphology, subaerial levees form differently than subaqueous levees and contain a contrasting sediment sequence. They result from deposition out of suspension during flood conditions and are therefore finer grained than their subaqueous counterparts. These linear fining-upward sediment accumulations are characterized by small ripple cross-stratification and disrupted laminations due to disturbance by vegetation.

Interdistributary bays and marshes

Although less complex than the distributary channel environments, the interdistributary regions comprise the bulk of the area on the deltaic plain. By way of contrast, it represents the lowest-velocity environment on the upper delta surface except for abandoned channels. The interdistributary areas are shallow, nearly devoid of relief, and accumulate sediment slowly. Depths range from a few meters in the open bays to supratidal in some of the marsh areas. Sediment is introduced primarily as the result of flooding by the river or by storm activity (Coleman, 1976).

Most of the sediment is silt and clay with scattered shell debris from shallow-water benthic organisms. Local and thin sand or shell lenses may be present as the result of reworking by storm conditions (Wright, 1978). Parallel laminations and burrow mottling are widespread.

The nature and distribution of interdistributary areas is largely tied to the tidal

Figure 9-8 Oblique aerial photograph of delta plain on the Mississippi Delta, showing interdistributary bays and marshes. (Courtesy of J. Coleman.)

regime on the delta. Small tidal ranges result in rather large, open interdistributary bays with marshes developed along the margins, such as on the Mississippi Delta (Figure 9-8). On the other hand, high tidal ranges are associated with broad, unvegetated tidal flats in the interdistributary areas (Figure 9-9), such as on the Ord Delta (Coleman and Wright, 1975).

Events associated with flood conditions cause crevasse splays to develop along

Figure 9-9 Photograph of extensive tidal flats on the delta plain of the Ord Delta, Australia. (Courtesy of J. Coleman.)

distributary channels and provide a mechanism for supplying interdistributary bays with relatively large pulses of sediment. These splays are also analogous to those in the fluvial environment but may be extremely large on the broad, flat delta. They sometimes represent small deltas in that they have a small distributary system (Figure 9–10) and may be repeatedly activated (Coleman and Gagliano, 1964; Coleman, 1976).

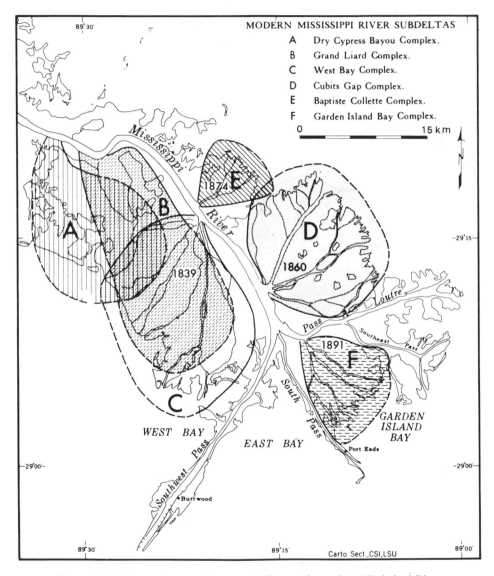

Figure 9-10 Large crevasse-splays or subdeltas on the modern Mississippi River delta. (After Coleman, 1976.)

Mudlumps

Diapiric masses of deformed muds have intruded the bar-finger sands (Fisk et al., 1954) on the Mississippi Delta. They form as the result of thick overburden on top of thick and unstable muds. Some of these clayey masses show no surface expression, whereas many break through the bar sands and may rise significantly above sea level. The mudlumps are proximal to the mouths of Mississippi River passes, where they have been known for more than a century. As yet they are unreported from other deltas.

The mudlump islands generally encompass less than a square kilometer, are elongate, and are sinuous (Morgan, 1951). The clayey sediments are typically inter-bedded with silt and are relatively wave resistant. The mudlumps rise 3 or 4 m above sea level and may display wave-cut cliffs along the margins. The outer margins are highest, with step faults and related subsidence in the central portion of the feature (Morgan, 1951).

An excellent summary of the sediments and sedimentary structures of several delta plain environments on the Mississippi Delta was compiled by Coleman et al. (1964) and is presented in Figure 9–11.

Delta Front

The seaward portion of the topset delta in the shallow subtidal zone is generally called the delta front or delta-front platform (Allen, 1970b). This zone is typically no more than a few kilometers wide and extends to a depth of about 10 m or less. It is characterized by sand-size sediment with grain size decreasing seaward, where some mud

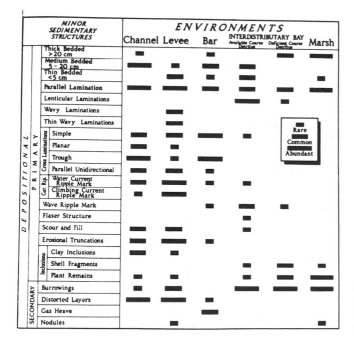

Figure 9–11 Summary chart of sedimentary structures on various delta plain environments. (From Coleman et al., 1964.)

is mixed with the sand. This zone is dominated by marine processes such as waves and longshore currents, with tidal influence important along coasts with high tidal range. The delta front is primarily a sheet sand accumulation with local sandbars (Wright, 1978). In many respects it is analogous to **shoreface** deposits of the inner continental shelf areas, where deltas are absent.

Sediment brought to the river mouth and debouched into the basin is typically entrained by waves and longshore currents which distribute it in a shore-parallel fashion. The relatively high energy causes sorting of the sediment and winnowing away of the fines. These sands display cross-stratification in a variety of scales and orientations, with only minor zones of bioturbation (Allen, 1970b). The latter result from periods of quiescense and a general slowing of rates of deposition. Except for crests of river mouth bars and beaches developed on the delta, the delta-front environment represents the highest-energy environment of the delta (Allen, 1970b).

Although progradation seaward is common, there are times when sediment supply is low and places where sediments are scarce along the delta front. Under these circumstances the delta-front sands are commonly reworked into low-lying beach ridges along the outer subaerial margin of the delta (Figure 9–12). This is typical of the destructional phase of deltaic sedimentation.

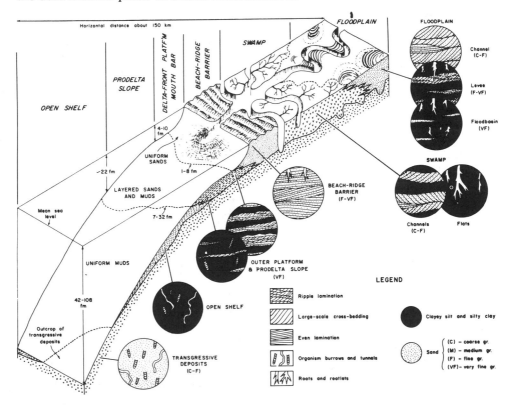

Figure 9–12 Schematic of typical section across the Niger Delta and adjacent marine shelf, showing sediment types and distribution as well as morphology. (From Allen, 1970b.)

Elongate or crescentic sand bodies are common on the delta front. Distributary-mouth bars actually occupy the delta-front region but are dominated by riverine processes; they were considered in the preceding section. Under macrotidal conditions, elongate tidal sand ridges (Figure 9–2) replace distributary mouth bars in the upper delta-front area. These features are composed primarily of bidirectional cross-bedded sands (Wright et al., 1975). **Distal bars** are located near the outer limits of the delta front and consist of interbedded fine sand and mud (Wright, 1978). They may be several meters high and are characterized by stair-step features which have been determined to be rotational slump blocks. Coleman et al. (1974) have concluded that these slump blocks are due to oversteeping caused by high sedimentation rates.

Prodelta

The most widespread and homogeneous of the delta environments is the prodelta. It represents the bulk of the volume of deltaic sediments and is dominated by silt and clay, with the latter being dominant. Prodelta sediments fine seaward and grade into shelf muds. Some deltas also contain thin sand beds in the prodelta muds (Figure 9–12). Deposition is from suspension and gives rise to thin laminations which may not be visible to the naked eye but show up well on **x-radiographs.** Burrowing organisms may destroy these laminations when sedimentation rates are low and benthic organisms colonize this environment. Shells and tests scattered throughout the prodelta muds are of normal marine organisms, indicating the absence of fresh-water influence in this environment.

Some convolutions in bedding and small-scale, soft-sediment faulting suggest slope instability. Pods of sand may be transported via the previously mentioned slumping and are incorporated in the shallow prodelta muds (Wright, 1978).

DELTA MORPHOLOGY

Deltas exhibit a wide variety of areal configurations, depending on the interrelationships between sediment supply and the major processes operating in this dynamic environment. Although this discussion is concerned primarily with the constructional aspects of deltaic sedimentation, much of the overall delta morphology is also due to destructional processes. Primary effects that must be considered are the riverine processes and their related sediment load, wave energy flux, longshore currents, and tidal energy flux. Succinctly put, delta morphology is a response to the interaction of the processes acting in the river mouth region.

Classification of Deltas

An uncomplicated, yet meaningful simple model of the process-response mechanisms operating along deltaic coasts is that presented by A. J. Scott (Fisher et al., 1969). He shows three basic delta morphologies which are the result of the interactions between riverine and marine processes (exclusive of tides). Dominance by riverine processes results in an elongate delta with distinct digitate features related to the major distributaries (Figure 9–13). An excellent example of such a delta is the modern Mississippi Delta. At the opposite extreme, where wave energy and long-shore currents dominate, are the cuspate deltas, characterized by a lack of major distributaries and a series of arcuate strandplain ridges (Figure 9–13). The São Francisco Delta of Brazil is an example. An intermediate form, which has a well-developed network of distributaries but a generally smooth outline, is the lobate delta, exemplified by the Niger Delta in Africa (see Allen, 1970b).

Scott's model can be subdivided into essentially two major types of deltas: constructive and destructive. Constructive delta systems are river dominated, contain a large and well-developed drainage system which supplies a large volume of sediment, and are characterized by distinct phases of construction and destruction (Fisher et al., 1969). Sediments are dominated by mud-size particles. Such deltas are common on broad, shallow shelves where wave activity is low. Destructive phases are limited to distal portions of the delta and are secondary in importance.

Highly destructive deltas are divided into wave-dominated and tide-dominated varieties. Wave-dominated deltas may service only a small local, drainage basin with sporadic input of only a moderate sediment volume (Fisher et al., 1969). Constructional phases are subordinate, with wave energy reworking sediments and forming a smoothed outline. A relatively high percentage of sand is present due to removal of fine particles by waves. Well-developed beaches and strandplains paralleling the delta outline are common (Figure 9–14).

Tide-dominated deltas contain an abundance of linear features oriented about perpendicular to the overall trend of the shoreline and parallel to tidal flow. Sand and distinct sand shoals are common. In some respects this general sediment type is similar to wave-dominated deltas, but the sand bodies are arranged differently, creating a markedly distinct general morphology (Figure 9–14).

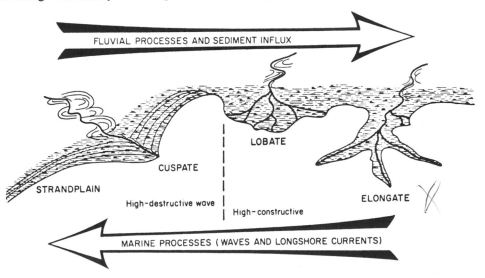

Arrows point in direction of increasing influence.

Figure 9–13 Relative influence of fluvial and marine processes on general configuration of deltas. (From Scott, 1969, in Fisher et al., 1969.)

Part Three   Transitional Environments

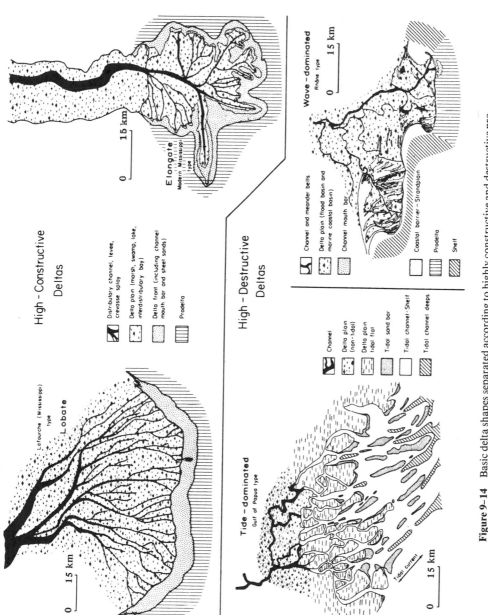

High – Constructive
Deltas

Lobate
Lafourche (Mississippi)
type

Elongate
Modern Mississippi
type

0   15 km

Distributary channel, levee,
crevasse splay

Delta plain (marsh, swamp, lake,
interdistributary bay)

Delta front (including channel
mouth bar and sheet sands)

Prodelta

High – Destructive
Deltas

Tide – dominated
Gulf of Papua type

Wave – dominated
Rhône type

Channel and meander belts

Delta plain (flood basin and
marine coastal basin)

Channel mouth bar

Coastal barrier – Strandplain

Prodelta

Shelf

Channel

Delta plain
(non-tidal)

Delta plain
tidal flat

Tidal sand bar

Tidal channel - Shelf

Tidal channel deeps

Tidal Current

0   15 km

Figure 9–14 Basic delta shapes separated according to highly constructive and destructive processes. (From Fisher et al., 1969.)

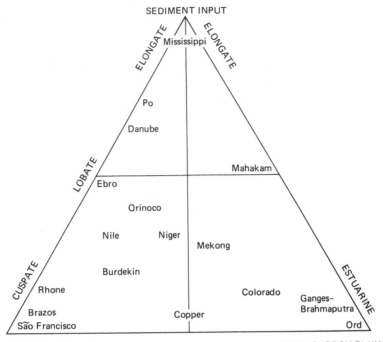

Figure 9–15 Classification of deltas based on processes and morphologic response. (After Galloway, 1975.)

A good morphological classification of deltas was presented by Galloway (1975) incorporating some of the initial concepts of Fisher et al. (1969). Using a ternary diagram with sediment yield (riverine processes), wave energy, and tidal energy as end members, he classified many of the world's major deltas as to shape (Figure 9–15, Table 9–1). Such a ternary approach enables one to visualize the

TABLE 9-1 CHARACTERISTICS OF DELTAIC DEPOSITIONAL SYSTEMS

	Fluvial-dominated	Wave-dominated	Tide-dominated
Geometry	Elongate to lobate	Arcuate	Estuarine to irregular
Channel type	Straight to sinuous distributaries	Meandering distributaries	Flaring straight to sinuous distributaries
Bulk composition	Muddy to mixed	Sandy	Variable
Framework facies	Distributary mouth bar and channel fill sands, delta margin sand sheet	Coastal barrier and beach ridge sands	Estuary fill and tidal sand ridges
Framework orientation	Parallels depositional slope	Parallels depositional strike	Parallels depositional slope

SOURCE: Galloway, 1975.

relative significance of each of the three major processes by examining the various examples illustrated on the classification scheme.

By adding littoral drift and the offshore slope to the process variables noted above, Coleman and Wright (1973) have presented six major delta morphologies which characterize virtually all of the world's major deltas. In addition to the incorporation of additional variables, this scheme shows the relative size, location, and geometry of sand bodies incorporated into the models. Because of the important role played by deltas as sites for the accumulation of petroleum, this classification is of great value to the exploration geologist in the search for oil and gas.

In this classification the river-dominated variety (Type I) corresponds closely to the morphology presented by Fisher et al. (1969) and also by Galloway (1975). Each of the subsequent delta types is based on variations in the aforementioned variables. The size and orientation of the sand bodies is related largely to the intensity and orientation of marine processes (Figure 9–16). For example, the combination of high wave energy and strong littoral drift gives rise to large sand bodies which parallel the trend of the coast.

SOME MODERN DELTAS

The previous sections of this chapter considered various elements of the modern deltaic environment. To grasp fully the nature of the complex delta system, one must consider not only the interrelationships between the many elements as they presently exist, but must consider how changes through time are incorporated into the three-dimensional prism of deltaic sediments. For purposes of discussion the modern deltaic environment will be extended back through Holocene time.

Each of a few examples of modern deltas is discussed briefly to provide the reader with a modest spectrum of the great variety of deltas that presently exists. Such discussions will provide a basis for recognizing ancient analogs of modern deltas.

Mississippi Delta

In addition to being among the largest of the world's deltas, the Mississippi Delta is probably the most intensely studied. A glance at the significant literature on deltas shows that most geological studies of this environment prior to the early 1960s were on the Mississippi (LeBlanc, 1975). Among these are the pioneering efforts of H. N. Fisk and his colleagues (1944, 1947, 1955, 1960, 1961). More recent important work has been done by J. M. Coleman and his associates at the Coastal Studies Institute, Louisiana State University (Coleman et al., 1964; Coleman and Gagliano, 1964; Coleman and Wright, 1971).

Seven discernible lobes of the Mississippi Delta have been active during the past 5500 years, with the present active lobe being about 600 years old (Kolb and Van Lopik, 1966). The combination of rapidly migrating depositional environments coupled with a tremendous rate of sediment influx has resulted in a thick and diverse wedge of sediment which forms the modern delta. Considering a relatively simplified

1. Low wave energy; low littoral drift; high suspended load

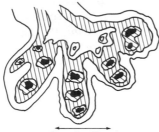

Shoreline Trend

2. Low wave energy; low littoral drift; high tide

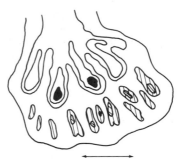

3. Intermediate wave energy; high tide; low littoral drift

4. Intermediate wave energy; low tide

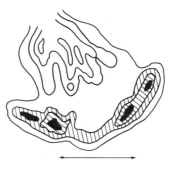

5. High wave energy; low littoral drift; steep offshore slope

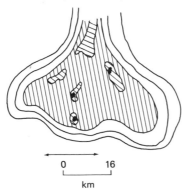

0   16
km

6. High wave energy; high littoral drift; steep offshore slope

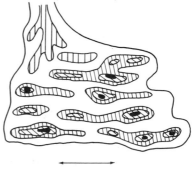

Figure 9–16 Classification of deltas based on dominant processes and showing sand distribution. (From Coleman and Wright, 1973.)

300

diagram of this delta, there are several sedimentary facies present (Figure 9–17). Pro-delta muds are overlain by delta front silts and sands which are locally interrupted by linear, bar-finger sands (Fisk et al., 1954). This schematic representation is rather generalized, in that such environments as natural levees, crevasse splays, and marsh deposits are not depicted stratigraphically.

A similar but more comprehensive representation has been produced by Frazier and Osanik (1969). This sequence of diagrams (Figure 9–18) shows the development of the active delta with its contained sedimentary environments and facies. The Mississippi Delta represents a classic example of the elongate, fluvial-dominated delta (Figure 9–14). It falls into the type I model (Figure 9–16) of Coleman and Wright (1973).

Because most geologists work with only stratigraphic sections, it is appropriate that consideration be given to such a section for the Mississippi Delta. A generalized sequence from Scruton (1960) shows transition and bottomset beds of sand and clay overlain by a thick sequence of prodelta muds (foreset beds). Topset beds are represented by delta front silt and sand with some organic-rich marsh deposits, all of which are cut by sandy channel deposits (Figure 9–19).

An idealized but more detailed stratigraphic column illustrates 10 distinct environments and their corresponding facies (Figure 9–20). Although the probability of encountering the entire sequence in any single core is low (Wright, 1978), the order and relative thicknesses of the facies is typical of the Mississippi Delta.

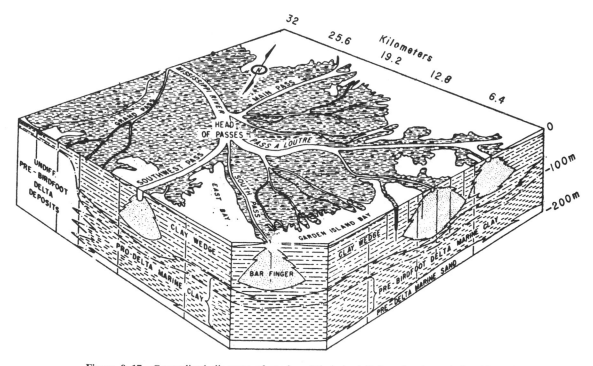

Figure 9–17 Generalized diagram of modern Mississippi Delta, showing relationships of sedimentary facies. (Simplified from Fisk et al., 1954.)

A. INITIAL PROGRADATION

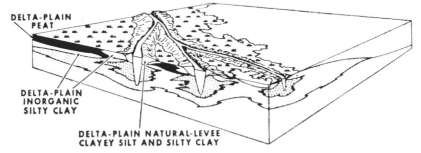

NATURAL LEVEE

FRESH-WATER
STREAM-MOUTH MARSH

DELTA-FRONT
DISTRIBUTARY-MOUTH-BAR
SILTY SAND

DELTA-FRONT
SILTY SAND
AND SILTY CLAY

PRODELTA
SILTY CLAY

B. ENLARGEMENT BY FURTHER PROGRADATION

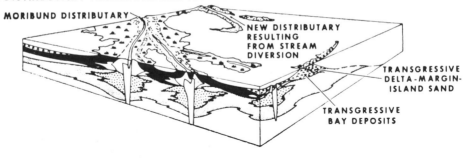

DELTA-PLAIN
PEAT

DELTA-PLAIN
INORGANIC
SILTY CLAY

DELTA-PLAIN NATURAL-LEVEE
CLAYEY SILT AND SILTY CLAY

C. DISTRIBUTARY ABANDONMENT AND TRANSGRESSION

MORIBUND DISTRIBUTARY

NEW DISTRIBUTARY
RESULTING
FROM STREAM
DIVERSION

TRANSGRESSIVE
DELTA-MARGIN-
ISLAND SAND

TRANSGRESSIVE
BAY DEPOSITS

D. REPETITION OF CYCLE

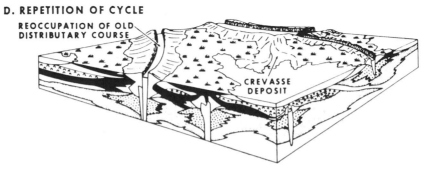

REOCCUPATION OF OLD
DISTRIBUTARY COURSE

CREVASSE
DEPOSIT

Figure 9–18 Sequence of delta development and associated facies. (From Frazier and Osanik, 1969.)

302

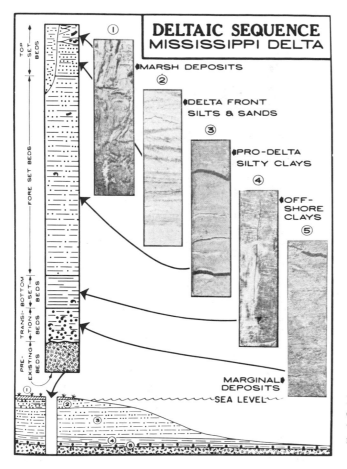

Figure 9-19 Generalized stratigraphic column through the Mississippi Delta with representative cores from the respective environments. (From Scruton, 1960.)

Niger Delta

Another well-known delta of a contrasting type to the Mississippi is the Niger Delta on the west-central coast of Africa along the Gulf of Guinea. The Niger is significantly smaller than the Mississippi and shows only about half the discharge. It experiences a mean tidal range of about 1.5 m and a modest wave climate. Both of the latter characteristics are considerably higher than the Mississippi (Coleman, 1976). As a consequence the Niger is about intermediate between the type III and IV categories (Figure 9-16). In the classification of Galloway it falls near the middle of the ternary classification (Figure 9-15). Briefly summarized, the Niger is about equally balanced between wave and tidal domination, with significant influence from the river. This gives rise to a rather smooth outline with several subequally spaced distributaries (Figure 9-21).

Extensive research on the Niger Delta and vicinity has been carried out by J. R. L. Allen (1963, 1964a, 1965a, 1970b), whose work has provided the data for the following brief discussion.

Prominent meandering of the many distributaries has resulted in extensive

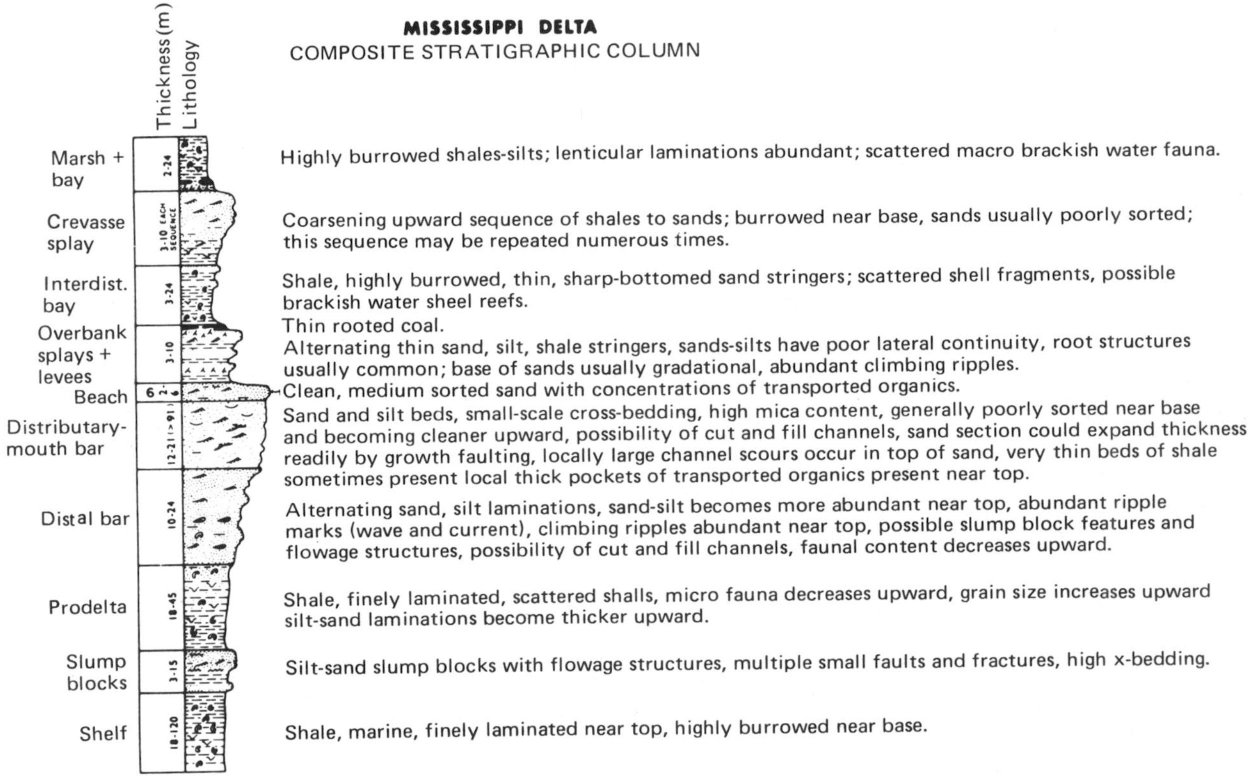

MISSISSIPPI DELTA
COMPOSITE STRATIGRAPHIC COLUMN

	Thickness (m)	Lithology	Description
Marsh + bay	2-24		Highly burrowed shales-silts; lenticular laminations abundant; scattered macro brackish water fauna.
Crevasse splay	3-10 EACH SEQUENCE		Coarsening upward sequence of shales to sands; burrowed near base, sands usually poorly sorted; this sequence may be repeated numerous times.
Interdist. bay	3-24		Shale, highly burrowed, thin, sharp-bottomed sand stringers; scattered shell fragments, possible brackish water sheet reefs.
Overbank splays + levees	3-10		Thin rooted coal. Alternating thin sand, silt, shale stringers, sands-silts have poor lateral continuity, root structures usually common; base of sands usually gradational, abundant climbing ripples.
Beach	2-6		Clean, medium sorted sand with concentrations of transported organics.
Distributary-mouth bar	(16<)12-21(>16)		Sand and silt beds, small-scale cross-bedding, high mica content, generally poorly sorted near base and becoming cleaner upward, possibility of cut and fill channels, sand section could expand thickness readily by growth faulting, locally large channel scours occur in top of sand, very thin beds of shale sometimes present local thick pockets of transported organics present near top.
Distal bar	10-24		Alternating sand, silt laminations, sand-silt becomes more abundant near top, abundant ripple marks (wave and current), climbing ripples abundant near top, possible slump block features and flowage structures, possibility of cut and fill channels, faunal content decreases upward.
Prodelta	18-45		Shale, finely laminated, scattered shells, micro fauna decreases upward, grain size increases upward silt-sand laminations become thicker upward.
Slump blocks	3-15		Silt-sand slump blocks with flowage structures, multiple small faults and fractures, high x-bedding.
Shelf	18-120		Shale, marine, finely laminated near top, highly burrowed near base.

Figure 9-20 Idealized stratigraphic column of the Mississippi Delta sequence. (From Coleman and Wright, 1975.)

304

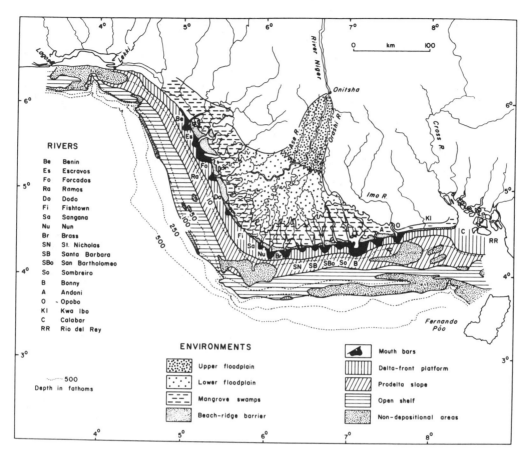

RIVERS

Be	Benin
Es	Escravos
Fo	Forcados
Ra	Ramos
Do	Dodo
Fi	Fishtown
Sa	Sangana
Nu	Nun
Br	Brass
SN	St. Nicholas
SB	Santa Barbara
SBo	San Bartholomeo
So	Sombreiro
B	Bonny
A	Andoni
O	Opobo
KI	Kwa Ibo
C	Calabar
RR	Rio del Rey

-----500
Depth in fathoms

ENVIRONMENTS

Upper floodplain

Lower floodplain

Mangrove swamps

Beach-ridge barrier

Mouth bars

Delta-front platform

Prodelta slope

Open shelf

Non-depositional areas

Figure 9–21 Major sedimentary environments of the Niger Delta. (From Allen, 1970b.)

swamp and channel deposits at the expense of the interdistributary bay environment, which is essentially absent. The wave climate provides sufficient energy to develop a rather pronounced beach–ridge barrier complex seaward of which are prominent distributary-mouth bars (Figure 9–21). From the outer delta-front platform to the open shelf there is the typical decrease in sand content and an increase in bioturbated marine muds.

A typical stratigraphic section of the Niger Delta would show older sands successively overlain by open shelf facies, prodelta muds, delta platform deposits including bar-mouth facies, beach-barrier facies, and swamp facies, culminating in floodplain deposits of the riverine portion of the delta (Allen, 1963).

São Francisco Delta

The São Francisco Delta of Brazil represents a classic example of a wave-dominated delta (Figures 9–13 and 9–15) and falls into the Type V category of Coleman and Wright (1973) (Figure 9–16). It is small in area and low in discharge compared to both the Mississippi and Niger. Although tidal range is near 2 m, the dominant process is the waves; it is subjected to about 100 times the wave power of the Mississippi and about 15 times that of the Niger (Coleman, 1976).

The result is a delta dominated by sand accumulations and high-energy environments. Muds are moved away from the delta by waves and wave-generated currents. Marsh and floodplain deposits are local and interdistributary bays are absent (Figure 9–22). The dominant sedimentary facies is the beach-ridge barrier sand, which covers most of the modern delta (Wright, 1978). The outer margin of the delta plain contains a well-developed dune system (Figure 9–22), unusual for a deltaic system.

An ideal stratigraphic section through the São Francisco Delta would consist of a basal muddy and bioturbated shelf-prodelta unit overlain by bar sands, beach ridges, fine-grained swale fill, dune sand, and capped by tidal plain deposits (Figure 9–23).

Ganges-Brahmaputra Delta

Much of the coast of Bangladesh is occupied by the Ganges-Brahmaputra Delta complex. This very large delta system represents a good example of the tide-dominated delta (Figure 9–15) or the Type II delta (Figure 9–16). In addition to being very large, the Ganges-Brahmaputra has a discharge twice that of the Mississippi, low wave

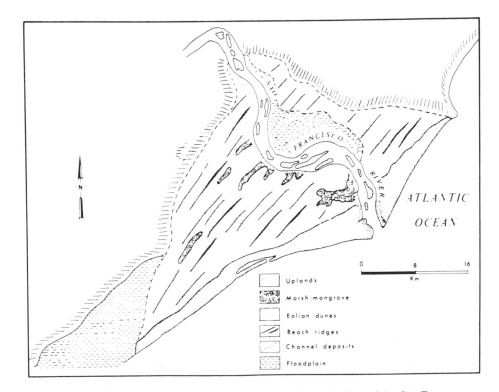

Figure 9-22 Major sedimentary environments and morphology of the São Francisco Delta, Brazil, a wave-dominated delta. (From Coleman and Wright, 1973.)

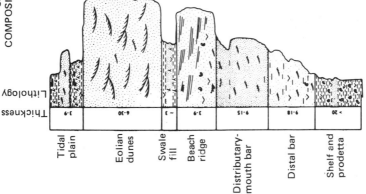

SÃO FRANCISCO DELTA

COMPOSITE STRATIGRAPHIC COLUMN

Lithology

Thickness

Tidal plain — 3-9

Alternating shale, silt, and sand layers, shale highly burrowed; sands thin and display scoured bases; small-scale x-strat, common in silts and sands; occasional thin peat stringers.

Eolian dunes — 6-30

Clean, well-sorted sands displaying large-scale festoon x-bedding; small-scale tangential x-bedding common, occasional distorted sand layer; high rooting on top of sand.

Swale fill — ~1

Poorly sorted silts and shales displaying abundant root burrowing.

Beach ridge — 3-6

Well-sorted, clean sands displaying low-angle dips, parallel beds extremely abundant; small-scale x-stratification common near base, some sand beds display excellent sorting; thin parallel layers of heavy minerals; occasional shell beds.

Distributary-mouth bar — 15

Alternating sands and silt with thin occasional shale parting; small-scale x-stratification abundant; occasional scour base on sand beds; grain size increases upward.

Distal bar — 9-18

Alternating silt, sand, and shale layers; sand layers display graded bases; small scale x-stratification common in sands and silts; scattered shells; lenticular shale lenses near top.

Shelf and prodelta — >20

Marine shale with thin silt and sand stringers; highly burrowed near base; scattered shell; shale layers thin upward.

Figure 9–23 Idealized stratigraphic column for the São Francisco Delta. (From Coleman and Wright, 1975.)

307

energy, and a mean tidal range near 4 m. This gives rise to rather striking morphology and sediment accumulations.

The definitive work on this delta is by Coleman (1969), who conducted an extensive investigation of the sediments, structures, and processes of the modern Ganges-Brahmaputra complex.

In many respects this delta is similar to a large braided stream complex that suddenly meets the ocean. During much of the year discharge is not great, as compared to the monsoon season when extreme floods are commonplace. The result is a braided stream type of situation with the large sand accumulations mobilized during these dramatic events. Natural levees are large and widespread, as are tidal flat environments. In place of the interdistributary bay environment we find shallow flood basins which are innundated during high discharge periods; water is trapped and fine sediments settle out of suspension. Tidal influence is dramatic in that it produces large, widespread tidal sand ridges. In addition, tidal influence extends up the rivers during periods of low discharge and generates distinctly bimodal cross-bedding orientations.

A generalized stratigraphic sequence from such a delta would include basal shelf and prodelta sandy muds overlain by distal bars and tidal ridges with channel filling, natural levees, splays and tidal flats, and flood basin muds. Out of context such a sequence would be rather similar to the braided stream sequence described in Chapter 8.

DELTAS IN THE ROCK RECORD

Recognition of ancient deltaic accumulations from the stratigraphic record is, and has long been, a major effort of the petroleum geologist who is engaged in exploration activities. Although the deltaic system is complex and contains many environments, it is rather easily recognized in sedimentary rock sequences. Stratigraphic associations are such that proper recognition of deltaic sequences is readily possible through cores or perhaps even using discontinuous samples such as cuttings. A recent paper describes statistical procedures for recognizing various deltaic environments from textural and mineralogical parameters (Ethridge et al., 1975).

A lengthy discussion of criteria for recognizing ancient deltas appeared in 1912 (Barrell, 1912), a time when little work of any kind had been done on the details of deltaic sedimentation. Barrell provided the first synthesis of important factors in deltaic sedimentation using Gilbert's (1885) delta as a conceptual model. It is important to observe that he recognized that other delta types existed. Barrell also realized the value of stratigraphic relationships between deltaic strata and sediments which accumulated in adjacent or related environments. The latter is quite an important point because of the possibility of deltaic development in nonmarine basins as well as along the marine coast.

Deltas through Geologic Time

Although the best known deltaic sequences from the rock record are those of the Pennsylvanian and Tertiary Systems, deltas have been recognized from nearly all portions of the rock record. The Taconic uplift gave rise to a series of coalescing deltas in the Ordovician of the central Appalachians (Horowitz, 1966). Devonian deltas from England (Allen, 1962) and the Appalachians (Barrell, 1913, 1914; Friedman and Johnson, 1966; Humphreys and Friedman, 1975) are somewhat similar and related to tectonic events in the source areas. A Late Devonian–Early Mississippian delta from the southeastern Appalachians Basin has been described by Walls (1975).

Late Paleozoic deltas are widespread and are of great economic importance because of their coal accumulations, especially in the Pennsylvanian of the United States and the Carboniferous of Europe. Details of a few examples are discussed in the following section.

The Cretaceous is also a period during which deltaic sedimentation was widespread. The literature contains discussions of examples from Brazil (Murphy and Schlanger, 1962), England (Taylor, 1963), and the United States (Pryor, 1960, 1961; Weimer, 1970; Hubert, 1972).

Tertiary deltas are widespread and have received much attention due to their potential for oil and gas. A detailed discussion of some Eocene deltaic accumulations is presented in a following section.

Tertiary Deltas

Tertiary time was characterized by considerable tectonic instability throughout much of the world. Such instability and its related rather rapid relative changes in sea level provide a striking contrast to that of the Pennsylvanian time. One might think, therefore, that deltaic systems would not be developed under such conditions. There are, however, some large and quite thick deltaic sequences from the Eocene of Oregon (Dott, 1966), where there was extensive volcanic activity and rather high relief, and from the Eocene of Texas (Fisher and McGowen, 1967, 1969), where deltas accumulated on a subsiding margin adjacent to a well-developed coastal plain.

Oregon

During Eocene time a narrow and accurate coastal plain bordered a volcanically active upland. This coastal plain terminated seaward in a discontinuous barrier system which was interrupted by deltaic systems (Dott, 1966). The subject of discussion here is the Coaledo Formation, a sequence of terrigenous sands and muds several hundred meters in thickness. Detailed study of this and adjacent units has been completed by Dott (1966), whose interpretations are summarized below.

The Coaledo is comprised of complex interbedding of cross-bedded, convoluted, and horizontal bedded sandstones, many of which show channeling, with mudstones and siltstones. Much compacted bedding, piercement of mudstones by sandstone dikes, and large mudstone clasts characterize the sequence (Figure 9–24). Thin coal beds are also present. In general, the lower part of the unit shows more channeling than is shown in the upper part, where coals are more common (Dott, 1966).

A comparison of this stratigraphic sequence with modern deltas indicates that the lower Coaledo is comparable to delta-front and prodelta environments where accumulation is rapid, with distal bars and river mouth bars being common [Figure

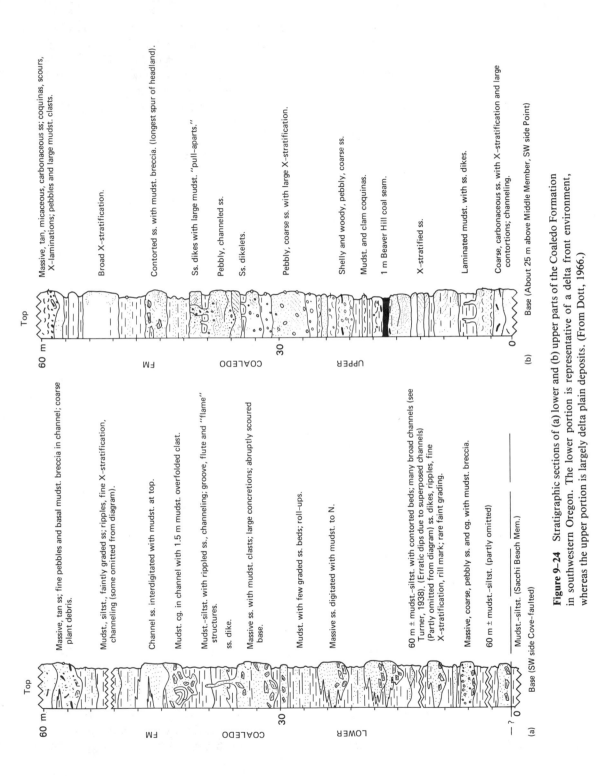

Figure 9–24 Stratigraphic sections of (a) lower and (b) upper parts of the Coaledo Formation in southwestern Oregon. The lower portion is representative of a delta front environment, whereas the upper portion is largely delta plain deposits. (From Dott, 1966.)

(a)

Top

60 m

30

LOWER COALEDO FM

Massive, tan ss; fine pebbles and basal mudst. breccia in channel; coarse plant debris.

Mudst., siltst., faintly graded ss; ripples, fine X-stratification, channeling (some omitted from diagram).

Channel ss. interdigitated with mudst. at top.

Mudst. cg. in channel with 1.5 m mudst. overfolded clast.

Mudst.-siltst. with rippled ss., channeling; groove, flute and "flame" structures.
ss. dike.

Massive ss. with mudst. clasts; large concretions; abruptly scoured base.

Mudst. with few graded ss. beds; roll-ups.

Massive ss. digitated with mudst. to N.

60 m ± mudst.–siltst. with contorted beds; many broad channels (see Turner, 1938). (Erratic dips due to superposed channels) ss. dikes, ripples, fine X-stratification, rill mark; rare faint grading.

Massive, coarse, pebbly ss. and cg. with mudst. breccia.

60 m ± mudst.–siltst. (partly omitted)

———— Mudst.–siltst. (Sacchi Beach Mem.)

Base (SW side Cove-faulted)

(b)

Top

60 m

30

0

FM

COALEDO

UPPER

Massive, tan, micaceous, carbonaceous ss; coquinas, scours, X-laminations; pebbles and large mudst. clasts.

Broad X-stratification.

Contorted ss. with mudst. breccia. (longest spur of headland).

Ss. dikes with large mudst. "pull-aparts."

Pebbly, channeled ss.

Ss. dikelets.

Pebbly, coarse ss. with large X-stratification.

Shelly and woody, pebbly, coarse ss.

Mudst. and clam coquinas.

1 m Beaver Hill coal seam.

X-stratified ss.

Laminated mudst. with ss. dikes.

Coarse, carbonaceous ss. with X-stratification and large contortions; channeling.

Base (About 25 m above Middle Member, SW side Point)

9–24(a)]. The presence of **clastic dikes** and many gravity-generated structures indicates an environment of rapidly accumulating sediments on an unstable surface. Much of the upper Coaledo Formation indicates various delta plain environments with interdistributary bays (laminated muds), marshes (coal), and distributaries (channel sands and overbank deposits) [Figure 9–24(b)].

This deltaic complex was probably lobate in nature and rapidly prograding. Wave and current processes caused development of some beach and barrier sands. Dott (1966) compares it to the modern Nile, Niger, or Orinoco Delta, making it fall between the Type III and Type IV deltas of Coleman and Wright (1975).

Texas

Deltaic systems are numerous in the Tertiary of the Texas Gulf Coast; however, this discussion will center on the Wilcox Group (Eocene), which contains sequences representing both high-constructive deltas and high-destructive deltas (Fisher et al., 1969). Most of this discussion is summarized from the excellent work of W. L. Fisher and J. H. McGowen (Fisher, 1969; Fisher et al., 1969; Fisher and McGowen, 1967, 1969).

As in the case with much of the stratigraphy of the Gulf Coast, most data are from the subsurface, through various types of logs, cores, and cuttings. The first interpretation of deltaic deposition in the Wilcox is attributed to Culbertson (1940), who conducted a regional study. A generalized cross section normal to the depositional strike (dip section) was prepared by Echols and Malkin (1948). This section shows the intertonguing of marine and nonmarine facies but contains no lithologic details of sedimentary facies (Figure 9–25).

Detailed dip sections showing various sedimentary facies were prepared by Fisher and McGowen (1967) for the fluvial-deltaic systems of the Wilcox (Figure 9–26). The Rockdale delta system represents a high-constructive lobate delta system which is characterized by interdistributary bays, channels, and marshes with numerous lignite deposits. Thin destructive units representing wave- and current-worked sandy sediments are interbedded with thick delta-front muddy sands. Seaward of this are thick muds of the prodelta environment (Figure 9–26). A paleogeographic reconstruction of the high-constructive Rockdale delta system is actually quite similar to the present-day Texas Gulf Coast if the barrier island system of the latter is ignored.

In the Upper Wilcox the delta system is of the high-destructive type (Fisher, 1969). The subsurface data for this portion of the Wilcox show significantly different geometry and extent of the major sand bodies. They are large and elongate parallel to the coast, indicating considerable wave and current influence. These sand bodies represent the beach-barrier or strandplain facies (Fisher, 1969) and would be comparable to the modern São Francisco Delta (Figure 9–22). This would be similar to the Type V delta of Coleman and Wright (1975).

Pennsylvanian Deltas

The cyclic and deltaic sequences of the Pennsylvanian System in North America and the analogous Carboniferous strata in other parts of the world have been studied in

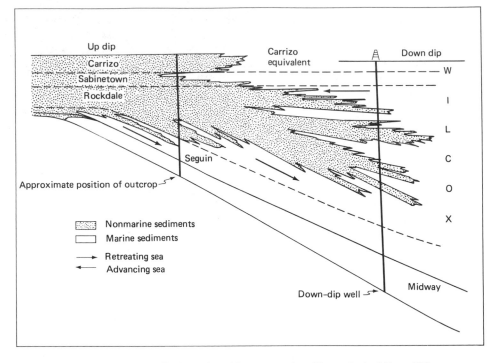

Figure 9-25 Generalized stratigraphic cross section (dip section) of Texas Wilcox strata. (From Echols and Malkin, 1948.)

detail. Not only has much of the world's coal supply been produced from these deposits, but they have also been significant producers of petroleum. Excellent studies have been conducted in Scotland (Greensmith, 1966; Belt, 1975), England (Collinson, 1968), and Spain (van de Graaf, 1975) as well as throughout the United States. In the following discussion Pennsylvanian deltaic systems of the Appalachian Basin, the mid-continent, and central Texas are considered. Although each of these areas represents a different geologic setting in terms of physiography and tectonics, there is a common theme found in the deltaic accumulations.

Appalachian Basin

During Middle Pennsylvanian time the Appalachian Basin in the Ohio–Pennsylvania–West Virginia area was receiving rather thick accumulations of terrigenous sediments, including sands and muds. Organic accumulations (now coals) and some carbonate sediments were associated with these terrigenous sediments. These comprise the Allegheny Formation, a stratigraphically complex unit generally less than 100 m in thickness. Through the careful and extensive efforts of John C. Ferm together with his colleagues and students (Ferm and Williams, 1965; Ferm and Caravoc, 1968; Ferm, 1970, 1974) we can now recognize these rocks as representing deltaic systems. Furthermore, it is possible to recognize the various environments of deposition within the delta system.

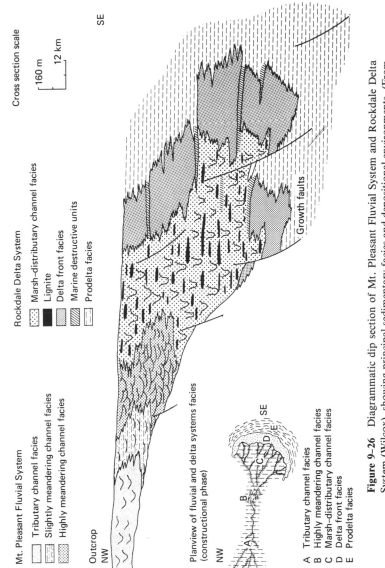

Mt. Pleasant Fluvial System

☐ Tributary channel facies

▨ Slightly meandering channel facies

▨ Highly meandering channel facies

Rockdale Delta System

▨ Marsh–distributary channel facies

■ Lignite

▨ Delta front facies

▨ Marine destructive units

▨ Prodelta facies

Outcrop

NW   SE

Cross section scale

[160 m

[12 km

Planview of fluvial and delta systems facies (constructional phase)

NW

SE

A

B

C

D

E

A Tributary channel facies

B Highly meandering channel facies

C Marsh–distributary channel facies

D Delta front facies

E Prodelta facies

Growth faults

Figure 9–26 Diagrammatic dip section of Mt. Pleasant Fluvial System and Rockdale Delta System (Wilcox), showing principal sedimentary facies and depositional environments. (From Fisher and McGowen, 1967.)

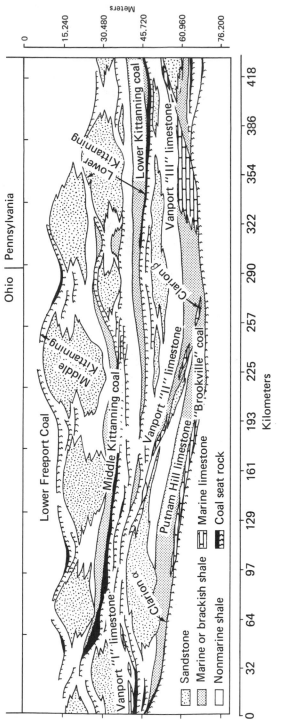

Figure 9-27 Generalized cross section of lower and middle Allegheny strata from the western portion of the Allegheny Basin. (From Ferm, 1970.)

314

Figure 9–27 shows a detailed stratigraphic cross section along the northwest side of the Allegheny outcrop belt, a distance of more than 400 km. It is important to note the continuity of the various lithofacies as well as their apparent geometry. The thin coal units and the thicker but tabular shales are typically continuous and widespread compared to the pod-shaped sand bodies. To gain a proper understanding of how these sediment bodies probably developed, it is necessary to consider a relatively simplified method of a prograding, fluvial-dominated delta system (Figure 9–28). In this model, Ferm (1970) depicts an active lobe with a well-developed distributary network adjacent to a peat-covered, inactive lobe. As one deltaic lobe is abandoned and another activated, there is a shift in the site of rapid accumulation [Figure 9–28(b)], with the delta wedge prograding over the peat which was accumulating on the delta plain of the older lobe below. The result is an accumulation of lithofacies that mimics the situation in the Allegheny Formation (compare Figures 9–27 and 9–28).

More details of a portion of this Allegheny sequence are shown in Figure 9–29, which depicts the facies and depositional environments of the upper and lower delta plain deposits (Ferm, 1974). These diagrams give a good feeling for both the geometries of the facies and their interrelationships. Observe the general trends for stratigraphic variation in grain size; in the delta plain in general there is the characteristic coarsening upward trend, whereas thick channel fill accumulations show a fining-upward trend (Figure 9–29).

A paleogeographic reconstruction of a part of Allegheny time is shown in Figure 9–30, taken from Williams et al. (1964). Numerous deltas developed in the shallow Pennsylvanian sea adjacent to a low-lying alluvial plain.

Eastern Interior Basin

During Pennsylvanian time the Eastern Interior Basin included what is now southern Illinois, southwestern Indiana, and northwestern Kentucky. It was dominated by fluvial-deltaic environments which were accumulating primarily terrigenous sediment with some carbonate and coal (Pryor and Sable, 1974). Like the Appalachian Basin, this area was the site of considerable coal accumulation, which has caused significant stratigraphic research, especially by the respective state geological surveys. Much of our knowledge of depositional environments and paleogeography of this basin and the western adjacent Forrest City Basin is attributed to the efforts of the late H. R. Wanless and students during the 1960s.

Numerous well-preserved cyclic sequences of deltaic deposition are preserved in this area. These cyclic sequences range in thickness from a few to several tens of meters. The complete sequence has a basal destructive and shelf-type marine portion succeeded by prodelta muds and overlain by delta plain units (Figure 9–31). The uppermost unit in the sequence is the coal representing the interdistributary marsh or swamp (Pryor and Sable, 1974).

Detailed environmental mapping of these cyclic sequences (Wanless et al., 1963, 1969) has permitted reconstruction of the paleogeography for various intervals (Figure 9–32). Although by necessity somewhat generalized, these maps are markedly similar to analogous modern delta environments.

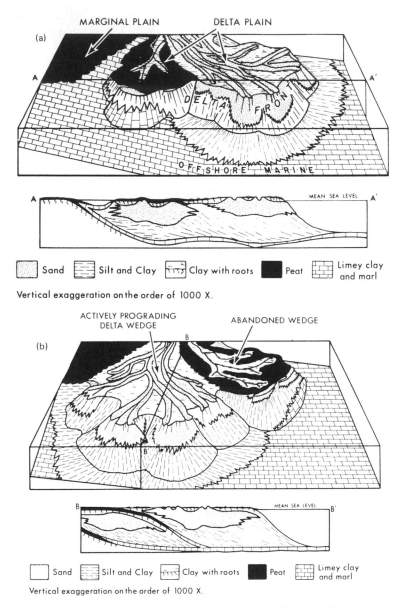

Figure 9–28 Diagrams of idealized prograding delta, showing how lobes are abandoned and new ones developed. Note the cross sections, which show how relationships between sand bodies and peats evolve. (From Ferm, 1970.)

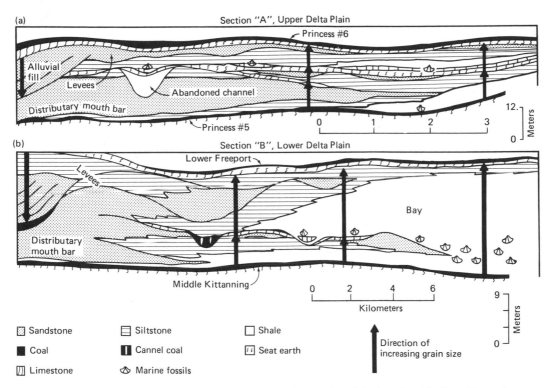

Figure 9-29 Cross sections of upper and lower delta plain deposits, Allegheny Formation. Note the stratigraphic trends in grain size. (From Ferm, 1974.)

North-Central Texas

Extensive sequences interpreted as fluvial-deltaic in origin were accumulating on the Eastern (Concho) Shelf of the Midland Basin in Texas during Pennsylvanian time. These sequences comprise part of the Strawn, Canyon, and Cisco Groups and have been of great interest due to their widespread petroleum production. Most of the detailed research on these strata has been carried out by L. F. Brown, Jr. and his associates at the Bureau of Economic Geology, the University of Texas, Austin (Brown, 1969a, 1969b; Brown et al., 1973).

More than 300 m of sediments in the Cisco Group alone can be ascribed to at least strong deltaic influence (Brown et al., 1973). A schematic cross section across a portion of the Eastern Shelf shows a complex network of shales, limestones, coal, and various sand bodies (Figure 9-33) within the Cisco Group. Sections for the Strawn and Canyon Groups are somewhat similar. By careful examination of these rocks, their textures, sedimentary structures, and stratigraphic relationships, it is possible to assign depositional environments. Following that, one can proceed to work out the sequences of environments through both space and time.

One of the basic tasks to be accomplished is that of isolating the various cyclic sequences and then assigning depositional environments to the individual facies. For the Pennsylvanian of Texas the sequences are like those of the Eastern Interior and

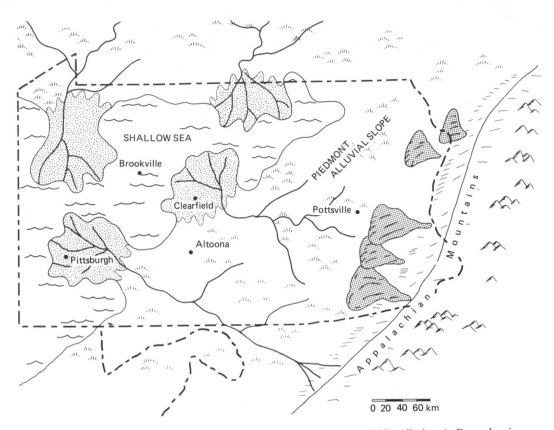

Figure 9-30 Paleogeography of lower Kittanning time (Middle Allegheny), Pennsylvania. (From Williams et al., 1964.)

Appalachian basins. Brown et al. (1973) have developed stratigraphic sequences representing deltaic cycles (Figure 9-34). Note that their cycles are comparable to those of Pryor and Sable (Figure 9-31) except that the cycles are not in phase. Brown et al. (1973) designate the prodelta muds as the base of the cycle, whereas Pryor and Sable (1974) choose the destructive phase as the base.

The work by Brown et al. (1973) goes a step further than most in that it recognizes and differentiates between elongate delta sequences and lobate delta sequences (see Figures 9-13 and 9-14). The primary difference between the two types shows up in the delta-front strata, where elongate deltas are characterized by contorted sand overlain by trough cross-beds, whereas lobate deltas have soft sediment faulting with overlying strata of generally horizontal bedding. This distinction is caused by the development of delta-front sheet sands due to wave and longshore current action on lobate deltas, in contrast to extensive channel mouth bars in the more fluvially dominated elongate deltas.

Pennsylvanian deltaic systems have caused some concern among geologists who try to compare them to modern deltas such as the Mississippi Delta. Due to extreme tectonic stability, the Pennsylvanian sequences are thin and laterally extensive. Brown et al. (1973) make an excellent point in recognizing this false comparison and

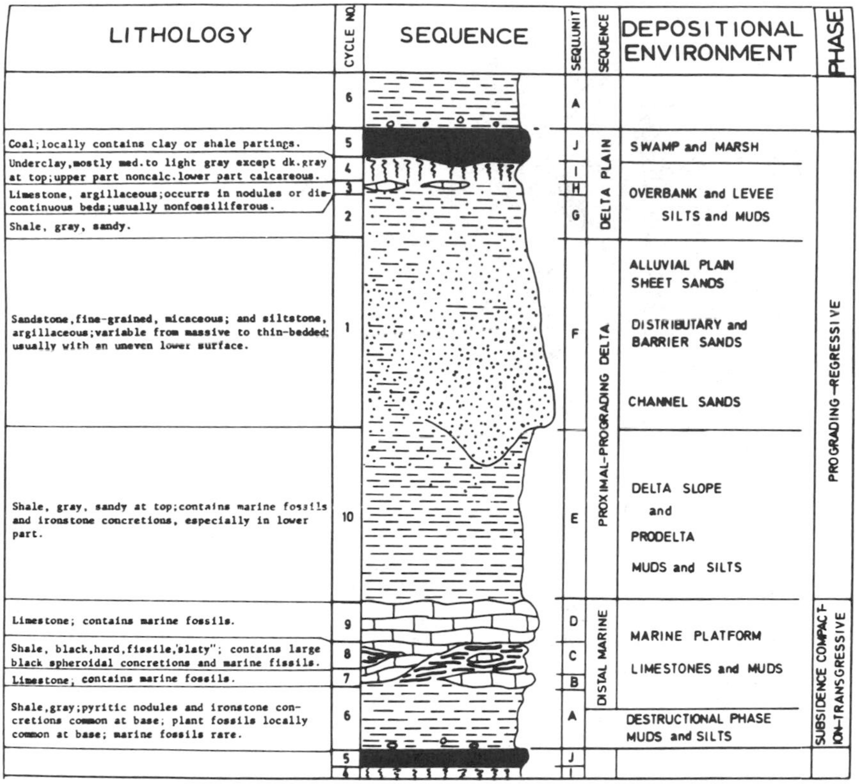

LITHOLOGY	CYCLE NO.	SEQUENCE	SEQ UNIT	SEQUENCE	DEPOSITIONAL ENVIRONMENT	PHASE
	6		A			
Coal; locally contains clay or shale partings.	5		J	DELTA PLAIN	SWAMP and MARSH	PROGRADING—REGRESSIVE
Underclay, mostly med. to light gray except dk. gray at top; upper part noncalc. lower part calcareous.	4		I			
Limestone, argillaceous; occurrs in nodules or discontinuous beds; usually nonfossiliferous.	3		H		OVERBANK and LEVEE SILTS and MUDS	
Shale, gray, sandy.	2		G			
Sandstone, fine-grained, micaceous; and siltstone, argillaceous; variable from massive to thin-bedded; usually with an uneven lower surface.	1		F	PROXIMAL-PROGRADING DELTA	ALLUVIAL PLAIN SHEET SANDS DISTRIBUTARY and BARRIER SANDS CHANNEL SANDS	
Shale, gray, sandy at top; contains marine fossils and ironstone concretions, especially in lower part.	10		E		DELTA SLOPE and PRODELTA MUDS and SILTS	
Limestone; contains marine fossils.	9		D	DISTAL MARINE	MARINE PLATFORM LIMESTONES and MUDS	SUBSIDENCE COMPACTION-TRANSGRESSIVE
Shale, black, hard, fissile, "slaty"; contains large black spheroidal concretions and marine fissils.	8		C			
Limestone, contains marine fossils.	7		B			
Shale, gray; pyritic nodules and ironstone concretions common at base; plant fossils locally common at base; marine fossils rare.	6		A		DESTRUCTIONAL PHASE MUDS and SILTS	
	5		J			

Figure 9–31 Ideal deltaic sequence relating depositional environments to sedimentary facies. (Modified from Pryor and Sable, 1974.)

suggest that systems such as the Guadalupe (Donaldson et al., 1970) or Colorado (Kanes, 1970) Deltas in Texas are better counterparts. These deltas are constructive and build into a shallow bay behind a barrier system on a tectonically stable coastal plain (Brown et al., 1973).

SUMMARY OF CHARACTERISTICS OF SEDIMENT/SEDIMENTARY ROCK BODIES REPRESENTING THE DELTAIC SYSTEM

The deltaic system is one of the more easily defined of the coastal depositional systems. It represents a complex interaction of riverine and marine processes with a great quantity and diversity of sediment. Consequently, there are numerous sedimentary environments and resulting facies.

Tectonic setting. Theoretically, deltas may form along coasts in any tectonic setting; however, they tend to be most abundant along coastal plains or trailing-edge coasts. Such conditions cause the development of large drainage networks carrying much sediment and water. In addition, they tend to have gently sloping continental

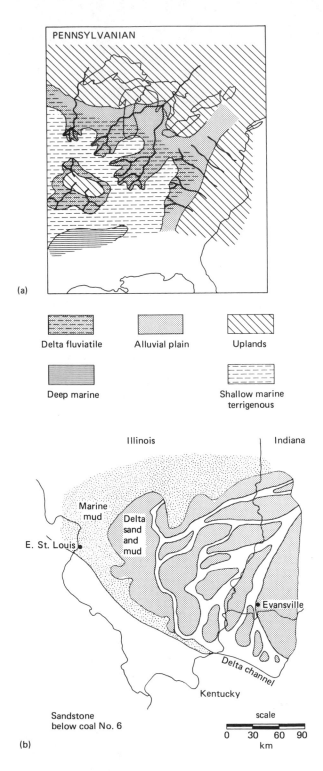

(a)

Delta fluviatile   Alluvial plain   Uplands

Deep marine   Shallow marine
                         terrigenous

Illinois   Indiana

Marine
mud
          Delta
          sand
          and
E. St. Louis mud

                                    Evansville

                          Delta channel

Kentucky

Sandstone   scale
below coal No. 6   0   30   60   90
                              km
(b)

Figure 9-32 (a) Paleogeography of the
Appalachian and Eastern Interior basins
during Pennsylvanian time with (b) example of a delta from the Eastern Interior Basin. (Modified from Pryor and
Sable, 1974; Wanless et al., 1969.)

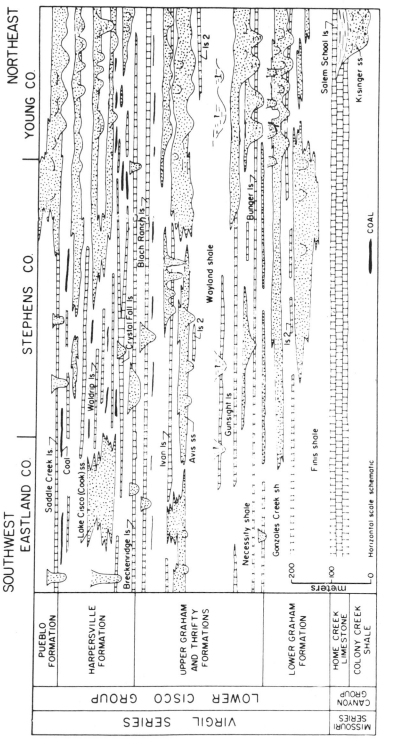

Figure 9–33 Schematic cross section of the Cisco Group, north-central Texas, with thin limestones and thicker channel sandstones. (From Brown et al., 1973.)

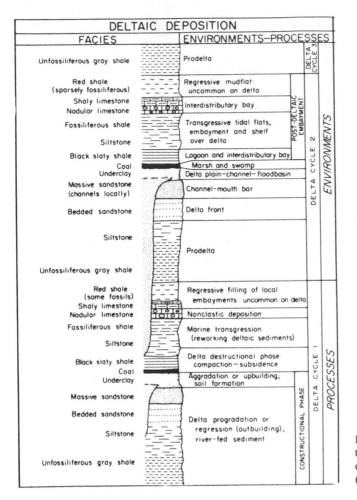

DELTAIC DEPOSITION

FACIES		ENVIRONMENTS–PROCESSES			

Unfossiliferous gray shale — Prodelta — DELTA CYCLE 3

Red shale (sparsely fossiliferous) — Regressive mudflat: uncommon on delta

Shaly limestone / Nodular limestone — Interdistributary bay

Fossiliferous shale — Transgressive tidal flats, embayment and shelf over delta

Siltstone

Black slaty shale — Lagoon and interdistributary bay

Coal — Marsh and swamp

Underclay — Delta plain–channel–floodbasin

Massive sandstone (channels locally) — Channel-mouth bar

Bedded sandstone — Delta front

Siltstone — Prodelta

Unfossiliferous gray shale

Red shale (some fossils) — Regressive filling of local embayments: uncommon on delta

Shaly limestone / Nodular limestone — Nonclastic deposition

Fossiliferous shale — Marine transgression (reworking deltaic sediments)

Siltstone

Black slaty shale — Delta destructional phase: compaction–subsidence

Coal

Underclay — Aggradation or upbuilding, soil formation

Massive sandstone

Bedded sandstone — Delta progradation or regression (outbuilding), river-fed sediment

Siltstone

Unfossiliferous gray shale

POST-DELTAIC EMBAYMENT — DELTA CYCLE 2 — ENVIRONMENTS

CONSTRUCTIONAL PHASE — DELTA CYCLE 1 — PROCESSES

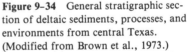

Figure 9-34 General stratigraphic section of deltaic sediments, processes, and environments from central Texas. (Modified from Brown et al., 1973.)

shelves which provide a site for accumulation of the delta and which attenuate waves before they reach the delta. By contrast, rugged coasts of a leading-edge setting have small drainage systems and steep continental margins. The result is little sediment being carried to the coast and large waves to remove the sediment that is deposited there; thus few deltas form.

Shape. The delta or triangular areal shape of these systems is a characteristic, although there is much variation, especially in fluvial-dominated or tide-dominated types. Deltas are commonly wedge shaped, or they may be very thick lens-shaped bodies.

Size. A great range exists in both areal extent and thickness. Deltas may be up to thousands of square kilometers in area and thousands of meters thick or they may be quite small.

Textural trends. A typical prograding delta displays a coarsening upward trend and a decrease in grain size in a seaward direction. These are, of course, gross generalities depending on the type of delta and the specific site of the stratigraphic section. In general, the delta plain is a mixture of sand and mud, the delta front is dominantly sand, and the prodelta is mud.

Lithology. Little can be stated about the composition of deltaic sediments. It varies widely and is generally complex because of the diverse source materials within the drainage basin. There are specific constituents that tend to be characteristic, however. Included are the presence of coal or peat seams in delta plain deposits, abundant micaceous particles, and numerous wood fibers dispersed through the sediment.

Sedimentary structures. No structures are diagnostic of deltas or environments within the delta. Some sediment deformation of prodelta muds is common, such as convolute bedding, diapiric features, loading structures, and so on. Specific facies, of which there are several in a delta system, may have a characteristic suite of structures.

Paleontology. Although there are not specific taxa which are diagnostic of deltas, the paleontologic trends in combination with lithologic trends may be indicative of a deltaic system. A freshwater–brackish water–saltwater transition would be indicative if related to sediment types and distinguished from an estuarine embayment.

Associations. The delta system itself would be expected to lie between continental deposits such as fluvial strata and open marine shelf strata. Within the delta system there is variety in various facies associations. Delta plains are complex and contain numerous sedimentary environments with their respective facies. Marsh, interdistributary bay, tidal flat, and various distributary environments all produce distinctive facies which may be in many stratigraphic arrangements. Delta-front facies show great variation depending on the relative strength of riverine, tidal, and wave-generated processes. As a result, their role in the overall delta sequence shows much variation. Prodelta muds may have many similarities with open shelf muds and certainly show gradual transition with them. The greater thickness and rate of accumulation of deltaic muds and the accompanying phenomena of deltas serve to distinguish them from shelf sediments.

ADDITIONAL READING

BROUSSARD, M. L. (ED.), 1975. *Deltas, Models for Exploration.* Houston Geological Society, Houston, Tex., 555 p. The most comprehensive and up-to-date volume of deltas available at this time. This book is the second excellent delta volume published by the Houston Geological Society (see Shirley, 1966, below). It contains sections on general topics, modern deltas, and ancient deltas with numerous articles by leading researchers.

COLEMAN, J. M., 1976. *Deltas: Processes of Deposition and Models for Exploration.* Continuing Education Publication Co., Champaign, Ill., 102 p. This is an excellent synthesis of sedimentation of modern deltas by the world's foremost authority. A well-illustrated, well-written, and concise treatment of the subject.

FISHER, W. L., BROWN, L. F., JR., SCOTT, A. J., AND McGOWEN, J. H., 1969. *Delta Systems in the Exploration for Oil and Gas—A Research Colloquium.* University of Texas, Austin, Tex., Bur. Econ. Geol., 102 p. and 168 illus. A superb synthesis of modern deltas and their rock-record counterparts by the people who coined the term "depositional systems." Numerous excellent illustrations of various deltaic sequences; no halftones are included, but the volume suffers little from the omission.

MORGAN, J. P., AND SHAVER, R. H. (EDS.), 1970. *Deltaic Sedimentation: Modern and Ancient.* Soc. Econ. Paleontologists and Mineralogists, Spec. Publ. No. 15, Tulsa, Okla., 312 p. One of the many excellent topical volumes eminating from the Society's annual research symposium. Both modern and ancient deltas are included. Emphasis is on morphology and sediments with little on processes.

SHIRLEY, M. L. (ED.), 1966. *Deltas in Their Geologic Framework.* Houston Geological Society, Houston, Tex., 251 p. This is the first of several good delta volumes. A collection of fine articles on both modern and ancient deltas.

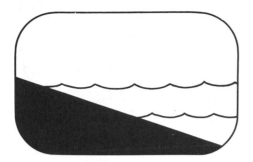

10 Intertidal Environments

Consideration of coastal depositional systems necessitates a reasonable and meaningful organization. The coastal zone is extremely diverse and is quite an important part of the geologic record of sedimentary strata. All of those environments that occupy this narrow zone are influenced by the terrestrial or continental environments as well as by marine environments. There is generally little topographic relief in the coastal zone on trailing-edge coasts, so that changes in sea level cause pronounced effects in displacing environments either in a landward or in a seaward direction.

It is relatively easy to separate the delta system from the remainder of the coastal zone. Other depositional environments that are important along the coast include intertidal environments, coastal bays, beaches, dunes, and inlets. All may be components of a barrier island system, but most of them do occur alone. This suite of environments may be in the carbonate or the terrigenous realm; in fact, some books separate coastal environments on that basis (e.g., Reading, 1978). Although there are important differences between carbonate and terrigenous environments, there are at least as many similarities. Consequently, this chapter and the two that follow are organized so as to include similar environments from the standpoint of processes and morphology rather than that of sediment composition.

This complex of coastal environments may be considered as the strandplain depositional system (Fisher et al., 1969). To discuss this entire system in one chapter would result in either a nearly endless chapter or would require extensive deletions and brevity. As a result, the strandplain system will be subdivided for discussion purposes into naturally similar portions. This chapter deals with intertidal environments such as marshes and tidal flats, including coastal sabkhas. Chapter 11 considers the adjacent subtidal environments; coastal bays, including estuaries and lagoons. The final chapter on coastal environments, Chapter 12, discusses the barrier island system: dunes, beaches, inlets, and washover fans.

DEFINITION AND SCOPE

This chapter considers intertidal environments which border coastal bays or which front open coasts but does not include beaches or those intertidal environments subjected to appreciable wave attack. The term intertidal zone is used here in the broad sense, and includes supratidal. In other words, all those depositional environments which are sometimes submerged and sometimes exposed will be discussed regardless of whether or not there is a regular cyclicity to the processes.

The intertidal environment is commonly considered as quite a narrow band along the strandline of the coast. This is not necessarily the case, especially in regions where extreme tidal ranges are combined with very low relief, such as on the northern coast of Australia, particularly the Gulf of Carpenteria. In these areas the intertidal zone may be several kilometers wide.

Another consideration that affects sedimentation in the intertidal environment is the tidal climate of a particular area. Included are not only such characteristics as tidal range and tidal species but also lunar variations and atmospheric affects. Those cyclic and predictable tidal datum levels which serve as reference horizon are listed in Figure 10-1. Observe that only relative positions are shown; any scale may be applied. It should also be observed that the middle of the tide range is termed the **half-tide level.** It is commonly and mistakenly equated with mean sea level. The midtide level is generally significantly above mean sea level along coasts. For sedimentologic purposes mean tide level (close to midtide level) is more important as a datum.

Aperiodic events that change water-level fluctuations from predicted levels may play a very important role in intertidal or peritidal sedimentology. Such phenomena are related typically to meteorologic events and may cause either higher- or lower-than-normal water levels; a single event would not produce both conditions. The aperiodic water-level change that is most often significant in terms of sediment transport and accumulation is the rise of water level induced by setup produced by onshore-moving waves and winds. Intense storms and hurricanes may cause water levels to rise more than 5 m above normal levels. If this **storm surge** comes during the high-tide portion of the cycle, especially a spring high tide, then considerable movement of sediment will take place because the entire intertidal environment will be inundated. Although negative or extremely low tides also occur, typically associated with offshore winds, their effect on sedimentation is not so dramatic. Exposure may cause desiccation or may result in killing benthic organisms but does not result in extreme sediment movement.

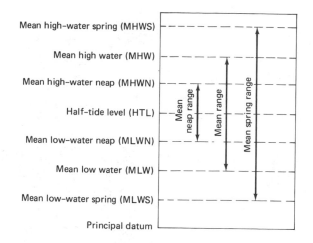

Figure 10-1 Tidal datum levels used in coastal hydrography. (From Frey and Basan, 1978, p. 109.)

SALT MARSHES

Salt marshes or coastal wetlands are vegetated intertidal areas and they occur along the coasts of most landmasses. These marshes may border estuaries, lagoons, and other protected strand environments, or they may occur along an open coast if energy conditions permit establishment of vegetation. Excellent examples of open coast marshes occur along the west Florida coast beginning east of the Appalachicola Delta and north of Florida Bay in the Ten-thousand Islands Area. Whereas the more traditional concept of salt marshes is one of grass vegetation, some would also include the mangrove environment under the definition, even though they are characterized by trees rather than grass; such will be the case in the following discussion. The examples mentioned from Florida include both grassy salt marshes and mangrove swamps or marshes (Figure 10-2).

Marshes have not been extensively studied by sedimentologists; most of the research has been conducted by biologists or paleobiologists. A recent excellent summary of marsh sediments and sedimentation has been written by Frey and Basan (1978). The reader is referred to this article as the best single source on the marsh as a sedimentary environment.

In considering the marsh as a sedimentary environment, one must keep in mind that our definition of the term places certain restrictions on the distribution of marshes through geologic time. Recall from Chapter 8 that the advent of land plants in the Devonian is considered as a factor in explaining the coincident abundance of meandering streams. Although these Devonian plants were not grasses and true

(a)

(b)

Figure 10–2 Intertidal vegetated environments are (a) *Spartina* marsh and (b) mangrove swamps.

grasses did not appear until the Tertiary, one can assume that some types of coastal vegetation produced at least a marshlike environment by the late stages of the Paleozoic Era.

Zonation

Salt marshes are a tide-stressed environment (Basan and Frey, 1977) which grades from nearly terrestrial to nearly marine in character. Relief is typically low, but there are numerous subtle variations in elevation or other factors which cause distinct zonation of vegetation in the salt marsh environment. Even slight changes in the tidal character of the marsh will cause changes in zonation. Variables that control this marsh zonation include nutrient flux, physicochemical conditions, waves, currents, and sediment parameters as well as tidal regime. Two levels of zonation are suggested by Frey and Basan (1978); at the ecosystem level there are changes in plants which result from the transition from terrestrial to marine, and at the habitat level there are changes which are reflected in the growth forms, or other changes in a given species.

Quite subtle or local changes in relief or general morphology are reflected in the zonation of plant species and in their growth forms. The substrate composition and stability are also important considerations. The marsh is typically subdivided into two major components: the high marsh and the low marsh. In addition to the obvious basis of the position in the tidal regime, there is correspondence to plant taxa and forms and to substrate (Figure 10–3).

The low marsh, which represents the younger portion of the marsh, is characterized by *Spartina* (cordgrass) throughout and it generally is floored by a mud substrate. The low marsh vegetation extends from about mean high-water neap (MHWN) to near mean higher high water (MHHW) (Figure 10–3). This part of the marsh is therefore flooded fairly regularly.

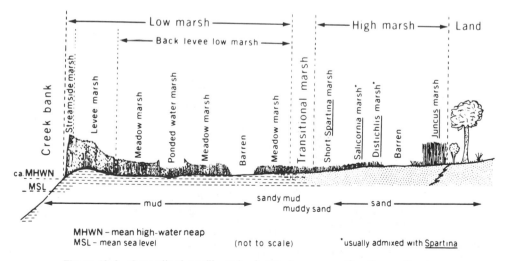

Figure 10–3 Generalized profile and salt marshes representing the southeastern coast of the United States. All low and transitional zones are characterized by *Spartina*. (After Edwards and Frey, 1977.)

By contrast, the high marsh consists of short forms of *Spartina* (cordgrass), *Salicornia* (glasswort), *Distichlis* (salt grass), and *Juncus* (needle rush). Most commonly, *Juncus* is dominant in the high marsh. Sand or muddy sand is likely to be the substrate and the high marsh is inundated only during spring tides or storm-related high-water conditions. A narrow transition zone between the high and the low marsh may be present (Figure 10-3).

It should be emphasized that there is variation in flora from one climatic belt to another and substrates also vary (Gallagher, 1977). Mangroves may occupy the marsh niche in low-latitude areas. Although the term "swamp" is used typically when trees are present (mangrove swamp) and "marsh" is reserved for grassed areas (Reid, 1961), there is little difference in the overall system. Mangrove types show a zonation of three species as one proceeds from low water through the intertidal zone. The roles of mangroves and marsh grasses in sedimentation have many common aspects.

General Morphology

Although features of the marsh surface are subtle and small in the vertical scale, they are quite important in sedimentation. If the tidal creeks and channels are excluded, and they can be due to their lack of vegetation, relief over the entire marsh may be measured in tens of centimeters. Highest elevation on the marsh is associated with the natural levees along the creek banks (Figure 10-3). These levees originate just as do those in any fluvial environment. The relatively high elevation causes *Spartina* to grow tall along the levees (Figure 10-4). The levees, together with their related tidal

Figure 10-4 Oblique aerial photograph showing subtle topography of marsh with (a) tidal channels and (b) natural levees.

creeks, provide the only relief on the marsh which is not inherited. There may be small topographically high areas around which the marsh developed.

The bulk of the salt marsh is a broad, flat expanse of vegetation with local barrens, sometimes called rotten spots. Local highs or lows on the marsh surface may be reflected by a change in vegetation or by an absence of vegetation. The delicate balance between vegetation and position in the tidal range may be illustrated by an example from the Delmarva Peninsula of Virginia along the margins of Wachapreague Bay. This bay is rapidly filling in with fine sediment and contains extensive intertidal flats and marshes. At low tide, rather large megaripples with small wave heights may be observed. These tidal flats and their contained bedforms are positioned such that some of the megaripples have *Spartina* growing on the crest but the troughs are barren. A rather complete gradation can be seen from tidal flat to completely colonized marsh. The relief on these features is only about 20 cm.

Development

Although salt marshes are extensive and represent an important ecological niche, they are one of the shortest lived of all sedimentary environments. Coastal marshes are, by definition, tied closely to a rather narrow position with respect to sea level. Changes in sea level can therefore cause great impact on the marsh environment. A rise in sea level may actually prolong marsh existence because it provides renewed sites for sediment accumulation and related colonization by marsh vegetation. A lowering of sea level can cause the reverse effect because marsh niches are left high and dry. Regardless of whether or not the sea level fluctuates or if so, which way it fluctuates, there must be a rather delicate balance between sea-level position and sediment influx to prolong the lifetime of a given marsh.

Marshes are sediment sinks; that is, they are sites of accumulation. It is rather rare for the marsh to be subjected to erosion by tidal currents except along the tidal creek. Sediment is derived from runoff of land, from washover of barriers, and from the shelf, being transported landward by tidal currents. As sediment is deposited along the shallow margins of the land masses there eventually comes a time when the substrate is in a proper position with respect to the tidal range so that marsh vegetation will establish. This is true both for grasses and for mangroves. Continued infilling of these shallow sediment sinks will result in a change in vegetation and the eventual destruction of the marsh by vertical accretion to a level above high tide.

The primary mechanism for providing sediment to the marsh is by overbank flooding of channels, much like the fluvial system. Sediments are carried largely in suspension and consist of fine organic detritus and mineral particles. During high spring tides and especially during storm tide conditions, the marsh is flooded by this sediment-laden water. There are numerous phenomena related to the marsh grasses that permit the suspended sediment to accumulate on the marsh. The marsh vegetation acts as a baffle, causing attenuation of waves and currents, thus permitting suspended sediment to settle. According to Pryor (1975), settling from suspension is not adequate to explain the volume of accumulation that occurs on most marshes. Flocculation of clays may enhance sedimentation, and organisms can be extremely important.

The obvious effects of grasses as baffles may provide only a small portion of the overall influence of the biota. The plants themselves physically collect sediment, especially in the dense lower levels. Animals may also play a role, in that suspension feeders take in large quantities of sediment, which is then excreted as pellets that readily accumulate on the substrate due to their relatively large size (Kraeuter, 1976).

Also like the fluvial system, marsh channels may be subjected to avulsion and neck cutoff. Breaching of levees and development of crevasse splays is a means of providing large quantities of sediment to the marsh, although it is local. Washover of sediment provides large pulses of sediment in a shape somewhat like the splays. In both situations the grasses are completely or partially buried but recover quickly (Godfrey and Godfrey, 1973).

As sediment continues to accumulate in the marsh, a maturation process occurs. This is best explained by the relative amounts of high marsh compared to low marsh. Frey and Basan (1978) have considered marshes to belong to three major categories based on this relationship (Figure 10–5). Other characteristics can then be related to each stage.

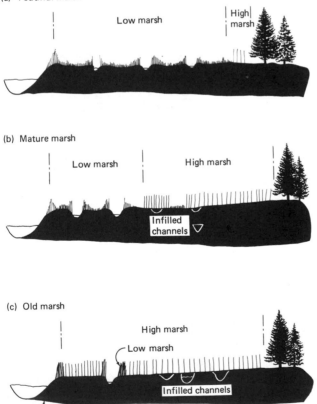

(a) Youthful marsh

Low marsh   High marsh

(b) Mature marsh

Low marsh   High marsh

Infilled channels

(c) Old marsh

High marsh

Low marsh

Infilled channels

Regular intertidal areas restricted to tidal creeks

Figure 10–5 Skematic diagram of three stages in development and maturation of salt marshes. (From Frey and Basan, 1978, p. 114.)

The youthful stage [Figure 10-6(a)] is characterized by a dominance of low marsh with *Spartina alterniflora*. Drainage is well developed and sedimentation rates are relatively high. At the outset, accretion is both lateral and vertical, but as development proceeds vertical accretion dominates. The mature marsh is about evenly divided between high and low marsh [Figure 10-6(b)]. Drainage is still well developed but concentrated on the low marsh; high-marsh channels are at least partially filled. Much meandering of tidal creeks occurs. Sedimentation is slow and dominated by vertical accretion. The old-age stage of marsh development is dominated by high marsh with much *Juncus* and *Salicornia* [Figure 10-6(c)]. Few tidal channels remain open and sedimentation is quite slow. Innundation by seawater occurs during storms only.

Sediments

The general nature of salt marsh sediment accumulations is quite unlike that of other depositional environments. It is commonly a subequal mixture of fine mud and plant debris with small amounts of shell material, sand-size terrigenous particles, and large plant fragments. Some authors (Bouma, 1963; MacDonald, 1977) have designated marshes as containing the finest sediments of all marine environments. This finding is not always true, especially for those marshes developed on washover deposits or flood tidal deltas associated with barriers; most of these are dominated by sand-size sediments.

Facies

The coastal marsh accumulates a distinctive combination of sediment, structures, geometry, and biogenic features. When placed in its proper stratigraphic context it is easy to identify. Although there is some nearly universal similarity among marsh sequences, there may be striking contrasts. Most marshes accumulate much plant debris and typically develop peats, eventually perhaps coals. The grain size is typically finer than that in adjacent tidal flats but may range from sand to clay size.

Bedding may show a broad spectrum of characteristics. Some marsh deposits are well bedded with wavy undulations [Figure 10-7(a)]. Bedding is typically due to alternations of plant debris and terrigenous sediment with undulations due to disturbance by plants (Bouma, 1963). By way of contrast, some marsh deposits are completely lacking in stratification [Figure 10-7(b)] due to bioturbation (Basan and Frey, 1977).

Small channel fill accumulations may occur in the marsh facies representing the tidal creek environment. Oyster shells (*C. virginica*) or other marine and brackish organisms may provide skeletal material. Channel deposits are typically coarse relative to the marsh itself, and cross-stratification is common.

Details of marsh paleogeography may be provided through microscopic study of the peats (Cohen and Spackman, 1974; Allen, 1977). Identification of the plant debris may permit reconstruction of ancient zonation patterns on the marsh.

Figure 10–6(a) Photograph of a youthful marsh (Sandy Neck, Barnstable Harbor, Mass.) Compare with Figure 10–5. (Courtesy of Joe R. Wadsworth, Jr.)

Figure 10–6(b) Photograph of a mature marsh (St. Catherines Id., Georgia) Compare with Figure 10–5. (Courtesy of Joe R. Wadsworth, Jr.)

Figure 10–6(c) Photograph of an old-age marsh (Nanticoke, Md.) Compare with Figure 10–5. (Courtesy of Joe R. Wadsworth, Jr.)

(a)

Figure 10–7 Photographs showing (a) well-bedded marsh deposits and (b) those which lack bedding.

(b)

MANGROVE SWAMPS

Mangrove swamps develop in the same niche as marshes except in more tropical regions; they may be mixed in subtropical areas. One of the most extensive areas of mangrove development is in the Ten-thousand Islands and Florida Bay areas of southern Florida. Mangroves display a typical zonation related to position in the intertidal zone; however, the swamp may extend throughout the intertidal zone (Chapman, 1976; Gallagher, 1977) to slightly below mean low tide. There are regions where mangroves are restricted to high in the intertidal zone. Apparently, anywhere within the intertidal zone where seeds can take hold and develop is where the mangrove swamps extend. Mangrove swamps are passive in development and become established only in areas where current and wave energy are low (Wanless, 1974).

Mangroves do exert a similar but probably less marked effect on sedimentation than that of marsh grass. The dense root systems act as baffles and enhance sediment accumulation. In addition, the roots provide a hard substrate for sessile benthos organisms such as barnacles and oysters to attach. Upon expiration and disarticulation, these organisms will contribute biogenic debris to the system.

Mangroves develop in terrigenous, carbonate, or mixed sediment and have similar effects in all these types. Peats may accumulate in these areas and tend to occupy the same stratigraphic position as marsh peats in the coastal sequence. According to Wanless (1974), development and preservation of mangrove peats is associated with (1) a substrate near tidal level, (2) protection from high-energy impact, and (3) only minor influx of detrital sediment. Mangrove peats typically overlie freshwater or terrestrial deposits and are in turn are overlain by or interbedded with brackish-water or tidal flat deposits (Scholl, 1969).

TIDAL FLATS

Tidal flats are low-relief environments which contain unconsolidated and unvegetated sediments that accumulate within the intertidal range, including the supratidal zone. Although one typically associates tidal flats with terrigenous mud, that is not a proper generality. There are extensive tidal flats of sand-size particles, and carbonate tidal flats also occur. The latter are considered in the following section.

Tidal flats have received attention from sedimentologists for a long time and this environment was among the first where process-related studies were conducted (Klein, 1977a). The general accessibility of tidal flats and their extensive development along the North Sea coasts of England and Europe has undoubtedly contributed to the early studies. At the present time there have been numerous research papers on the subject.

Tidal flats do contain a number of rather distinctive characteristics which enables their recognition in ancient stratigraphic sequences. It should be observed, however, that many features of tidal flats also occur in subtidal but tide-dominated environments. A recent book (Ginsburg, 1975) is devoted to consideration of criteria that characterize modern tidal deposits and their recognition in the rock-record.

General Morphology

The most common position for tidal flats to occupy is the perimeter of the terrestrial environment. They also occur along open coasts or in embayments where the flat is bounded by subtidal environments.

There is generally a direct relationship between the areal extent of tidal flats and tidal range. As examples, the North Sea coast of Europe experiences 2.5- to 4-m tides and is characterized by tidal flats up to 7 km wide, and some coastal areas of the

Yellow Sea possess 5-m tides with tidal flats up to 25 km wide. There are places with ranges of less than a meter but with broad tidal flats; such conditions exist along some estuary margins of the Gulf coast of the United States. A general dearth of sediment or rather high wave energy may restrict tidal flat development in regions of high tidal range.

Tidal flats commonly display little relief except for tidal channels, which may contain subtidal floors. Lateral migration of tidal channels and associated lateral sedimentation follows the pattern of marshes and that previously described for meandering fluvial systems. Maximum relief on the upper surface of the tidal flat is likely to be dependent on the size of the bedforms produced, and this may in turn be related to grain size.

Process-Response Phenomena

Tidal currents dominate the various processes that affect channels in tidal flats, but waves may play a more important role on the flats themselves. In addition, biogenic processes such as bioturbation and pelletization are of widespread importance. The rise and fall of tides on a diurnal or semidiurnal basis commonly creates a great deal of sediment movement due to the flooding and ebbing currents created by the passage of successive tidal waves. Tidal currents are generally thought of as moving in opposite directions. Bedforms generated by bed load transport and their contained cross-stratification would reflect these directions, the result being the typical herringbone pattern (Figure 10–8). The implication is that flooding tides move up the gentle slope of the tidal flat and that during ebb conditions the opposite motion takes place.

In many situations the flood and ebb currents show distinctly different velocities at a given location, and in addition there is considerable variation in tidal current velocity at any location throughout the tidal cycle. Slack water conditions may permit fine, suspended sediment particles to settle out, whereas throughout much of the flood and ebb cycle, bed load transport of sand-size particles is dominant and mud is kept in suspension.

The alternation of bed load and suspended load accumulation produces bedding characteristics which are commonly associated with tidal flats or at least with

Figure 10–8 Bidirectional (herringbone) cross stratification on tidal flat. Trench is 30 cm high. (Courtesy J. C. Boothroyd.)

tide-dominated environments. An idealized tidal bedding model has been proposed by Reineck and Wunderlich (1968b) based on field experiments along the North Sea tidal flats of Germany. They found that a cyclic sequence of sand and mud was produced during each tidal cycle and that this sequence was arranged into four distinct strata, reflecting conditions during the rise and fall of the tide. During or near slack water, mud accumulated from suspension, whereas sand was deposited during the higher velocities of ebb and flood currents (Figure 10-9). This style of bedding has been observed in modern tidal flats and is preserved in the rock record as well (Klein, 1977a).

Actually, flaser and lenticular bedding (see Chapter 3) are probably more commonly preserved in tidal accumulations than is tidal bedding. Ripples cover most tidal flat environments, and the accumulation of suspended mud or pellets may be restricted to the troughs between ripple crests. Sediments accumulated during successive tidal cycles will produce flaser bedding. The general dearth of sand and development of discontinuous ripples will give rise to the formation of lenticular bedding (Reineck and Wunderlich, 1968a).

The phenomenon of **time-velocity asymmetry** (Postma, 1967) is important in affecting sediment transport over tidal flats and also in other tide-dominated environments. The common situation is that of maximum flood current velocity being

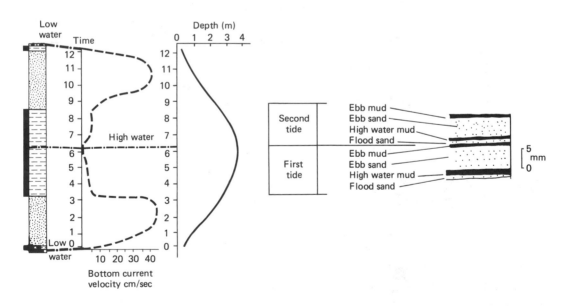

Figure 10-9 Development of tidal bedding. Marker beds at high and low tide bracket bed load deposition of sand (stipple) and deposition of mud from suspension (dashes and black bar). During two tidal cycles, four couplets of sand and mud, comprising a tidal bed, are deposited. (From Klein, 1977, p. 46; modified after Reineck and Wunderlich, 1968.)

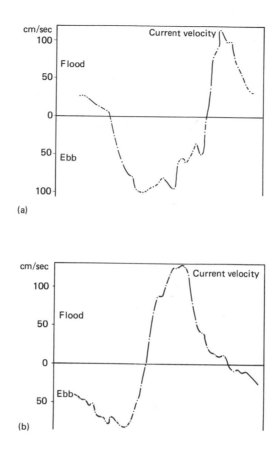

Figure 10–10 Time-velocity curves, showing typical asymmetry patterns that might occur. (After Postma, 1961.)

earlier than midtide and maximum ebb current much later on tidal flats (Figure 10–10). In addition, it is common for a particular location on an intertidal or tide-dominated surface to experience domination by the flooding phase or by the ebbing phase of the tidal cycle. It is rare that there is an equal balance between the two.

A consequence of this type of time-velocity asymmetry is the development of reactivation surfaces (Klein, 1970). This surface is one that truncates cross-stratification of bedforms but is inclined in the same direction (Figure 10–11). Large bedforms develop in sand-size sediment during the dominant tidal phase; it may be flood or ebb. This represents the constructional aspect of the cycle. During the subordinate and destructional phase these previously developed bedforms are partially destroyed, but the subsequent constructional phase causes a reforming or reactiva-

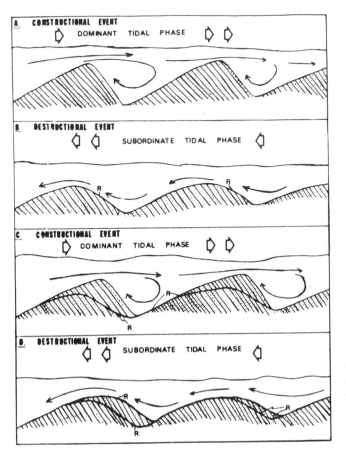

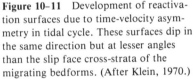

Figure 10–11 Development of reactivation surfaces due to time-velocity asymmetry in tidal cycle. These surfaces dip in the same direction but at lesser angles than the slip face cross-strata of the migrating bedforms. (After Klein, 1970.)

tion of the bedform, thus generating the reactivation surface. Although reactivation surfaces are commonly associated with tide-dominated environments, they may also occur in areas of braided streams (Collinson, 1969) and meandering streams (Jackson, 1976).

The previous discussion considered sediment transport processes which acted essentially in opposite directions. There are conditions that may result in currents and therefore in sediment transport occurring at right angles and producing bedforms so oriented. This has been termed **late-stage emergence** (Klein, 1977a). It is caused by crests of large bedforms, which are generated by high-velocity tidal currents, acting as channel boundaries when these bedforms are exposed during ebb conditions. The result is a channelization of the late-stage flow parallel to these crests and the formation of small-scale bedforms at right angles to large ones (Figure 10–12).

Sediments and Sedimentary Structures

Tidal flats may be considered as being essentially of two types: muddy and sandy. Much attention has been aimed at the extensive mud flats bordering the Wadden Sea coast of Europe (Van Straaten and Kuenen, 1957; Postma, 1961). Mixed sand and

Figure 10-12 Photograph of bedforms with crests at right angles formed during late-stage emergence. Megaripples were generated by flooding currents, and smaller ripples were formed by currents confined to troughs of the large bedforms as the water level was lowered.

mud flats occur along the German Bight (Reineck, 1967, 1972; Reineck and Singh, 1980). Sand-dominated tidal flats are commonly restricted to mesotidal and macrotidal coasts such as the Bay of Fundy (Klein, 1970; Knight and Dalrymple, 1975). Gravel-size sediment is uncommon on tidal flats and is restricted to shell material, generally from organisms that live in this environment, or to terrigenous particles from nearby sources.

There is a characteristic grain size distribution across the tidal flat trending from relatively coarse near low tide to relatively fine near high tide (Figure 10–13). The reason for this distribution is the spectrum of energy conditions across the environment and the relative time of submergence. Near low tide, wave energy is relatively high and tidal currents are prolonged, so that winnowing of fine particles takes place. These small particles are either transported up onto the tidal flat or they settle out of suspension in the calmer subtidal environment. Near the high-tide area both wave and tidal current energy are low, thus permitting accumulation of mud. Under conditions of equal flood and ebb energy on the upper part of the tidal flat there is a net accumulation of sediment. The explanation for this is found in Hjulstrom's curve (Figure 2–8), which demonstrates that greater flow strength is required to erode the muddy sediment than to deposit it. Van Straaten and Kuenen (1951, 1958) refer to this phenomenon on tidal flats as scour lag.

Tidal flats tend to display a wide variety of bedforms and commonly contain many infaunal organisms which produce bioturbation. The variety of bedforms covers the spectrum of lower-flow-regime types and may include upper plane beds.

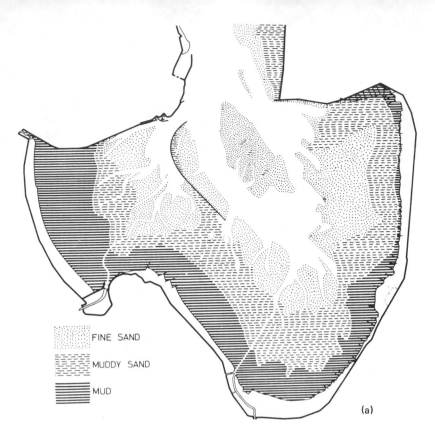

FINE SAND

MUDDY SAND

MUD

(a)

Aerial variation in the crest patterns is also pronounced (Figure 10–14). It is common to have one style or size of bedform superimposed upon another; one may result from flood currents and the other from ebb currents. The consequence of this spectrum of bedforms is an equally broad spectrum of cross-stratification patterns (Klein, 1970; Evans, 1965). Above mean high tide and in the supratidal zone, desiccation features are the most common physical structures. Mudcracks (Figure 10–15) and wrinkled algal mats are typically present.

Bioturbation may be present throughout the tidal flat sediments but is more likely in the upper portion, where physical energy is relatively low. Infaunal organisms produce pellets which may accumulate on the tidal flat surface and give rise to flaser bedding or wavy bedding. As much as 20 to 40% of the sediment may be pellets (Anderson, et al., 1981).

Marsh and Tidal Flat Sequences

Like most of the environments previously discussed, tidal flats tend to display a fining-upward stratigraphic trend. Although individual tidal flat sequences may be only a few meters thick, it is possible to recognize the various positions in the tidal zone as represented by their characteristic textures and sedimentary and biogenic structures.

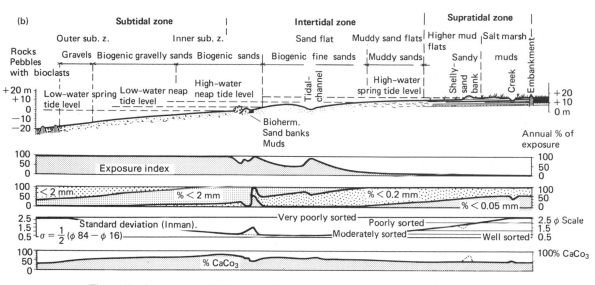

Figure 10-13 Zonation of tidal flat environments from areas of extensive development along the North Sea coast of Europe: (a) aerial view of Jade Bay, West Germany; (b) profile showing zonation in the Bay of St. Malo on the Brittany Coast of France. [(a) from Reineck and Singh, 1980, p. 432; (b) from Larsonnieur, 1975, p. 24.)]

Figure 10-14 Patterns of ripples and megaripples on tidal flats exposed at low tide.

Figure 10-15 Mud cracks developed in cohesive fine-grained sediments in the supratidal environment, Lake Reeve, Victoria, Australia.

The Wash, England

Extensive intertidal flats are present on the embayed coast of eastern England (Lincolnshire). This region, known as The Wash, experiences neap tidal ranges of 3.5 m and spring ranges of 7 m, with tidal flats reaching a width of 3.5 km (Evans, 1965).

General morphology and sedimentological conditions on intertidal environments of The Wash are representative of the typical marsh–tidal flat complex (McCave and Geiser, 1979). A look at a schematic summary delineating sediment facies shows six intertidal depositional environments (Figure 10-16). Observe that there are both higher and lower mud flats in this system and that certain taxa of pellet-producing organisms are restricted to specific positions within the intertidal zone.

Progradation of this tidal flat complex will produce a characteristic sequence which ideally would have a thickness about equal to that of the tidal range. Such an idealized sequence includes some bioturbation and skeletal material in the sands of the lower portion (Figure 10-17). Flaser bedding in ripple and megaripple cross-stratified sands is also present at this horizon (Evans, 1975). The overlying facies is interbedded ripple cross-stratification and wavy mud beds, which represent the lower mud flat (E) (Figure 10-16). The sand facies above is bioturbated and ripple cross-stratified, grading into the higher mud flat facies, which may display algal mats or mud cracks. The top of the sequence is organic-rich mud with many roots and represents the marsh (Figure 10-17).

Similar sequences have been described from the German Bight area of the North Sea along the coast of the Netherlands (Van Straaten, 1954; Van Straaten and Kuenen, 1957) and Germany (Reineck, 1967), and on the Brittany coast of France (Larsonneur, 1975), and are summarized by Klein (1971).

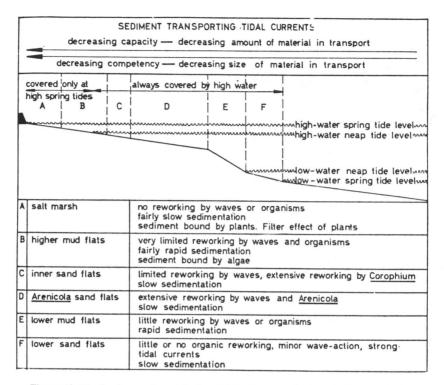

SEDIMENT TRANSPORTING TIDAL CURRENTS		

decreasing capacity — decreasing amount of material in transport

decreasing competency — decreasing size of material in transport

covered only at high spring tides	always covered by high water	
A B	C D E F	high-water spring tide level
		high-water neap tide level
		low-water neap tide level
		low-water spring tide level

A	salt marsh	no reworking by waves or organisms fairly slow sedimentation sediment bound by plants. Filter effect of plants
B	higher mud flats	very limited reworking by waves and organisms fairly rapid sedimentation sediment bound by algae
C	inner sand flats	limited reworking by waves, extensive reworking by Corophium slow sedimentation
D	Arenicola sand flats	extensive reworking by waves and Arenicola slow sedimentation
E	lower mud flats	little reworking by waves or organisms rapid sedimentation
F	lower sand flats	little or no organic reworking, minor wave-action, strong· tidal currents slow sedimentation

Figure 10–16 Surface sediment facies distribution over the intertidal zone of The Wash, eastern England. (From Evans, 1965.)

Bay of Fundy, Canada

The extensive tidal flats of the Bay of Fundy in Nova Scotia (Canada) experience some of the highest tidal ranges in the world; spring values are 15 m, with mean range near 12 m. The morphology and dynamics of these sandy tidal flats have been studied extensively (e.g., Swift and McMullen, 1968; Klein, 1970; Knight and Dalrymple, 1975). Of particular interest are the large bedforms and their complex geometries.

A hypothetical sequence of this environment has been developed by Knight and Dalrymple (1975) and is fairly typical of sand-dominated, macrotidal tidal flats. The progradational sequence fines upward and is about 15 m thick, reflecting spring tidal range. More than one-half of the sequence is comprised of well-sorted sand which displays large-scale, bidirectional cross-stratification (Figure 10–18). It represents migration of megaripples and sand waves by tidal currents. Some tidal channel deposits with characteristic shape and lag deposits may be scattered throughout. Above is a thin facies of sand containing flasers, current, and ripple cross-stratification (Figure 10–18) which represents a lower-energy environment than the underlying sand. The upper intertidal zone is dominated by a mudflat facies which display thin bedding with scattered ripples and bioturbation. Pellets are also com-

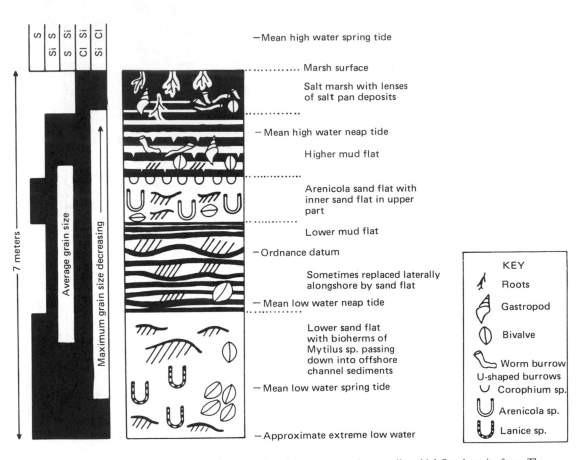

Figure 10–17 Generalized stratigraphic sequence of prograding tidal flat deposits from The Wash. (From Evans, 1975, p. 17.)

mon. The dominant mode of accumulation is by settling from suspension (Knight and Dalrymple, 1975). The top of the sequence is the salt marsh, which is primarily plant debris and mud. There is some terrigenous contribution from tidal input and from runoff from the adjacent land. Stratification is essentially absent.

CARBONATE INTERTIDAL ENVIRONMENTS

The general morphology and many aspects of the process-response system in the carbonate intertidal system are similar to those features of the terrigenous intertidal environment. There are, however, some important differences which justify a separate discussion. Well-developed intertidal carbonate environments of the world are located on the west side of Andros Island in the Bahamas, in Abu Dhabi along the

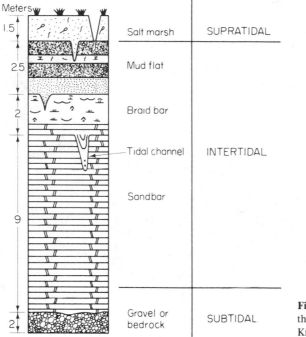

Figure 10-18 Intertidal sequence from the Bay of Fundy, Canada. (From Knight and Dalrymple, 1975, p. 54.)

southern Persian Gulf, and in Shark Bay, Western Australia. All three areas have been studied thoroughly and each is somewhat different from the others.

Shark Bay, Australia

Shark Bay represents what is probably the best locality in the world for studying modern algal stromatolites (Logan et al., 1964, 1970, and 1974b) because of the presence of large stromatolite heads which are common in the rock record. Unlike the other two areas, Shark Bay tidal flats are dominated by sand, most of which is biogenic debris. Tidal ranges in the areas of tidal flat development range from 0.60 to 1.70 m (Logan and Cebulski, 1970). Width of the flats may be up to a few kilometers and slopes are less than 40 cm/km.

The most striking feature of most of the Shark Bay intertidal and supratidal zone is the widespread development of algal stromatolites. The lower portion of the intertidal zone is characterized by smooth algal mat layering with mounds or ridges commonly developed by scour of tidal currents (Figure 10–19). The middle and upper intertidal zone develops a pustular mat, which is an irregular or disrupted algal mat. The supratidal zone supports a film mat which is comprised of carbonate mud draped over the top of the mound or column (Logan et al., 1974a). The stromatolite sequence produced is then analogous to a prograding tidal flat sequence in that each vertical section through the stromatolite contains recognizable accumulations from various portions of the tidal flat environment.

Figure 10-19 Algal stromatolite forms in Hamelin Pool, Shark Bay, Western Australia. Each stromatolite body is 30–40 cm in height. (Courtesy of M. A. H. Marsden.)

Abu Dhabi, Persian Gulf

The southern Persian Gulf is bounded by extensive carbonate intertidal and supratidal flats or coastal sabkhas. These areas have received considerable attention from geologists (Kinsman, 1964; Evans et al., 1969; Kendall and Skipwith, 1968) and the rock record appears to contain numerous sequences representative of the sabkha environment. The best known and largest of the Persian Gulf sabkhas extends for about 200 km along the coast of Abu Dhabi (Figure 10-20).

The intertidal flats seaward of the sabkha are inundated by a neap tidal range

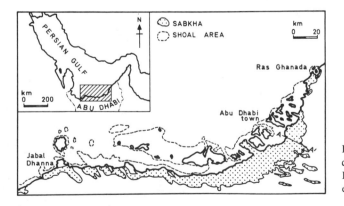

Figure 10-20 Location and geographic distribution of coastal sabkha in Abu Dhabi on the Persian Gulf. (From Evans et al., 1969, p. 146.)

near 1 m and a spring tidal range of about 2 m (Schneider, 1975). The intertidal zone contains two rather distinct facies: a lower one characterized by burrowed lime mud, fecal pellets, and some lithified crust with black mangroves scattered about, and the upper intertidal zone, which is dominated by algal mats and has scattered bushes of *Arthrocnemum*. Tidal channels are common throughout the flats. Vertical accretion takes place by the settling-lag effect described previously and is enhanced by the sticky surface of the blue-green filamentous algae which comprise the mats (Bathurst, 1975).

The sabkha or supratidal portion of this coastal complex is more extensive than the intertidal flats. Deflation typically removes unconsolidated sediment to the level of the water table; however, cementation and development of an algal mat cover also serve to prevent wind erosion. The surface of the sabkha is essentially planar and free of vegetation.

The sabkha develops as the result of both deposition of lime mud and biogenic debris, and also by early diagenetic phenomena. Although this text is not aimed at a consideration of diagenetic processes in general, the sabkha is an exception. The nucleation and growth of evaporite minerals, particularly gypsum and anhydrite, play an integral role in the development of the sabkha sequence.

Groundwater under the sabkha surface becomes saturated, and at a salinity of 117‰ gypsum begins to precipitate (Bathurst, 1975). Other evaporite minerals also precipitate; halite is particularly common on the surface, and anhydrite forms throughout, as does dolomite (Hsu, 1973; Purser, 1973).

The sabkha sequence has been referred to as a shoaling-upward sequence (James, 1977). There are numerous variations that might develop but all begin with an intertidal portion and are overlain by progressively landward portions of the sabkha (Figure 10–21). The general section consists of (1) laminated or bioturbated subtidal lime muds and algal mats, some of which may show desiccation; (2) gypsum mush, dolomite, and cerithid shells; and (3) nodular anhydrite, dolomite, and halite, all typically bedded (Schneider, 1975; Kendall, 1979).

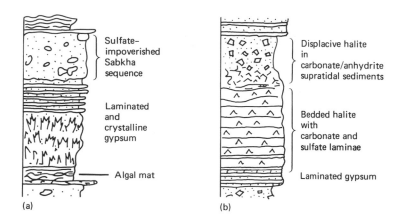

Figure 10–21 General shoaling-upward sabkha sequences showing various combinations of evaporites and carbonates. (After Kendall, 1979, p. 153.)

Andros Island, Bahamas

The western margin of Andros Island on the Great Bahama Bank is one of the largest complexes of tidal carbonate flats in the world. This complex is several kilometers wide and extends for over 150 km in a roughly north-south orientation. This coast is characterized by low wave energy, and astronomical tides range from 17 cm (neap) to 41 cm (spring) (Ginsburg and Hardie, 1974). Wind or storm tides frequently exceed astronomical tides and produce a pronounced effect on the tidal flats.

A low beach ridge separates the shallow, subtidal marine platform from the tidal flat complex. This complex includes both intertidal and supratidal environments, with tidal channels, natural levees, and polyhaline ponds (Figure 10-22). Most of the sediment landward of the beach ridge is lime mud or pelleted mud, with lesser amounts of biogenic skeletal sand.

Tidal channels are subjected to currents that may reach 1 m/sec; however, the water is generally completely clear, as is the water over the flats themselves. Due to cohesion of the lime mud and the tendency for rapid precipitation of carbonate cement in this environment, there is essentially no sediment transport taking place under so-called normal conditions (Ginsburg and Hardie, 1975). Abnormally high tides and waves which are generated by the passage of frontal systems cause considerable sediment to be transported, and it is as a result of these conditions that this environment experiences accumulation. Bioturbation is widespread, especially in the subtidal and the intertidal environment; these organisms cannot survive the prolonged exposure of the supratidal flat.

The combined result of the features described above is that additional detrital sediment is transported to tidal flats via flooding of the tidal channels. This gradual vertical accretion produces a slowly prograding tidal environment that is in many

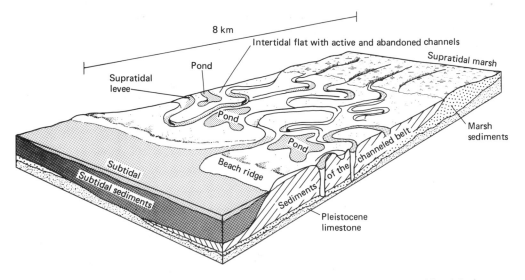

Figure 10-22 Skematic diagram of extensive tidal flat complex on the west side of Andros Island, Bahamas. (After Shinn et al., 1969, p. 1226.)

Part Three   Transitional Environments

ways analogous to the tidal flat–marsh complex described earlier in the chapter. The intertidal muds are pelleted and burrowed. Well-developed natural levees rise up to 15 cm above the adjacent flats and are comprised of laminated mud. These natural levees are the site of formation of modern dolomite (Shinn et al., 1969).

The intertidal portion of this complex may have small supratidal mounds which support a tropical marsh vegetation and which contain shallow ponds that display a wide variation in salinity depending on rainfall, flooding by storms, and rate of evaporation. The adjacent supratidal area is analogous to a marsh but has scattered grasses and mangroves with widespread development of algal mats (Figure 10–23).

An excellent summary of the Andros Island tidal complex is shown in Figure 10-24, which utilizes the concept of exposure index to present a generalized stratigraphic section (Ginsburg and Hardie, 1975). The exposure index is simply the percentage of time a given zone or environment is subjected to subaerial exposure. In this complex the levee crest is nearly always exposed, the lower marsh is exposed about 60 to 70% of the time and the channel floor is almost never exposed. This exposure index then can be viewed in terms of the stratification style, desiccation features, and burrow types (Figure 10–24). As a result it is convenient to consider the prograding carbonate tidal sequence as one of shoaling-upward or greater exposure upward.

ANCIENT TIDAL ENVIRONMENTS

The term **tidalites** (Klein, 1971) has been proposed to refer to sediments that have been deposited by tidal currents. Such sediments may accumulate in intertidal or subtidal environments. The discussion here, as in the preceding sections, is concerned only with the intertidal (including supratidal) environment.

A conference was convened in February 1973 to assess the state of the art of tide-dominated sedimentary environments and their counterparts from the rock record. Modern and ancient examples of both terrigenous and carbonate environments were considered (Ginsburg, 1975). The consensus was that whereas it is relatively easy to recognize tidalites, it is commonly not possible to distinguish between tidalites which are intertidal and those which are subtidal. In general, one can do a better job of interpreting carbonate environments than terrigenous ones because of better exposure indicators and the presence of distinctive growth forms of algal stromatolites. In terrigenous environments one can be certain of the position of marshes relative to sea level because of the excellent datum provided by this environment.

Cohansey Sand (Miocene-Pliocene), New Jersey

The Cohansey Sand is a medium to coarse quartz sandstone which crops out throughout most of the New Jersey coastal plain. Thickness of exposures is typically 6 m but ranges from 2 to 13 m (Carter, 1978). A detailed analysis by Carter (1975, 1978) of textures, sedimentary structures, and stratigraphy has shown that two sequences, comprised of a total of nine facies, characterize the Cohansey Sand. One is

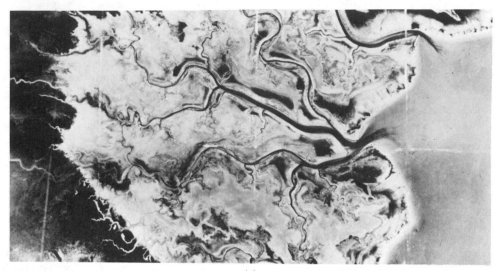

(a)

(b)

Figure 10–23 Aerial photographs of extensive carbonate intertidal and supratidal flats on the west side of Andros Island; (a) vertical view; (b) oblique view.

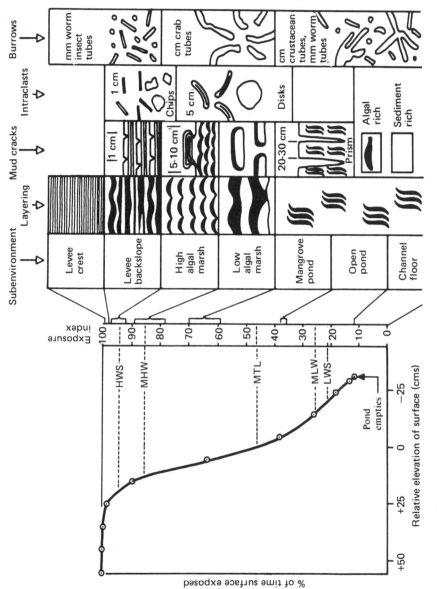

Figure 10–24 Diagram showing exposure index of various environments in the intertidal complex. (From Ginsburg and Hardie, 1975.)

353

interpreted as a barrier island deposit and the other as a tidal deposit or barrier-protected sequence (Figure 10–25). At numerous locations the barrier-protected sequence overlies the barrier sequence.

Four facies characterize the barrier-protected sequence: cross-stratified sand, burrowed cross-stratified sands, burrowed massive sand, and interbedded sand and mud facies (Carter, 1978). The basal, relatively thick, cross-stratified sand facies contains both trough and planar cosets with abundant reactivation surfaces. Although most cross-strata dip in one distinct direction, a few dip in opposition to this direction [Figure 10–25(a)]. Distinct channels are evident in laterally continuous exposures. The burrowed and cross-stratified facies is similar to the channel, cross-stratified facies except that *Ophiomorpha* burrows are widespread. This generally vertical burrow with a diameter of a few centimeters is the trace fossil produced by *Calianassa* (ghost shrimp). Flaser bedding is also present in this facies. The burrowed sand facies has abundant *Ophiomorpha* burrows near the base. There is a fining-upward trend in mean grain size, with mud lenses and thin concentrations of heavy minerals near the top of the facies. Interbedded mud and sand is the fourth facies. It consists of thin layers of alternating texture with some ripple cross-strata in the sands. Deposition in

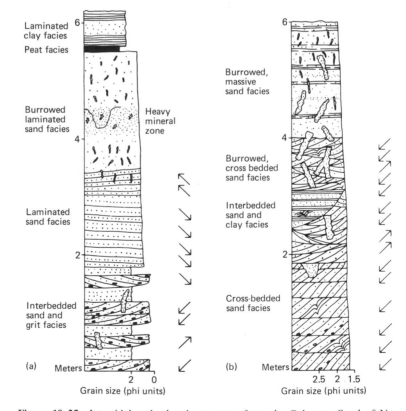

Figure 10–25 Intertidal and related sequences from the Cohansey Sand of New Jersey: (a) a barrier sequence; (b) a barrier-protected sequence. (From Carter, 1978, pp. 935, 936.)

low-energy channels is indicated by its relation to the channel, cross-stratified sand, [Figure 10–25(a)] and its long, sinuous geometry (Carter, 1978).

The interpretation of depositional environment for this sequence as described by Carter (1975; 1978) is similar to the modern complexes bordering the North Sea (Van Straaten and Kuenen, 1957; Reineck, 1967). The cross-stratified unit represents a subtidal but tide-dominated environment with megaripples and some channelization. Time-velocity asymmetry was responsible for the reactivation surfaces and the dominance of a single direction of the cross-strata. The interbedded sand and mud facies represents deposition in protected or abandoned tidal channels, and the burrowed and cross-stratified facies indicates a general decrease in depth, perhaps intertidal, and represents a sandy flat environment generally devoid of channels. The intensely burrowed unit was deposited under intertidal conditions. Thin layers of heavy minerals are lag concentrates. The sequence is therefore generally shoaling upward and represents a progradation of a tidal complex.

The other sequence from the Cohansey Sand [Figure 10–25(b)] also contains a tidal component but is largely a barrier island deposit. The upper meter or so of peat and laminated mud represents a marsh environment (Carter, 1978) analogous to many of the back-island marshes that are developed along the modern Atlantic and Gulf Coast barrier island systems.

Lower Coal Series (Lower Jurassic), Bornholm, Denmark

Two recent papers by Sellwood (1972, 1975) describe what is interpreted as an intertidal sequence from the island of Bornholm near the mouth of the Baltic Sea on the south flanks of the Baltic Shield. A general lithofacies map of northern Europe and vicinity from the Early Jurassic time shows the island of Bornholm to have been covered by marginal and marine facies with carbonates to the southwest (Figure 10–26). Stratigraphically, the Lower Coal Series rests on the Keuper marls (Triassic) and is itself overlain by the Marine Lias. The Lower Coal Series is comprised primarily of fluvial sandstones, with some coals. The sequence of interest here is the upper portion of this unit, which is interpreted as a transition between the underlying fluvial facies and the marine strata above (Sellwood, 1972).

Description

Most of Sellwood's interpretations for a tidal origin are based on an extensive exposure nearly 400 m long and about 15 m thick (Sellwood, 1972). All or most of the strata in the basal 5 m consist of cross-stratified, fine to medium sandstone which displays a variety of large-scale cross-strata varieties (Figure 10–27). Some herringbone cross-stratification is present. Bedding styles indicate upper-flow-regime conditions at the base but that this gave way to the lower flow regime with small ripples (Sellwood, 1975). Reactivation surfaces are present on megaripples, which also display clay drapes. Individual bedding surfaces have several burrows descending from them. Shallow channels are cut into the sand and are filled with lateral accretion deposits.

Most of the remainder of the exposure is dominated by well-preserved tidal bed-

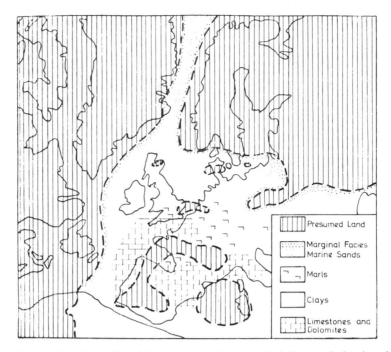

Figure 10-26 Generalized lithofacies and paleogeography in Europe during deposition of the Lower Coal Series of the Jurassic. (From Sellwood, 1972, p. 95.)

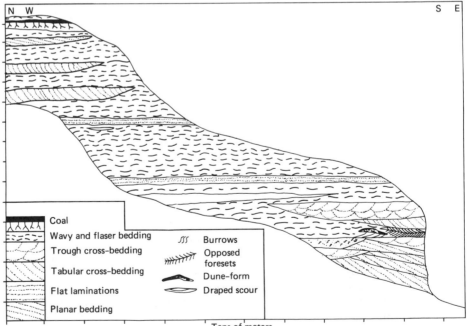

Figure 10-27 Lithofacies in exposure of Lower Coal Series at Bornholm (Jurassic). (From Sellwood, 1972, p. 97.)

Figure 10-28 Tidal bedding from the Jurassic in Bornholm, Denmark. (Courtesy of B. W. Sellwood.)

ding (Figure 10-28) with well-developed flaser and lenticular types. The dominant texture is sand, which shows both wave and current ripples draped with clay. Lateral accretion sets with considerable lateral continuity are common; some extend for 100 m (Sellwood, 1975). Small megaripple bedding is present in the lower portion of this tidal sequence. Coal seams up to 30 cm thick are present at the top of the sequence. They have rootlets extending from the coals and disturbing the underlying flaser bedding.

Interpretation

Texturally, the sequence at Bornholm shows a distinct fining upward trend and gives the distinct impression of a shoaling-upward sequence. A systematic analysis confirms the latter. The cross-bedded sand with herringbone patterns represents megaripples with reactivation surfaces and the planar bedding represents upper-flow-regime conditions. This relatively high energy set of circumstances was tide dominated but may have been either subtidal or intertidal. Channels were also present on this sand flat environment.

The tidal bedding with abundant flasers and wavy bedding (Figure 10-28) is almost exactly like the tidal bedding described and figured by Evans (1965) from The Wash and by Reineck (1972) from the Jade Bay area of the North Sea. This bedding style accumulated on an intertidal flat with ripples and scour lag which gave rise to the vertical accretion. Coals near the top of the sequence represent a marsh-type environment in the upper intertidal and supratidal zone.

Manlius Formation (Lower Devonian), New York

The Lower Devonian Helderberg Group of eastern and central New York State has been investigated in detail by Laporte (1967, 1969, 1971, 1975) from the standpoint not only of the physical nature of the rocks but also by careful analysis of its contained fauna. The Helderberg is a sequence of fossiliferous limestones about 100 m in thickness, representing a situation of general westward transgression (Laporte, 1967, 1969). The Manlius Formation is the basal unit and rests on the Rondout Dolomite (Silurian–Devonian). It is only 7 to 15 m thick (Laporte, 1975) but contains a rather complete sequence of tidal and tide-related depositional environments.

Description and interpretation

Three distinct facies are present within the Manlius: subtidal, intertidal, and supratidal (Laporte, 1967). The stratigraphic and geographic distribution of the individual facies is complicated. In general, the basal Manlius is intertidal facies throughout, with most of the upper portion being subtidal. The supratidal facies is limited in extent (Laporte, 1967).

The subtidal facies is mostly pelleted lime mud (pelmicrite) with an abundant and diverse fauna. Burrows are common and bedding is generally lacking. Lack of physical structures indicates that biologic activity was dominant.

A thinly stratified pelmicrite which is interlayered with pelsparite and biosparite characterizes the intertidal facies (Laporte, 1967). Some algal stromatolites and oncolites are present, as are lenses of oolite. Intraclasts are associated with horizons displaying desiccation features. Only a modest fauna is present. Evidence of reworking is suggested by mud-filled shells in sparite, and preferred orientation indicates at least modest currents. Small cut-and-fill structures are also present. The interpretation is one of a classic intertidal sequences with only a small tidal range. The latter is inferred by absence of cross-stratification, bedforms, and other features associated with strong tidal curents.

Thinly stratified, unfossiliferous dolomite units located within the Manlius are considered as representing the supratidal environment. Pellets are present and some of the layering is interpreted as of an algal mat origin (Laporte, 1967). Dolomite rhombs are common in the micrite layers; some appear to be filling mudcracks. **Birdseye structure** in the form of irregular vugs is also present and probably represents desiccation features.

Reconstruction of environments

The three facies recognized and interpreted by Laporte display a stratigraphic distribution that indicates numerous minor fluctuations in sea level, probably coupled with some small-scale relief on the depositional surface (Laporte, 1967). This would cause the apparently sporadic stratigraphic position of the supratidal facies. A general cross section of the Manlius environment shows a situation much like that on the west side of Andros Island, Bahamas (see Figures 10–22 and 10–29).

Walker and Laporte (1970) have reconstructed the various components of the Manlius environments, including both sediments and the associated communities. The subtidal facies contains a diverse marine fauna, including various corals, brachiopods, snails, and calcareous algae. These occupy a burrowed lime mud which contains skeletal fragments. These authors have divided the intertidal facies into a lower and upper community based on the exposure index. Organisms in the lower intertidal community consist of brachiopods, oncolites, some ostracods, and a variety of burrowing deposit feeders. The substrate rarely shows desiccation features and is

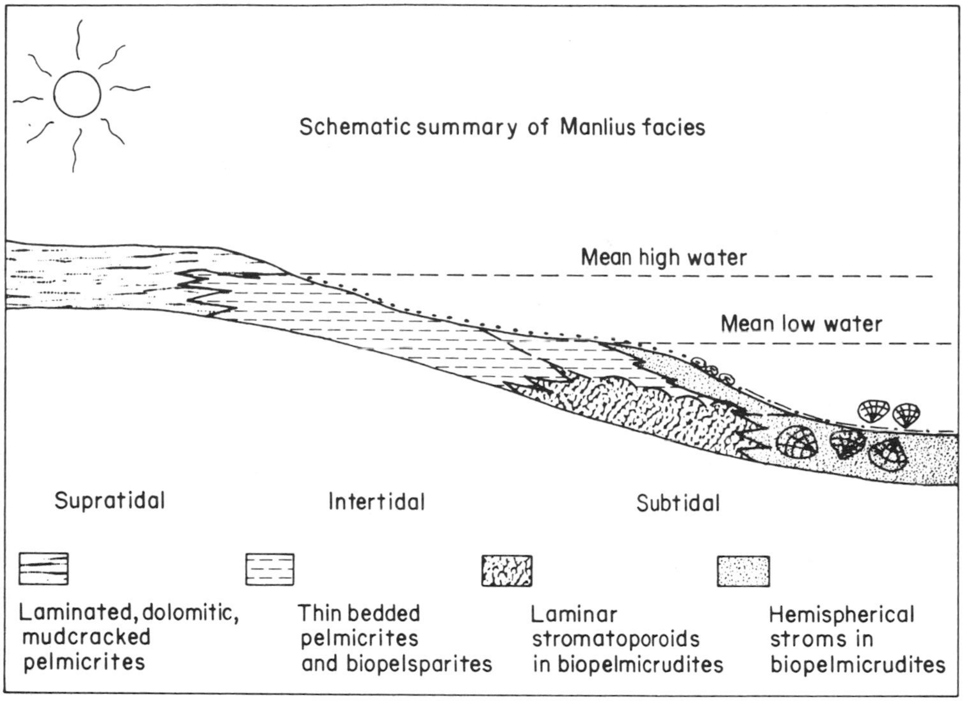
Schematic summary of Manlius facies
Mean high water
Mean low water
Supratidal
Intertidal
Subtidal
Laminated, dolomitic, mudcracked pelmicrites
Thin bedded pelmicrites and biopelsparites
Laminar stromatoporoids in biopelmicrudites
Hemispherical stroms in biopelmicrudites

Figure 10–29 Interpretation of Manlius depositional environments during Devonian time. (From Laporte, 1967, p. 91.)

interlayered pelleted mud and biogenic skeletal debris. By contrast the upper intertidal environment was dominated by burrowing deposit feeders and filter feeders, by algal mats, and had some mudcracks in the dominantly pelleted mud. The supratidal community had scarse ostracods and was covered with desiccated algal mats.

Wood Canyon Formation (Precambrian), Nevada

Much effort aimed toward establishing models for intertidal sedimentation in terrigenous sediments has been expended. Considerable work in this regard has been carried out by Klein (1970, 1971, 1972a, 1972b, 1975, 1977a), who has proposed a stratigraphic model for determination of paleotidal range.

The Middle member of the Wood Canyon Formation represents the uppermost Precambrian unit in the Great Basin of Nevada (Klein, 1975). Thickness ranges from 150 to 550 m, with numerous sequences that have been interpreted as intertidalites.

Description of intertidal features

A total of 10 different structures is listed by Klein (1975) as being characteristic of the Middle Wood Canyon. Included are bedform features such as herringbone cross-stratification, interference ripples, and reactivation surfaces. All are common on intertidal surfaces but may also occur in subaqueous environments. Tidal bedding, including flasers, wavy bedding, and lenticular bedding, is also present. Desiccation cracks and conglomerates of mud clasts also occur. There are burrows similar in form to those in modern intertidal flats.

Although individually these structures may be found in environments other than intertidal flats, their collective presence is strongly indicative of the intertidal environment. Probably some of these features were developed, at least in part, in the subtidal environment.

Paleotidal model

Detailed stratigraphic analysis of Middle Wood Canyon has led Klein to propose a model sequence which can be used to establish paleotidal range during accumulation (Klein, 1970, 1971, 1972b). This fining- and shoaling-upward sequence progresses from a relatively thick, lower intertidal flat through the high tidal flat and supratidal zone.

The base is characterized by rather thick herringbone cross-strata with some interference ripples (Figure 10–30). The latter are interpreted as representing late-stage emergence runoff. Above this sand sequence is typically a thin sandy layer with flaser bedding. The flaser unit grades upward into tidal bedding, with a general upward decrease in sand-size material. The sequence is capped by burrowed mudstone with

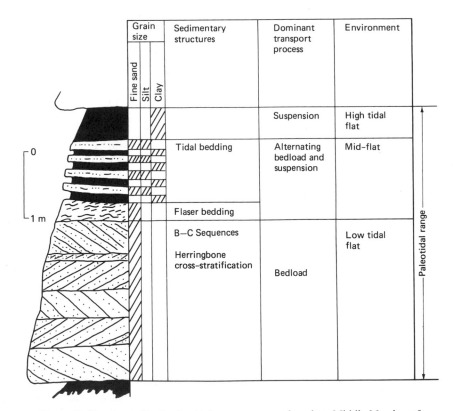

Figure 10–30 Generalized paleotidal range sequence based on Middle Member of the Wood Canyon Formation, Nevada. (From Klein, 1972a.)

desiccation cracks, representative of the upper tidal flat or supratidal zone (Figure 10–30).

The mode of sediment transport ranges from bedload dominated at the base to suspended mode for the upper tidal flat. Klein (1971, 1972b) proposed that the thickness of such a sequence is representative of the paleotidal range for the time of accumulation. In many respects this seems reasonable, but it should probably be interpreted as the maximum tidal range. The reason for this is the extreme difficulty of determining the lower limit of the intertidal zone as contrasted to the subtidal but tide-dominated zone. Another problem is the completeness of the section or lack thereof. Certainly, several sequences in the same general stratigraphic horizon should be examined before making paleotidal interpretations. Nevertheless, this model is at least as good as anything else available and warrants consideration when one makes paleoenvironmental reconstructions.

SUMMARY OF CHARACTERISTICS OF SEDIMENT/SEDIMENTARY ROCK BODIES REPRESENTING INTERTIDAL ENVIRONMENTS

Both carbonate and terrigenous intertidal environments present a unique set of characteristics that may be preserved and recognized in the stratigraphic record. Although there are some features common to intertidal environments in general, separation of vegetated from nonvegetated and terrigenous from carbonate types is appropriate for most considerations.

Tectonic setting. Although there is no specific relationship between tectonics and the development of tidal flats and marshes, the more extensive intertidal environments are relatively common on broad stable coasts compared to leading-edge coasts of high relief. Broad intertidal environments tend to be most prominent boardering large shelf seas; however, tidal range and sediment availability are also important factors.

Shape. Intertidal deposits are tabular in shape. However, some are not extensive perpendicular to the paleo-shoreline orientation, thus yielding an elongate configuration to the sediment body.

Size. Typically, intertidal deposits are local in extent and only a few meters in thickness. This is largely a result of the special conditions necessary for their formation. Such phenomena as rapid sediment influx or changes in sea level may cause this zone to become subtidal or elevated above direct influence of the sea. There are some very thick sequences interpreted as having accumulated, at least in part, in the intertidal environment. These sequences typically have cycles that include marine and nonmarine components with tidalites incorporated in the section. Such sequences occur in both carbonate and terrigenous systems.

Textural trends. Sequences deposited in the intertidal environment do tend to show distinct textural sequences with a general fining-upward trend. This reflects progradation over the intertidal environment which displays a decrease in grain size from low tide to high tide. This trend commonly holds true for both terrigenous and carbonate tidal flat accumulations.

Lithology. For purposes of characterizing the sediment composition of intertidal environments, it is appropriate to consider three distinct categories: terrigenous tidal flats, carbonate tidal flats, and marshes. The most varied lithology is found in terrigenous tidal flat deposits, which reflect the composition of related sediments or rocks. Typically, quartz, feldspar, rock fragments, and clay minerals are dominant, with some biogenic materials, generally carbonate. Although carbonate tidal flat sequences are characterized by biogenic debris of calcite and aragonite composition, some dolomite can be expected, especially in the high natural levee and supratidal deposits. Marshes are generally a combination of plant debris and terrigenous particles, with some biogenic carbonate. The terrigenous particles may have various compositions, but quartz and clay minerals are dominant.

Sedimentary structures. Three distinct categories should be considered when characterizing sedimentary structures. Marshes may display some thin bedding but generally only various types of bioturbation, primarily by roots and invertebrates burrowing into the marsh substrate. Terrigenous tidal flat sequences may also display various bioturbation structures, especially those produced by mollusks and worms. Small-scale ripple cross-stratification, flaser bedding, wavy bedding, and lenticular bedding are common, with desiccation features in the muds near the top of the sequence. Carbonate tidal flat deposits display a similar sequence but with less abundance of physical structures. Algal mats are common in the upper, supratidal portion of carbonate sequences.

Paleontology. The fauna and flora of intertidal environment is abundant and relatively diverse. Marsh sequences would be expected to contain much plant material, generally as peat or coal. In addition, oysters and burrows of crabs and worms are common. Tidal flat sequences may contain some scattered plant debris in the upper strata but would be expected to include various burrow structures, gastropods, and pelecypods as dominant fossils.

Associations. Because of their deposition in the strand zone, intertidal deposits may occur between marine and nonmarine accumulations, most generally in progradational sequences. Thin-bedded intertidal zone sediments commonly overlie sandy, cross-stratified tidal delta deposits or subtidal estuarine deposits. Changes in sea level may result in cyclic sequences of subtidal and intertidal deposits.

ADDITIONAL READING

FREY, R. W. AND BASAN, P. B., 1978. Coastal salt marshes *in* Davis, R. A. (ed.), *Coastal Sedimentary Environments.* Springer-Verlag, New York, 101–169. Probably the best summary on coastal marshes from the point of view of the geologist. Contains an extensive reference list.

GINSBURG, R. N. (ED.) 1975. *Tidal Deposits.* Springer-Verlag, New York, 428 p. The best single book on tidal sedimentation. Forty-five short articles on both modern and ancient tidal flat deposits in terrigenous and carbonate environments by the recognized experts of the world.

KLEIN, G. DEV. (ED.), 1976. *Holocene Tidal Sedimentation.* Benchmark Papers in Geology, v.

30. Dowden, Hutchinson & Ross, Stroudsburg, Pa., 423 p. A collection of seventeen reprinted papers representing the best work on modern tidal environments. Only two are from carbonate terrains.

KLEIN, G. DEV. 1977. *Clastic Tidal Facies.* Continuing Education Publication Co., Champaign, Ill., 149 p. A concise general treatment of tidal sedimentation which emphasizes the author's point of view but covers the topic well. A handy, short reference text.

11 Coastal Bays: Estuaries and Lagoons

The various styles of embayments or semi-enclosed, relatively small water bodies that occupy the coastal zone are here called **coastal bays**. This term is used in the broad sense regardless of origin, hydraulics, salinity, or sedimentation style. For purposes of depositional sedimentary environments it is practical to consider two primary types of coastal bays: estuaries and lagoons. An **estuary** is a coastal bay that is subjected to freshwater runoff and experiences regular influence from astronomic tides. By contrast, a **lagoon** is a coastal bay that receives no significant freshwater runoff and is restricted in its circulation with the open ocean.

It is readily apparent that all coastal bays are not included under these two categories, even if each is used in the broad sense of the definitions. Some coastal bays are simply coastal embayments with no special attributes regarding sediment discharge from land, salinity, or hydrodynamics. Such coastal bays are not placed in a special category, and because of their similarity with the adjacent open marine environment, these bays do not present a unique sedimentary depositional environment. As a consequence, the presence of such an environment preserved in the rock record would probably contain the same or similar features as the shallow open marine system.

It should be observed that in the several excellent symposia volumes on estuaries and lagoons (e.g., Lauff, 1967; Castanares and Phleger, 1969; Nelson, 1972; Cronin, 1975) there has been a tendency to use the terms "estuary" and "lagoon" loosely. As an example, Phleger (1969) considers virtually all coastal bays as being lagoons. The following discussions adhere to the definitions given above.

ESTUARIES

There is great variety in morphology, hydrodynamics, and sediment distribution in estuaries, but there are broad factors which are common, at least to most estuaries (Schubel, 1971). As a result, it is possible to recognize the estuarine environment in the rock record.

General Characteristics

According to Pritchard (1967) and Dyer (1973), there are four main morphologic types of estuaries: drowned river valleys, fjords, bar-built estuaries, and tectonically produced estuaries. The widely distributed, irregularly shaped estuaries common along the coastal plains of the world are the result of drowning river valleys [Figure 11–1(a)] as sea level rose during the Holocene. Both the Chesapeake Bay and Delaware Bay of the eastern United States are examples. This type of estuary is the site of rapid sediment accumulation and has rather high preservation potential. Fjords are deep, steep-sided estuaries carved by glaciers. They are commonly characterized by poor circulation and slow sediment accumulation. Fjords are small and typically are developed along tectonically active coasts with high relief. They possess a low preservation potential because of their geologic setting.

As estuaries develop and coastal processes begin to transport sediment along shore it is common for spits or barriers to develop across the estuary mouth. These sediment bodies may only partially close the estuary mouth or they may close it almost completely, with only tidal inlets interrupting an otherwise continuous barrier [Figure 11–1(b)]. The large estuaries behind the Outer Banks of North Carolina represent such estuary types.

Tectonically produced estuaries are generally confined to leading-edge coasts, where faulting and subsidence create embayments. San Francisco Bay in California is such an estuary.

Estuarine Processes

The estuary is a focal point for a variety of interactive processes. Some of these are regular and predictable and some are not. The processes in question may result from freshwater runoff, from tidal-generated forces, or from a combination of the two. Wind may also affect estuarine dynamics. There are varying degrees to which each process affects a given estuary. Numerous classifications of estuaries have been developed utilizing a variety of parameters. Because of the complexity of the estuarine system and the complete gradation through the spectra of conditions, the reader should realize that boundaries between categories are gradational and difficult to define.

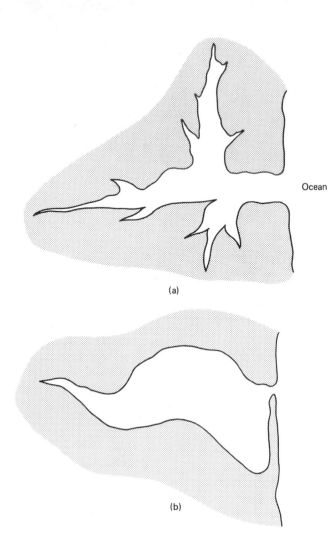

(a)

Ocean

(b)

Figure 11-1 (a) Drowned estuary configuration, showing typical complex shoreline pattern; (b) bar-mouth estuary configuration.

Tides

By definition, estuaries are subjected to tidal current activity, and these currents, together with riverine discharge, are the two most important estuarine processes. Probably the best approach to categorizing estuaries by tidal range is by following Davies (1964), who designated microtidal ($<$ 2 m), mesotidal (2 to 4 m), and macrotidal ($>$ 4 m) range coasts (see Chapter 2). Hayes (1975) has applied this classification to morphologic models of estuaries.

In general, tidal currents in estuaries are directly related to tidal range, but this relationship may not be valid at some locations. Estuaries that exhibit broad mouths with little or no constriction for entering or exiting tidal currents [Figure 11-1(a)] follow the expected relationship well. By contrast, the mouths of those estuaries that contain spits or other constrictions [Figure 11-1(b)] do not necessarily follow this relationship. The inlets or other openings between the estuary and the open sea represent the primary pathway for exchange of water during rising and falling tides. It is through these pathways that the **tidal prism** must pass during each tidal cycle. The tidal prism is essentially a water budget and is comparable to stream discharge except that it flows both ways.

As a result, a large estuary that shows only a small tidal range may generate extremely rapid tidal currents at its mouth because of the tidal prism (area of estuary times tidal range) that must be transmitted during each tidal cycle. Examples include Aransas Pass on the Texas coast, John's Pass on the west coast of Florida, and Oregon Inlet on the Outer Banks of North Carolina. In situations where freshwater runoff is high, there is a much greater ebb tidal prism than flood tidal prism. It is also possible for circulation cells to become established between adjacent inlets such that one might be flood dominated and the other ebb dominated (Price, 1963; Lynch-Blosse and Davis, 1977) with respect to tidal prism.

Rapid currents through inlets cause entrainment of sediment. The sudden decrease in current speed at the landward or seaward side of the inlet causes accumulation of sediment. These large sediment bodies are called **tidal deltas** because of their origin and shape. **Flood deltas** are on the landward side of the inlet and are deposited by flood tidal currents, and **ebb deltas** accumulate on the seaward side and are the result of ebb tidal deposition (Figure 11–2). These important sediment bodies are treated in detail later in the chapter.

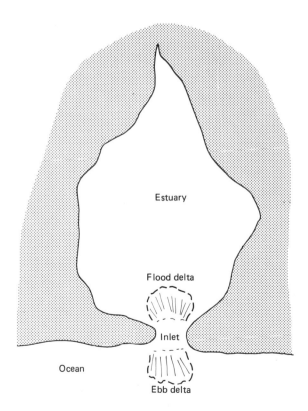

Figure 11-2 General configuration of tidal inlet and associated tidal deltas.

Tidal Currents. As tide ebbs and floods, one observes the currents generated by this process and also the change in water level that is associated with it. It would seem logical to assume that the tide level or vertical tide would be in phase with tidal currents. That is, high slack water and high tide would coincide in time, and vice versa. Under some conditions, particularly in small, shallow estuaries, the expected situation prevails. Large and deep estuaries, however, show a distinct phase lag of from several minutes to more than an hour (Postma, 1967). It is actually rather rare for tidal stage and currents to be both in phase and symmetrical.

Most commonly, the peak current velocities occur somewhat before or after midtide level under flood and ebb conditions, respectively (Figure 11–3). Because of the lack of symmetry in current velocities through time and in estuary configuration, the vertical tide may be asymmetric. This phenomenon of time-velocity asymmetry (Postma, 1961) is very important in sediment transport. In many estuaries strong currents are flowing when the tide turns. This causes a segregation of flow such that some channels are dominated by ebbing currents and others by flooding currents. It is especially important in inlets where ebb flow is maintained after rising tide is initiated. Flood currents are initiated along the margins of the inlet, whereas ebb currents persist in the throat. Time-velocity curves for each of these channels show distinct differences (Figure 2–19). The marginal channels are distinctly flood dominated, with ebb flow confined to the early stages of the ebb cycle.

Estuarine circulation

As a tidal wave of salt water moves up an estuary and as the freshwater exits from the estuary, there is a zone where these quite distinct water masses interact. The types of interaction that occur may be important to sedimentation in the estuary and represent another method of classifying estuaries. There are three primary estuary types, based on the nature and distribution of the zone where fresh water and seawater come together.

A highly stratified or salt-wedge estuary is one where there is little mixing between fresh and salt water and a distinct density stratification persists [Figure 11–4(a)]. Such a situation exists where the river discharge is the dominant process in the estuary (Pritchard, 1955). The only mixing occurs by vertical advection in the shear zone between the two opposing water masses (Biggs, 1978). Sediment carried to the estuary from streams may settle into the salt-wedge layer and be transported landward, eventually being deposited at the tip or landward extent of this water mass.

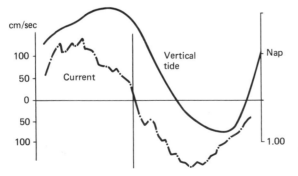

Figure 11–3 Tidal current and tide level over a tidal cycle. (From Postma, 1967, p. 163.)

Part Three   Transitional Environments

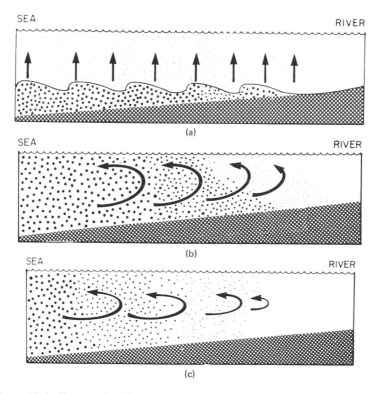

Figure 11-4 Estuary classification based on hydrography includes (a) a well-stratified type, (b) one that shows partial mixing, and (c) a well-mixed type. (From Biggs, 1978, pp. 89, 90, 91.)

Coarse sediment particles generally settle out landward of the salt wedge. These phenomena produce a zone of relatively high sediment accumulation just beyond the tip of the salt wedge.

The major distributaries of the Mississippi River represent excellent examples of this type of circulation. A similar circulation type is present in the Vellar estuary on the southeast coast of India (Dyer, 1973).

Well-stratified estuaries display a complicating circulation phenomenon which is related to the Coriolis effect. During flood tides the interface between water masses is tilted upward on the right side of the estuary in the northern hemisphere as one looks landward, and during ebb tides it is tilted upward on the left side. As a result, there is a circular component to circulation in a well-stratified estuary, with a null point or **amphidromic point** in the center.

Partially mixed estuaries are those where tidal influence plays an important role in determining circulation and in mixing of the water masses. The turbulence created by tidal action causes downward movement of fresh water as well as upward movement of seawater (Pritchard, 1955). The result is that there is a gradual increase in salinity in the seaward direction at any level of the water column and that at any location the surface water is less saline than the bottom water [Figure 11-4(b)]. Isohalines drawn on a longitudinal section through a partially mixed estuary would be nearly

Chapter 11   Coastal Bays: Estuaries and Lagoons

parallel and inclined upstream. The Coriolis effect is also apparent in partially mixed estuaries, in that salinity is highest on the right side of the estuary as the observer looks landward.

Suspended sediment tends to concentrate in the area of the turbidity maximum, which is located just downstream from the landward limit of seawater intrusion (Biggs, 1978). Chesapeake Bay on the Atlantic coast of the United States and the Thames estuary in England are good examples of partially mixed estuaries.

When tidal activity is at least as much an influence on mixing as the river discharge, a vertically homogeneous estuary is developed [Figure 11-4(c)]. Wide-mouthed estuaries where wave action can also cause mixing tend to be the best examples of vertically homogeneous or totally mixed estuaries. Mixing of the water masses results in isohalines that are nearly vertical and increasing seaward.

The Coriolis effect plays a role in circulation and sediment transport in totally mixed estuaries. As flood tide occurs in the northern hemisphere, the isohalines are displaced landward on the right side as one looks up the estuary, and on the ebb tide they are displaced seaward on the left side from the same point of view. Sediment tends to follow the same pattern, with marine sediments concentrated on the right side of the estuary with the concentration of riverine sediments on the opposite side (Biggs, 1978).

Although the foregoing classification of estuaries is useful, one must be aware that there is complete transition from well-stratified to totally mixed estuaries; no natural boundaries exist. It is also important to realize that, depending on freshwater discharge, one estuary can change types.

Sediments

Estuaries are sediment sinks because the estuarine environment in general acts as a sediment trap. Unless there are tectonic events or abrupt sea-level changes, estuaries are geologically short lived. It is obvious that much sediment is carried to the estuary via freshwater runoff from land, but there are other sources as well. It is rather difficult to generalize about estuarine sediments because each estuary is different from the rest. One general characteristic that does seem to have some control over the sediments that accumulate is the relative roles of marine processes compared to riverine processes. This is similar to the situation in deltas (see Chapter 9) and also to the classification of estuaries mentioned above. Marine-dominated estuaries are subjected to strong tidal currents and may also experience wave action that interacts with sediments. Such estuaries are commonly characterized by sand-size sediment and distinct sediment body morphology. By contrast, the riverine-dominated estuary is a low-energy environment typically dominated by mud and having thin, nearly horizontal sediment layers. There are, of course, combinations of both.

Provenance

The origin of sediments that accumulate in estuaries is at least as complex as the circulation system that operates therein. There has been considerable lack of agreement among researchers as to the source of estuarine sediment. There are several possibilities and certainly the most obvious is the runoff from land via streams. In addition, sediment may be carried into the estuary from the open ocean. Most people would probably agree that these two sources contribute the bulk of estuarine sediments; however, there is no agreement about the relative proportion of each. Other sources of estuarine sediments are biogenic detritus from estuarine organisms, sediment particles derived from erosion of the estuarine margins, and biodeposition of sediments by pellet-making organisms.

There is greatest understanding of the sediment carried to the estuarine environment via streams as compared with other sources (Meade, 1972). The amount is relatively high but varies greatly in both space and time. In many estuaries there may be more than one stream providing sediment. The heavy mineral assemblages accumulating in the estuary may be used to trace the source of sediments and determine relative contributions of each stream (Guilcher, 1967). Other more subtle criteria such as grain shape (Erlich and Weinberg, 1970) and grain size may be utilized.

Offshore sediment represents a potentially large source for estuaries where tidal circulation is the dominant mechanism for transport. It is generally believed that most of this sediment supply is in the form of bed material transport. Overall clarity of the water during flooding tides testifies to the lack of wash load. Estuaries that are barred may be characterized by littoral drift transporting much sediment to the inlet or estuary mouth. It may then be entrained by tidal currents and be transported into the estuary. Additionally, barred estuaries receive great quantities of sediment from blowover or washover phenomena. These are commonly discussed in relation to high-energy events, where transport over the barrier and into the estuary is abrupt and on a large scale. Onshore winds may provide considerable sediment to the estuary under nonstorm conditions.

The sediments and bedrock exposures along the estuary margin may be an important sediment contributor or they may not be. Large estuaries with at least a modest fetch will contain waves that cause coastal erosion similar to that in the open marine environment. A good example of this is along many parts of Chesapeake Bay (Rosen, 1979). Tidal currents may be strong enough to erode the bottom or sides of the estuary. Many inlets which serve estuaries contain channels scoured tens of meters deep, even into bedrock, such as along the Atlantic and Gulf coasts. Some of this eroded material will accumulate in the estuary. A good example of underlying and adjacent bedrock serving as a source of sediment is in the Bay of Fundy, where most of the modern sediment is derived from the Triassic red beds into which the estuary is carved (Klein, 1963c).

Organisms contribute to the origin of sediment in estuaries from two avenues. The most obvious is the contribution made by skeletal debris provided as organisms that live in the estuary die. Predominant are such mollusks as oysters (*Crossostrea*), mussels (*Mytilus*), other pelecypods, and gastropods. The amount and the distribution of these biogenic sediment particles varies widely over the spectrum of estuaries. The other contribution by organisms is commonly overlooked. It is the formation and excretion of pelleted mud. Estuaries support large populations of suspension feeders, which in turn produce great quantities of sand-sized pellets (Haven and Morales-Alamo, 1972). In many estuaries the accumulation of the sand-size aggregates appears to be the primary mechanism for deposition of suspended particles.

Rate of accumulation

Rates of sediment influx into the estuarine environment vary with both time and space. Similarly, the relative amounts of wash load and bed material load contributions to the estuarine environment also show great variation. Because of these variations it is difficult to determine meaningful sedimentation rates for estuaries. Data from various sources show values of 15 cm per century for estuaries on the Texas coast (Shepard and Moore, 1960), 20 cm per century for Delaware Bay (Oostdam and Jordan, 1972) and 50 to 80 cm per century for the upper Chesapeake Bay (Schubel, 1972). Projections of such rates indicate the relatively short existence of these environments in geologic time.

The data cited above are based primarily on sediment accumulations from suspension in a low-energy, river-dominated estuary. Sediment accumulation patterns are much different in tide-dominated estuaries or at the constricted seaward entrances to low-energy estuaries. An excellent example of the latter is Chesapeake Bay, which contains a tide-dominated mouth characterized by much movement of sediment as bed material (Ludwick, 1975).

Distribution

A combination of riverine and tidal circulation together with sediment availability tends to control most aspects of sediment distribution. For purposes of sediment distribution, one can divide estuaries into two types or at least into two distinct portions: one of high energy and one of low energy. Low-energy estuaries or at least the low-energy portions of estuaries are dominated by the accumulation of mud with biogenic debris and minor amounts of sand. These estuaries may have much riverine influence and generally only slight tidal current activity except at narrow constrictions at the mouth. They would most likely fall into the microtidal estuary category of Hayes (1975).

In both mesotidal and macrotidal estuaries, tidal currents are the dominant process and sand bodies are prominent in the estuarine system. Mesotidal estuaries contain tidal sand bodies near the mouth and point bars in the system of tidal creeks. Mud dominates the intervening shallow bays and tidal marshes. Macrotidal estuaries are not common and are dominated by tidal sand bodies (Hayes, 1975). In many respects the macrotidal estuary and the tide-dominated delta (Chapter 9) cannot be distinguished.

Calm-water facies

The quiescent portions of estuaries show great variety in their surface sediment facies. Sluggish currents from tidal or riverine sources together with waves provide mechanisms for distribution of sediments. Only very broad statements are possible insofar as generalizing about these environments. Sediment is typically coarser at or near the margins than in the center, as it is in shallow lakes (Chapter 5). Rates of accumulation are highest adjacent to river mouths or constrictions, where tidal transport may occur.

These estuarine environments exhibit low relief, they are typically without channels, and the primary physical structure is nearly horizontal stratification.

Bioturbation may be widespread locally depending on salinity and rates of accumulation. Biogenic accumulations such as oyster reefs may be a significant contributor to these sediments.

One of the best areas as far as the variety of coastal bays is concerned is the central Texas coast. There are numerous estuaries, **polyhaline bays**, and lagoons along the reach of coast and their sediments are well known (e.g., Shepard and Moore, 1960). There are several shallow and large estuaries along this coast, with Copano, Aransas, and San Antonio bays rather typical (Figure 11-5). Among these three bays are three different types. San Antonio Bay is an estuary with a rather large delta developed at the mouth of the Guadalupe River. Copano Bay is also a brackish estuary, but no large river enters it. Aransas Bay is generally considered to be a polyhaline bay, in that it usually shows salinities below open marine levels, although during the dry season it may exceed $35^0/_{00}$.

A broad subdivision of these bays into four sedimentary facies has been made by Shepard and Moore (1960). These are the bays near rivers, oyster reefs, deep central bays, and bays near narrows (Figure 11–6). The bulk of these environments consists of the deep central bay facies, which is mostly mud with some terrigenous sand and biogenic debris. The oyster reef facies contains various sand constituents but is dominated by about two-thirds coarse shell material (Figure 11-6). The facies that is proximal to the rivers is very much like that of the central bay facies except that the terrigenous sand fraction is about twice as great in the areas near rivers. Near constrictions where tidal currents are relatively swift, about one-half of the sediment is terrigenous sand, with some mud being carried away to calmer waters. One should remember that in these large estuaries with low tidal ranges, there is a large tidal

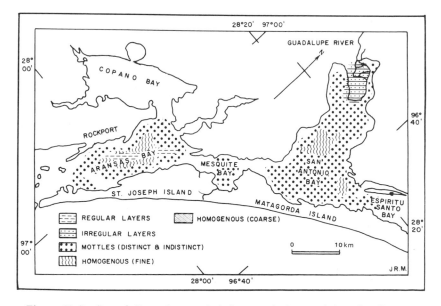

Figure 11-5 Central Texas bays and their general characteristics of sediment distribution. (From Shepard and Moore, 1960, p. 131.)

Chapter 11 Coastal Bays: Estuaries and Lagoons 373

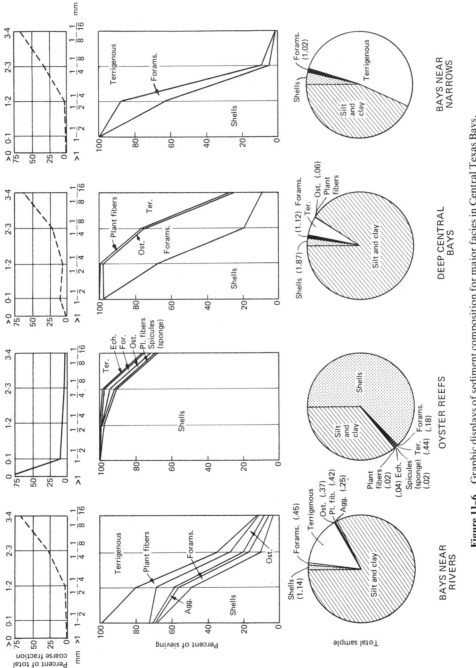

Figure 11-6 Graphic displays of sediment composition for major facies in Central Texas Bays. (From Shepard and Moore, 1960, p. 130.)

prism that must pass through the narrow areas, thus creating these sand-dominated facies.

High-energy facies

Mesotidal and macrotidal estuaries typically display extensive sand-size sediment accumulations. The bulk of these is comprised of the tidal-generated sand bodies associated with mesotidal inlets (Hayes, 1975). Macrotidal sand bodies are typically dominated by elongate and nearly parallel bars which are generated by the strong tidal currents in these generally funnel shaped estuaries (Figure 11–7). These sand bars are about at right angles to the trend of the coast and may be intertidal or subtidal (Hayes, 1975; Greer, 1975; Wright, 1978). Large bedforms develop on the bars, and channels tend to be separated as to flood or ebb dominance (Figure 11–7).

Mesotidal estuaries and inlets contain extensive tidal sand bodies. The tidal channel systems that are distal to the mouth of the estuary resemble the anastomasing patterns of streams. More apparent organization of the tidal sand bodies may occur in those estuaries where well-developed point bars are present in the tidal creeks which service extensive marsh and tidal flats systems, such as along the Georgia–South Carolina coast (Ward, 1978).

Tidal Deltas. By far the most common sand bodies associated with estuaries are tidal deltas. These rather fan shaped accumulations occur in the narrow mouths of estuaries or in inlets that service the estuaries. They are probably best developed in mesotidal estuaries (Hayes, 1975) but are also common in some microtidal environments. Tidal deltas represent primarily sand-size sediment accumulations which are formed by the interaction of flood and ebb currents with the morphology of the estuary mouth. They are called deltas because of their general shape and because

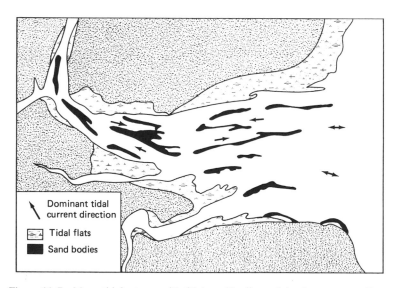

Figure 11–7 Macrotidal estuary with tidal sand bodies and dominant current directions. Compare with tide-dominated river deltas (Chapter 2). (From Hayes, 1975, p. 21.)

their mode of formation is similar to that of riverine deltas. A large volume of sediment accumulates due to rather sudden loss of competence of the tidal currents. Each estuary or inlet contains a flood delta, generated by flood tides, and an ebb delta, which is generated by ebb currents (Figure 11-8). An excellent summary of tidal deltas is presented by Boothroyd (1978).

Flood deltas are generally well developed and are molded by tidal action only. They form behind barriers or in estuaries where there is no significant wave action. This contrasts to ebb deltas which develop seaward of the mouth of the estuary and are exposed to open-water wave attack. The result is that some ebb deltas are small, with an arcuate bar at the terminus due to wave domination of the sand body [Figure 11-9(a)]. The opposite situation is where the tide is dominant and the ebb delta extends seaward in the form of large, shore-normal sand bodies [Figure 11-9(b)]. A scheme of ebb delta configurations depending on the interrelationships between tides, waves, and longshore currents has been developed by Oertel (1975) (Figure 11-10).

The morphologic model used for flood deltas follows that proposed by Hayes (1975). Its shape is somewhat reminiscent of a horseshoe crab (Figure 11-11). The tidal delta contains a broad low-relief ramp and typically shallow, bifurcating flood channels near the central area. Ebb channels are commonly deep and are deflected around the ebb shield at the landward limit of the delta (Figure 11-11). Spillover lobes, which are similar to splays, may be formed as ebb flow leaves the channel and floods the tidal flat. Ebb flow also tends to develop spits along the seaward margin of the flood delta.

Hayes's (1975) ebb delta model consists of a broad and deep main ebb channel flanked by relatively shallow, lateral flood channels (Figure 11-12). These channels are separated by elongate, channel-margin bars which are essentially shore-normal. In the tide-dominated ebb delta these are the primary sand bodies of the delta [Figure 11-9(b)]. The seaward extent of the ebb delta may have an arcuate terminal lobe which is formed primarily by wave action. In the wave-dominated ebb delta this sand body is well developed [Figure 11-9(a)], but it is not so in the tide-dominated version. Small intertidal swash bars are common near the seaward end of the linear bars (Figure 11-12).

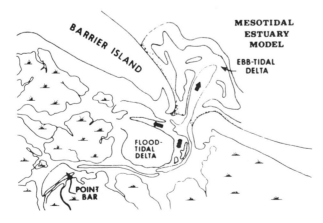

Figure 11-8 Generalized map of coastal complex, showing the configuration of tidal deltas and their relationships to the mesotidal coast. This diagram is based largely on the Massachusetts and South Carolina coasts. (From Hayes, 1975, p. 11.)

(a)

(b)

Figure 11-9 Oblique aerial photographs of a wave-dominated ebb delta (Matanzas Inlet, Florida) (a) and a tide-dominated ebb delta, Fripp Inlet, South Carolina (b).

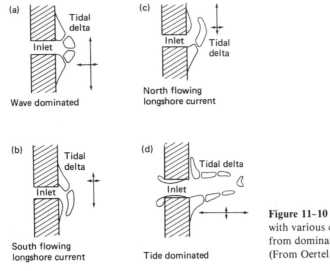

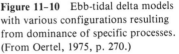

Figure 11-10 Ebb-tidal delta models with various configurations resulting from dominance of specific processes. (From Oertel, 1975, p. 270.)

These tidal deltas contain a very complex system of bedforms on the intertidal sand bodies and in the deeper channels. They range widely in size and in areal configuration, depending largely on the flow strength and direction (Figure 11-13). The typical range is from linear ripples up to plane beds which are in the upper flow regime; antidunes are isolated and rare on tidal deltas. Bedforms and resulting cross-stratification have orientations that reflect the dominant flow at a particular location on the tidal sand body. Consequently, large and small, flood- and ebb-oriented bedforms may coexist. Such situations give rise to very complex sequences of cross-

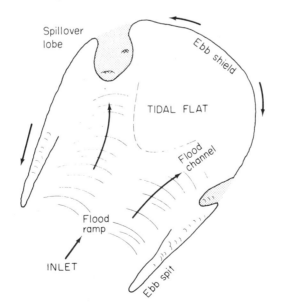

Figure 11-11 Flood-tidal delta model, showing major morphologic components. (From Hayes, 1975, p. 17.)

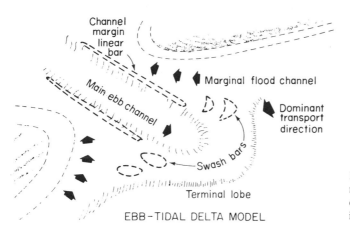

Figure 11-12 General morphologic model for mesotidal ebb delta with a combination of tidal current and wave influence. (From Hayes, 1975, p. 13.)

stratification which are discussed below. Excellent illustrations of estuarine bed-forms are found in Boothroyd (1978).

Estuarine Sequences

The relatively short lived geologic lifetime of estuaries tends to produce a variety of stratigraphic sequences. These sequences exhibit a rather high preservation potential and are fairly easily recognized in the rock record as being of estuarine origin. The sequences produced in this environment tend to fall within three general themes, each of which has variations. One of these is the sequence that represents the low-energy,

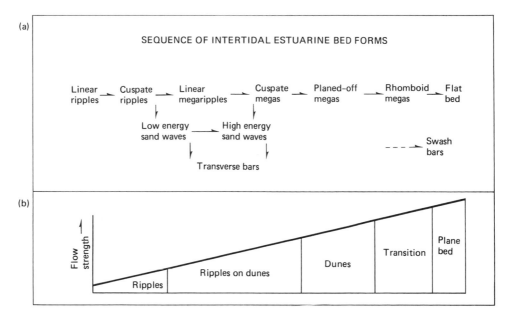

Figure 11-13 Sequence of bedforms that develop on intertidal sediment bodies in estuaries and on tidal deltas. (From Boothroyd, 1978, p. 304.)

muddy estuary which is commonly dominated by riverine processes. The relatively high energy estuarine sequences may be characterized by two rather distinct sequences; one represents the tide-dominated estuary, which is essentially funnel shaped and may be in the mesotidal or macrotidal range, and the other represents the tidal delta complex. The latter system may be present in any tidal category and may be a part of other types of estuarine systems.

It should be observed that the juxtaposition of the estuary to other coastal environments, such as marshes, tidal flats, deltas, and others, dictates that these environments be stratigraphically related to estuaries. Also important in the interpretation of estuarine sequences is the contained fauna. This is especially important because of salinity variations and contrasts with related environments.

Calm-water estuaries

Regardless of shape, the calm-water estuary accumulates a fairly characteristic sequence of sediment. In most cases the salinity distribution may be the most important feature in determining sedimentary facies and depositional environments. The James River in Virginia is an elongate estuary which is a major branch of Chesapeake Bay. In the region near the mouth of the river, salinity is essentially marine. The sedimentary facies are distributed in a nearly linear fashion from the fluvial-estuarine to the marine (Figure 11–14). There is a general trend of decreasing organic content, decreasing stratification, and increasing grain size from the fluvial to the marine

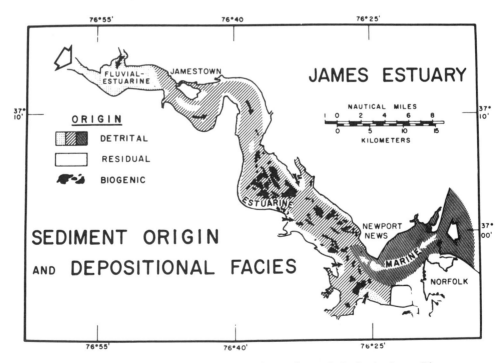

Figure 11–14 General distribution of surface sediment facies in the James River estuary, a tributary of Chesapeake Bay. (From Nichols, 1972, p. 209.)

areas of this estuary (Nichols, 1972). The intermediate estuarine facies has much mottling, abundant biogenic debris, and highest rates of accumulation. The prograding sequence in such a relatively simple estuary would be one characterized by a fining-upward texture and increasing organization of stratification upward. The fauna would reflect a decreasing salinity upward.

Good core data are available from some of the central Texas estuaries, especially San Antonio Bay (Figure 11–15). These estuaries were formed as the sea level rose and flooded coastal plain drainage systems. The result is that the Holocene estuarine sequence has accumulated on Pleistocene fluvial sediments (Shepard and Moore, 1960). The cores taken from San Antonio bay show no marked trends in texture or composition. Grain size is in the clay or silty clay ranges, and thick accumulations of oyster debris occur in most cores. The cores near bay margins are coarser (Figure 11–15). Mottling is widespread except near the Guadalupe Delta, where stratification persists.

Tide-dominated estuaries

Estuaries that are somewhat funnel shaped and experience at least a mesotidal range tend to accumulate sequences which have marked contrast with the low-energy variety described above. Good examples of these estuarine types are along the Georgia–South Carolina coast of the United States and in the German Bight area of Europe.

Estuarine facies along the Georgia coast tend to show an increase in grain size and cross-stratification in a seaward direction (Howard et al., 1975). Bioturbation is widespread except in the upper reaches of the estuaries (Howard and Frey, 1975). A recent study by Greer (1975) develops a good conceptual model for an estuarine sequence which can be used to characterize the funnel-shaped, mesotidal estuary (Figure 11–16). This model is presented in the progradational mode.

The coarse tidal channel lag accumulations form the base of the sequence. Overlying channel bar and channel bottom deposits show diversity consisting of planar foresets, laminated muds, some flaser and wavy bedding, and small-scale ripple cross-stratification, any of which may display bioturbation (Figure 11–16). The shallower portions of the estuary which overly these units consist of ripple cross-strata, flaser bedding, and bioturbation. These lower-energy environments include tidal flats. The cap of the sequence represents the salt marsh and consists of mud and peat with abundant root structures (Figure 11–16).

Holocene sequences representing the tidal portion of a river and an estuarine channel have been described by deRaaf and Boersma (1971) from the coast of the Netherlands. One of these sections consists of a 5-m fining-upward section of bidirectional, cross-stratified sands with flasers and lenticular beds in the upper portion (Figure 11–17). This sequence is interpreted by the authors as transitional between fluvial and estuarine. The lower few meters are interpreted as a point-bar deposit in a tidal stream, with the bottommost portion being ebb dominated. Herringbone cross-strata in the middle portion are indicative of bimodal tidal influence. The muddy upper section indicates lower-energy conditions but is presumably subtidal (deRaaf and Boersma, 1971).

The Haringvliet estuary just inland from the North Sea shows an accumulation

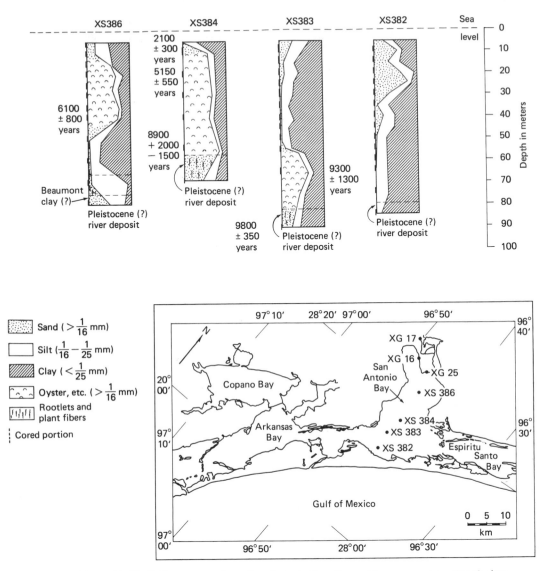

Figure 11–15 Sediment cores from San Antonio Bay, Texas. These cores show that the bay persisted for several thousand years and that its sediments overlie river deposits. (From Shepard and Moore, 1960, p. 139.)

of three estuarine lithofacies (deRaaf and Boersma, 1971; Terwindt, 1971). These facies are (1) large-scale cross-stratification, (2) flaser bedding with some horizontal and ripple cross-strata, and (3) lenticular bedding. Previous research in this estuary suggested that a fining-upward succession of facies 1 through 3 would be expected (Oomkens and Terwindt, 1960). Nearly 200 cores from the estuary and adjacent tidally influenced areas show no particular pattern of accumulated sediments. Many cores were characterized by an alternation of fining-upward and coarsening-upward

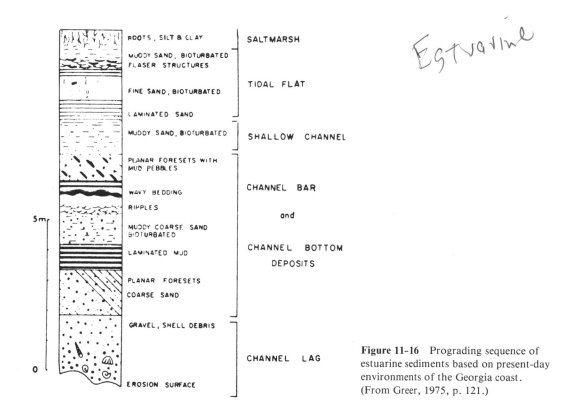

Estuarine

Figure 11-16 Prograding sequence of estuarine sediments based on present-day environments of the Georgia coast. (From Greer, 1975, p. 121.)

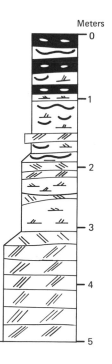

Mud

／／／ Cross-strata

⌣⌣ Flaser bedding

Figure 11-17 Estuarine sequence from Barendrecht along the Netherlands coast. There is a general fining-upward trend, with lenticular beds at the top. (From deRaaf and Boersma, 1971, p. 486.)

sequences, each of which ranged from about 30 to 100 cm. The reasons given are the changes in tidal hydraulics resulting from wind and from neap to spring tidal cycles (Terwindt, 1971). The variability exhibited by estuary processes should not be expected to produce regular and predictable sedimentary sequences.

Tidal deltas

Research efforts dealing with the coring of tidal deltas and the incorporation of tidal delta accumulations into stratigraphic sequences have only recently been undertaken. LeBlanc (1972), Kumar and Sanders (1974, 1975), and Hubbard and Barwis (1976) have all considered tidal deltas as they are incorporated into inlet sequences as barrier spits migrate laterally alongshore. This approach is discussed in Chapter 12.

The only detailed stratigraphic sequence described for the tidal delta is one that considers the flood deltas as preserved in a regressive system (Boothroyd, 1978). The sequence would be expected to reach about 10 to 12 m in thickness for a mesotidal coast and is dominated by large-scale cross-stratification (Figure 11-18). Ebb channel deposits show the expected ebb-oriented cross-strata, and most of the remainder of the sequence is dominated by flood-oriented cross-strata. Spillover lobes and ebb spits (Figure 11-11) provide the exceptions, with ebb-oriented megaripple cross-strata (Figure 11-18). The flood delta sequence is capped by a muddy sand tidal flat succeeded by marsh deposits.

This sequence is an idealized one which may be interrupted by extreme events such as hurricanes or may be halted in its development by encroachment of an adjacent barrier.

Ancient Examples of Estuarine Deposits

A search through the current literature on modern estuaries as environments of deposition will reveal a wealth of research publications from throughout the world. A

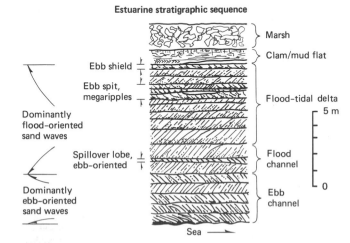

Figure 11-18 Stratigraphic sequence of mesotidal flood-tidal delta and related channels. (After Hayes and Kana, 1976.)

similar effort to locate examples of estuarine deposits from the rock record results in a surprisingly small number of papers. The nature of estuaries and their contained deposits is such that they enjoy a generally high preservation potential. One must conclude therefore that there are many ancient estuarine sequences which are unrecognized as such. Probably, they are so similar to shelf and some deltaic deposits (Chapters 9 and 13) that estuarine sections are falsely interpreted as other environments.

Eocene, London Basin (England)

An excellent exposure of the Lower Bagshot Beds in the Eocene of the London Basin shows accumulations of estuarine and related sediments. These beds are overlain by the marine Bracklesham Beds (Bosence, 1973). The Bagshot Beds are here comprised of two distinct lithofacies. Facies 1 consists of sands with ripple cross-strata showing bidirectional orientation normal to depositional strike. Climbing ripples and symmetrical sets are present. Some mud drapes and flasers are present but become increasingly abundant upward [Figure 11–19(a)]. The flasers give way to a thin lenticular sequence which then coarsens upward, reversing the lower portion of this section. The cyclic sequence is about 50 to 70 cm thick and is repeated in facies 1.

Facies 2 is more widespread in the Bagshot Beds and consists of sands and gravels with abundant scour and fill configuration [Figure 11–19(b)]. Individual channel-fill accumulations show fining-upward trends and contain flakes of mud derived from facies 1. Cross-strata indicative of megaripple migration are common and contain pebbles. Although the complete facies displays a bimodal current direction compatible with that of facies 1, each channel-fill unit displays a single direction of cross-stratification. Burrows of *Ophiomorpha,* which represent ghost-shrimp activity, are present. They are relatively common in facies 1 but rare in facies 2.

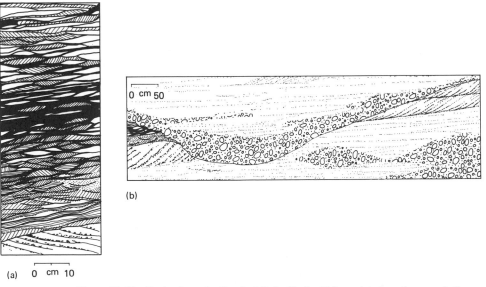

(a) 0 cm 10

(b)

Figure 11–19 Facies from the Bagshot Beds. Facies 1(a) consist of medium sand, flasers, and lenticular beds. Facies 2(b) is sand and gravel with channeling. (From Bosence, 1973, p. 65.)

A comparison of the sequence from the Eocene of the London Basin described above with the sequence previously described from the Haringvliet estuary in the Netherlands shows remarkable similarities, especially in facies 1. This finer facies is interpreted as representing accumulation along the sides of tidal channels or point bars in the estuary system. The sequence indicates a subtidal origin and bidirectional currents. Facies 2 represents more of a scour and fill environment typical of the channel floor (Bosence, 1973). The cyclic nature of facies 1 may represent energy changes related to lunar tidal cycles. The overlying Bracklesham Beds are rich in glauconite and have extensive bioturbation. They are interpreted as accumulating in the inner marine shelf beyond the estuary. Bioturbation in the estuarine sequence is not abundant, due to extensive sediment mobility by tidal currents.

Fall River Formation (Cretaceous), Wyoming

The Lower Cretaceous Fall River Formation crops out in the Black Hills of northeastern Wyoming. The formation overlies fluvial deposits of the Lakota Formation, which in turn are on top of the dinosaur-bearing Morrison Formation (Jurassic). Much of the Fall River Formation consists of thin-bedded sandstone and mudstone, which has been interpreted as accumulating in a tidal flat environment.

Campbell and Oaks (1973) have subdivided the Fall River into four major lithofacies—the fluvial, upper estuarine, lower estuarine, and shallow-shelf facies—each of which may contain strata that accumulated in various environments. Only the middle two will be discussed in detail. These estuarine facies are closely associated with tidal flat, marsh, and tidal channel deposits.

The sand bodies in question are scoop shaped with the narrow ends landward (Campbell and Oaks, 1973, p. 770). The thickness is several meters, width is 1 km or more, and length may be up to 5 km. Bedding surfaces tend to be wavy and there is a general thinning of the units in a seaward direction. The facies interpreted as being representative of the upper estuarine environment ranges from very fine to coarse sand and consists primarily of unidirectional, high-angle, large-scale planar cross-strata. They may represent sand wave migration. The azimuth of dip is in the seaward direction, which suggests domination by ebb tidal currents. Lenticular scour fills are scattered throughout this unit. They exhibit both flood- and ebb-oriented cross-stratification (Campbell and Oaks, 1973).

The lower estuarine facies is distinctly finer in grain size than its upper estuarine counterpart. There is abundant ripple cross-stratification and some flaser and lenticular bedding (Campbell and Oaks, 1973). The fossils preserved are rare and not diagnostic. Although all features presented by the authors do occur in the estuarine environment, they occur elsewhere. The lack of diagnostic sequences and structures makes the interpretations reached subject to some question.

Campbell and Oaks present a reasonable conceptual model for the origin and development of this estuarine complex. Streams cut through coastal marshes and tidal flats to the open marine environment (Figure 11–20). Scour caused by tidal currents provided the sites for the scoop-shaped sediment bodies that dominate this portion of the Fall River Formation. As time passed these scoop-shaped deposits overlapped one another, producing the complex facies relationships as they presently exist in the rock record.

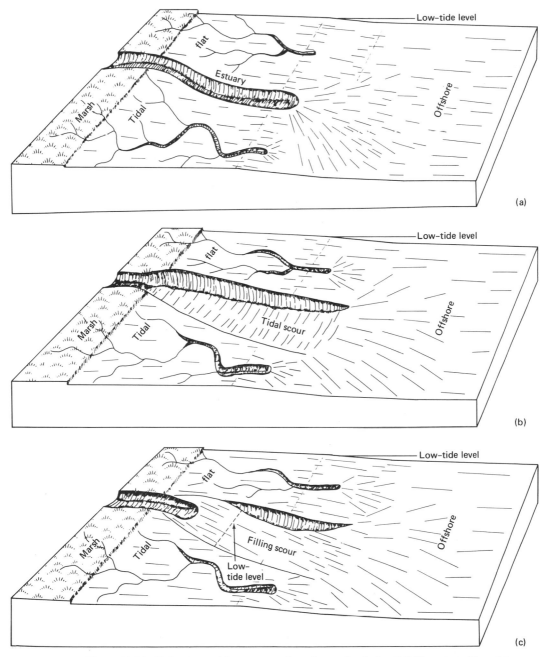

Figure 11-20 Paleogeographic reconstruction of estuaries during Fall River deposition: (a) a stream cuts the coastal marsh, (b) tidal scour generates a site for (c) sediment accumulation. (From Campbell and Oakes, 1973, p. 775.)

Breamrock Formation (Devonian), Ireland

A sequence of Upper Devonian rocks exposed in the south of Ireland near Cork represent a transgressive system overlying the Old Red Sandstone (de Raaf and Boersma, 1971). This unit of nearly 300 m in thickness contains numerous sedimentary structures, which indicate accumulation under tidal current conditions. Two general facies are distinguished: one with even stratification and minor scouring at the base, and the other which has much flaser and lenticular bedding (Kuijpers, 1971).

The even-layered sandstones comprise only about 20% of the Breamrock. This facies has current lineation and some large-scale cross-stratification where channeling is present. The other facies shows lots of small-scale foreset stratification with distinctly bimodal orientations, although there is a distinct domination by one mode. All data point toward a definite tidal origin for this unit (Kuijpers, 1971; deRaaf and Boersma, 1971) and probably an estuarine setting. Either ebb or flood current appears to have dominated, but previous work has not determined paleogeographic details that would permit this designation.

COASTAL LAGOONS

Most definitions of the term lagoon are nonspecific. Typically, they are indicated as being parallel to the coast and separated from the open marine water by a natural barrier, but little or nothing is mentioned about runoff from land or circulation with the open marine environment. In the following discussion, lagoons are considered as being coastal bays with no significant runoff from land and with restricted circulation with the open marine environment. Such an environment is sedimentologically distinct from estuaries and from normal marine conditions.

Lagoons display various shapes, but most are probably elongate parallel to the coastline and virtually all are separated from the open marine environment, commonly by a barrier island but in some places by a reef. This environment is generally other than normal marine salinity and is not the site of rapid detrital sediment influx. Chemical precipitates may be widespread. The fauna and flora are typically distinct because of the restricted conditions.

Climate is a factor in the development of lagoons; they are generally low- to mid-latitude features and are associated with semiarid or arid conditions. More rainfall would produce streams emptying into the coast and therefore estuaries would be formed, such as those along the eastern coast of the United States.

General Characteristics

Coastal lagoons may form in a variety of geologic settings as long as there is some embayment from the open marine environment and a mechanism for isolating it, such as development of a barrier or spit. They may develop along high-relief areas such as the Pacific coast of Mexico, but most lagoons are developed along coastal plains. Typically, lagoons are shallow and are low-energy systems. Size limits fetch, so that waves are small and currents are not strong except adjacent to inlets.

Salinity

Lagoons are commonly **polyhaline** or **hypersaline**, at least in the broad senses of these terms. Polyhaline water bodies are those which display great change in salinity from brackish to hypersaline, generally in response to seasonal rainfall or at least some cyclic phenomena. Some shallow coastal lagoons fall into this category. A rather extreme example is Lake Reeve in Victoria, Australia, which is a long, shallow lagoon (Figure 11–21). During the wet season it is almost fresh, whereas in the dry season it nearly dries up and becomes quite saline. This type of salinity pattern would characterize lagoons of the midlatitudes with strong seasonality to the precipitation.

Hypersaline lagoons are those in which salinities are continually above normal marine concentrations. They characterize the semiarid and arid coastal areas, where little or no freshwater influx occurs. Salinity gradients commonly increase away from the connections with the open sea, if any are present. Laguna Madre in Texas and Shark Bay in Western Australia are examples.

Lagoonal Sediments

The low level of physical processes operating in coastal lagoons precludes any large quantity of detrital sediment from reaching the lagoon under normal conditions. Extreme events such as hurricanes or typhoons may present exceptions. The shallow, hypersaline, quiet conditions of the lagoons are conducive to chemical precipitation. The overall rate of sediment accumulation is low, however.

Detrital sediments

The reader may notice that the term "detrital" is used to describe the non-chemical portion of the sediment accumulating in a lagoon, whereas "terrigenous" has been used in a similar context for previous environments. Detrital is here invoked

Figure 11–21 Photograph of Lake Reeve, Victoria (Australia), which is an ephemeral lagoon.

in order to include locally generated biogenic debris together with terrigenous particles, the combination of which is the nonchemical portion of lagoonal sediments.

Biogenic components of the detrital sediments are much the same as in estuaries: skeletal material and pellets. Most or even all of the lagoonal sediment may be comprised of these sources. It is particularly common to have all skeletal sediments in tropical coastal lagoons, such as in the Persian Gulf. Pellets may comprise much of the detrital component in some lagoons. Filter- and detritus-feeding benthic organisms injest sediment from the relatively calm lagoonal waters and the underlying substrate. The Coorong in South Australia and Lake Reeve in Victoria accumulate pelleted muds produced by the gastropod *Batillaria* as the most important detrital sediment in the lagoon.

Terrigenous detritus, primarily in the form of sand, is rather patchy in its distribution within the lagoon. It is restricted to the area adjacent to the barrier between the lagoon and the open sea. Essentially, three mechanisms of sediment transport into the lagoon are available. The most obvious is the tidal delta sand bodies associated with inlets. Flood deltas are sediment sinks and provide a large quantity of sediment, although it is typically only local in extent (Figure 11–11).

The other two mechanisms are related in origin and distribution. Prevailing winds and many coastal storms have significant onshore components and they also generate onshore moving waves. Much sediment is blown from the barrier beach and related dunes into the lagoon, commonly producing a rounded and sorted sand-size mode in the lagoonal muds. In some areas, such as Laguna Madre (Texas) and along the Persian Gulf, there is much large-scale aeolian transport of sand into lagoons (Figure 11–22). Washover of barrier sediment during storms is also an important mechanism of transporting terrigenous detrital sediment into lagoons. This

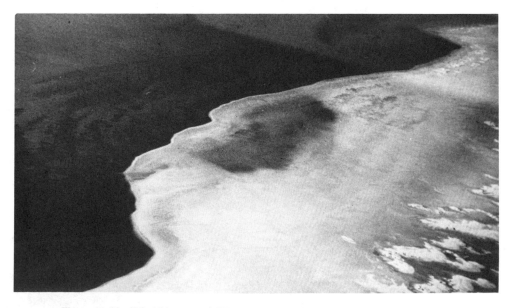

Figure 11–22 Wind-blown sand filling Laguna Madre, Texas, behind Padre Island.

phenomenon takes place locally and as rather large sediment pulses, in contrast to the more gradual and steady transport by aeolian means. Washover features are considered in detail in Chapter 12. Both of the mechanisms described above result in sediment transport into lagoons from the seaward direction.

Chemical sediments

In many respects the coastal lagoon system is like playa lakes. The general absence of continued detrital sediment influx and changing or extreme salinity in both are conducive to chemical precipitation of carbonates and evaporite minerals.

Oolites are a fairly widespread type of sediment particle in many lagoons, such as Laguna Madre (Rusnak, 1960) and Shark Bay, Australia (Logan et al., 1970). These spherical grains are most common in purely carbonate environments where water is agitated. In Laguna Madre, oolites have been found in grassy areas which are characterized by calm water (Freeman, 1962).

Most chemical sediments in lagoons are precipitates of carbonate mud or evaporite minerals such as gypsum and sometimes halite. Under conditions of excess evaporation over precipitation, more dense, high-salinity water sinks to the bottom and density stratification may occur. Most coastal lagoons are shallow and if there is development of any density stratification, it is shortlived or has only a modest vertical gradient.

The precipitation of carbonate mud is widespread in modern coastal lagoons. Recently, dolomite has also been interpreted as forming from direct precipitation (von der Borch, 1976). The latter is restricted to hypersaline lagoons. Gypsum will precipitate as large selenite crystals in lagoonal muds where pore water reaches salinities of 200‰ or more, and it also forms as crusts in the wind tidal flats or upper intertidal flats along lagoon margins. Some local conditions may permit salinities to reach high enough levels for halite precipitation, but this would be restricted to isolated pockets or ponds in the lagoon complex.

Examples of Modern Lagoons

Because virtually each modern lagoon is unique in at least some of its important sedimentologic attributes, selected examples will be considered to cover a spectrum of lagoonal situations. The three areas chosen are all well known and have been discussed extensively in the literature. Mugu Lagoon, California, is small and has nearly normal marine salinity due to its size and the presence of a tidal inlet. Laguna Madre, Texas, is probably the best known of all coastal lagoons. It has a wide range of salinities and accumulates a combination of terrigenous and chemical sediments. Coastal lagoons along the Trucial Coast of the Persian Gulf contain no terrigenous component and represent the arid coastal lagoon environment.

Mugu Lagoon, California

Mugu Lagoon is located between Santa Barbara and Los Angeles on the southern coast of California. It is only a few square kilometers in area and contains extensive salt marsh and tidal flats in its northeastern portion (Figure 11–23). The barrier developed from a spit and contains a tidal inlet which is open most of the

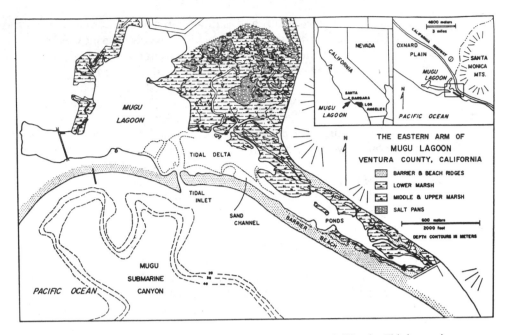

Figure 11-23 Map of eastern portion of Mugu Lagoon, California. This lagoon is receiving terrigenous sediment from the seaward direction and has normal marine salinity. (From Warme, 1969, p. 139.)

time. Mugu Lagoon receives no significant runoff; however, circulation with the Pacific Ocean through the tidal inlet in combination with an annual rainfall of 25 to 50 cm keeps salinities essentially at normal marine levels. The lagoon has been studied extensively by Warme (1967, 1969, 1971; Warme et al., 1977), who was particularly interested in sediment-organism relationships.

The barrier that forms Mugu Lagoon developed on a blanket of shallow marine sands (Warme, 1969, 1971). Numerous beach ridges are present and the inlet contains a well-developed flood tidal delta (Figure 11-24). The subtidal lagoon includes three components: tidal channels, which are related to the tidal delta; ponds; and tidal creeks, which service the intertidal flats and marsh (Warme, 1971). Early maps show that in the mid-nineteenth century the lagoon extended much farther to the northwest along the coast and covered an area several times that of the present Mugu Lagoon (Warme, 1971, map 3). Dredging has modified the western, deeper part of the lagoon.

Sediments. There is little question that detrital sediments accumulating in the lagoon are derived from seaward of it. The tidal currents in the inlet and washover phenomena provide the transport mechanisms. Sand-size sediment is provided by both these mechanisms, and mud is carried largely by tidal currents. A small quantity of mud is undoubtedly also carried to the lagoon by runoff from the adjacent hillsides (Warme, 1971). Virtually all of the present lagoon complex is underlain by inner shelf sands which accumulated prior to formation of the barrier.

Except for much of the marsh, the lagoonal complex is dominated by sand-size sediment. This includes tidal channels, tidal creeks, and subtidal ponds. Skeletal car-

Figure 11-24 Aerial photograph of Mugu Lagoon, California. (Courtesy of J. Warme.)

bonate material is present essentially throughout but only as a few percent. Bioturbation and distinct burrow structures are also widespread. Most physical structures are in the form of thin stratification, but cross-strata are developed as bedforms migrate in tidal creeks and channels (Warme, 1971).

Salt ponds are present in some supratidal areas around the lagoon. Sediments are dominantly mud and are well stratified, with disruptions only by desiccation features. Gypsum crystals may be several millimeters in diameter and increase in abundance landward in the salt ponds. As sediment dries, halite may precipitate.

Desiccated mud curls up and becomes susceptible to wind transport. Much of this mud is blown as sand-size aggregates into adjacent small dunes which are called "clay dunes" (Huffman and Price, 1949). They are restricted in their development to lagoon or playa margins.

Rates of sediment accumulation in various environments of the lagoon show a wide range (Figure 11-25). Immediately landward of the barrier the washover deposits accumulate nearly a meter per year. Rates comparable to that are found in tidal creeks and channels. The subtidal ponds, marshes, and tidal flats range between 20 and 50 cm per year (Warme, 1971). All of these rates are relatively high for coastal

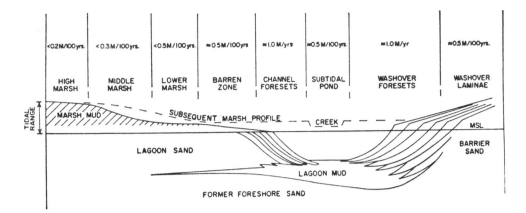

Figure 11-25 Diagram of transverse section across Mugu Lagoon, California, showing various depositional environments and rates of accumulation. (From Warme, 1971, p. 51.)

lagoonal systems. The primary reasons are the small size coupled with a high-energy coast, and the good tidal circulation.

Laguna Madre, Texas

One of the largest and best known coastal lagoon systems lies along the southern coast of Texas between Corpus Christi and the border with Mexico (Figure 11-26). Adjacent Baffin Bay is physiographically distinct from Laguna Madre because of its origin as a drowned river valley; however, it is presently similar in most other respects and is included in this discussion. This area is characterized by a semiarid climate and is bounded in part by an extensive eolian plain (Figure 11-26). Evaporation typically exceeds precipitation by over 50 cm per year, giving rise to hypersaline conditions in the lagoon (Miller, 1975).

Baffin Bay is nearly 2 m in depth throughout much of its extent, whereas Laguna Madre displays more variety but is typically about 1 m deep. Adjacent areas are low-lying and gently sloping, especially on the landward side of Padre Island. The lagoon is not affected by significant lunar tides but does experience wind tides. Extreme events may cause tides to rise over 1 m locally and to have an important effect on the wind tidal flats adjacent to the lagoon (Miller, 1975).

The chemistry and hydrography of the lagoon display variations and play extremely important roles in sedimentation. Circulation between Laguna Madre and the Gulf of Mexico is quite restricted. Two small, natural and ephemeral inlets are present at the northernmost end and actually serve Corpus Christi Bay more than Laguna Madre (Davis et al., 1973). About 90 km south of Corpus Christi Bay is Mansfield Pass, a man-made, structured inlet which has been closed naturally for lengthy periods (Hayes, 1967a). Brazos Santiago Pass at the southernmost portion of the lagoon is the only other source of water exchange with the Gulf (Figure 11-26). Additionally, there is little circulation within the lagoon except that provided by wind. The salinity reflects this lack of circulation and excess evaporation. Although there is variation throughout Laguna Madre, it is hypersaline throughout except im-

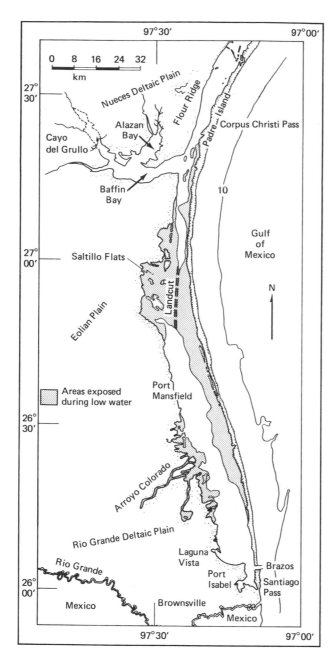

Figure 11-26 General physiographic map of Laguna Madre, Texas. (Modified after Rusnak, 1960, p. 155.)

mediately adjacent to the inlets. Salinities range from near 40‰ at each end to 65‰ or more in the central area and Baffin Bay (Rusnak, 1960). Small local areas and shallow ponds may reach values of over 100‰.

Extreme events may cause severe changes in the chemistry, biota, and even the sediments. Hurricanes occur with a frequency of 40 per century along this coast

(Price, 1958). These events cause great temporary change in salinity patterns as the result of high storm tides and extreme rainfall associated with the storm. The result is a marked decrease in salinity and accompanying large-scale killing of the hypersaline organisms which commonly populate this lagoonal system. Runoff related to extreme rainfall causes density stratification with fresh surface water. This may persist for up to several weeks until wind-derived circulation causes mixing (Behrens, 1969).

Sediments. Laguna Madre sediments come from a variety of sources and accumulate slowly except during extreme storm events. Terrigenous sediments include wind-blown and washover sand from the adjacent barrier. This source is the largest and may show very rapid accumulation, especially along the seaward margin of the lagoon. Mud-size terrigenous sediment reaches the lagoon complex via inlets and occasional runoff. Biogenic debris is common and may comprise up to 30% of the sediment (Rusnak, 1960). Only a few species are present due to the hypersalinity of the environment.

Chemical precipitates are also common and widespread, showing some relation to salinity patterns. Oolites are present throughout much of the lagoon, especially near beaches and other relatively agitated locations (Rusnak, 1960; Freeman, 1962). Carbonate muds including dolomite have been identified, especially in Baffin Bay (Behrens and Land, 1972). In addition to carbonates, local areas of extreme salinities and wind tidal flats may contain gypsum and even halite crystals if salinities are extremely high (Figure 11-27).

Sedimentary structures are virtually absent from the sediments which accumulate away from the shores of the lagoon. Cores to depths of several meters exhibit little or no bioturbation and display thin, well-stratified fine layers (Behrens, 1969). Wave ripples and local cross-stratification are confined to lagoon-margin sediments.

Trucial Coast, Persian Gulf

The central region of the Trucial Coast contains a number of large lagoons, the largest of which is Khor al Bazam (Figure 11-28), although the best known are in the vicinity of Abu Dhabi (Purser, 1973). Lagoons along the Trucial Coast are restricted in their circulation but not to the extent of Laguna Madre. Tidal circulation in Khor al Bazam is generally shore-normal. In both cases there are restrictions which when coupled with extreme rates of evaporation produce salinity levels of 60 to 70‰ (Kinsman, 1969). There is a general increase in salinity from west to east (Purser and Evans, 1973). The barriers that help to form the lagoon are dominated by coral-algal reefs and related sabkhas (Kendall et al., 1968; Purser and Evans, 1973). These barriers are interrupted by wide inlets which have well-developed tidal deltas composed primarily of ooids and skeletal sands.

Sediments. There are no areas of abundant terrigenous sediments along the Trucial Coast. The sediment present is produced largely within the general area and is carbonate except for some evaporites in adjacent sabkhas. Within the Khor al Bazam Lagoon proper, mean grain size is dominantly medium sand except immediately landward of the barrier, where mud is abundant (Kendall et al., 1968). Sediments are nearly all biogenic, with a variety of phyla being represented. Some quartz sand is

(a)

(b)

Figure 11–27 (a) Areal photo of wind tidal flat (Laguna Madre); (b) core from this environment (Courtesy of J. Miller).

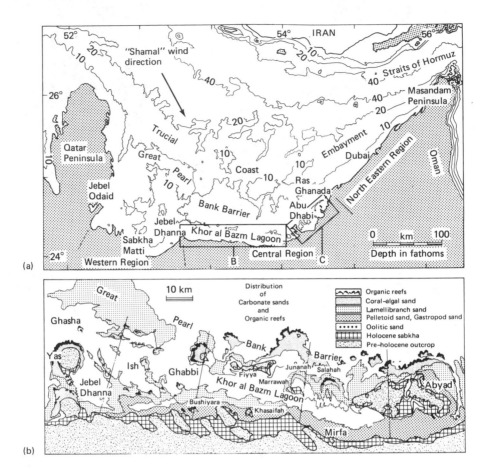

(a)

(b)

present and is derived from Cenozoic rocks and blown into the lagoon (Evans and Bush, 1969).

Sediments in lagoons of the eastern Trucial coast near Abu Dhabi are typically finer grained and contain abundant foraminifera and mollusk debris. Subtidal algal stromatolites are present to depths of 3 m (Kinsman and Park, 1976). Aragonite mud is produced both by physicochemical precipitation and indirectly by photosynthesis taking up large amounts of carbon dioxide. Pelleted mud is common around the shallow low-energy regions of the lagoons and is produced by cerithid gastropods which graze on the surface. This facies is widespread in lagoon margins around the world (Coorong and Lake Reeve, Australia; Laguna Madre, Texas).

Ancient Lagoonal Sequences

Although numerous references are made to the lagoonal depositional environment in discussions of ancient coastal barrier systems, most of the sequences in question are in fact estuarine according to the definitions used in this chapter. It is important to make this distinction between lagoon and estuary in the rock record because of the

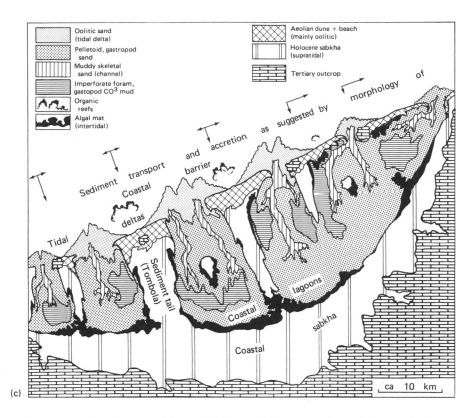

Legend:
- Oolitic sand (tidal delta)
- Pelletoid, gastropod sand
- Muddy skeletal sand (channel)
- Imperforate foram, gastopod CO_3 mud
- Organic reefs
- Algal mat (intertidal)
- Aeolian dune + beach (mainly oolitic)
- Holocene sabkha (supratidal)
- Tertiary outcrop

Figure 11-28 (a) Index map of the Trucial Coast with lagoon complexes of the central region. (b) Surface sediment facies of Khor al Bazm and (c) lagoons near Abu Dhabi show marked similarities. (From Purser and Evans, 1973, pp. 212, 216, 221.)

important differences in reconstruction of the paleogeography, depending on which type of coastal bay was present.

Specific criteria that can be used for recognition of lagoons include evidence for restricted circulation, lack of prominent tidal influence, and a slow rate of terrigenous influx. The presence of evaporites, a fauna lacking diversity, and abundant burrow mottling would favor a lagoonal setting.

Lower Silurian, southwest Wales

A well-documented transgressive barrier sequence has been described from Lower Silurian sections exposed on the southwestern tip of Wales (Bridges, 1976). The barrier sequence is associated with volcanic strata, including basalts, tuffs, and rhyolites. The barrier horizons are contained within the Skomer Volcanic Group and consist of three facies: (1) lagoonal, (2) barrier island, and (3) open marine [Figure 11-29(a)].

The lagoonal deposits are intensely bioturbated muds, although some thin horizons displayed a well-stratified character. Faunal diversity is low, with the phosphatic brachiopod *Lingula* being the most common taxon. A few other sessile

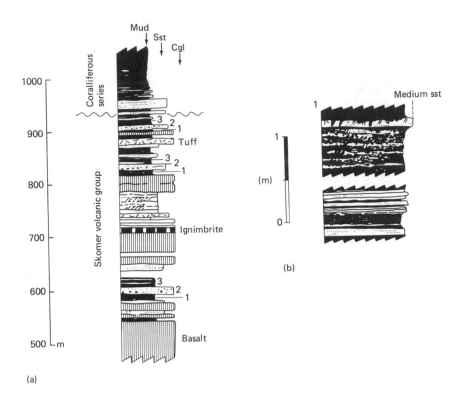

Figure 11-29 (a) General stratigraphic section of the Lower Silurian of southwest Wales, showing volcanics interstratified with sedimentary rocks. Barrier sequences include (1) lagoonal, (2) barrier island, and (3) open marine. The lagoonal portion (b) is dominated by burrowed mud with thin fine-sand layers. (Modified from Bridges, 1976.)

brachiopods and gastropods are present, some of which are phosphatized. The extensive burrowing is probably due to annelids and crustaceans (Bridges, 1976).

Thin sandstone beds are present in the lagoonal sequences [Figure 11-29(b)] and are interpreted as being generated by washover processes on the barrier. All data indicate that this muddy facies accumulated in an environment of restricted circulation and therefore elevated salinity, probably along a semiarid coast (Bridges, 1976).

SUMMARY OF CHARACTERISTICS
OF SEDIMENT/SEDIMENTARY ROCK BODIES
REPRESENTING COASTAL BAYS

Coastal bays in general are protected environments which are also sediment sinks. Because of this combination, coastal bays have a high preservation potential. The most widespread types of bays, estuaries and lagoons, have several features that are distinct to each type, although they also have many in common.

Tectonic setting. Coastal bays may be developed in any tectonic setting, although there are certain conditions that enhance sediment accumulation and preservation in the stratigraphic record. Wide coastal plains on the trailing edges of plates tend to give rise to large estuaries and lagoons. These coastal bays receive sand and mud and are commonly preserved by transgression of the sea.

Shape. No specific shapes characterize sediment bodies that accumulate in coastal bays, although most are somewhat tabular. Most lagoons tend to be elongate parallel to the coast and produce sediment bodies that do likewise.

Size. The thickness of individual coastal bay sequences is typically less than 10 m; however, numerous sequences may be stacked in the stratigraphic record. The extent of these sequences is typically limited to tens or a few hundreds of square kilometers.

Textural trends. Sequences that accumulated in coastal bays commonly display a fining-upward textural trend. This is especially true for estuarine sequences in microtidal environments. Tide-dominated estuaries and lagoons do not show any particular stratigraphic trends in grain size.

Lithology. Muds composed largely of clay minerals are the most abundant lithology in coastal bay deposits. Sand-size material may be comprised of almost anything, but quartz and rock fragments are most abundant, with gravel-size carbonate biogenic particles common in many sequences. Lagoonal sequences may contain evaporite minerals such as gypsum, anhydrite, and halite.

Sedimentary structures. A broad spectrum of sedimentary structures may be present in coastal bay sequences. Estuarine sequences show structures that are different from those of lagoons. In addition, tide-dominated estuaries have a somewhat different suite of structures from those of estuaries where tidal currents are not prominent. Lagoons display parallel bedding and some burrowing, but few structures generated by waves or currents because of the quiescent conditions of this environment. Estuaries tend to display much cross-stratification, flaser bedding, wavy bedding, and lenticular bedding together with bioturbation. Those which accumulated in tide-dominated estuaries have much megaripple cross-stratification. These sequences will have distinctly bidirectional cross-stratification formed by the alternating tidal currents.

Paleontology. Fossils are a common constituent of coastal bay deposits. Lagoons tend to have a restricted fauna generally with little diversity, reflecting the generally hypersaline conditions of this environment. Estuarine sequences contain abundant brackish to normal marine organisms. Bioturbation may leave distinctive burrow structures some of which contain organisms in living position.

Association. Coastal bay sequences may be stratigraphically adjacent to continental strata or marine strata. Most commonly they are associated with barrier island sequences, especially in a transgressive mode.

ADDITIONAL READING

CASTANARES, A. A., AND PHLEGER, F. B. (EDS.), 1969. *Coastal Lagoons—A Symposium.* Universidad Nacional Autónoma de México/UNESLO, Mexico City, 686 p. Somewhat old but this volume is the only one devoted solely to lagoons; mostly modern but also includes papers on ancient analogs. The term "lagoons" is interpreted in the loose sense in that some coastal bays described above as estuaries are included.

CRONIN, L. E. (ED.), 1975. *Estuarine Research*, v. 2: *Geology and Engineering*. Academic Press, New York, 587 p. An excellent compilation of papers dealing primarily with sedimentology of mesotidal estuaries.

LAUFF, G. H. (ED.), 1967. *Estuaries*. Amer. Assoc. Adv. Sci., Publ. No. 83, Washington, D.C., 757 p. This is the original comprehensive volume on estuaries with numerous classic papers. It is somewhat out of date by today's standards.

NELSON, B. W. (ED.), 1972. *Environmental Framework of Coastal Plain Estuaries*. Geol. Soc. Amer., Mem. No. 133, 619 p. Series of papers dealing with a broad spectrum of estuarine studies; dominantly concerned with the east coast of the United States.

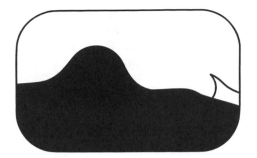

12 The Barrier Island System

The barrier island complex is one of the most diverse of all depositional systems. Modern barriers are extremely valuable for habitation and recreation, and in protecting the fragile environments on their landward side. Ancient barriers have been extremely productive reservoirs for oil and gas.

Origin

Barriers are elongate accumulations of sediment which are parallel to the coast and separated from it throughout at least most of their extent. The origin of barriers has received much discussion in the literature. Since the pioneering report by de Beaumont (1845) there have been three primary mechanisms considered for the origin of barriers: (1) emergence of shallow bars by wave processes (de Beaumont, 1845); (2) long spits developed by longshore currents and then breached, forming inlets (Gilbert, 1885); and (3) drowning of coastal ridges by rising sea levels (McGee, 1890). More recent authors have each chosen one of these origins to champion (Hoyt, 1967, 1968; Fisher, 1968; Otvos, 1970). There seems to have been a tendency for each of these researchers to consider that barriers have but a single origin. A more broad-minded approach is advocated by Schwartz (1971), who suggests a multiple causality for their origin.

A look through a world atlas quickly shows that barrier islands exhibit markedly different morphology, size, and relationships to nearby landmasses. In addition, barriers occur in a wide variety of settings and wave climates. It seems logical, therefore, that a single mechanism is an unlikely explanation for these features. The origin and development of small barriers has actually been monitored along the coast of Florida, where they are emerging from subtidal shoals (Davis et al., 1979).

General Setting

Barriers comprise between 10 and 13% of the coastline of the major landmasses of the world (Schwartz, 1973). A systematic check of the physiography and geology of the landmasses adjacent to barriers indicates that they are most widespread in trailing-edge regions adjacent to coastal plains (Glaeser, 1978). The Atlantic and Gulf coasts of the United States are among the world's largest barrier chains and are certainly the best known geologically. There are also extensive barriers in the Arctic on the Alaskan coast, along the North Sea, in Brazil, and in many other places. Virtually every landmass at least has barrier spits which have developed from headlands across embayments.

Barrier islands as they now appear are very young geologically. Rapid sea-level changes during the early Holocene transgression probably did not permit the development of well-formed barriers. The slowing of the transgression about 7000 years B.P. permitted barriers to develop and to migrate landward, such as along the middle Atlantic coast of the United States (Fischer, 1961; Kraft, 1971). This same period of decreased sea-level rise also permitted barriers to develop on prograding coasts. The classic example of that situation is Galveston Island, Texas (Bernard and LeBlanc, 1965). Most of the present-day barriers are less than 5000 years old, reflecting the relative stillstand in sea level during this period.

In addition to the obvious role of sea level and wave action combined with longshore currents, the tides may play a significant role in barrier island development and morphology. Using the tidal classification of Davies (1964), a relationship has been developed between tidal range and barrier systems (Hayes 1975). Microtidal coasts are characterized by well-developed and nearly continuous barriers such as Padre Island in south Texas. The barriers of the mesotidal coast are typically short with tidal inlets common, such as are present on the Georgia–South Carolina coast. Coasts that experience macrotidal ranges generally do not contain barriers. These tide-dominated coasts are subjected to such extreme tidal currents, oriented essentially perpendicular to the coast, that wave action is not effective and longshore currents are practically absent. The German Bight region of the North Sea shows nearly the complete spectrum of these conditions. The tidal range increases toward the Elbe River, and nearby estuaries and the barriers show the expected related trend. The reader must be aware, however, that there are exceptions to this generalization, primary along microtidal and mesotidal coasts.

Morphology and Major Depositional Elements

The complete barrier island system actually includes everything from the nearshore marine environment to the mainland coast. Although each of the several depositional

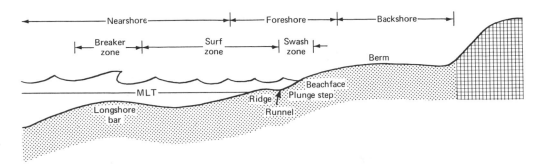

Figure 12-1 Diagram showing major environments in the beach and nearshore zone.

environments included may be separated from one another by certain characteristics, each is also an integral part of the barrier system because of the interaction of processes between adjacent environments. A typical barrier system includes numerous linear environments which parallel the trend of the barrier system and also a few which are more local in extent.

The most seaward environment is the nearshore zone, which is the zone of high wave action and of longshore bars and troughs. It is comparable to the inner shoreface of some investigators. The beach and nearshore environments meet at the low-tide line. The beach is the zone of unconsolidated sediments that extends from this line to the uppermost limit of wave action. The latter is commonly associated with an abrupt change in slope and morphology. The intertidal beach may consist of a **ridge and runnel** lying just seaward of a concave-upward storm beach (Figure 12-1), or it may consist of a relatively steep, seaward-dipping **foreshore** and a nearly horizontal **backshore** (Figure 12-1). The latter profile is characteristic of accreting conditions.

Landward of the beach are coastal dunes (Figure 12-2) which vary greatly in size and extent from one barrier to another. These dunes may occur in several rows or only a single one; they may be only slightly recognizable or as much as 100 m above water level. Their development depends primarily on sediment supply but also on the wind conditions. The next sedimentary environment varies from island to island but is typically nearly flat and only slightly above sea level; commonly, it is supratidal. These back-island flats may contain terrestrial grasses, marsh grasses, or be free of vegetation.

The tidal range in the estuary or lagoon which separates the barrier from the mainland determines the extent of intertidal zones that border it. Barren tidal flats,

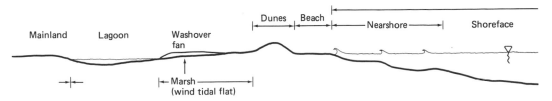

Figure 12-2 Profile diagram of barrier island complex, showing major environments.

Chapter 12 The Barrier Island System

marshes, or mangrove swamps may be present in a variety of combinations. Although they surround these coastal bays, the relative distribution of each environment as well as the overall extent range widely throughout the various barrier island systems.

The distinct environments described above are all essentially continuous and parallel along the barrier system (Figure 12–2). There are some local sedimentary environments which are also integral components of the barrier system. The inlets that disrupt the continuity of the barrier and their associated tidal deltas play vital roles in barrier island sedimentology. Storm conditions commonly cause dune ridges to be breached or overtopped, resulting in the accumulation of **washover fans** (Figure 12–3). These thin fan-shaped pulses of sediment represent a primary mechanism for landward transport of barrier sediments.

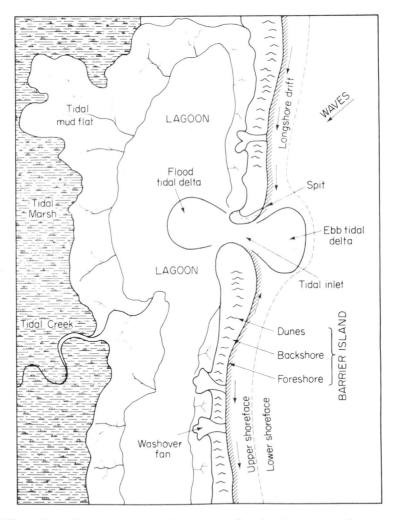

Figure 12–3 Sketch map of a barrier island complex, showing various sedimentary environments present. (From Blatt et al., 1980, Figure 19-14.)

Several of the major sedimentary environments listed above as constituent elements in the barrier system have already been discussed as individual entities. These include tidal flats and marshes (Chapter 10) and coastal bays and related inlets (Chapter 11). The primary aims of this chapter are twofold. One is to discuss major elements of the barrier system, which are dominated by seaward-derived processes such as the nearshore, beach, coastal dunes, and washover fans. The other is to consider the entire barrier island system as it occurs in the stratigraphic record.

BEACH SEDIMENTATION

Marine processes that play roles in barrier island sedimentary environments are restricted to three types: wind, waves, and currents. In most cases there is a great deal of interaction among these, and the interactions are very complicated. Waves are dominant most of the time in the nearshore zone and in the foreshore portion of the beach. Wind is dominant on the backshore most of the time and essentially always in the dunes. Currents may be wave-generated, such as longshore and rip currents. These are extremely important in the zone of wave action and under some circumstances may be the dominant process.

By this time the reader is probably wondering why tidal currents have not been included as playing a significant role in barrier island sedimentation. In actuality, tidal currents are only of direct significance very near and inside inlets (Davis and Fox, 1981), where they move sediment. Away from inlets tidal range does play an indirect role in coastal processes (Davis et al., 1972). As tide level rises and falls, the effects of waves and currents are moved from place to place in the beach and nearshore zone.

Nearshore Zone

As waves move into shallow water and begin to be interfered with by the bottom, there are some phenomena that take place which are critical to an understanding of nearshore sedimentation. A good general summary of these phenomena for a non-barred coast has been constructed by Clifton et al. (1971), who recognize three environments within the nearshore. In the absence of longshore bars, the outer limit of the nearshore zone will be considered as the point where waves become peaked and approach a solitary form. Seaward of that point, waves are nearly sinusoidal in shape and generate asymmetric ripples. These are the combined-flow ripples of Harms (1969).

In the zone of buildup, waves achieve a solitary shape with rather peaked crests. The asymmetry of the orbital velocity causes landward currents which generate lunate megaripples (Figure 2–24). The landward margin of the zone of buildup is at the breaker line, where waves reach instability. In the surf zone there is extreme turbulence and energy, which produces plane bed sediments. A narrow zone of rough beds separates the outer plane beds from the inner plane beds or swash zone deposits (Figure 2–24). What occurs, therefore, is a sequence of bedforms which represent the transition from the lower part of the lower flow regime up to plane bed deposition in the upper flow regime. In fact, it is not uncommon for antidunes to develop in the swash of gently sloping, fine-grained beaches (Figure 3–11). The plane beds and antidunes that develop in the swash zone are generated in a zone where backwash with seaward motion dominates, whereas all other zones display a landward motion (Figure 2–24).

The model described above is representative of high-energy, nonbarred coasts such as are common along the west coast of the United States. Another coastal situation to be considered is the barred coast, which is typically lower in wave energy and which characterizes the east and Gulf coasts of the United States.

In a study of a nearshore environment adjacent to barriers in the Gulf of St. Lawrence, Davidson-Arnott and Greenwood (1974, 1976) constructed a model comparable to that of Clifton et al. (1971). The addition of longshore bars presents a more complicated profile, and when this profile interacts with waves it produces a correspondingly more complicated set of processes and resulting structures. The profile in question contains two longshore bars which are asymmetrical with the steeper side shoreward (Figure 12–4). Detailed study of box cores as well as surface features enabled the investigators to establish sedimentary facies which are characteristic of the various subenvironments in the nearshore (Davidson-Arnott and Greenwood, 1976).

The seaward slope facies is located in the area of shoaling waves just seaward of the breaker zone. It is characterized by wave-generated asymmetric ripples and plane beds, depending on the location relative to breakers and on the wave size. The bar crest is the location of breakers; however, under certain conditions waves may be breaking only on the inner bar. Structures developed there are lunate megaripples and plane beds. The cross-stratification produced by the megaripples may be seaward in orientation as well as landward. Seaward-flowing currents interfere with wave motion, causing seaward migration of these bedforms (Davidson-Arnott and Greenwood, 1976). The landward slope facies is reminiscent of the seaward slope except for the attitude of the plane beds. This facies displays significant differences between the inner and outer bar (Figure 12–4). Primary among them is the avalanching of sediments in a landward direction as wave-generated currents move over the shallow inner bar. Waves begin reforming over the trough. In this area, longshore currents

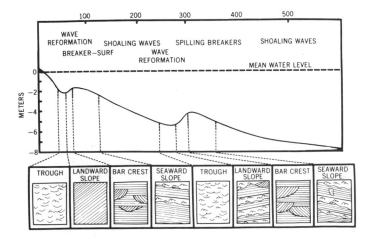

Figure 12–4 Surface sediment facies on nearshore barred topography, showing sedimentary structures. (After Davidson-Arnott and Greenwood, 1976, Figure 4, p. 154.)

Part Three   Transitional Environments

have their greatest influence and may dominate wave effects. Ripples are widespread and commonly have crests perpendicular to shore, testifying to the influence of longshore currents.

There is a total of four sedimentary facies in the bar and trough system, with each bar and trough having one complete set. Although the study by Davidson-Arnott and Greenwood (1976) included only two bars, it seems reasonable to extrapolate their characterization to systems with more bars.

Both of the nearshore systems described above considered only the relatively simple two-dimensional profile. It is now appropriate to consider the third dimension: the alongshore variation in topography and its influence on the process-response mechanisms that operate in the nearshore. Such three-dimensional consideration is most likely to show variety on a barred coast. The most dramatic variability is the presence of rip channels through the bar crests. These channels are rather subtle in their topographic expression, but they are important in the circulation scheme in barred coasts (see Chapter 2). Earlier workers (Cook, 1970; Davis and Fox, 1972) demonstrated that rip currents transported sediment as bed load but did not define the rip channel facies as did Davidson-Arnott and Greenwood (1976). The dominant feature of this facies is seaward-dipping cross-stratification with sets up to 20 cm thick. They are the result of seaward rip current flow which produces small megaripples.

Previous consideration of the nearshore zone covered only physical processes and their responses. Biological processes, specifically bioturbation, may also be significant. There is an obvious and indirect relationship between the level of physical energy and the bioturbation in sediments. A detailed study of benthic organisms and bioturbation in the nearshore zone off Padre Island, Texas, has shown that many organisms occupy this dynamic zone (Hill and Hunter, 1976). Some benthic invertebrates, such as surf clams (*Donax*) and amphipods, are abundant across the entire bar and trough system. By contrast, the sand dollar (*Mellita*) is more abundant in the relatively low energy of the trough and does not occur landward of the second bar (Figure 12–5).

Intertidal Beach

The beach is not only the most dynamic coastal environment, it is probably also the most studied. One of the first generalizations that was made about beach dynamics is that the beach represents a cyclic process-response system. The beach cycle is related to wave energy variations. Storms cause much beach erosion and typically take place over only a short period of time, whereas the rebuilding of the beach between storms may take at least an order-of-magnitude more time.

The storm or erosional beach is characterized by a seaward slope throughout with a slightly concave-upward profile (Hayes, 1969). The intertidal portion of this beach consists of a ridge and runnel system (King and Williams, 1949), which is an ephemeral bar and trough formed as the result of erosion in the higher portion of the beach. The accretional beach has a well-developed berm with a rather steep forebeach and a horizontal or slightly landward sloping backbeach area (Figure 12–1). The processes and morphologic responses that take place during the cycle from one beach

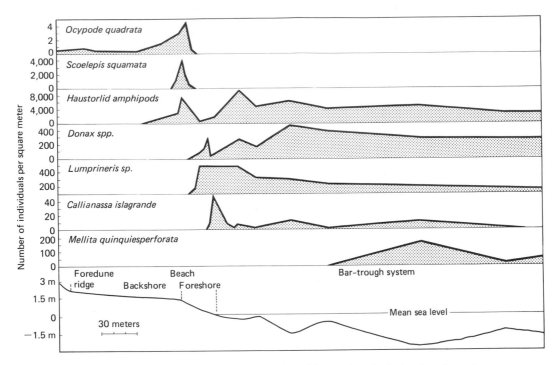

Figure 12-5 Distribution and zonation of benthic organisms in the beach and nearshore area, Padre Island, Texas. (From Hill and Hunter, 1976, Figure 4, p. 174.)

configuration to another are predictable and generate a characteristic sequence (Hayes, 1972).

Early studies of ridge and runnel features served to formulate the model for this cycle (Hayes, 1969). After subsidence of a storm, the beach exhibits the characteristic profile (Figure 12-1). The ridge is initially somewhat symmetrical but rapidly becomes asymmetrical, with the steep side landward (Figure 12-6). Small waves create currents that pass over the ridge crest and also transport sediment in a landward direction. This not only gives the bar its asymmetric shape but causes it to

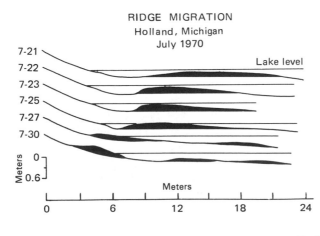

Figure 12-6 Sequential change in shape and location of storm-generated ridge as it migrated onto storm beach in Lake Michigan. (From Davis et al., 1972, p. 416.)

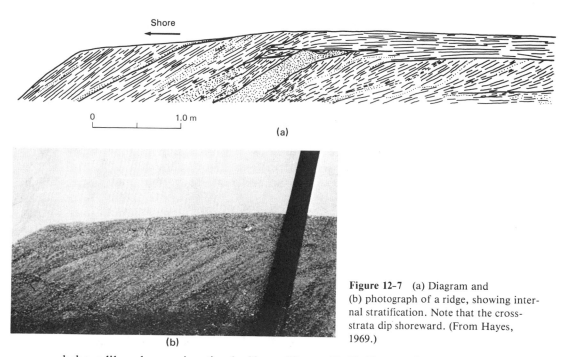

Figure 12-7 (a) Diagram and (b) photograph of a ridge, showing internal stratification. Note that the cross-strata dip shoreward. (From Hayes, 1969.)

behave like a large migrating bedform (Figure 12–7). Eventually, the ridge will weld to the storm beach and thus serve to repair it (Figure 12-8). The rate of migration varies considerably and is inversely related to tidal range (Davis et al., 1972). The reason for this inverse relationship is simply the time within each tidal cycle that the ridge is being moved by wave-generated currents. In areas of high tidal range the ridge is exposed for much of the cycle, whereas on microtidal coasts it is under wave motion for much of the tidal cycle.

The three-dimensional nature of this ridge and runnel system is much like the shallow longshore bar and trough. The ridge is cut by rip channels and there are com-

Figure 12-8 Photo showing ridge as it begins to weld to the storm beach. (Courtesy of M. O. Hayes.)

plex currents. As the tide floods, the rip channel and runnel slowly fill with water. During high tide and ebb tide, bedforms are generated on the ridge, in the runnel, and in the rip channels. The ridge surface is typically plane bed deposition, although antidunes have been reported (Wunderlich, 1972; Hayes and Kana, 1976). The runnel contains a bed of **ladderback ripples** developed due to processes oriented at right angles to one another (Figure 12–9). There are shore-parallel ripples formed by waves which pass over the ridge, and there are ripples perpendicular to the shore which are generated by ebbing currents flowing over the runnel toward a seaward outlet (rip channel). The circulation pattern is much like the nearshore rip system. An additional feature of the runnel area is the accumulation of fine sediment or pelletized fines in the troughs of ripples. This provides the mechanism for formation of flaser bedding (Figure 3–15).

Migration of the ridge over the runnel and eventually welding to the storm beach causes a rather abrupt change in sedimentation conditions. As the ridge moves landward it experiences some vertical accretion such that it causes waves to cease overtopping it. Instead, waves move up and back in a typical swash zone fashion, reworking the ridge sediments into seaward-dipping foreshore deposits (Hayes, 1969).

These morphologic features and related sedimentary structures permit the development of a typical transgressive sequence for intertidal beach deposits. The basal unit is the storm beach deposits, which dip seaward at 2 to 5° and are characterized by zones of lag deposits of heavy minerals (Figures 12–10 and 12–11). The overlying unit represents the runnel and contains ripple cross-stratification with directions pointing landward and alongshore. Flasers may be present. Rather large-scale, avalanche foresets which indicate landward transport characterize the ridge unit (Figure 12–10). The reworking of welded ridge deposits creates seaward-dipping foreshore deposits with dips of 5 to 15°. The next unit that might be encountered in a

Figure 12-9 Photo of runnel, showing ladderback ripples. One set is formed by waves and the other by currents moving parallel to shore. (Courtesy of M. O. Hayes.)

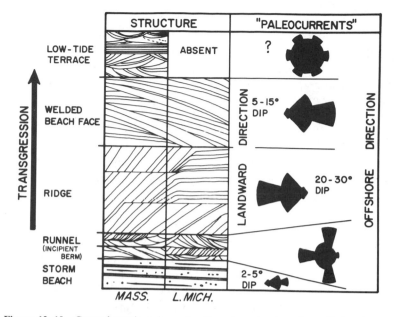

Figure 12–10 General stratigraphy and sedimentary structures in cyclic beach sequence. (From Davis et al., 1972, Figure 11.)

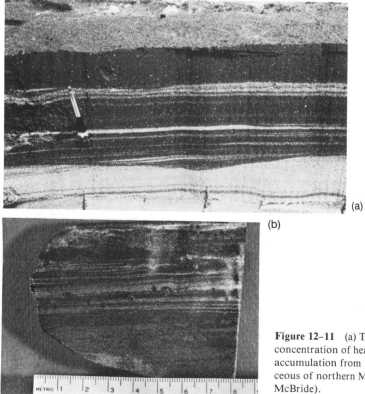

(a)

(b)

Figure 12–11 (a) Trench in beach, showing concentration of heavy minerals; (b) similar accumulation from Difunta Group (Cretaceous of northern Mexico) (courtesy of E. F. McBride).

413

complete transgressive sequence would be a low-tide terrace with plane and cross-stratification oriented in various directions due to wave and current orientations beyond the surf zone (Figure 12-10). Studies of numerous coasts have confirmed that this process-response system is essentially universal, although the scale and rate of change may differ (Davis et al., 1972).

Subaerial Beach

The backshore portion of the beach is also complex but is sometimes overlooked by investigators of beach processes. Generally, the backshore is subaerial except during spring tides or storm conditions, when it becomes inundated. The backshore may not exist on some beaches at some times due to storm erosion.

Winds generate ripples, sand shadows, and small shadow dunes in the backshore area. Stratification is commonly nearly horizontal and conforms well with the surface. Bioturbation is widespread in this zone and may at least partially destroy stratification. The ghost crab (*Ocypode*) is the most characteristic taxon in this area (Figure 12-5), but rove beetles are also present (Hill and Hunter, 1976).

The preservation potential of the backshore environment is extremely low, due to its ephemeral nature and the difficulty of burying it in a sequence without first causing its destruction.

COASTAL DUNES

Dunes that form along the coast or on a barrier island are fundamentally like those of inland deserts. They are alike in their processes, sedimentary structures, and the morphology of individual dunes. Primary among the differences are the overall shape of the dune complex and vegetation, which is common on coastal dunes. Coastal dunes are not related to climate; they occur on arid coasts of Baja California, Mexico, and along the humid coasts of the Pacific in the United States.

Some coastal dunes are free of vegetation and migrate rapidly (Figure 12-12), whereas many are vegetated and are relatively fixed in their position. Vegetated dunes are most common (Goldsmith, 1978) and generally take the form of well-defined ridges parallel to the shore. These dune ridges are called **foredunes**. They may occur as multiple ridges (Figure 12-13) and they provide evidence of progradation of the barrier on which they occur. They may also develop on migrating spits.

The primary requisites for the formation of coastal dunes are an abundance of sediment and a mechanism for accumulating it in dune form. Meteorologic phenomena along the coast are such that there is almost always some onshore component of wind which carries sediment landward from the backshore. Dunes tend to begin development as wind shadow deposits in the lee of vegetation or other obstructions. Growth of such incipient dunes was monitored in the Cape Cod, Massachusetts, area by Goldsmith (1978), who measured increases in elevation at a rate of 0.3 to 0.5 m per year.

Although many coastal foredunes are relatively immobile because of the

Figure 12–12   Large shoreward-migrating coastal dune displaying well-developed slipface, Coorong National Park, South Australia.

presence of vegetation, there still exists the typical internal cross-stratification associated with aeolian dunes. Roots may locally disrupt the stratification and the dunes may be slow in moving, but their internal nature is still characteristic. The most obvious feature is the somewhat planar, large-scale cross-stratification. Studies of several foredune areas show both bimodal directions (Goldsmith, 1973; McBride and Hayes, 1962) and unimodal directions (Land, 1964). Differences result from seasonal

Figure 12–13   Aerial photograph showing multiple parallel foredune ridges, 90-mile Beach, Victoria, Australia.

changes in prevailing wind and from prevailing versus predominant winds, both of which may be recorded in the dune accumulation. Dip values range widely but are generally between 10 and 35° (Goldsmith, 1978). Small-scale cross-stratification may be common and may take many forms. It is of ripple origin but displays a variety of climbing ripple and ripple-drift bedding (Hunter, 1977a, 1977b).

The stratification types discussed above are generated by traction transport of sediment over a rippled or smooth dune surface. Grainfall and grainflow deposition may also occur. The latter shows distorted stratification and high dip angles (Hunter, 1977a).

Eolianites

Coastal areas in tropical environments are dominated by carbonate sediment, largely of biogenic origin. The wind accumulates the sand-size carbonate particles in dune forms along the backshore. The combination of climate and calcium carbonate causes rapid lithification of these dune sediments, forming **eolianites.** Resistant ridges of eolianite are present in many parts of the world, including Bermuda, Andros Island (Bahamas), Israel, and Africa. Multiple ridges in Australia near the Coorong indicate former positions of the shore during the Pleistocene.

The internal nature of eolianite is similar to noncarbonate dune accumulations except that cross-stratification may be more regular and ripple cross-strata are uncommon (Figure 12–14). Large sets of planar foresets are widespread, but lenses of cross-strata occur as well.

Figure 12–14 Pleistocene aeolianite deposit along the southern Victoria coast, Australia.

WIND TIDAL FLATS

The supratidal zone on the landward side of a barrier is commonly broad, nearly devoid of relief, and at some locations lacks vegetation. This environment is called the **wind tidal flat** because of its susceptability to inundation by even slight wind or storm tides. The Padre Island area of south Texas contains a well-developed wind tidal system and is well known (Hayes, 1967a; Miller, 1975). Modest winds may cause wind tides of 30 to 50 cm and thereby cover more than 50 km^2 in this area.

The wind tidal flat is somewhat like the backshore beach in that it experiences both eolian and subaqueous processes. It differs in that the storm tides bring about deposition on the wind tidal flats, whereas they cause erosion on the backshore. During elevated tidal levels, water with suspended sediment covers the flat. Commonly, there is little wave or current energy and fines settle onto the wind tidal flat. These fines are trapped by the blue-green algal mats that cover the environment (Figure 12–15), and vertical accretion takes place. Occasionally, waves contain sufficient energy to tear up some of the algal mats, but they recover rapidly.

Eolian processes on the wind tidal flats prevail because inundation by water is occasional. In most respects the wind tidal flat is considered as a surface of bypassing. Erosion by the wind is prevented by the continuous algal mat, which acts as an armor. Accumulation of sand derived from the barrier is rather temporary. Where

Figure 12–15 Well-developed algal mats covering the wind tidal flat surface and a trench showing layering of older mats and sand, Padre Island, Texas. (Courtesy of J. A. Miller.)

there is little sand available, it typically is carried across the wind tidal flat and deposited in the lagoon. Some areas possess much available sediment, and transverse dunes that migrate rapidly are developed (Figure 12–16).

The residence time of the mud that accumulates during inundation may be short. Rapid drying and desiccation provide flakes of mud aggregate which are blown landward. A significant portion of this may be incorporated in the clay dunes landward of Laguna Madre (see Chapter 11).

Aeolian processes play an important role in the landward transport of sediments and even barrier islands. This is particularly evident in the Padre Island barrier system, where a large area of Laguna Madre has been completely filled by sand transport by wind (Figure 11–26). There are also numerous places where large sediment accumulations have encroached on the lagoon (Figure 11–22).

STORM EFFECTS ON THE BARRIER SYSTEM

Impingement of intense weather systems on the coastal zone is a widespread and geologically very important phenomenon. Barrier island systems are particularly susceptible to storm-related events because of their relative vulnerability compared to other coasts. The dominant processes that cause coastal changes are the waves and the storm tides. The latter are generally a passive factor in that their biggest role is to provide access to wave action.

Figure 12–16 Scattered transverse dunes on wind tidal flats behind Padre Island, Texas.

Basic Meteorological Phenomena

The passage of cyclonic systems or weather fronts over the coastal zone is the primary mechanism for generating high-wave-energy conditions that alter the coast. The cyclone is a counterclockwise-circulating low-pressure system that is reponsible for most of the large-scale coastal changes.

Cyclonic systems are especially important in the mid-latitude regions, where low-pressure systems move from west to east. Extremely intense storms may be spawned in the low latitudes over water and move into higher latitudes. These are variously called hurricanes, cyclones, and typhoons, but all are very intense low-pressure systems which can cause great change to coastal sedimentary environments.

In addition to the obvious increase in wave energy associated with low-pressure systems, the wind may create significant storm tides, which may be of great importance in barrier systems. Although there is an obvious relationship between the intensity of the storm and storm surge, an equally important factor is the rate of movement of the storm itself. High wind speed and a rapidly moving storm may cause less storm tide and less damage than a hurricane that has winds just above hurricane level but moves slowly. Intense rains which may accompany hurricanes are also important in sedimentation.

Washover Phenomena

Larger waves in concert with storm tides may break or overtop the crest of the barrier. This is especially important if the storm intensity peaks at high tide or if the storm occurs during spring tide conditions. Regardless of the specific situation, large quantities of sediment and water are carried through a rather small saddle or excavated channel in the crest of the island and then spread out in a fan shape over the back-island flats (Figure 12–17). These fans may be small or can cover several square kilometers (Andrews, 1970). Large fans have recognizable distributary systems and may be reactivated during successive storms.

Internal stratification of washover fans shows two different styles. Most of the transport of this sediment is by sheet flow; upper plane bed conditions prevail. Some situations may cause the washover sediment to move into a water body such as a lagoon or an estuary (Schwartz, 1975). In the event that this happens, foreset stratification develops in the distal part of the fan (Figure 12–18). Marsh grasses possess the ability to rapidly overcome burial by washover fans and grow up through the sand (Godfrey and Godfrey, 1973; Hosier and Cleary, 1977).

Most barriers exhibit the subtle morphology of barren fans which are recognizable as washover features. The phenomenon of washover fan production has been recognized as the dominant mechanism for landward migration of barriers in a transgressive situation. Schematic models have been developed to demonstrate this (Godfrey and Godfrey, 1973; Hosier and Cleary, 1977). Such models are cyclic and show washover features which develop at the expense of the beach.

CENTRAL PADRE ISLAND

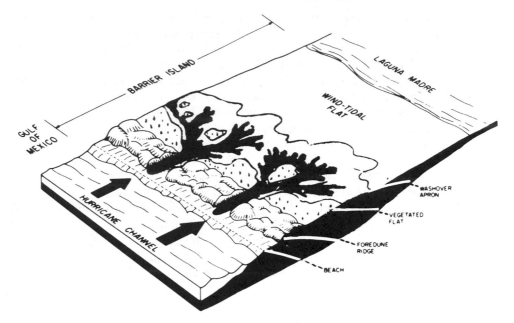

Figure 12–17 Generalized diagram of barrier system, showing washover deposits and their relation to other barrier environments. (After Scott et al., 1969.)

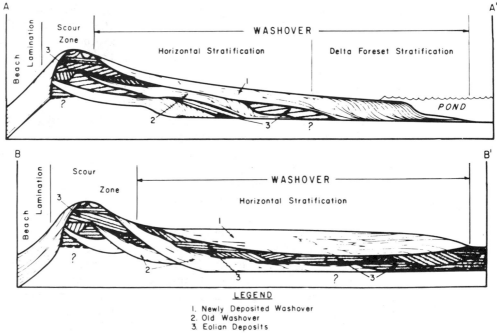

LEGEND
1. Newly Deposited Washover
2. Old Washover
3. Eolian Deposits

Figure 12–18 Schematic cross sections through washover fans, showing (a) Gilbert-type delta prograding into the water and (b) horizontal stratification of sheet flow deposition. (After Schwartz, 1975.)

Inlets and Hurricane Passes

Intense meteorologic events cause not only storm tides of as much as a few meters but also cause great changes in the tidal prism which is exchanged between the open water and the back barrier coastal bays. Existing inlets may be modified or new pathways may be generated to accommodate this large and abrupt increase in tidal prism.

One typically considers that large waves and high storm tides which accompany storms generate changes largely through landward transport of water. Certainly, there are situations where this is the case. What is not commonly realized is that the extreme discharge through passes in the barrier system is equally important. Washover phenomena create small channels which may be enlarged to the point that tidal currents are carried through them. These hurricane passes (Hayes, 1967a) typically generate a sediment body similar to a flood delta. They are generally sealed off from tidal circulation by littoral drift within only months of their formation (Figure 12-19). It is also common for storms to cause considerable modification of flood deltas on tidal inlets.

The storm effects described above are easily recognized and show readily discernible lasting effects. The great discharge of water and sediment in a seaward direction across the barrier may display its effects shortly after the storm but they are then destroyed by nearshore waves and currents. There are three factors that may contribute to this seaward flushing of sediment-laden water from the bays. The most

Figure 12-19 Small hurricane pass with mouth sealed off due to littoral drift, Padre Island, Texas.

obvious is the generally great quantity of precipitation and runoff that commonly accompanies a storm. Another is the relatively rapid release of onshore energy when the storm subsides. A hydrostatic head has developed between the bays and open water. The third factor is that, depending on location of the storm, there may be a shift from onshore to offshore winds as the center of the cyclonic system passes over the coast. Any or all of these will result in much sediment being carried through various pathways in the barrier and out onto the shallow shelf.

These phenomena were first documented in detail by Hayes (1967a), who found extensive graded units on the inner continental shelf of Texas following Hurricane Carla in 1961. Numerous hurricane channels carried the ebbing surge out over the shelf (Figure 12–20). Mansfield Pass, a structured inlet that had been closed due to littoral drift, was open to tidal circulation after Hurricane Carla (Hayes, 1967a). A similar but more pronounced ebb surge effect took place at Corpus Christi Pass dur-

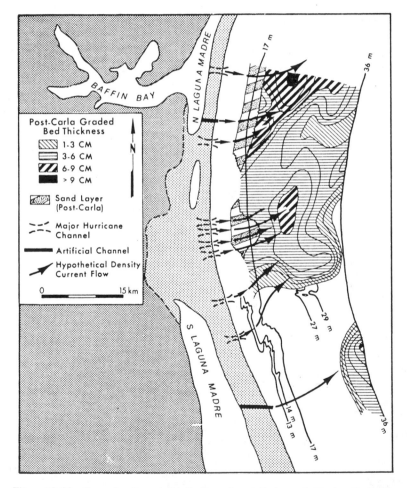

Figure 12–20 General sediment distribution of post-Carla graded beds adjacent to Padre Island, Texas. (From Hayes 1967a, p. 24.)

ing Hurricane Beulah in October 1967. This small natural inlet was closed off and nearly filled with washover and eolian sand prior to the storm, but was totally flushed of sediment afterward (Figure 12–21). This seaward transport of great quantities of sediment also produces linear sand bodies normal to the coast. They are similar to the form taken by ebb tidal deltas in tide-dominated areas, but these storm-generated ebb deltas are quickly dispersed by wave action.

Cheniers

Elongate, smoothly arcuate, and subparallel ridges of sand or shell debris located along the margin of a coastal plain are called **cheniers**. In the rock record cheniers bear some likeness to barrier island sediments, but they are markedly different in origin. Cheniers are located on a generally muddy, marshy coast and are not

(a)

(b)

Figure 12–21 Oblique aerial photos showing Corpus Christi Pass (a) prior to Hurricane Beulah in 1967 and (b) after the storm had completely flushed the pass of sediment. (From Davis et al., 1973, p. 125.)

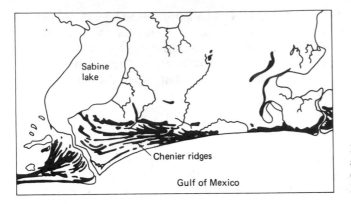

Figure 12-22 Map of chenier ridges along the Texas-Louisiana coast of the Gulf of Mexico. (After Gould and McFarlan, 1959.)

separated from the mainland by water. These low-lying ridges are abundant in many areas of the western and southern Gulf coast, such as near the Sabine River (Figure 12-22).

The similarities in general morphology compared to barriers, and the necessity of high-wave-energy conditions for formation, make this an appropriate place to discuss cheniers. These ridges are features of prograding coasts. Prime requisites for formation of cheniers are an abundance of sediment and a coast of rather low wave energy. Commonly, chenier plains are developed adjacent to river mouths and are similar in general appearance to wave-dominated deltas with numerous beach ridges (compare Figures 9–22 and 12–23). Rivers supply abundant fine sediment, producing prograding mudflats and marshes. A reduction in sediment supply or a period of elevated wave energy causes the removal of the fine sediment, with shell debris and coarse terrigenous sediment concentrated in the form of a ridge (Gould and

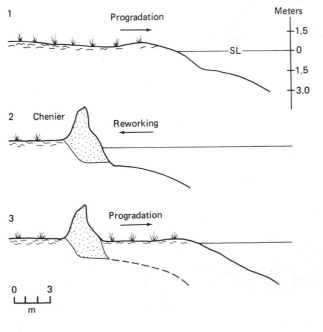

Figure 12-23 Sequence showing chenier formation by alternation of progradation and wave reworking. (From Hoyt, 1969.)

McFarlan, 1959; Hoyt, 1969). A rather cyclic continuation of these processes produces the chenier plain (Figure 12-22).

Cheniers reach only a few meters above sea level but are continuous for many kilometers. The chenier plain may be up to 15 km wide. Internal stratification of cheniers is complex, with directions both seaward and landward depending on swash or washover origin (Hoyt, 1969). There is a general trend of increasing-upward grain size (Byrne et al., 1959).

BARRIER SEDIMENTS

The spectrum of sediments contained in the barrier system is extremely broad; this includes both texture and mineralogy. If one includes the adjacent coastal bays, an even broader spectrum is present. Essentially all sediments are terrigenous or are biogenic debris. The variety of terrigenous mineralogy is dependent on the source(s) of sediments available, but is generally dominated by quartz with lesser amounts of feldspar and rock fragments. Accessory or heavy minerals typically account for 1 or 2%. Biogenic debris is dominantly molluscan, but a broad spectrum of taxa is commonly represented.

Nearshore Zone

Numerous studies of the beach and nearshore zone have included textural parameters of the sediments. Many authors have attempted to characterize specific environments by textural parameters, but as yet no widespread agreement has been reached.

The topography of the barred nearshore zone is such that energy variations may be pronounced. This is commonly reflected in the texture of the bottom sediments. For purposes of discussion we consider four subenvironments within the nearshore: offshore, bar, trough, and surf zone. Absolute values for grain size, sorting, and other parameters are not very meaningful, but relative values may show patterns. Unfortunately, some studies present opposite comparisons with others (Komar, 1976).

In a general way, sediment is coarsest and least sorted at the **plunge point** and fines in both the landward and seaward directions. This is best illustrated in tideless environments (Fox et al., 1966; Davis and McGeary, 1965), where the plunge point remains stationary. Although the rising and falling tide shifts the plunge point, it is also coarsest along marine coasts (Visher, 1969; Greenwood and Davidson-Arnott, 1976; Miller and Zeigler, 1964).

Some researchers have found generally finer and better-sorted sediments on the longshore bar compared to the trough (Greenwood and Davidson-Arnott, 1976; Hunter et al., 1979). This is the relationship most commonly found by the present author. By contrast, the reverse relationship has also been found (Figure 12-24). The same disagreement exists within the breaker zone, where some authors find poor sorting (Inman, 1953) and others record the best sorting in the nearshore profile (Miller and Zeigler, 1964). The very complex interactions among waves, currents, and the sediment on the bottom vary greatly and are the cause of these apparent disagreements. These data are testimony to the problems of using sediment texture to

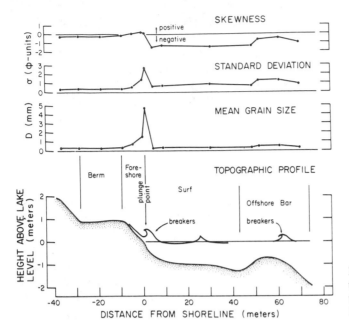

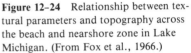

Figure 12-24 Relationship between textural parameters and topography across the beach and nearshore zone in Lake Michigan. (From Fox et al., 1966.)

identify environments. There is agreement on skewness, which is negative across the nearshore profile (Figure 12-24).

Beach

The intertidal beach or foreshore serves as an effective boundary for marine processes during normal energy levels. The swash zone, where water rushes up and back as each wave breaks at the plunge point, is relatively easy to characterize and compare to other beach and nearshore environments. The lag concentrations of heavy minerals due to storm erosion are fairly diagnostic (Figure 12-25). In addition, textural trends are rather universal in nature. Grain size is coarser than all but the plunge point and sorting is correspondingly poor. Both show a distinct shoreward trend (Figure 12-24). These characteristics have been observed in a broad spectrum of tidal range and wave energy (Fox et al., 1966; Greenwood and Davidson-Arnott, 1976; Hunter et al., 1979).

The backbeach is typically subjected to rather uniform conditions and displays textural characteristics which reflect this. Sediments are fine, well sorted, and show slight positive skewness (Fox et al., 1966; Dickinson, 1971; Moiola and Spencer, 1973).

Dunes

Textural parameters of dunes seem to have more general uniformity than other environments of the barrier. It should be realized that these generalizations are statistically based and that there may be significant differences within a single barrier

Figure 12-25 Veneer of heavy minerals on a recently eroded beach.

island (Goldsmith, 1978). Dunes typically exhibit mean grain size which is finer than adjacent seaward or landward environments. Differences may be only 0.10ϕ or less (Mason and Folk, 1958) or they may be many times that, especially along coasts with coarse beaches.

Dunes are typically the best sorted of barrier island environments and display a positive skewness (Mason and Folk, 1958; Moiola and Spencer, 1973; Goldsmith, 1978). The lagging behind of coarse particles as dunes accumulate is thought to be the reason for the positive skewness and better sorting of dune sediments as compared to neighboring areas (Mason and Folk, 1958).

Dunes exhibit better rounding and sphericity than do adjacent environments, due to their eolian nature. This is due to rounding within the eolian environment by impact and also to selective transport of spherical grains from the beach to the foredune area (Williams, 1964). Surface textures such as frosting have also been utilized to characterize dune sediments. Although this may be true in part, it is also well established that the frosting can be formed by chemical as well as mechanical means (Kuenen and Perdok, 1962). Detailed recent studies of grain textures by use of the scanning electron microscope have provided what many people feel is the most definitive means for recognizing depositional environments by grain surface textures (Krinsley and Doornkamp, 1973).

Back-Island Flats

Compared to other barrier island environments, the landward portion of the dunes has received little study. It is also a diverse environment as compared to others within the barrier complex, except perhaps for the nearshore zone. The presence or absence of vegetation and combined eolian and aqueous processes serve to make generalizing about back-barrier sediments a difficult task.

As a consequence there is much variation in data reported from back-barrier environments. Mason and Folk (1958) found that these sediments were slightly finer grained than the beach but coarser than the dunes. By contrast, Dickinson (1971), working only tens of kilometers away, found back-island sediments coarser than those of the beach. Both writers concur that sorting is poorer than that in beaches and dunes, but Dickinson's study produced negatively skewed sediments and Mason and Folk (1958) found skewness to be positive. The decrease in sorting is probably due to the combination of aeolian and aqueous processes, giving a small fine mode in addition to the dominant coarser mode.

BARRIER ISLAND MORPHOLOGY

Although barriers contain common environments both in the transverse and longitudinal direction, there is some variety to general barrier shape. These differences are due to a combination of coastal processes, including waves, longshore currents, and tidal currents, with sediment supply. Hayes has classified barriers as microtidal and mesotidal (Hayes and Kana, 1976), with the contention that the shape of the barrier reflects tidal range and that macrotidal coasts do not permit development of barriers. Although this approach does have some merit, there are numerous exceptions. The present writer prefers to consider morphology irrespective of tidal range.

With this context in mind it is appropriate to think of barriers as being (1) rather long, narrow, and only occasionally interrupted by inlets, or as being (2) short, relatively wide, and with inlets being common along the coast. Generally, a rather extensive reach of coast is characterized by one type or the other, not a mixture. For example, the Gulf Coast of Texas and the Outer Banks of North Carolina fall into the "long, skinny" type, whereas the Gulf Coast of peninsular Florida, South Carolina, and Georgia are of the "short, fat" variety.

Long, Narrow Barriers

The relatively continuous barriers with few inlets tend to be developed along wave-dominated coasts where longshore transport of sediment is great. The Texas coast, including Mustang and Padre Islands, is an excellent example. Here the tidal range is low and waves are generally small except during extreme storm events. Nevertheless, longshore transport is great. Although storms generate hurricane passes (Hayes, 1967a) and man has dredged passes, they are rapidly closed to tidal currents by sediment from littoral drift. This results in washover and blowover phenomena being the dominant mode of landward transport of sediment. In addition, there is little or no seaward transport mechanism except flushing of sealed passes after storms. This sediment is then rapidly dispersed along the coast by longshore currents.

Short, Wide Barriers

Some extensive reaches of coast are characterized by short barriers up to 15 to 20 km in length and by numerous tidal inlets. These barriers commonly are wide or at least display a complex depositional history via multiple sets of beach ridges. Although tidal range may be important locally, it is more important for the coast to be at least tide influenced. That is, the tidal currents which enter and exit the inlets exert considerable influence or even dominate the coastal processes. Most commonly, it is a combination of tide- and wave-generated processes which prevail.

The presence of numerous tidal inlets and the development of ebb tidal deltas exerts a marked influence on the overall configuration of barriers. The close proximity of inlets to each other permits tidal circulation cells to develop, with some inlets being ebb dominated and some being flood dominated. Ebb-dominated inlets may produce some rather extensive ebb deltas and also make much sediment available to the open marine coastal system.

This type of barrier-inlet relationship generates islands that tend to prograde at one end through beach ridge accretion and migrate landward at the other end via washover (Figure 12–26). Such islands have been called **drumstick barriers** by Hayes (1975). They display a prograding end sheltered by the ebb delta and characterized by longshore current reversal and formation of linear ponds as beach ridges develop. The downdrift end shows spit development, and washovers are common (Figure 12–27). Although Hayes has designated drumstick barriers as mesotidal, there are several good examples on the west coast of peninsular Florida where tides are 70 to 80 cm.

Figure 12-26 Oblique aerial photo of Caladesi Island, a good example of a drumstick barrier.

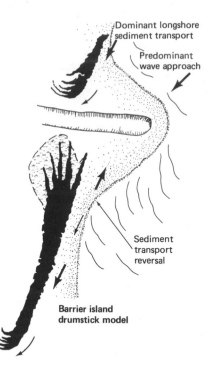

Dominant longshore
sediment transport

Predominant
wave approach

Sediment
transport
reversal

**Barrier island
drumstick model**

Figure 12-27 General morphologic model of drumstick barrier island. (From Hayes and Kana, 1976.)

BARRIER SEQUENCES

As sediments accumulate and sea levels change, sequences of facies representing various barrier island environments accumulate. There is a wide variety of barrier sequences because of the diversity of styles of barrier complexes and the differing relative sea-level situations. Sequences discussed here are representative of Holocene conditions. In general, barrier sequences of this age are less than 5000 to 6000 years old, with development of the barrier related at least partially to the slowing of sea-level rise at about that time.

Regardless of the details of any style of barrier island sequence, the general stratigraphic character of the facies is determined by the relative influence of two important variables: sea-level change and sediment supply (Curray, 1964). Certain of the barrier environments have a high preservation potential during a prograding or regressive situation, whereas others are benefited by transgressive conditions.

It becomes convenient to categorize barrier sequences into regressive (prograding), transgressive, and inlet-migration models. Although these models are widely applicable and relatively easy to recognize, one must be acutely aware that one barrier only a few kilometers in length may produce all three of the sequence models. This realization is critical to proper interpretation of sequences in the rock record, especially in the subsurface.

Regressive (Prograding) Barrier

Under conditions of high sediment supply relative to sea-level change, there is a tendency for the barrier to prograde seaward. In older, classical writings, regression implied a lowering of sea level; however, these sequences may actually develop during rising sea level. The latter situation has taken place along Galveston Island, Texas, a classic example of the prograding barrier (Bernard et al., 1962). Cores in the island display a coarsening-upward stratigraphic trend, with time lines reflecting the depositional surface (Figure 12–28). Indications are that the age of the island is about 4000 years B.P.

Davies et al. (1971) described a stratigraphic sequence from Galveston which can serve as a model for prograding barriers (Figure 12–29). The lower half or so represents the shoreface, or inner continental shelf. This is the fine-grained portion and consists of muddy fine sand with much bioturbation. Grain size increases and bioturbation decreases upward in the section. Some laminations and cross-stratification appears in upper shoreface sediments.

The upper shoreface or nearshore zone begins at the horizon where bioturbation becomes absent and physical structures dominate (Figure 12–29). Structures are primarily low-angle planar stratification, some ripple cross-stratification, and some shell concentrations. This sequence represents the bar-and-trough region through the foreshore beach. Above is a trough cross-stratified zone, well sorted, with some root structures. This represents the eolian environment.

Much more detailed stratigraphic sequences of the intertidal and subaerial barrier are presented by Barwis (1976). Using cores from prograding beach ridges on Kiawah Island, South Carolina, he constructed two sequences: one from the seaward edge and one from the landward edge (Figure 12–30). Both sequences are developed over marshes. The seaward one goes through a welded ridge (beach-face), runnel, backshore (berm crest), and dunes. Cross-stratification patterns are complicated but important to the recognition of each environment. The landward sequence is similar except that washover fan deposits and wind-tidal flats underlie the dune deposits

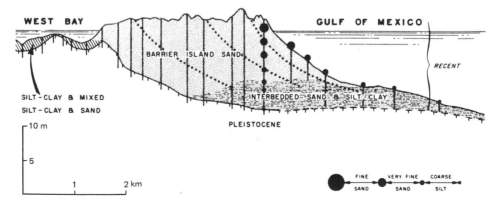

Figure 12-28 Stratigraphic model of regressive barrier, Galveston Island, Texas, displaying a coarsening-upward trend. (Modified from Bernard et al., 1962.)

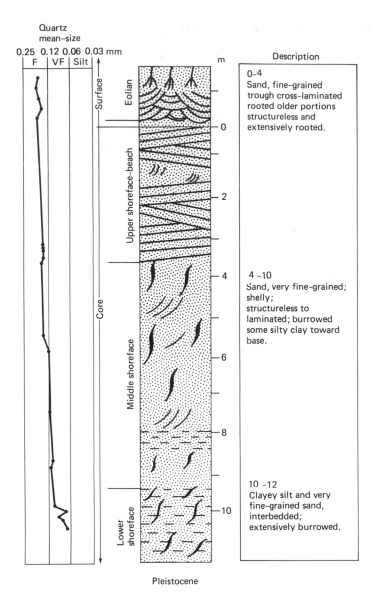

Quartz
mean-size
0.25 0.12 0.06 0.03 mm

| F | VF | Silt |

Surface

Eolian

Core

Upper shoreface-beach

Middle shoreface

Lower shoreface

m

Description

0-4
Sand, fine-grained
trough cross-laminated
rooted older portions
structureless and
extensively rooted.

4 -10
Sand, very fine-grained;
shelly;
structureless to
laminated; burrowed
some silty clay toward
base.

10 -12
Clayey silt and very
fine-grained sand,
interbedded;
extensively burrowed.

Pleistocene

Figure 12-29 Core from Galveston Island, Texas, showing characteristic structures and representative depositional environments. (From Davies et al., 1971, Figure 6.)

(Figure 12-30). These sequences, developed by Barwis (1976), fit into the upper portion of the Galveston Island section of Davies et al. (1971).

Transgressive Barrier

The landward translation of the barrier system is considered to be transgressive and typically implies a rising sea level. John C. Kraft of the University of Delaware and

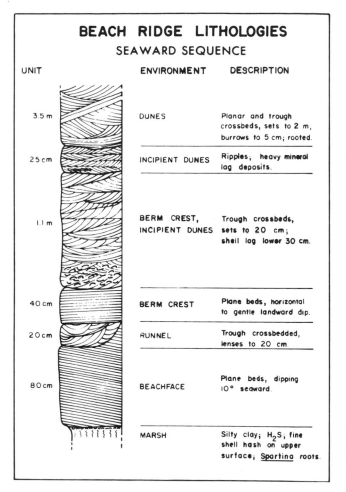

BEACH RIDGE LITHOLOGIES

SEAWARD SEQUENCE

UNIT	ENVIRONMENT	DESCRIPTION
3.5 m	DUNES	Planar and trough crossbeds, sets to 2 m, burrows to 5 cm; rooted.
25 cm	INCIPIENT DUNES	Ripples; heavy mineral lag deposits.
1.1 m	BERM CREST, INCIPIENT DUNES	Trough crossbeds, sets to 20 cm; shell lag lower 30 cm.
40 cm	BERM CREST	Plane beds, horizontal to gentle landward dip.
20 cm	RUNNEL	Trough crossbedded, lenses to 20 cm.
80 cm	BEACHFACE	Plane beds, dipping 10° seaward.
	MARSH	Silty clay; H₂S; fine shell hash on upper surface; _Spartina_ roots.

Beach Column

Figure 12–30 Stratigraphic sequence in beach ridge from Kiawah Island, South Carolina. (From Barwis, 1976, p. II–117.)

his colleagues have provided excellent documentation of these transgressive sequences, particularly along the mid-Atlantic coast. The rate of sea-level rise has been shown to be important in preservation of the barrier (Kraft, 1971). The more rapid the rise in sea level, the more likely it is that the barrier will be preserved. During a stillstand or very slow rise, all of the barrier may be removed by coastal erosion. This generalization assumes a finite sediment supply in the barrier system. Kraft (1978) stated that present conditions of the mid-Atlantic coast cause about one-half of the coastal sequence to be preserved.

The present Delaware barrier coast serves as a good example of a transgressive barrier system. Marsh deposits on the Pleistocene erosion surface may be more than 10,000 years B.P. (Figure 12–31). These are overlain by extensive lagoonal muds. The widespread marsh and lagoon deposits are the site upon which the barrier itself develops and across which it migrates. The barrier and shoreface deposits are continually reworked as washover, blowover, and nearshore erosion occurs (Figure

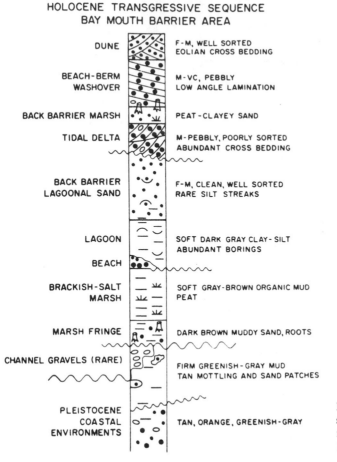

HOLOCENE TRANSGRESSIVE SEQUENCE
BAY MOUTH BARRIER AREA

DUNE	F-M, WELL SORTED EOLIAN CROSS BEDDING
BEACH-BERM WASHOVER	M-VC, PEBBLY LOW ANGLE LAMINATION
BACK BARRIER MARSH	PEAT-CLAYEY SAND
TIDAL DELTA	M-PEBBLY, POORLY SORTED ABUNDANT CROSS BEDDING
BACK BARRIER LAGOONAL SAND	F-M, CLEAN, WELL SORTED RARE SILT STREAKS
LAGOON	SOFT DARK GRAY CLAY-SILT ABUNDANT BORINGS
BEACH	
BRACKISH-SALT MARSH	SOFT GRAY-BROWN ORGANIC MUD PEAT
MARSH FRINGE	DARK BROWN MUDDY SAND, ROOTS
CHANNEL GRAVELS (RARE)	FIRM GREENISH-GRAY MUD TAN MOTTLING AND SAND PATCHES
PLEISTOCENE COASTAL ENVIRONMENTS	TAN, ORANGE, GREENISH-GRAY

Figure 12–31 Stratigraphic section showing transgressive barrier sequence from the Delaware coast. (From Kraft, 1971.)

12–31). The complex interrelationships of these processes produce a typical transgressive barrier sequence.

The sequence for a transgressive barrier consists of fringing marsh, lagoonal, and back-island marsh deposits in the basal portion (Figure 12–32). Tidal deltas may be present depending on the location of the specific sequence. Washover deposits and back-island flats underlie dune deposits. The sand units display various structures dominated by cross-strata. Washovers show planar stratification or Gilbert-type delta development, tidal deltas contain bimodal cross-stratification, and dunes exhibit a generally onshore, low-angle cross-stratification.

Barrier-Inlet Sequence

Barrier inlets may migrate laterally due to spit accretion and related downdrift erosion, or they may migrate landward in a diagonal direction as the barrier transgresses. In either situation the barrier-inlet system may become incorporated into the stratigraphic record. This portion of the barrier complex presents a complex stratigraphy and also shows several variations. Recent publications by Kumar and

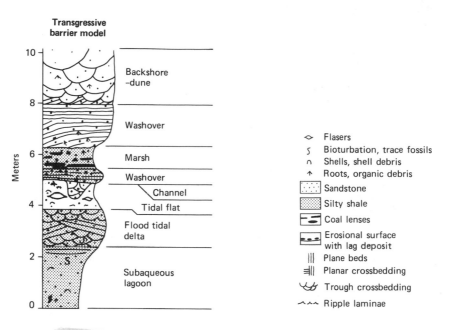

Figure 12–32 General stratigraphic model of transgressive barrier. (From Reinson, 1979, Figure 24, p. 70.)

Sanders (1974, 1975), Hayes and Kana (1976), and Reinson (1979) present good syntheses and summaries on the topic.

Regardless of which variation may accumulate, the basal portion of the section generally consists of a coarse channel lag overlain by deep channel deposits (Figure 12–33). They are characterized by large-scale, ebb-oriented cross-strata in sand, with flood-oriented reactivation surfaces. This sequence represents the main ebb channel (Figure 11–12). Above may be channel-margin bar, spit platform, or shallow channel facies. Although all are characterized by sorted, cross-stratified sand, the foresets may dip in a variety of directions, with both ebb and flood components reflecting migration of megaripples. The welded ridge sequence typical of beaches (Figure 12–10) is commonly present in the barrier-inlet sequence. As the spit moves downdrift it experiences many episodes of welding of ridges in the accretion process. The top of the sequence is commonly comprised of beach deposits with typical low-angle foreshore deposits (Figure 12–33). The sequence may be capped by overlying dunes.

ANCIENT BARRIERS

The recognition of barrier island systems in the rock record involves the collection and synthesis of many data. These data may come from outcrops, cores, chip samples, or various types of logs. The proper interpretation and reconstruction from the data requires knowledge of, and experience with, modern barrier systems as well

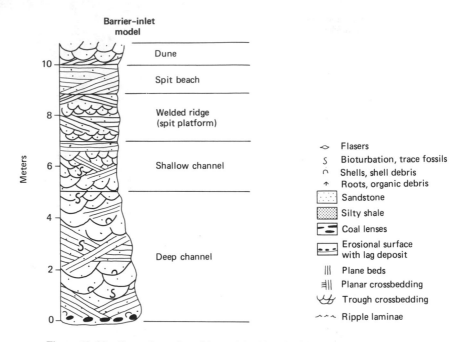

Figure 12–33 General stratigraphic model of barrier-inlet sequence. (From Reinson, 1979, Figure 24, p. 70.)

as the ancient record. Excellent summaries of the various criteria that may be used are presented by Dickinson et al. (1972) and Davies et al. (1971).

In conducting this type of large-scale synthesis it is first necessary to be able to recognize and interpret the small-scale features, such as composition, textures, sedimentary structures, sediment-organism relationships, and other components of the rocks. These data, coupled with the geometry of the rock body, permit designation of the various depositional environments. Studying the environments in their three-dimensional setting then permits interpretations regarding the entire depositional system. The geologist who is working only with geophysical logs is limited to the geometry of the rock bodies and some interpretations as to rock type and texture. From these limited data the same interpretations must be made.

The first well-documented interpretation of an ancient barrier island system was that by Bass (1934), who studied the now famous Bartlesville shoestring sands (Pennsylvanian) in Oklahoma. This interpretation was based largely on sand body geometry, a criterion that is widely applied. It should be observed that barrier islands, longshore bars, cheniers, spits, beaches, and offshore marine sandbars all exhibit a similar geometry and contain similar sediments and sedimentary structures. The distinction between these depositional environments is not easy (Dickinson et al., 1972).

The following examples represent but a few of the many barrier island systems that have been recognized from the rock record. They have been selected so as to include only well-documented examples which cover a spectrum of geologic time. In most instances, one research paper serves as the source of information for the example chosen.

Wilcox Group (Eocene), Texas

The discussion of ancient deltas in Chapter 9 included excellent examples from the Wilcox Group (Eocene) of east Texas (Fisher and McGowen, 1967, 1969). It so happens that this extensive coastal plain complex also includes a good example of a barrier island system in the south-central portion of the state.

Approach to the study

Various types of well records serve as the data base from which this study is derived. Spacings range widely, but more than 100 wells were used in the area, which covers a few thousand square kilometers. Outcrop data were also incorporated into the study. Although all data available were utilized, including fossils and sedimentary structures, most of the interpretations are based on sedimentary facies distribution and **isolith maps** (Fisher and McGowen, 1969).

Sedimentary facies and their distribution

In this area of south-central Texas, the Wilcox Group contains two rather distinct lithofacies: a mudstone or sandy mudstone which locally contains dolomite, glauconite and gypsum; and a sandstone which is fine to medium, well sorted, and also contains glauconite. Lack of core data prohibits specific information on sedimentary structures.

The mudstone facies is the Indio Formation. It is typically less than 100 m thick and contains varying amounts of sand. In general, the sand content increases down depositional dip or to the southeast (Figure 12–34). The sandstone unit is collectively much thicker, reaching over 300 m. This unit, known as the Cotulla Formation, is composed of numerous elongate sandstone bodies oriented along depositional strike (Fisher and McGowen, 1969) with individual bodies up to 30 m thick.

Distribution of these facies shows a generally elongate system with the long axis parallel to depositional strike. Sand/mud ratios and the isolith map indicate not only the general texture and sediment body geometries, but also depict smaller sediment bodies with a shore-normal orientation (Figure 12–34). These small sandy bodies are documented from both subsurface and outcrop data.

Depositional environments

An interpretation of a barrier island and adjacent lagoon or estuary is logical from the data presented. Orientation and geometry of the sediment bodies and their spatial relationships to one another are much like those seen on the present Texas coast only a few hundred kilometers to the southeast.

More subtle evidence for such an interpretation is also present. The maps shown in Figure 12–34 show irregular contours and contacts on the landward side of the barrier compared to the seaward side. This is due to the smoothing action of

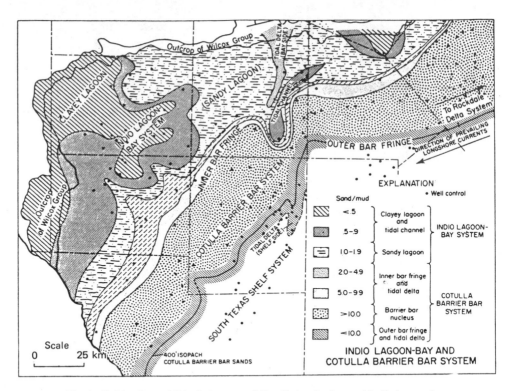

Figure 12-34 General lithofacies map of Cotulla barrier-bar and Indio lagoon-bay systems, Wilcox Group (Eocene), Texas. (From Fisher and McGowen, 1969, Figure 8, p. 45.)

waves and longshore currents in open waters relative to the lack of these smoothing processes along the landward side. No direct evidence is present to indicate the specific nature of the back-barrier bay. The presence of tidal channels, apparent terrigenous runoff, and nearby deltas suggest that it may be a polyhaline bay similar to those behind barriers of the modern Texas coast.

The tidal channel and tidal deltas depicted in Figure 12-34 can be substantiated by data from numerous wells. More closely spaced data and abundant core data would probably reveal more of these ancient tidal inlet systems. Again, the combination of sediment body geometry, textures, and relationships to other units make the interpretation a reasonable one. Combined outcrop and subsurface data permit reconstruction of a flood tidal delta in the Indio bay system (Figure 12-35). A great variety of facies can be recognized, including channels and tidal flats. Both large megaripples and plane beds were widespread on this system with bioturbation also common. The abundance of bioturbation, flood-dominated cross-strata, and the mud-filled main channel suggest that this tidal delta may have been a storm-generated feature that was subsequently not flushed by continuing tidal currents and eventually had its main channel plugged.

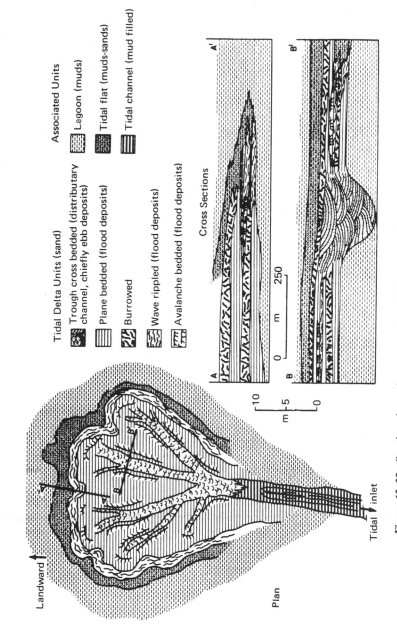

Figure 12-35 Stratigraphy and reconstruction of flood tidal delta complex, Indio system, Wilcox Group, Texas. (From Fisher and McGowen, 1969, Figure 9, p. 46.)

Tidal Delta Units (sand)

Trough cross bedded (distributary channel, chiefly ebb deposits)

Plane bedded (flood deposits)

Burrowed

Wave rippled (flood deposits)

Avalanche bedded (flood deposits)

Associated Units

Lagoon (muds)

Tidal flat (muds-sands)

Tidal channel (mud filled)

Cross Sections

Landward

Plan

Tidal inlet

Muddy Sandstone (Lower Cretaceous), Wyoming and Montana

The Muddy Sandstone in Wyoming and Montana has been studied for some time because of its petroleum-producing history and potential. Of the group of studies on the Muddy, three are of special interest here because of the data and interpretations presented (Berg and Davies, 1968; McGregor and Biggs, 1968; Davies et al., 1971).

General stratigraphy

The area of study is the Powder River Basin in northeastern Wyoming and southeasternmost Montana. In this area, the Muddy Sandstone displays a maximum thickness of about 30 m and is overlain by the widespread Mowry Shale and rests on the equally extensive Skull Creek Shale. To the southeast is the thick and extensive Dakota Sandstone, of which the Muddy is thought to represent a distal tongue (Berg and Davies, 1968).

The Bell Creek Field in Montana provides most of the data for the interpretation of depositional environments of the Muddy. About 300 wells have been drilled in this region on 40-acre centers, providing a dense network of data (McGregor and Biggs, 1968). Cores were studied from 9 of these wells in addition to the interpretation of geophysical logs from many of the others.

The Muddy Sandstone is here dominated by fine quartz sandstone which is fairly well sorted. Minor constituents include small percentages of orthoclase, mica, and rock fragments with 6% matrix in the sandstone. This places it just on the boundary between immature and submature in Folk's maturity classification.

Sedimentary facies and their distribution

Four distinct facies comprise the Muddy. The basal meter or so is comprised of mudstone with irregular or wavy laminations. Burrows are common and are parallel to bedding. Some lenses of very small cross-stratification are present (Davies et al., 1971). This basal unit grades into an overlying sandstone in about 25 cm of section. This sandstone is homogeneous, fine grained, and slightly mottled. It reaches near 2 m in thickness (Figure 12–36). Homogeneity may reflect intense bioturbation, with the mottling being the only visible evidence of it. The third facies is also the thickest, reaching near 4 m. It is a well-stratified fine sandstone with rare burrows. Strata are gently inclined and in sets about 30 cm thick. Small cross-strata are also present. The top facies is less than a meter thick. It is a homogeneous, well-sorted, fine sandstone with many root structures (Figure 12–36). The overall trend in grain size is one of coarsening upward except for a slight reversal near the top of the section.

Depositional environments

The sequence of facies described above represents the typical nearshore and barrier sequence. The lowermost facies represents the lower shoreface, probably beyond the influence of waves except during storms. The next unit represents the

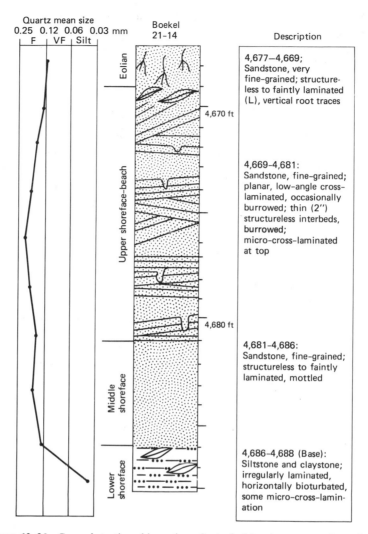

Figure 12-36 General stratigraphic section of a typical barrier sequence from the Muddy Sandstone (Cretaceous), Bell Creek field, Montana. (From Davies et al., 1971, Figure 6, p. 555.)

shallower shoreface, where wave action is active, sorting and winnowing out fines. The suggestion is, however, that energy was not so high that benthic infauna could survive.

The thick coarsest unit represents the beach and surf zone area, where waves, longshore currents, and swash are prominent. Low-angle cross-strata probably represent foreshore deposition. The finer-rooted sandstone is interpreted as an eolian deposit (Davies et al., 1971).

This sequence is actually only the sandstone or barrier portion of the barrier island system. The related shales which serve as trap rock in the Bell Creek field are the bay and marsh environments. The authors recognized a lagoonal facies with interbedded sandstone and mudstone, and a marsh facies. A homogeneous and burrowed thin sandstone in this mud-dominated sequence was interpreted as a washover deposit (Davies et al., 1971).

Pennsylvanian of the Appalachian Basin

The detailed stratigraphy of coal-bearing deposits of Pennsylvanian age in the Appalachian Basin as well as in other basins of the same age and character has been known for some time. Reconstructions of depositional environments for these strata have generally invoked some type of deltaic system and have been agreed upon as such. More recently J. C. Ferm, J. C. Horne, and their associates have found that at least some of these strata represent a barrier island depositional system (Ferm, 1974; Hobday, 1974; Horne et al., 1978). The bulk of these studies has been conducted in eastern Kentucky and adjacent West Virginia.

General stratigraphy

Everyone who has driven along highway I-69 in western Kentucky or who has read any of the literature on these coal-bearing and related Pennsylvanian sequences knows of their complex stratigraphy. Individual lithic units are typically thin and may be extensive or may be lenticular. Thick units of several meters generally are only local in extent. Additionally, there are numerous lithofacies present, representing a wide variety of depositional environments (Figure 12–37). Both terrigenous and carbonate units are present along with coal.

A typical section may show fossiliferous biosparites at the base, grading into shale with carbonate lenses and scattered fossils. Burrowed and rippled sandstone facies follow and are overlain by a relatively thick sandstone unit with large, low-angle cross-stratification and plane beds. Lenticular sandstones with coarse basal gravel are above the thick sandstone (Figure 12–37).

Adjacent or overlying the foregoing sequence may be another type of section, in this case dominated by mudstone and coal with only thin sandstone units. The various lithofacies may recur and be arranged in a variety of orders. There are four basic lithofacies present: (1) a coarsening upward, burrowed mudstone; (2) a dense clay with overlying coal; (3) thin quartz sandstone; and (4) lenticular cross-stratified, fining-upward sandstone (Figure 12–38).

Interpretation and reconstruction of depositional system

The general sequence represented by the sections described above is one of a prograding or regressive barrier system. The basal carbonate unit represents open marine deposition beyond the influence of terrigenous influx. Overlying mudstones and burrowed sandstones indicate outer and inner shoreface deposition. Sandstones that display well-developed stratification and cross-strata were deposited in the beach and nearshore zone. Lenticular sand bodies with gravel lag deposits and cross-stratification represent a tidal channel–tidal delta complex (Figure 12–38).

The mudstone-dominated section represents essentially a back-barrier com-

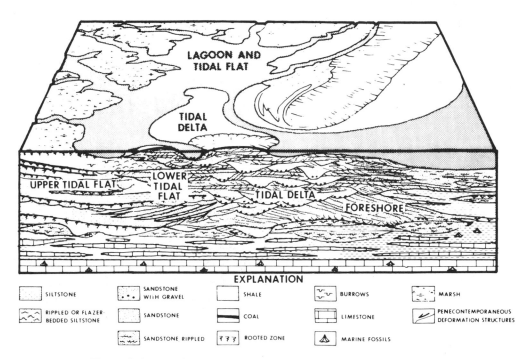

EXPLANATION

SILTSTONE		SANDSTONE WITH GRAVEL		SHALE		BURROWS		MARSH
RIPPLED OR FLAZER-BEDDED SILTSTONE		SANDSTONE		COAL		LIMESTONE		PENECONTEMPORANEOUS DEFORMATION STRUCTURES
SANDSTONE RIPPLED			ROOTED ZONE			MARINE FOSSILS		

Figure 12–37 Detailed lithofacies relationships in a barrier island depositional system in the Pennsylvanian of eastern Tennessee. (From Ferm et al., 1972.)

plex, with the lagoons (or estuaries) accumulating clay and silt, in some cases coarsening upward. Swamps or marshes are represented by dense clay called seatrock with coal overlying it. Coarse facies consist of washover fans that display planar stratification, in some cases dipping slightly landward. The washover deposits may occur anywhere in the back barrier. Local, thick, lens-shaped cross stratified sandstone represents flood tidal deltas (Horne et al., 1978).

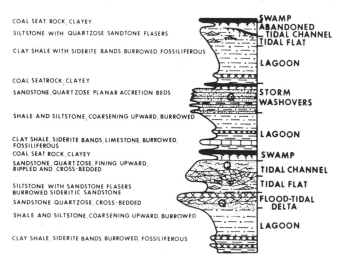

COAL SEAT ROCK, CLAYEY — SWAMP
SILTSTONE WITH QUARTZOSE SANDTONE FLASERS — ABANDONED TIDAL CHANNEL / TIDAL FLAT
CLAY SHALE WITH SIDERITE BANDS, BURROWED, FOSSILIFEROUS — LAGOON
COAL SEATROCK, CLAYEY
SANDSTONE, QUARTZOSE PLANAR ACCRETION BEDS — STORM WASHOVERS
SHALE AND SILTSTONE, COARSENING UPWARD, BURROWED
CLAY SHALE, SIDERITE BANDS, LIMESTONE, BURROWED, FOSSILIFEROUS — LAGOON
COAL SEAT ROCK, CLAYEY — SWAMP
SANDSTONE, QUARTZOSE, FINING UPWARD, RIPPLED AND CROSS-BEDDED — TIDAL CHANNEL
SILTSTONE WITH SANDSTONE FLASERS BURROWED SIDERITIC SANDSTONE — TIDAL FLAT
SANDSTONE QUARTZOSE, CROSS-BEDDED — FLOOD-TIDAL DELTA
SHALE AND SILTSTONE, COARSENING UPWARD, BURROWED
CLAY SHALE, SIDERITE BANDS, BURROWED, FOSSILIFEROUS — LAGOON

Figure 12–38 Generalized vertical sequence through back-barrier deposits in the Pennsylvanian of eastern Kentucky and southern West Virginia. (From Horne et al., 1978.)

Chapter 12   The Barrier Island System   **443**

Keyser Limestone (Upper Silurian), Virginia

Although only a few barrier-inlet accumulations have been well documented from the pre-Pleistocene rock record, they are probably not rare. The recent development of facies models for such a sequence and their local extent have been major factors in this general dearth of literature. A recent detailed study of such a barrier-inlet sequence from the Silurian of Virginia (Barwis and Makurath, 1978) serves as a good example of this type of depositional system.

General stratigraphy

The Keyser Limestone is exposed throughout much of the Valley and Ridge Province in the Central Appalachians. In the outcrop belt the thickness ranges from 30 to 85 m. The unit consists of a variety of carbonate rock types as well as quartz sandstone. The Keyser overlies the Tonoloway Limestone, which has been interpreted as representing a supratidal flat environment (Figure 12–39). The overlying New Creek Limestone was deposited on the open marine shelf.

A representative section of the Keyser shows three distinct lithofacies within it: (1) a basal terrigenous sandstone called the Clifton Forge Sandstone Member (Barwis and Makurath, 1978); (2) a thick unit of fossiliferous carbonate, mudstones, and sandstones; and (3) an upper fossiliferous limestone.

Depositional environments

The uppermost carbonate unit is considered to represent a shallow lagoon, and the middle unit is interpreted as being deposited on shallow to supratidal mudflats and sandbars. These largely tide influenced environments were adjacent to, and integrated with, the subtidal lagoon.

Of primary interest is the Clifton Forge Member. How might a relatively thick, mature, cross-stratified quartz sandstone become incorporated in a dominantly carbonate sequence? Careful examination of textures, composition, sedimentary structures, and the overall sequence led the authors to invoke a tidal inlet, channel-fill environment for the Clifton Forge. Their evidence includes the fining-upward sequence, bimodality of the cross-strata, mature nature of the sediment, and the relationships with adjacent units (Barwis and Makurath, 1978). One of the most difficult questions to be answered is touched on only briefly. Although the depositional environment suggested seems proper, how did the terrigenous sediment reach its inlet location, and where did it come from? The authors suggest a geomorphically complex setting, allowing for quartz sand to be reworked. The coherence of carbonates could allow low-energy currents or waves to transport quartz sand over them and become entrapped by tidal currents.

BARRIER SEQUENCE MODEL

As a result of their research on both modern and ancient barrier systems, Davies et al. (1971) have developed a general stratigraphic model for regressive barrier systems using the observed relationships of various sedimentary facies (Figure 12–40). The salient features of the model are as follows:

Barwis sequence 445

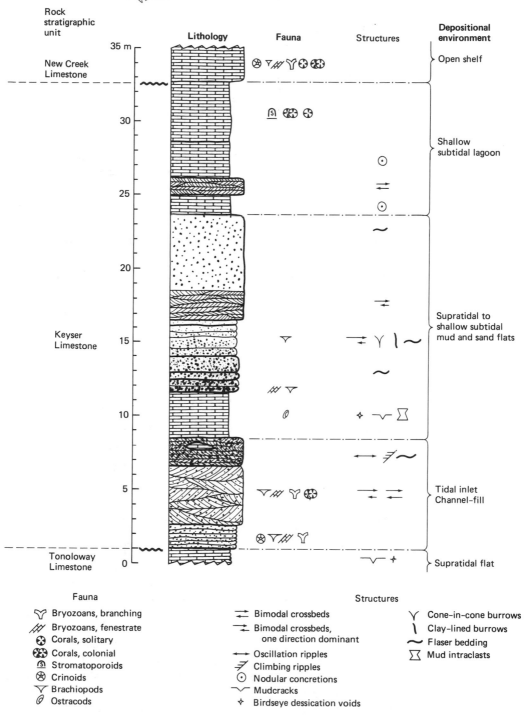

Figure 12-39 Lithologic sequence of Upper Silurian rocks at Gala, Virginia, which includes a tidal inlet sequence. (From Barwis and Mukrath, 1978, Figure 4, p. 67.)

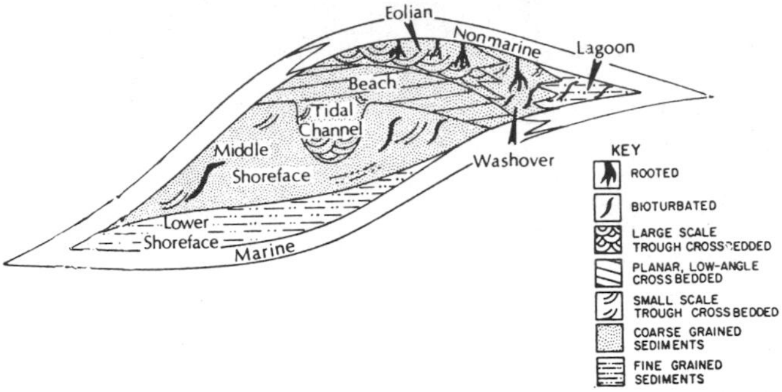

Figure 12–40 General facies model of prograding barrier island system. (From Davies et al., Figure 15, 1971.)

1. Sand bodies are lenticular, thinning both seaward and landward. They are elongate parallel to depositional strike.

2. Sand bodies of the barrier may be underlain by marine or nonmarine sediments or both.

3. The base of the barrier sand bodies is nonerosive except in inlet sequences.

4. The sand bodies display a sequence of sedimentary structures, representing upward transition from the shoreface environment to successively shallower environments.

5. Tidal channel or inlet deposits may interrupt the barrier sand sequence. These deposits have many similarities with fluvial channel deposits but show distinctly bidirectional cross-strata.

6. Grain size in the sand bodies increases upward from shoreface to beach deposits.

7. Barrier sand bodies are flanked by coastal bay sediments.

8. Barrier sand bodies may be capped by marine or nonmarine sediments.

9. The model is applicable to both terrigenous and carbonate systems.

SUMMARY OF CHARACTERISTICS
OF SEDIMENT/SEDIMENTARY ROCK BODIES
REPRESENTING THE BARRIER ISLAND DEPOSITIONAL SYSTEM

The recognition of the barrier island system in the stratigraphic record is typically rather easy; however, the distinction of the specific environments within the system may be difficult. There seems to be a general consistency to the stratigraphic record that accumulates in the barrier system.

Tectonic setting. Development of barrier island systems is fostered by stable, gently sloping coastal areas with at least a modest supply of sediment. Coastal plains such as those present on the trailing edge of plates represent the most likely geologic setting for barrier development.

Shape. The barrier system as a whole presents an elongate body in the stratigraphic record, with the long axis parallel to the depositional strike of the coast.

Within this system there may be local fan-shaped tabular bodies and channel deposits.

Size. The extent of barrier island deposits ranges widely but is generally only a few kilometers in width and several tens to even hundreds of kilometers long. Thickness also ranges widely, but tens of meters is typical. Distinct environments within the barrier system, such as washover fans, may display sedimentary bodies only a meter or so in thickness and covering less than a square kilometer. Tidal channels may be up to tens of meters thick and elongate but local in extent within the main barrier sand body.

Textural trends. The general trend in barrier systems is to display a coarsening-upward scheme both in transgressive and progradational sequences. Basal shoreface deposits are commonly fine to very fine sand, with an increase to medium or coarse sand taking place in nearshore and beach deposits which cap the progradation sequence. In transgressive barrier systems the basal portion is mud or very-fine sand of the coastal bays, and intertidal environments are overlain by barrier island sands. There are some pertubations in the textural trends. The beach and adjacent inner surf zone deposits may display a coarser grain size than the horizons stratigraphically above or below. Inlet channel fill sequences may be anomalously coarse.

Lithology. Barrier island deposits are typically mineralogically mature. Quartz is distinctly dominant with feldspar and rock fragments as secondary constituents. Biogenic carbonate fragments may be very abundant. The barrier island quartz sand sequence is generally isolated between offshore muds and coastal bay muds.

Sedimentary structures. No specific sedimentary structures characterize the barrier system; however, the individual environments may be, in part, characterized by their contained structures. The inner shoreface has a combination of bioturbation structures and small-scale ripple cross-stratification with various orientations. The surf zone deposits include a variety of ripple- and megaripple-generated cross-stratification with a shoreward direction of dip dominating over the seaward direction. Foreshore beach stratification overtops this portion of the section. Eolian trough cross-stratification may be the uppermost sand accumulation.

Local washover fans would display plane bedding with perhaps a slight landward dip. Inlet sequences would include a complex combination of ripple and megaripple cross-stratification with bidirectional dips of the cross-strata.

Paleontology. Diverse and generally abundant fauna characterize the open marine shoreface and nearshore facies of the barrier system. The subaerial facies contain only small skeletal fragments that are blown by winds or washed into that environment during storms. Back-island marsh, tidal flat, or bay facies contain various taxa, depending on salinity conditions of the individual barrier complex. They may range from brackish to hypersaline.

Associations. The barrier island sequence is associated with readily recognizable and distinct sequences which represent either seaward environments, such as the shoreface or inner shelf, or the landward, generally more variable coastal bay environments. These include tidal flats, marshes, estuaries, and lagoons. This combination of source beds and the reservoir properties of the barrier island deposits make this an ideal geologic setting for the exploration of oil.

ADDITIONAL READING

Davis, R. A., Jr. (ed.), 1978. *Coastal Sedimentary Environments*. Springer-Verlag, New York, 420 p. Several of the chapters are pertinent to barrier island systems. These include those on coastal dunes, beaches, inlets, and estuaries as well as a chapter on coastal stratigraphic sequences.

Davis, R. A., Jr. and Ethington, R. L. (eds.), 1976. *Beach and Nearshore Sedimentation*. Soc. Econ. Paleontologists and Mineralogists, Spec. Publ. No. 24, Tulsa, Okla., 187 p. State of the art papers on various aspects of beach sedimentation. Papers by Davidson-Arnott and Greenwood, and by Hill and Hunter are of particular interest to barrier island studies.

Hayes, M. O., and Kana, T. W. (eds.), 1976. *Terrigenous Clastic Depositional Environments*. Tech. Rept. No. 11-CRD, Coastal Res. Div., University of South Carolina, Columbia, S.C., I-131, II-184 p. Excellent summaries of all coastal environments in Part I and selected research papers in Part II. Complete bibliography on each environment.

Heward, A. P., 1981. A review of wave-dominated clastic shoreline deposits. Ear. Sci. Reviews, 17:223–276. Good synthesis of existing literature on this complicated coastal system.

Leatherman, S. P. (ed.), 1979. *Barrier Islands: From the Gulf of St. Lawrence to the Gulf of Mexico*. Academic Press, New York, 325 p. Several excellent papers on various aspects of barriers, but not a comprehensive treatment on the subject.

Schwartz, M. L. (ed.), 1973. *Barrier Islands*. Benchmark Papers in Geology. Dowden, Hutchinson & Ross, Stroudsburg, Pa., 451 p. An enormous number of the classic barrier papers from throughout the literature. Probably the most comprehensive coverage of barriers between two covers, but it is somewhat dated.

Part Four

MARINE

ENVIRONMENTS

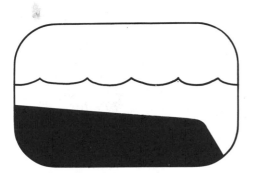

13 Terrigenous Shelves and Shallow Seas

The present continental margins of the world include extensive, relatively shallow continental shelves which extend seaward from the coast to the continental slope, where there is a marked increase in gradient of the marine bottom. These shelves extend to depths of 200 m or more and may be hundreds of kilometers wide. The slope of this region is generally much less than a degree and relief is low. Present sea-level conditions provide for more extensive shelf development now than during much of geologic time. These peripheral marine environments have also been called shelf seas or pericontinental seas.

During the geologic past and to a limited extent now, there have also been shallow marine seas which are nearly surrounded by continental land masses. Modern examples include Hudson Bay, the Yellow Sea, and the Baltic Sea. These epicontinental or epeiric seas (Figure 13–1) are similar in many respects to shelves. The major difference is that the epeiric seas receive sediment from essentially all sides, whereas shelves have sediment provided from only one side. In addition, epeiric seas may not be in direct contact with an open ocean basin, whereas a continental shelf is open to the sea throughout its extent. Tidal range was varied in ancient epeiric seas and the size and depth limited wave size.

This discussion emphasizes the shelves or shallow seas that are dominated by terrigenous sedimentation. To interface properly with Chapter 12, some overlap in coverage is neccessary. Treatment of this environment will extend from the outer

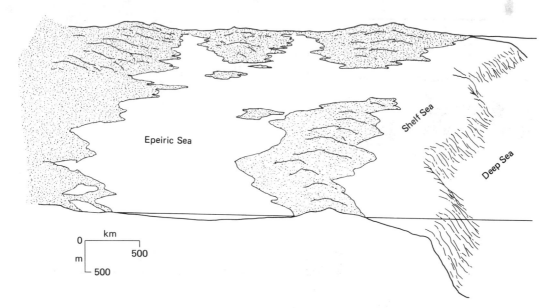

Figure 13–1 General setting of epeiric and shelf seas. Observe the somewhat different slope between the two types. (After Heckel, 1972.)

limit of bar and trough topography to the shelf-slope break; the shoreface will be included as a part of the shelf system.

Modern continental shelves present a situation that is not ideal for studying shelf sedimentation. Because of rapid and extreme changes in sea level during the past several thousand years, the modern shelf system presents a set of circumstances that makes relating it to the rock record somewhat difficult.

Holocene Sea-Level Changes

To discuss sedimentation in modern shallow marine seas in proper detail, it is necessary to consider the setting in terms of recent sea-level changes. At the time of the Late Wisconsinan glaciation, sea level was more than 100 m below its present level. Rapid initial melting caused very rapid rises in sea level as the Holocene transgression began (Figure 13–2). This extreme rate of sea-level rise lasted for only about 2000 years. Between 17,000 years B.P. and about 7000 years B.P., sea level rose rapidly, nearly a centimeter per year.

This rapid and large-scale rise in sea level produced the wide shelf seas that exist today. It also produced an odd pattern of shelf sediment distribution. Shoreline sediments were accumulating as the strandline rapidly transgressed the shelf. These included beach and nearshore sands, marsh peats, bay deposits, and other coastal sediments. As this transgression took place these coastal deposits rapidly became covered with several tens of meters of water. The transgression was too rapid to allow sediment from land runoff to be transported out on the shelf over these coastal deposits. The result is extensive relict (Emery, 1952) or palimpsest (Swift et al., 1971) sediments on the present continental shelf. **Relict** refers to sediments that have been

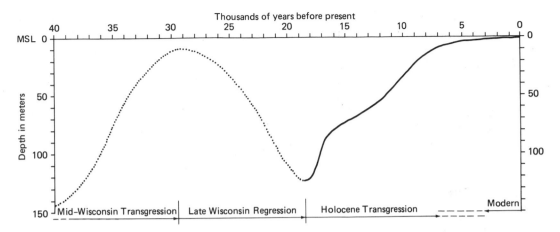

Figure 13–2 Late Pleistocene sea-level fluctuations, including generalized curve for Holocene transgression. Observe changes in rate of sea level rise at 17,000 and 7000 years B.P. (From Curray, 1965.)

deposited in a particular environment but now occupy a different one and are not in equilibrium with their present environment. **Palimpsest** refers to reworked relict sediments.

At about 7000 years B.P., sea level was approximately 10 m below its present position. From that time to the present, the rise in level has been markedly slower than it was earlier, only about 1 cm every 7 years. During this period much sediment has been transported from land to the shelf seas, producing a blanket of shelf sediments. The present shelf situation is one of about two-thirds relict and palimpsest sediments on the outer shelf, with modern shelf sediments on the inner shelf (Emery, 1968). It is only these modern, inner-shelf sediments which can be utilized for uniformitarian purposes in studying shelf environments in the rock record.

Related problems plague modern analogs of epeiric seas in that they were subjected to the same sea-level fluctuations and sediment supply problems. Compounding this is the fact that modern epeiric seas are rare.

CONTROLLING FACTORS IN SHELF SEDIMENTATION

Numerous models of shelf sedimentation have been proposed dating back to Douglas Johnson (1919). This and subsequent efforts have typically called on the same primary factors. Development of a stratigraphically based model (Swift et al., 1972; Swift, 1976) utilizes these factors as well. Primary controlling factors include rate of sediment yield (S), grain size of the sediment (G), rate of wave and current energy input (E), sea-level change (R), and slope (L). Relationships between these factors may be expressed as

$$\frac{SG}{E} - \frac{R}{L} \propto T$$

where T is the strandline movement (Swift, 1976).

Chapter 13   Terrigenous Shelves and Shallow Seas

The term SG/E is essentially the rate of coastal deposition. Both increase in sediment delivered and increase in grain size cause the rate of deposition to increase. The latter is important because coarser grains will come to rest rather than bypassing the shelf. The greater the energy input, the less sediment accumulates—thus the inverse relationship with E. The R/L term is the effective rate of sea-level change, but decreases as the slope, L, increases. The reason for this is that the steeper the slope, the more that sea level must be raised in order for the strandline to advance a given distance (Swift, 1976). The rate and sense of strandline movement, T, is dependent on the relationship between these two terms.

SHELF SEDIMENT

Varieties

A wide variety of sediment types exists in the shelf seas. These include detrital (terrigenous), biogenic, authigenic, volcanic, residual, and relict sediments (Emery, 1968). The detrital or terrigenous sediments comprise, by far, the greatest percentage of sediments which are presently accumulating on the shelves of the world. These sediments are carried to the coast by rivers. Although some rivers empty directly onto the shelf, many empty into estuaries. This causes the coarser material to become trapped in the estuaries and only the finer sediment, typically mud, is allowed to bypass these sediment traps. Most of the sand and gravel that is presently accumulating along the marine coast and shoreface is the result of reworking and redistribution of older coastal deposits. Most terrigenous particles are quartz with some feldspar and mica. Small amounts of stable accessory minerals such as zircon, garnet, rutile, and other heavy minerals are also commonly present. There may be great local variation in these due to composition of source areas. Terrigenous particles may be deposited by water, wind, or ice.

Shelf sediments contain variable amounts of biogenic debris, most of which is carbonate skeletal material (Swift, 1969). Some of the noncarbonate particles are phosphatic material from fish skeletal fragments and siliceous microorganism tests of radiolarians and diatoms. The carbonate particles range in size from coccolith tests only a few microns in diameter, to large coral heads which are meters across. A broad spectrum of taxa contributes to the biogenic carbonate sediment with mollusks, coelenterates, echinoderms, algae, and foraminifera being most abundant. The percentage of biogenic debris ranges from essentially zero to 100% (Ginsburg and James, 1974).

On a worldwide basis, authigenic sediments comprise only a small percentage of shelf sediments. Common species are phosphorite and glauconite. Phosphorite may be abundant locally and glauconite is spotty in distribution, with lack of detrital sediment accumulation favoring its formation. Authigenic calcium carbonate may also form and subtidal cementation by carbonate precipitation may be present. Volcanic shelf sediments, which can be locally abundant if an appropriate source is nearby, include particles of scoria, pumice, and basalt, mostly of sand size. Finer-grained particles are commonly carried beyond the shelf (Emery, 1968). Weathering in place has been documented but is difficult to distinguish from transported material. Some clay minerals have been formed in the shallow marine environment (Rex and Martin, 1966).

Relict sediments may theoretically be characterized by any composition or origin. Most are terrigenous, but biogenic and authigenic particles (ooids) may be common. Recognition of relict sediments may be by several means: the presence of coarser sediment seaward and deeper than finer sediment, iron oxide stains and pitting, oyster shells or mastadon bones on the outer shelf, oolites on the shelf, or indications of environments other than the existing one (Emery, 1968). Relict sediments on the present shelves are a Holocene phenomenon (Figure 13–3). Whereas the present shelf area is dominated by these sediments, the original deposition of shelf sediment was dominated by water-deposited detrital sediments.

General Distribution

Sediments described in the preceding section show a broad and predictable distribution which is closely tied to major ocean basin circulation patterns and to climate. Emery (1968) has summarized the general distribution as seen in an idealized ocean basin system (Figure 13–4). Primary sediment provinces are biogenic material near the equator, water-contributed terrigenous from the tropics to the arctic, and glacial sediments in the high latitudes. A fourth sediment type is in the authigenic zone in the midlatitudes on the east side of the ocean.

There are some important subtle deviations from symmetry which are due primarily to oceanic circulation. The biogenic area on the west side of the ocean extends to higher latitudes than on the east side because of the warming effect of currents. Similarly, the glacial deposits extend to lower latitudes on the west side of the basin compared to the east (Figure 13–4). Cold currents from the high latitudes cool

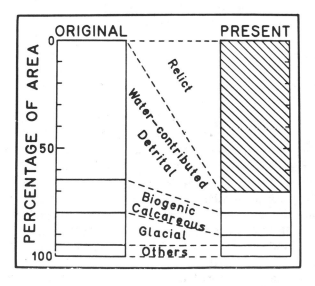

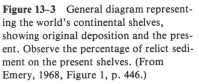

Figure 13–3 General diagram representing the world's continental shelves, showing original deposition and the present. Observe the percentage of relict sediment on the present shelves. (From Emery, 1968, Figure 1, p. 446.)

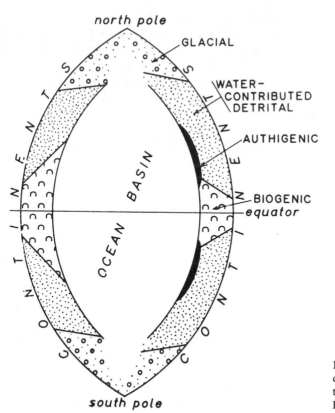

Figure 13-4 Idealized diagram of an ocean basin, showing major sediment types on continental shelves. (From Emery, 1968, Figure 2, p. 446.)

these regions. The authigenic zone is dominated by phosphorite, which is precipitated along the areas of upwelling on the east side of ocean basins such as off the California and Peru coasts. This theoretical model presented by Emery (1968) assumes that the sediments are in equilibrium with their environment.

A somewhat similar pattern of shelf sediment distribution is presented by Hayes (1967b), who related the observed distribution of sediments to climatic belts. Considering only the inner shelf and thereby eliminating relict sediments, he found that generalizations can be made about shelf sediments. Extremely high latitude areas were excluded.

Shell material, largely molluscan, is rather uniformly distributed and shows no relation to climate (Figure 13-5). On the other hand, coral shows an expected maximum in low latitudes where temperatures are high. Coral is restricted to below 35° of latitude. Rocky bottoms and gravel increase greatly from middle to high latitudes. This reflects glacial activity and the lack of sediment supply from land due to low rainfall and a short melting season. A somewhat opposite trend is present in mud distribution. It is maximum at the equator and drops to a minimum near 30°. This reflects regions of high rainfall and high temperature (Hayes, 1967b), both of which bring sediment to the shelf. Sand is just over 60% of the total near 30° of latitude and decreases to 30% in both high- and low-latitude directions (Figure 13-5). The maximum is in areas of moderate temperature and rainfall. Actually, much of the sand is

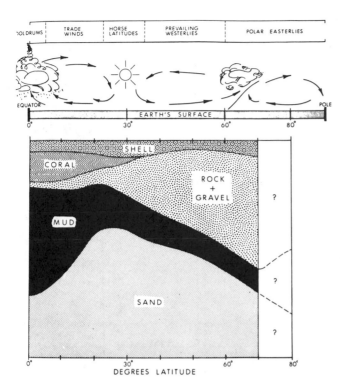

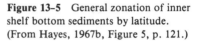

Figure 13–5 General zonation of inner shelf bottom sediments by latitude. (From Hayes, 1967b, Figure 5, p. 121.)

being reworked into the modern shelf from older shelf and coastal deposits; it is not coming directly from runoff via rivers.

The transportation and distribution of mud from present-day rivers onto the shelf takes on several general appearances. Throughout all coasts except muddy ones there is a zone of sand-dominated modern sediment (McCave, 1972). Such muddy coasts are in western Louisiana, Surinam, and the Orinoco Delta. In most areas however, the mud is found in three or four different zones on the shelf. Coasts protected from wave action display a belt of mud near the shore (Figure 13–6). Coasts exposed more to waves contain a sandy area on the inner shelf, mud in the middle, and relict sediments on the outer shelf. A few places such as the northern Gulf of Mexico show mud concentrated on the outer shelf (McCave, 1972). Adjacent to large deltas it is common for mud to cover the entire shelf sea floor. Mud is typically present beyond the shelf-slope break on all coasts (Figure 13–6).

SHELF PROCESSES

A broad variety of biological, chemical, and physical processes takes place on the shelf seas. Each has its own significant impact on sediment distribution. Benthic organisms ingest sediment, excrete it, and also rearrange it. Both biochemical and physicochemical reactions may take place. The more obvious physical processes involve the interactions between sediments and tides, waves, and currents.

Chapter 13 Terrigenous Shelves and Shallow Seas **457**

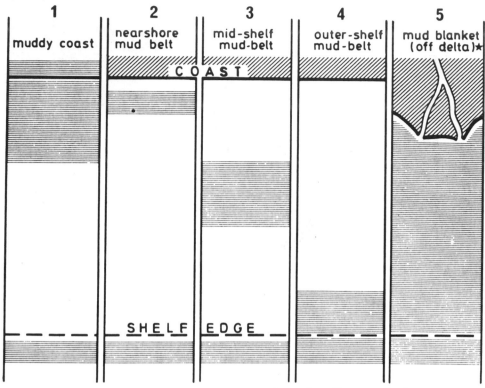

Figure 13–6 Schematic diagram of shelf mud in five different distribution types. (From McCave, 1972, p. 228.)

Sediment-Organism Interactions

The shelf and epeiric seas are typically one of the most densely populated of all depositional systems. The sands and muds of these shallow regions support a diverse and abundant epifaunal and infaunal community because of abundant food. Most epifaunal forms are vagrant, such as echinoderms and mollusks; however, sessile sponges and coelenterates may be common. Infaunal varieties include mollusks, echinoderms, various worms, and arthropods; both vagrant and fixed burrowers are present.

Typically, the shallower areas of the shelf are subjected to nearly continual physical processes which may prevent some taxa, especially epifauna, from inhabiting this region. Sediments in this zone tend to display a dominance of physical structures; however, once beyond the surf zone it is possible to find all physical structures or complete bioturbation, or any combination thereof. There is a general tendency for benthic organisms to increase as sediment particle size decreases. A number of factors contributes to this trend: lower physical energy and deeper water, more food available, and substrate preference. This trend parallels the decrease in physical structures and results in much bioturbation in outer shelf deposits (Figure

13–7). Howard and Reineck (e.g., 1972a, 1972b), utilizing box corers, have provided good information on the nature of the bioturbation features of the shelf.

Bottom dwellers may have effects on the sediment other than its rearrangement. They can affect the cohesiveness and therefore the stability of the shelf floor. The secretion of mucus or organic films on sediment increases cohesion and therefore stability; this requires dense populations of the particular taxon in question and is typically local. The opposite effect is more widespread. Flocculation and other sediment-dispersing phenomena may be produced by feeding activities (Rhoads, 1972).

The production of fecal pellets by suspension feeders can alter the grain size of bottom sediment. Ingestion of mud-size particles and production of sand-size pellets

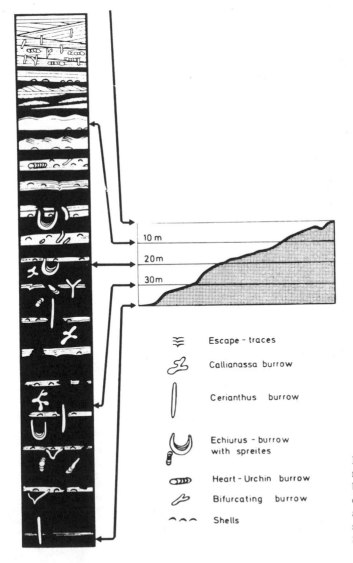

10 m

20m

30m

≋   Escape - traces

Callianassa burrow

Cerianthus burrow

Echiurus - burrow
with spreites

Heart - Urchin burrow

Bifurcating burrow

Shells

Figure 13–7 General stratigraphic representation of various deposits and bioturbation features across various shelf depths, showing a decrease in mud and an increase in physical structures in shallower deposits. (From Reineck and Singh, 1980, Figure 554, p. 402.)

by infauna is common. If pellets are coherent, they may then act as sand grains in terms of their response to physical processes. In some areas nearly all the mud is pelletized.

Chemical Processes

The most widespread and important chemical process in shelf or epeiric seas is the production of calcium carbonate. This topic is treated in detail in Chapter 14. Other shallower marine chemical phenomena which involve sediments tend to be masked by the physical or biological processes, which are readily apparent. The previously mentioned precipitation of phosphorite on the outer shelf as the result of upwelling is probably the most significant from the volume point of view. These deposits are currently being considered for commercial recovery and probably will be mined during the 1980s.

Some precipitation on the shelf is dependent on a very slow rate of terrigenous influx. Both chamosite (chlorite group) and glauconite (iron-rich illite, mica group) fall into this category. Glauconite has value in the rock record because it is only known to form in the marine environment.

Chemical reactions, especially submarine cementation, may be important in stabilizing the shelf floor (Hatcher and Segar, 1976).

Physical Processes

The transport and subsequent modification of shelf sediment is dominated by physical processes, including tides, waves, and currents. In the case of currents there are numerous types depending on the driving mechanism.

Tides

As the tidal wave progresses across the open ocean it is noticed only on the shores of oceanic islands and the range at these locations is typically low. As the tidal wave passes the outer edge of the shelf there is a gradual increase in elevation of the sea surface. Concurrently, currents are generated which have speeds of several centimeters per second, with the greatest speed near the midshelf area (Figure 13–8). This situation is one where the tidal wave approaches the margin perpendicular to it with no component parallel to the coast (Redfield, 1958). It should be observed that these tidal currents flow in an offshore as well as an onshore direction.

Actually, most shelf seas experience tidal waves that tend to move along the coast or at least show a shore-parallel component. These are called progressive tidal waves. In semienclosed shelf seas like the North Sea or Georges Bank, these currents may reach over 1 m/sec (Gorsline and Swift, 1977). It is on such shelves that much sediment entrainment takes place.

The shelf is important in controlling the tidal range at the coast and the tidal current velocity on the shelf (Redfield, 1958; Cram, 1979). There is a general and direct relationship between shelf width and tidal range on the coast. This has been documented along several extensive areas of the world. Examples of broad shelves

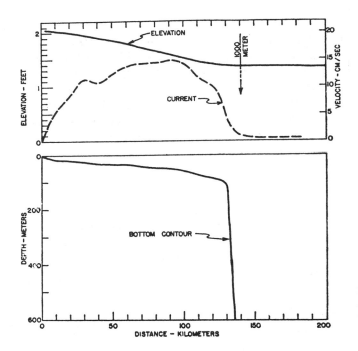
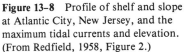

Figure 13-8 Profile of shelf and slope at Atlantic City, New Jersey, and the maximum tidal currents and elevation. (From Redfield, 1958, Figure 2.)

with high tidal ranges are the coasts of northwestern Europe, northwestern Australia, and southeastern South America, whereas western Africa and western South America contain narrow shelves and experience small tidal ranges (Redfield, 1958). An apparent exception is the western coast of the United States.

Waves

Although one tends to think of waves as having a major role in sediment transport in the surf zone, they are also quite important throughout the shelf. Under normal wave conditions the orbital motion tends to entrain sediment only to depths of about 10 m or so. Large waves such as those generated by storms may affect the entire shelf floor. The boundary layer for wave activity is the thin zone of sediment disturbance, generally less than 10 cm on the shelf (Gorsline and Swift, 1977). The oscillatory motion of these waves does not provide any net transport of sediment, but it does accomplish two things:

1. It brings sediment into the water column and thereby makes it available for transport by currents.
2. It also causes a reworking of sediment such that even if no sediment is moved away by currents, the resettling of particles entrained by the wave may take on a texture or fabric different from the previous one due to wave influence.

Internal waves are formed on the interface between water masses. Cacchione and Southard (1974) have speculated that some shelf sediment motion is induced by shoaling of these long gravity waves, particularly at the shelf edge.

Currents

In addition to currents generated by tidal waves (see above), there are also wind-induced currents, oceanic currents, and density currents which may interact with shelf sediments. The interactions of these various current systems with each other and their role in other processes are complex.

Wind-induced currents. Weather influences on the ocean are well known and are varied. The one under consideration here is the transfer of energy from the motion of the atmosphere to the hydrosphere and in turn to the sediment on the floor of the shelf. Some of these currents are continuing and predictable, whereas others are sporadic. They also cover a spectrum of scales in time and space:

1. Land-sea breeze effects which are local and diurnal
2. Passage of frontal systems, high- or low-pressure systems which are broadly predictable, have a cycle of a week or so and encompass hundreds of kilometers
3. Major seasonal weather systems which last for months and cover a major geographic region (Moers, 1976)

Some events, such as hurricanes, are rare and intense, causing much sediment transport. Although the currents generated by weather systems are generally more pronounced along the coast than on the shelf, they are nevertheless important and do cause sediment transport.

Storm-induced currents and related sediment transport have been studied in detail on the Atlantic shelf of North America (Lavalle et al., 1978; Swift et al., 1979; Field, 1980). If this shelf is representative of worldwide conditions, there is widespread transport of large quantities of sediment taking place on the present continental shelves. In studies of the New York Bight area it was determined that winter storms control sediment transport patterns (Lavalle et al., 1978). Various large bedforms, including megaripples and sand waves, are present on this shelf and are generated by "northeasters," which transport sediment to the southeast (Swift et al., 1979).

Large linear ridges on the inner shelf move from 2 to 120 m per year based on measurements over several decades. In a single storm in March 1962, a ridge moved 76 m (Field, 1980). It appears therefore, that storms and the currents that they generate may be the dominant force in sediment transport on modern continental shelves.

Both storm surge and wave surge are also a result of weather. These phenomena involve the net transport of water due to friction between strong winds and the water mass. The term **set-up** is commonly used to express this increase in water level, which may be pronounced along the coast. The only real significance the phenomenon has in sediment transport is that after the stress is removed, the return flow may move sediment along the coast or on the inner shelf (Hayes, 1967a; see Chapter 12).

Oceanic currents. The major, semipermanent current systems of the world ocean intrude onto the shelf at various times and places, exerting a pronounced effect on shelf sediments. Such places as the northwestern Gulf of Mexico, southeastern and Georges Bank shelf areas of the Atlantic margin of North America, and across the northern part of South America are subjected to currents with speeds up to 1 knot (Swift, 1969). Convergence of such currents off the coast of Texas produces a whirlpool-type circulation which Curray (1960) believes has a strong seaward-oriented return flow created by the pileup of water.

Density currents. Creation of density gradients between water masses by variations in salinity, temperature, and suspended sediment generates currents. Typically, these currents are slow and do not entrain sediment. Near major river mouths, however, the gradient is high, creating modest currents, which when coupled with the typical instability of the muddy bottom in this environment, will cause sediment transport. Recent studies off the Mississippi Delta area have shown mass movement of sediment, which is at least partially due to such currents (Prior and Coleman, 1978, Coleman and Prior, 1981).

SHELF SEDIMENTATION

Considerable attention has been devoted to the process-response systems currently operating on our present shelf environments. Symposia have been held, short courses offered, and special publications have been prepared on the subject (Stanley, 1969; Swift et al., 1972; Stanley and Swift, 1976; Gorsline and Swift, 1977). All data demonstrate that shelves are either dominated by storms or by tidal currents. During most of the time there is little sediment movement on storm-dominated shelves. Only when large storm waves or wave-generated currents are developed is there significant sediment movement. This typically is only a very small percentage of the time. By contrast, the tide-dominated shelf is continually experiencing currents of sufficient competency to move sediment. There are striking differences in sediment body type and distribution between the two shelf types.

Another fundamental distinction has been made among shelves by Swift (1974, 1976), who considers shelves to be either autochthonous or allochthonous in nature. **Autochthonous shelves** are those on which sediment already on the shelf is reworked and redistributed so as to equilibrate with existing conditions. **Allochthonous shelves** derive their sediment from other adjacent environments, typically via rivers. These major distinctions along with the dominant processes will be used to categorize shelf sedimentation in the following discussion.

Storm-Dominated Shelves

Most shelf seas in the world experience little in the way of tidal effects. Tidal currents are typically less than 25 cm/sec; the tidal range may vary. Under normal conditions waves do not feel bottom until reaching fairly shallow water. Along the west coast of the United States where waves are commonly large, the slope is rather steep, so that sediment entrainment does not take place until waves are close to the shore. Some shelves are very shallow, so that even under normal wave conditions there is breaking and sediment entrainment. The Georgia coast and the west coast of Florida, especially near the Appalachicola Delta, are examples.

The net result of these conditions is that except during intense weather conditions, much of the world's shelf environment is rather static insofar as sediment movement is concerned. Some of these storm-dominated shelves experience a seasonal pattern to the wave energy such as on the shelf off New England, where nor'easters develop in the winter, or on the shelf off Washington and Oregon. In the Gulf of Mexico the storm-related activity is typically related to tropical storms of the late summer. These hurricanes may occur sporadically and typically affect only a portion of the shelf rather than having a regional impact.

Autochthonous shelves

One of the best examples of an autochthonous, storm-dominated shelf is off the eastern coast of the United States in the western Atlantic. Although numerous rivers drain across the adjacent coastal plain, their sediment load is largely trapped in estuaries. Sediments presently on this shelf are reworked primarily from older and different depositional environments than the ones they now occupy.

A lengthy period of little or no sediment motion occurs during the summer and into both spring and fall. Numerous factors contribute to these situations. Tidal currents are slow, typically less than 20 cm/sec. This shelf is in the lee of prevailing winds and summer swell is too small to cause sediment movement until depths of < 15 m are reached (Swift, 1974). Some ripples have been observed to form under these nonstorm conditions (McClennen, 1973), but it is not common. Another factor in the prolonged low-energy conditions is the thermal stratification of shelf water during the warm seasons. This prohibits wind-driven currents from reaching the bottom and moving sediment.

The stormy months of late fall through early spring are accompanied by cooling of surface water. This, plus the increase in wave activity, causes the destruction of stratification (Swift, 1974). Most storms on this shelf are the result of low-pressure systems which stall over the western Atlantic and increase in intensity. These intense onshore winds from the northeast generate large waves and also create currents. A generalized diagram shows the simplistic distribution of processes along the shelf during such a storm (Figure 13–9). Note that both wave- and wind-driven currents are present. There is an obvious longshore component to these phenomena. In addition, there are currents along the bottom which have an offshore component. These currents are the result of set up against the coast and the resulting downwelling, which relieves this unstable situation. Monitoring studies in the New York Bight area have shown significant sediment movement by bottom currents which fluctuate in both direction and time (Lavelle et al., 1978; Gadd et al., 1978). Currents generated by storms and strong enough to move shelf sediment occur about once every 20 days with current speeds occasionally exceeding 60 cm/sec.

Morphology and sediment distribution on the northwest Atlantic shelf show a complicated situation. It is partly due to the combination of reworking of the early Holocene substrate and the stranding of numerous features as rapid transgression occurred (Field and Duane, 1976). A map of the Mid-Atlantic Bight, the region between Narraganset Bay and Cape Hatteras, shows several different morphologic features, including sand ridges, cuestas, channels, deltas, and **shoal retreat massifs** (Figure 13–10). Virtually all of the large features of the coast are relict or first-order features

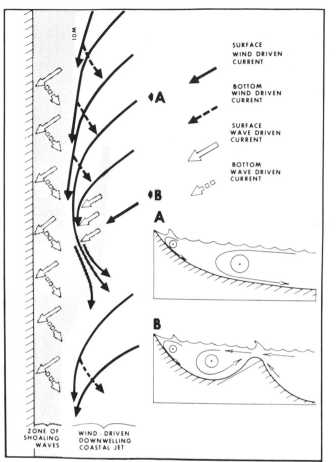

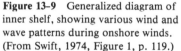

Figure 13-9 Generalized diagram of inner shelf, showing various wind and wave patterns during onshore winds. (From Swift, 1974, Figure 1, p. 119.)

(Swift et al., 1973). These include all but the elongate sand ridges, which have been interpreted as arising from modern processes of the shelf environment (Swift et al., 1973). Some authors have interpreted at least some of these linear shoals as relict barriers (Field and Duane, 1976). The ridges occur on and in between the large and older relict features (Figure 13–10) and their orientations are both nearly parallel and distinctly diagonal to the adjacent coastline. Such general observations suggest that both possibilities may exist.

These ridges may exhibit 10 m of relief with the shelf floor and they display a pattern of surface sediment textures. The ridges are typically coarser than are the troughs (Stubblefield et al., 1975). An explanation by Stubblefield and Swift (1976) accommodates both relict sand ridges and modern ones on the Atlantic shelf. Primary ridges are formed during rapid retreat of the shoreline during the Holocene transgression. Storms generate helical currents and wave surge, which cause a circulation pattern that scours existing troughs on the shelf. As a result of this scour there is modification of the topography such that modern ridges are created and

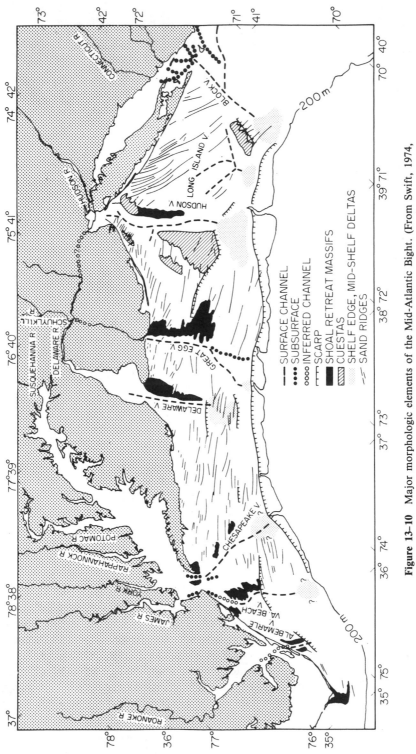

Figure 13–10 Major morphologic elements of the Mid-Atlantic Bight. (From Swift, 1974, Figure 3, p. 120.)

troughs are scoured down into older mud and sand layers (Figure 13–11). This process, coupled with some modification of the original ridges, produces the shelf topography as we now see it (Stubblefield and Swift, 1976) (Figure 13–10).

Large bedforms and lineations are commonly associated with these features and when cored they show cross-stratification. High-resolution seismic profiles and side-scan sonar devices also provide information on the geometry and large-scale internal stratification of these bedforms (Swift et al., 1973).

Allochthonous shelves

Shelves which are presently receiving a significant amount of sediment from runoff are distinctly different from autochthonous shelves, both in sediment texture and shelf morphology. Typically, such storm-dominated shelves have less relief and are finer grained than their autochthonous counterparts.

Virtually all sediment supplied to the shelf at present is carried by rivers; only about 2% results from the erosion of coastal bedrock or eolian sources. Additionally, only a few large rivers carry a large portion of the world's fluvial sediment to the sea, and most of these are in southeast Asia (Drake, 1976). It is also important to observe that virtually all of this fluvial sediment is transported to the sea in a suspended mode and that it comes to rest on the continental margin; only 10% reaches the deep ocean basin (Lisitzin, 1972).

Although storms are infrequent, sediment on these shelves is transported much more of the time than on autochthonous shelves, due to its fine nature. Even sluggish tidal currents, wind-driven currents, and intrusions of oceanic currents cause these suspended particles to be transported across the shelf.

General distribution patterns of modern sediment on these shelves show that extensive windows are present where relict sediments occur and the modern blanket of mud is discontinuous (Allen, 1964a; Curray, 1969). These windows are individually ephemeral but seem to persist on the shelf in general (McCave, 1972). An excellent example of such a shelf is adjacent to the Niger Delta in Africa (Figure 13–12).

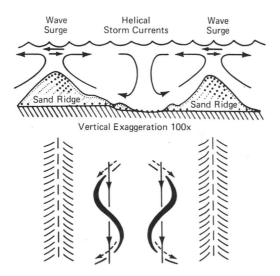

Figure 13–11 Schematic diagrams of helical flow and storm-induced wave surge associated with sand ridges on the eastern U.S. shelf. (From Swift, et al., 1973.)

Chapter 13 Terrigenous Shelves and Shallow Seas

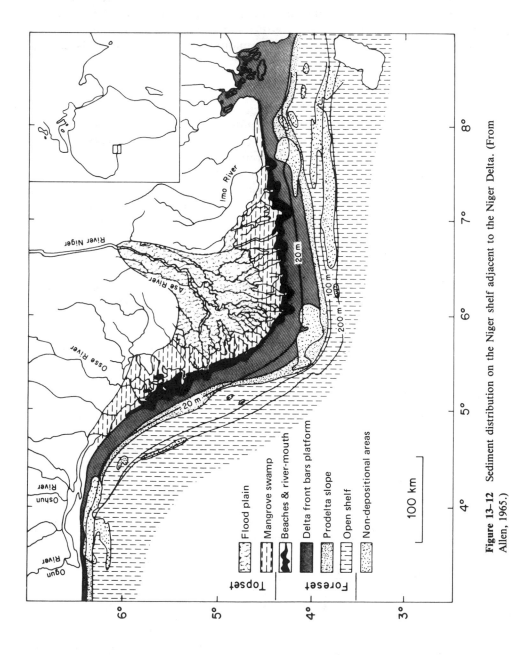

Figure 13-12 Sediment distribution on the Niger shelf adjacent to the Niger Delta. (From Allen, 1965.)

Topset
- Flood plain
- Mangrove swamp
- Beaches & river-mouth

Foreset
- Delta front bars platform
- Prodelta slope
- Open shelf
- Non-depositional areas

100 km

Osse River
Osse River
Oshun River
Ogun River
River Niger
Ase River
Imo River

20 m
20 m
100 m
200 m

Although much sediment is being supplied to the shelf, there are extensive shore-parallel bands of older sediments exposed through the younger muds. In some areas these "windows" cover as much or more area as the modern sediment; such a case is in the northwest Gulf of Mexico. These situations have prompted the generalizations formulated by McCave (1972) and shown in Figure 13–6.

Intense conditions on these shelves may occur in two general forms: (1) those which are cyclic and seasonal, and (2) those which are sporadic, such as hurricanes. The narrow shelf adjacent to Oregon and Washington is largely allochthonous, with significant sediment coming from numerous nearby rivers. Sediment supply and wave energy is seasonal, with maxima in both occurring in winter (Kulm et al., 1975). Similar data have been reported by Sternberg and Larsen (1976), who observed that nearly 10% of the time in a 260-day period, sediment was moved on the shelf in 80 m of water. Bottom sediment is homogenized and extensively rippled throughout the shelf as the result of these winter storms (Figure 13–13).

The Texas shelf is allochthonous in the broad sense of the term in that some sediment bypasses the numerous estuaries and the Mississippi River makes a significant contribution as well. Here, the high-energy events are hurricanes which occur nearly every year and which cause much sediment transport. Graded beds in the shoreface have previously been discussed (Figure 12–20). Based on existing wave data, Curray (1960) calculated the frequency of sand movement across this wide shelf. The range was from about once in 5 years at the outer margin to about once each year just beyond the shoreface. In between periods of physical transport by

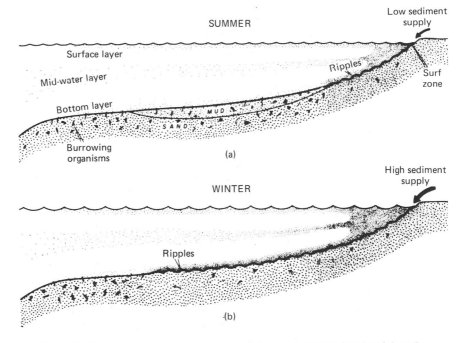

Figure 13–13 Sedimentation patterns on the Oregon shelf showing (b) high sediment supply during winter storm period, which contrasts with (a) the summer, when sediment supply is low and bioturbation is common. (From Kulm et al., 1975.)

Chapter 13 Terrigenous Shelves and Shallow Seas   **469**

waves, these sediments are subjected to bioturbation which shows variable distribution (Bouma, 1972).

Tide-Dominated Shelves

Many shelves whose adjacent coasts experience mesotidal and macrotidal ranges possess tidal currents which commonly range between 50 and 100 cm/sec. Such conditions give rise to considerable sediment transport over extensive periods of time and in a regular cyclic fashion. Shelves, where this occurs, include the North Sea, Korea Bay of the Yellow Sea, Gulf of Cambay (India), and north of Australia. Similar conditions exist off the mouths of high-discharge rivers such as the Ganges-Brahmaputra and the Amazon (Chapter 9) and also on carbonate shallow seas such as the Great Bahama Bank and Persian Gulf (Chapter 14). Large-scale sediment bodies are produced by the tidal currents on these shelves and are of significance to rock record interpretations of shelf deposits (Off, 1963).

The nature of tidal currents in two opposing directions thus produces sediment bodies which tend to be elongate parallel to tidal currents (Off, 1963). Some features may be elongate normal to transport, and the inequality of ebb flood currents can cause net migration of the sediment bodies and related bedforms.

At least one of the tide-dominated shelves mentioned above, the North Sea, is also subjected to intense storm wave energy, but the tides exert more of an influence on sediment transport and patterns (Figure 13–14).

Autochthonous shelves

Most of the extensive tide-dominated shelf regions in the world that are neither carbonate nor adjacent to rivers are autochthonous. These shelves tend to be dominated by sand-size sediment with some terrigenous and biogenic lag gravels but with little mud-size sediment except in areas sheltered from the tidal currents. Because of sustained and rapid currents passing over the shelf, the reworking of existing sediment is such that inherited morphology and sediment patterns are essentially nonexistent (Swift, 1976).

The most striking features of these shelves are the large, long ridges of sand. They were originally described in detail by Off (1963), who called them tidal ridges. Although some subsequent writers have also described these features (Robinson, 1966; Houbolt, 1968), others have referred to similar forms called sand ribbons (Kenyon, 1970; Kenyon and Stride, 1970). Tidal current ridges are 7–30 m in relief, up to 65 km long, and spaced a few kilometers apart (Off, 1963).

The initial detailed study of tidal ridges was by Houbolt (1968), who investigated the Southern Bight area between England and Holland. His study included not only processes and morphology but also coring the ridges themselves. Numerous large ridges are present with orientations essentially parallel to tidal flow (Figure 13–15). Houbolt (1968) considered these features to be tidally generated; however, subsequent writers have indicated that similar features on the Atlantic shelf of the United States appear to be relict (Swift, 1974, 1976).

The profile across the short dimension of the banks shows a distinct asymmetry, like a current-generated bedform. Cores from these ridges show gravel lag

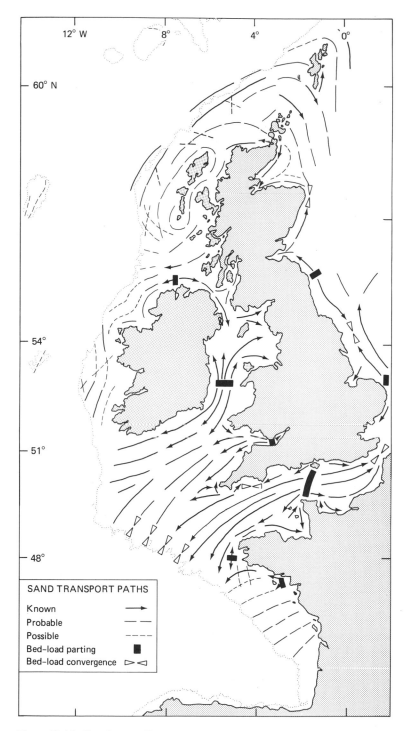

Figure 13–14 Dominant sediment transport patterns on the shelf areas of the North Sea and vicinity. (From Kenyon and Stride, 1970.)

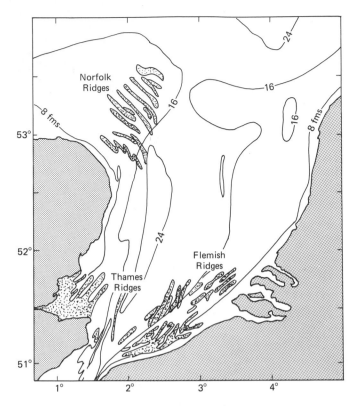

Figure 13–15 Major concentrations of sand ridges in the southern North Sea. (From Houbolt, 1968.)

deposits in swales between ridges, sorted sand with erratic cross-stratification on the gentle stoss slope, and well-developed foreset stratification on the steep slope, which is only inclined about 5° (Houbolt, 1968). These apparent incongruities are explained by Figure 13–16. The combination of tidal current data, morphology, and internal stratification suggests the following:

1. Sediment is moved along the steep face of the ridge by tidal currents.
2. The tide reverses and generates currents over the gentle side of the ridge as shown by megaripples (Figure 13–16).
3. Sediment then migrates up to the crest and moves down the steep side.

In this fashion the morphology, bedforms, and cross-stratification are explained. The net movement of the tidal ridge depends on which tidal direction dominates (Houbolt, 1968). These bank areas have flood- and ebb-dominated channels much like estuaries. It has been suggested that sand ribbons require currents of 100 cm/sec to maintain themselves. Below this speed the dominant features are sand waves (Swift, 1976).

Without a doubt the North Sea region is the best example of a tide-dominated

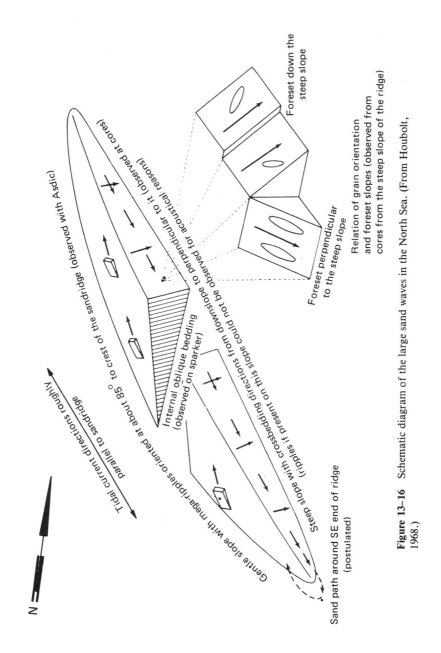

Figure 13-16 Schematic diagram of the large sand waves in the North Sea. (From Houbolt, 1968.)

autochthonous shelf (Figure 13-14). It has a high tidal current energy and displays great variety and complexity in its sediment patterns. It has also received considerable study by marine geologists. The tidal systems in the North Sea are somewhat self-contained and currents produced by the tidal wave may reach over 5 knots locally; 2-to 3-knot currents are widespread (Kenyon and Stride, 1970).

In addition to sand ribbons, discussed above, there are various bedforms present. Sand waves are a few meters high, over 100 m in wave length, and generally exhibit straight crests (Johnson, 1979). Except for the very large forms, they have well-defined avalanche faces and typically occur where tidal currents are asymmetric with respect to ebb and flood and where speeds are about 50 to 75 cm/sec. Small bedforms such as megaripples and ripples are nearly ubiquitous.

The North Sea shelf region includes essentially four characteristic bottom types: sand ribbons, sand waves, sand patches, and mud areas. The latter two are not characterized by large bedforms but may be extensively rippled, especially the sand areas. The sand patches probably represent a similar environment to that of the shoreface on a wave-dominated shelf. They are subjected to a combination of physical energy, which produces ripples, and infaunal organisms, which cause bioturbation (Johnson, 1979). The mud is commonly restricted to areas that are sheltered from both strong tidal currents and wave activity, that is, to the deeper parts of the shelf sea.

Allochthonous shelves

Tide-dominated shelves which receive their sediment from beyond the shelf itself are not common. If sufficient tidal energy is present to move the fine sediment presently supplied to the shelf, it is typically sufficient also to carry it past the shelf with deposition occurring on the slope and rise. One type of exception is the tide-dominated coast, which has a major river-estuary system adjacent to the shelf. Here sediment is abundant from the riverine source and tidal energy moves it readily. Off (1963) mentions the Gulf of Cambay (India) and Amazon Delta as good examples. An important difference in sediment body orientation exists between the previously discussed autochthonous shelves and these allochthonous examples. Those that are not influenced by rivers have tidal currents and sediment bodies which tend to be parallel or at least oblique to the coast, whereas tidal ridges on allochthonous shelves adjacent to rivers have an orientation that is nearly normal to the coast (Figure 13-17). It appears that the general nature of these ridges is like that of the North Sea, but detailed studies are not available. Note also that the allochthonous tide-dominated regime typically covers only the inner part of the shelf unless it is narrow and tidal currents dominate its entire width.

ANCIENT TERRIGENOUS SHELF DEPOSITS

The marine shelf or epeiric sea represents an environment where there is considerable homogeneity through both space and time compared to the environments of terrestrial or coastal areas. The extensive and tabular deposits with their contained sedimentary structures are rather difficult to recognize as having a shelf origin unless

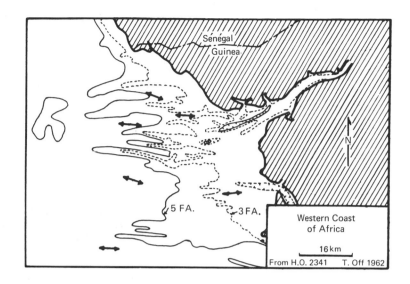

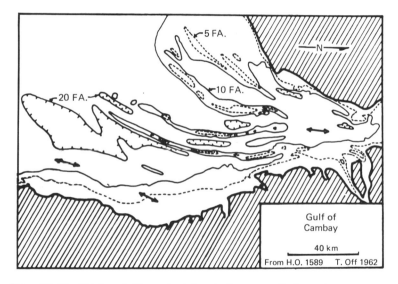

Figure 13–17 Tidal sand ridges oriented somewhat perpendicular to the coast due to extreme tidal currents and funnel-shaped coastlines. (From Off, 1963.)

additional criteria are utilized. Distinction between ancient shelf and epeiric sea deposits is difficult but frequently necessary in order to reconstruct the basin of deposition. In many ways the terrigenous shallow marine environment is more difficult to recognize and subdivide than is its carbonate counterpart (see Chapter 14).

Criteria for Recognition in the Rock Record

Designation of a sequence of strata as representing accumulation in a shelf sea requires utilization of all criteria typically called upon, such as sediment body geometry, lithology, textures, and sedimentary structures. These are used extensively in interpreting all ancient depositional environments, but in the case of open marine environments much attention should be focused as well on the biogenic component of the rocks. Fossils provide what is typically the most diagnostic criterion for recognition of ancient shelf deposits. Excellent summaries on recognition of these environments are presented by Heckel (1972), and Johnson (1979), both of whom stress the importance of fossils.

Overall lithology is not a good criterion for shelf sediments because virtually all sediment types occur in this environment. The minerals glauconite and chamosite may be useful marine indicators, especially the latter, which occurs only at shelf depths but is restricted to low latitudes (Heckel, 1972). Phosphorite is most abundant in shelf areas but occurs also in deep and coastal marine waters.

Sedimentary structures are widespread in shelf deposits but contain no dependable unique characteristics or distribution which distinguishes them from other subaqueous deposits, with the possible exception of hummocky cross-stratification, which appears to be restricted to the shoreface between normal and storm wave base. It has been well documented from ancient shelf deposits and is commonly associated with graded turbidite units (Hamblin and Walker, 1979; Bourgeois, 1980). There are biogenic structures which may be useful in recognizing shelf deposits. A summary of trace fossils from shelf environments indicates that most are formed by active scavengers searching for food (Crimes, 1975). Vertical or U-shaped burrows are rather scarce, but if present they are short. In early Paleozoic time with abundant trilobites, various traces were abundant. Burrows were common in later seas. Relatively deep areas of the shelf have higher amounts of organic matter in the sediment and burrowers become more common than in shallower and more oxygenated water, where there is less food (Crimes, 1975).

The fossil itself is the best indicator of the environment. The shelf environment is characterized by nearly homogeneous salinities, typically about 35‰. Organisms that live in this environment are not tolerant of salinity changes and normal marine environments contain a great diversity of organisms. Most of the common groups that are likely to be preserved live in the shelf environment (Heckel, 1972).

Depth of water is the other important factor in controlling the distribution of organisms with preservation potential. Unfortunately, a large number of these is not restricted to shelf depths, and many of those that are so restricted are typical of carbonate seas. With the exception of the modern horseshoe crab (*Limulus*), all major groups of organisms are either characteristic of carbonate seas or may extend to the slope and beyond (Heckel, 1972). There is a much greater abundance of these marine organisms on the shelf compared to other environments. Virtually any assemblage of marine fossils which displays typical growth form and size, and which contains abundant individuals, can be attributed to the shelf environment.

The examples described below represent a variety of shelf situations where most attention is directed to texture, sedimentary structures, and stratigraphic sequences. Recognition of a unit as a shelf deposit is fairly straightforward by using paleontology. The detailed analysis of the paleoenvironment is much more difficult and it is the latter toward which this discussion is directed.

Wave-Dominated Shelf Deposits

A lower Carboniferous sequence in County Cork, Ireland, has been interpreted as representing deposition in a wave-dominated shelf environment (deRaaf et al., 1977). Three main lithotypes are present: sandstone, mudstone, and a mixture of both. The mudstones are typically only millimeters thick, the mixed lithotype is decimeters thick, and the sandstone may exceed a meter. The total section is in excess of 500 m thick and consists of a complex stratigraphy of interlayered beds of all three lithotypes (Figure 13–18). The complete spectrum of stratification types (Figure 3–17) is present from cross-stratified sandstone through flaser, wavy, and lenticular stratification to mudstone. Within the section are also coarsening upward, fining upward, and random intercalation sequences (deRaff et al., 1977). Small-scale, low-angle cross-stratification is essentially ubiquitous through the section.

One of the biggest questions to be answered in the interpretation of this Carboniferous section concerns the origin of the cross-stratification. Previous studies of modern ripples (e.g., Newton, 1968) have shown that wave ripples may display unidirectional cross-stratification regardless of the external shape. Further, the distinction between wave- and current-generated cross-stratification was systematized by Boersma (1970), who presented characteristics for the wave-formed variety (Figure 13–19); cross-strata may be in opposing directions, the chevron and bundled upbuilding, and the undulating lower set boundary.

The great range in lithology over a narrow stratigraphic interval requires explanation. Cross-stratified sandstone is intercalated with mudstone. Storm or wave

W   E

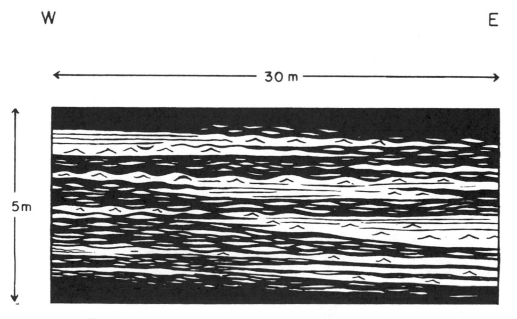
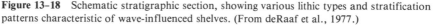

Figure 13–18 Schematic stratigraphic section, showing various lithic types and stratification patterns characteristic of wave-influenced shelves. (From deRaaf et al., 1977.)

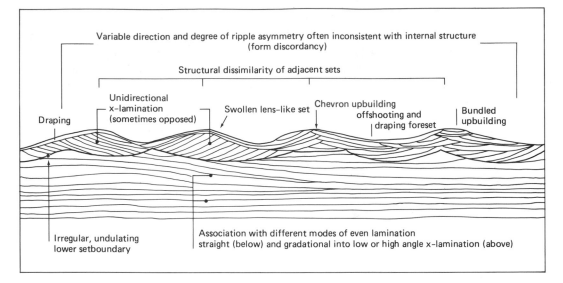

Figure 13–19 Diagnostic features of wave-generated cross-stratification. (From deRaaf et al., 1977.)

energy which winnows the fines and permits sand to accumulate can explain this type of succession. The authors (deRaaf et al., 1977) envision that deposition took place on a shallow shelf where large waves created shoaling and breaking conditions together with possible emergence (Figure 13–20). Such conditions gave rise to three somewhat different sand body types: incipient bar, submerged bar, and emergent bar. The latter experienced swash and backwash much like a beach (deRaaf et al.,

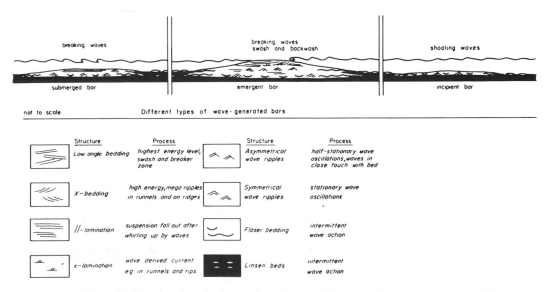

Figure 13–20 Three bar-shoal types, including stratification styles and wave processes. (From deRaaf et al., 1977.)

1977). This shelf sea may have been either pericontinental or epicontinental in nature. The apparent shallow nature suggests that an epicontinental origin is likely.

Tide-Dominated Shelf Deposits

Tide-dominated sediment transport during marine transgression has been interpreted for a number of Tertiary sequences in western Europe (England, Spain, and Switzerland) (Nio, 1977). The reason for studying these units is the presence of large-scale cross-stratification in well-documented marine strata. The study was designed to determine the origin of this cross-stratification and to compare it to possible modern analogs.

The entire sequence at each of the areas studied contains strata deposited in both marine and coastal environments. The standstone facies with the large-scale, sand wave-generated, cross-stratification is typically a few tens of meters thick. Although paleocurrents indicated by orientation of the cross-strata are unimodal, some horizons display a wide range in directions (Nio, 1977).

Stratigraphic position of the sand wave sequences is above tidal flat–estuarine deposits and is overlain by and interlayered with fossiliferous marine deposits of mudstone. The data presented have enabled the investigator (Nio, 1977) to propose a model for the origin of the sand wave sequences.

In this model the initial sand waves develop in or near estuaries as transgression begins. With greater tidal currents and increasing depths these sand waves double in size, and foresets steepen and become irregular. As the rate of sedimentation decreases, foresets become gently inclined and orientation becomes variable as current directions spread. Nio (1977) envisions this depositional environment to have been very much like that of the present-day North Sea.

Tide-dominated shelves and epeiric seas have also been suggested for Precambrian, Paleozoic, and Mesozoic depositional systems (Klein, 1977b; Klein and Ryer, 1978), although other investigators prefer to have these environments essentially tideless (e.g., Mazullo and Friedman, 1975). Data indicate that tides were important factors in sedimentation on ancient shelves and epeiric seas.

Combination Storm- and Tide-Generated Shelf Deposits

Cretaceous, Wyoming

The Upper Cretaceous strata of Wyoming represent a sequence of up to 3000 m of mostly terrigenous rocks deposited in nonmarine, coastal, and open marine environments. A large portion of these rocks has been interpreted as representing deposition in a shelf sea; in fact, it is considered as a "textbook example" of such depositional conditions (Asquith, 1970). Although many units and facies within the Upper Cretaceous of this area have been investigated, the following discussion will utilize only two related sandstone units: the Sussex and the Shannon members of the Cody Formation, a shale. These units are present in the Powder River Basin, where the Shannon is up to 30 m thick and the Sussex is about 15 m thick (Figure 13–21). An

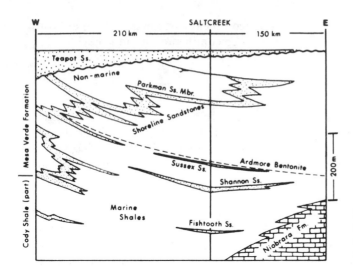

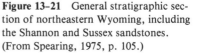

Figure 13-21 General stratigraphic section of northeastern Wyoming, including the Shannon and Sussex sandstones. (From Spearing, 1975, p. 105.)

extensive bentonite bed overlies the Sussex Sandstone and serves as an excellent time datum. Both units have been interpreted similarly: the Shannon by Spearing (1975, 1976) and the Sussex by Berg (1975) and Brenner (1978).

Each of these sandstone units is actually composed of multiple lithofacies. For example, the Shannon has silty mudstone, thin-bedded sandstone, and cross-stratified sandstone facies (Figure 13–22). A sequence of these three facies, in coarsening upward fashion, occurs in cyclic fashion (Spearing, 1975). The mudstone is burrowed, glauconitic, and contains phosphorite nodules at the contact. It is interpreted as representing essentially a nondepositional shelf surface. The overlying sandstone has a few burrows and some cross-strata, but it is dominated by wave-formed ripples. The cross-stratified facies contains abundant clay rip-ups incorporated in the bedding and is interpreted to represent large sand wave migration caused by storm events or tidal currents (Figure 13–22). Note that cross-strata orientations indicate opposing currents, although one dominates over the other.

The same type of coarsening-upward section with the same three facies has been described from the Sussex Sandstone (Berg, 1975; Brenner, 1978, 1980). An additional feature found in the Sussex is the presence of apparent channels which cut across the sand bodies. It should be remembered that although fossils are essentially absent in these sandstone units, the surrounding marine shales do contain a marine fauna.

The depositional model proposed for the Shannon and Sussex consists of migrating sand sheets which have their surfaces covered by sand waves. These produce sand bodies elongate parallel to shore at about the middle to outer shelf. The sequence which accumulates is coarsening upward (Figure 13–22), indicating continued shoaling. Large-scale cross-strata with mud clasts support this interpretation (Spearing, 1975). Both tidal and storm-generated currents contributed.

A reconstruction of a shelf on which Shannon and Sussex type of sediments accumulated shows the paleogeography that prevailed (Figure 13–23). The sand bodies were covered by large-scale bedforms and locally cut by storm-generated channels

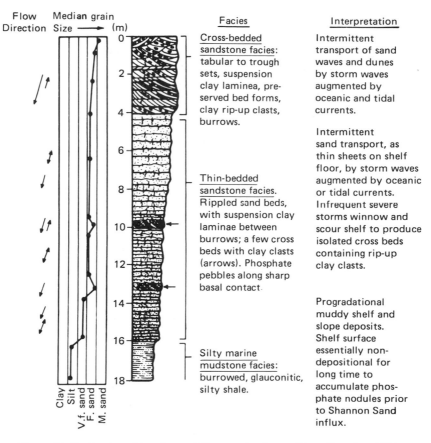

Figure 13–22 Stratigraphic column of Shannon Sandstone. (From Spearing, 1975, p. 108.)

(Figure 13–24), which produced fan-shaped subtidal deltas (Brenner, 1978). An almost exact duplication of this Cretaceous system has been described by Brenner and Davies (1973, 1974) from the Upper Jurassic in the Wyoming and Montana area. Particularly comparable is the coarsening-upward stratigraphic sequence, which displays similar facies. The Precambrian Jura Quartzite also represents deposition under combined tidal and storm current conditions (Anderton, 1976).

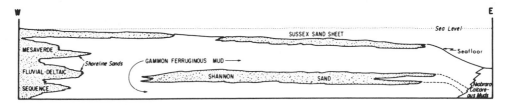

Figure 13–23 Paleogeographic cross section across Bighorn and Powder River Basins prior to deposition of Ardmore Bentonite Bed. Compare with Figure 13–22. (From Brenner, 1978, p. 186.)

Chapter 13 Terrigenous Shelves and Shallow Seas

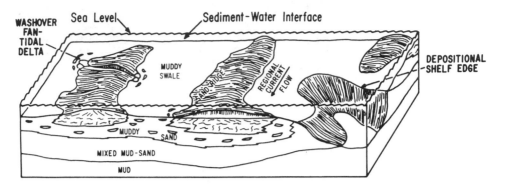

Figure 13-24 Depositional model for Sussex Sandstone, showing major morphologic features and stratigraphic relationships. (From Brenner, 1978, Figure 19, p. 198.)

Cambro-Ordovician, Upper Mississippi Valley

Studies of the physical and biostratigraphy of the Cambro-Ordovician sequence in the upper Mississippi valley are classics in their fields (Bell et al., 1952; Berg, 1954; Hamblin, 1961). These efforts established not only the nomenclature of the region but also demonstrated a shallow shelf sea as the overall environment of deposition. This sea was dominated by mudstone and sandstone with some carbonate deposition. Glauconite and trilobites are widespread constituents of the strata.

The Upper Cambrian units consist of planar bedded and cross-stratified mudstones and sandstones with clasts of mudstone reworked into the sandstone. Cross-strata are small scale, nearly symmetrical ripples are common, and bioturbation is prevalent (Berg, 1954). Extensive study of the cross-stratification has shown a consistent pattern to paleocurrent directions (Farkas, 1960; Hamblin, 1961; Emrich, 1966; Michelson and Dott, 1973). Although there is much variability, the regional dispersal pattern is northeast to southwest across what was probably a broad epeiric sea (Figure 13-25).

The nature of the sequences and the cross-strata suggest that both waves and shelf currents influenced sedimentation. There is a distinct and prevalent mode on the plot of cross-strata (Figure 13-25). The nature of the cross-strata sequences compared to those of Boersma (1970) and Raaf et al. (1977) indicate a wave-formed origin. In addition, there are bimodal and bidirectional cross-strata (Michelson and Dott, 1973), suggesting some tidal influence. The wide range may be due to a spectrum of wind and, therefore, wave direction.

Extreme energy events in these strata have been interpreted from strata exposed near the Precambrian Baraboo district which were islands in this shallow shelf sea. The structure and stratigraphic relationships in this area indicate islands of a few hundred feet of relief with steep slopes (Dott, 1974). Large rounded and sorted clasts up to several meters in diameter were derived from the Baraboo Quartzite and incorporated into the Cambrian sandstones. Cross-stratification in these rocks adjacent to the Baraboo Islands shows great range in direction (Dalziel and Dott, 1970).

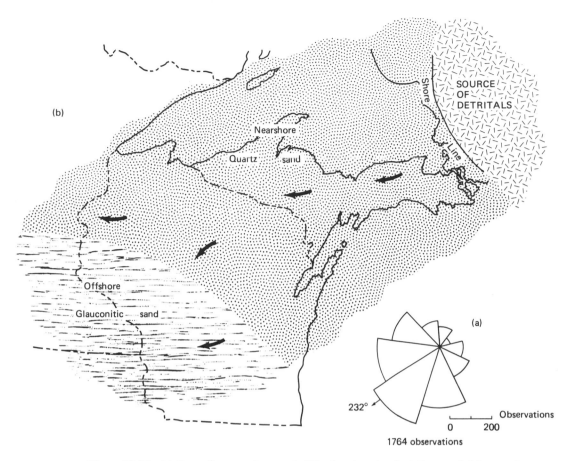

Figure 13–25 (a) Rose diagram of cross-stratification trough orientations and (b) general paleogeography of Cambrian Franconian strata. (After Michelson and Dott, 1973; Hamblin, 1961.)

SUMMARY OF CHARACTERISTICS OF SEDIMENT/SEDIMENTARY ROCK BODIES REPRESENTING TERRIGENOUS SHELVES AND SHALLOW SEAS

Terrigenous shelf and shallow sea deposits are among the most prevalent in the stratigraphic record. These sequences are typically rather easy to recognize because of the combined characteristics they exhibit.

Tectonic setting. Extensive shelf and shallow marine environments are favored by stable, trailing-edge margins. They tend to be absent or quite limited in extent along collision zones where plates converge.

Shape. Shelf sequences are tabular, with little variation on the theme of two long axes and one short axis.

Size. Shallow terrigenous marine strata are the most extensive preserved in the stratigraphic record. These sequences may cover thousands of square kilometers with a thickness up to hundreds of meters.

Textural trends. Shelf sequences show distinct patterns with fining-upward grain size trends, characterizing transgressive sequences and coarsening-upward sequences typifying progradational sequences. These trends prevail essentially without exception.

Lithology. The composition of shallow marine terrigenous deposits is dominated by quartz and clay minerals with secondary amounts of rock fragments and feldspar. Biogenic material of calcium carbonate is common and widespread but not abundant. Thin layers of concentrated shell material may be present and indicate storm conditions.

Sedimentary structures. There are some sedimentary structures and sequences of structures which are suggestive of a shelf or shallow sea environment. The most diagnostic structure is hummocky cross-stratification which occurs within the zone of storm wave base. There is also a predictable stratigraphic trend in the relative abundance of bioturbation structures and wave- or current-generated structures such as plane beds and ripple cross-strata. Bioturbation and abundance of mud reflect deeper-water deposition compared to physical structures and abundance of sand-size material.

Paleontology. Fossils preserved in shallow marine deposits are diverse in type, including virtually all types of stenohaline invertebrates plus some scattered phosphatic fish skeletal fragments.

Associations. Terrigenous shelf deposits may be stratigraphically above or below deposits such as barrier sands. Although less common, progradational shelf deposits may overlie deeper-water slope, canyon, or submarine fan deposits. Shelf deposits may be laterally adjacent to carbonate shelf buildups such as reefs.

ADDITIONAL READING

BURK, C. A., AND DRAKE, C. L. (EDS.), 1974. *The Geology of Continental Margins.* Springer-Verlag, New York, 1009 p. This is a compendium of more than 70 papers, mostly on specific continental margin areas of the world. The first section deals with general papers. Shelf sedimentation is only a minor portion of the book.

STANLEY, D. J. (ED.), 1969. *New Concepts of Continental Margin Sedimentation.* Amer. Geol. Inst. Short Course Notes, Washington, D.C. Fine collection of papers on the subject by several experts. This is the first of the several volumes on margins and contains most of the fundamental work on continental margins.

STANLEY, D. J., AND SWIFT, D. J. P. (EDS.), 1976. *Marine Sediment Transport and Environmental Management.* John Wiley, New York, 602 p. The title is misleading; the book is really on shelf sedimentation and it is the best book currently available on the subject. Included are physical and chemical aspects as well as geology.

SWIFT, D. J. P., DUANE, D. B., AND PILKEY, O. H. (EDS.), 1972. *Shelf Sediment Transport: Process and Pattern.* Dowden, Hutchinson & Ross, Stroudsburg, Pa., 656 p. An excellent collection of papers on sediment transport on the shelf environment.

VAN STRAATEN, L. M. J. U. (ED.), 1964. *Deltaic and Shallow Marine Deposits.* Developments in Sedimentology, v. 1. Elsevier, New York, 464 p. Excellent collection of papers on modern and ancient shallow marine deposits by world experts. The papers are somewhat dated, but many are classics.

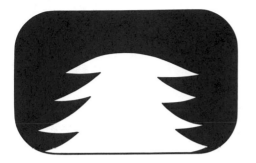

14 Reefs and the Carbonate Platform System

Carbonate-dominated shallow marine environments are restricted in their modern global distribution (Figure 14–1) compared to many periods in the geologic history of the earth. The present carbonate deposition on platforms is in two broad settings: one attached to landmasses such as the Persian Gulf, Florida Bay, and the North shelf of Australia, and one rising from oceanic depths such as the Bahama Platform and the many atolls of the Pacific (Blatt, et al., 1980). Extensive carbonate seas, such as those which existed during the Ordovician, Devonian, Mississippian, and Cretaceous periods, are not present today.

There are some fundamental differences between marine carbonate and terrigenous depositional systems. An excellent summary is the recent statement that "carbonate sediments are born, not made" (James, 1979a). In other words, whereas terrigenous sediments are products of previously existing rocks that have been transported to the basin of deposition (allochthonous), carbonate sediments are all formed in the basin of deposition (autochthonous). The material had its origin at or near the place of accumulation. This is true regardless of the origin of the carbonate particles. In the shallow carbonate factory (James, 1979a) there is great carbonate production and most of the sediment remains there. However, some does make its way landward, and some is carried over the shelf edge to the deep basin (Figure 14–2).

In general, carbonate-dominated platforms (Wilson, 1975) develop in areas where little or no terrigenous influx persists. This is largely the reason that there are

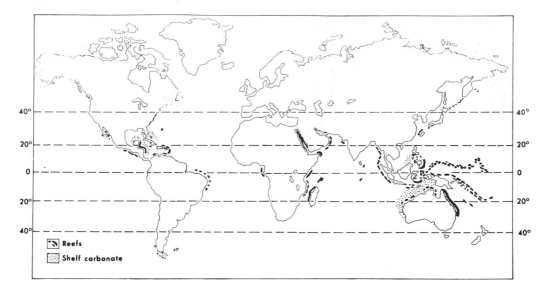

Figure 14-1 Distribution of modern carbonate sediments in shallow water. (From Wilson, 1975, p. 2.)

so few carbonate shelves today. The present influx of terrigenous sediments on most continental margins has the effect of stifling the luxuriant growth of carbonate-producing organisms, especially reef builders. As a result, carbonate buildups must occur on shelves where little terrigenous sediment is introduced, such as the southern Persian Gulf, or they must rise from the sea floor, where no terrigenous source exists, such as the Bahama Platform. In these situations there must be a dynamic equilibrium between carbonate deposition and subsidence to allow thick accumulations and to keep the depth shallow where carbonate production is favored.

Carbonate sediment production is typically in warm shallow water of low-latitude regions, but there are areas where much carbonate is produced in colder climates. Two broad skeletal grain assemblages have been recognized as having rela-

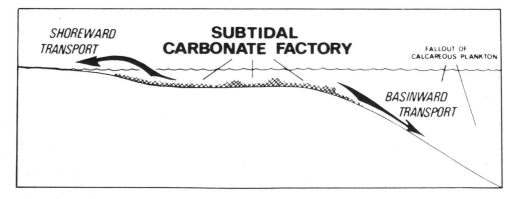

Figure 14-2 Generalized sketch of carbonate shelf, showing main regions of accumulation. (From James, 1979a, Figure 1, p. 106.)

tionships to latitude (Lees and Butler, 1972). The tropical assemblage is characterized by coral and green algae and the high-latitude foramal assemblage is dominated by mollusks and foraminifera, with other invertebrate phyla also present. Such phenomena as photosynthesis, respiration, evaporation, rainfall, and freshwater runoff may be involved in carbonate production (Bathurst, 1975). In addition, all the previously discussed physical processes, such as waves, tides, currents, bioturbation, and effects of benthic organisms, interact with carbonate sediments. Early submarine cementation may also occur and affect these processes.

SHALLOW MARINE CARBONATE ENVIRONMENTS

Sedimentary environments and facies associated with carbonate shelf or offshore bank deposits are markedly different from terrigenous shelves. To consider the shelf or bank complex in the proper context, it is best to first discuss each of the primary carbonate buildups individually. Using the dominant sedimentary facies, there are four primary buildup types to consider: organic framework reef, sediment piles, organic banks, and lime mud accumulations (Wilson, 1975). Only the subtidal region is considered; for related intertidal carbonate environments, see Chapter 10.

Organic Framework Reef

The term **reef** has been used somewhat loosely on occasion but is here defined as a wave-resistant organic framework. Related terms sometimes confused with reef are biostrome and bioherm. A biostrome is an accumulation of biogenic debris that is not necessarily in situ; the term **skeletal mound** is equivalent. A **bioherm** is generally considered to be an in situ accumulation of benthic organisms, although some geologists would apply the term to any mound-shaped accumulation of skeletal debris. Consequently, some bioherms are reefs and some are not, but all reefs are bioherms.

In the classification of carbonates discussed in Chapter 1, the term biolithite is used for in situ bioconstructed materials. An expansion of this depositional texture and fabric has been described by Embry and Klovan (1971). They subdivide biolithites into three types: bafflestone, bindstone, and framestone (Figure 14–3). **Bafflestones** include stalked organisms which trap sediment in between, such as branching corals; **bindstone** is used for tabular or encrusting organisms that bind sediment, such as encrusting coralline algae; and **framestone** includes the massive framework organisms, such as some corals.

In the reef there are two distinct styles of carbonate accumulation; the framework and the intervening unconsolidated biogenic debris. The relative amounts of each vary widely. It is not unusual to find that the framework portion is markedly less than 50% of the reef volume. Such an accumulation could be mud supported. Grainstones, which are better sorted and are washed, are also common. There is typically much pore space also. The reef core includes the in situ framework portion of the reef and related sediments. Adjacent to this is the reef flank, which is a distinctly different facies comprised of reef debris only, showing relatively steep deposi-

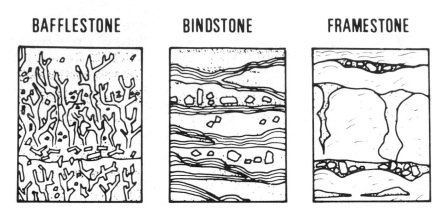

BAFFLESTONE **BINDSTONE** **FRAMESTONE**

Figure 14-3 Different fabrics and textures of reef material as recognized by Embry and Klovan (1971). (From James, 1979b, Figure 3, p. 122.)

tional dip (Figure 14–4). Beyond the reef flank may be a wide variety of sediments which are composed of both fine reef debris and other carbonates or even terrigenous sediment.

Reef types

Initial reef classification by Darwin was based on his many observations of Indo-Pacific reefs during the famous voyage of H.M.S. *Beagle* in the mid-nineteenth century. Darwin recognized three basic reef types—fringing, barrier, and atoll—based on their general shape and position relative to a related landmass. A **fringing reef** is one that is developed immediately against the land, lacking an intervening lagoon. In the situation where a lagoon separates the reef from land, the term **barrier reef** is applied. **Atolls** are somewhat circular reefs with a central lagoon. Darwin correctly interpreted that through subsidence there is a continual sequence from fringing to atoll in Pacific reefs.

Although Darwin's simplistic classification is a good one, the varied forms and conditions of reef development demand a more comprehensive classification. Probably the most widely used is that of Maxwell (1968), who considers two major categories: oceanic reefs and shelf reefs (Figure 14–5). The oceanic category closely follows the approach used by Darwin. Shelf reefs include a wide variety of forms, with wave and current regimes having primary influence.

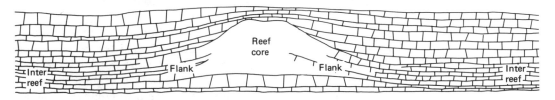

Figure 14-4 Generalized diagram showing major zones of reef and reef-related carbonate accumulations.

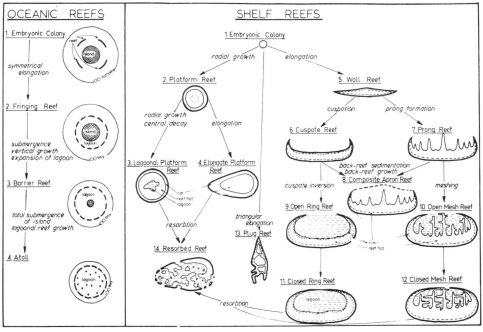

Classification of reefs.

Figure 14–5 Classification of reefs. (From Maxwell, 1968, p. 101.)

Reef environments

The reef is actually composed of numerous contrasting small environments which range from very low energy lagoons to the crashing surf. Although there is such a broad spread of conditions, there is also a reasonably predictable pattern to the position of the various environments (Purdy, 1974). An excellent example is Alacran Reef on the Yucatan shelf of Mexico (Hoskin, 1963). It is similar to the closed mesh type of shelf reef (Figure 14–5) in Maxwell's classification. Major subdivisions of the reef are the windward reef, lagoon, and leeward reef (Figure 14–6). The windward reef is where overall growth is greatest. It includes the slope, which consists largely of reef debris accumulating in steeply bedded, apronlike deposits. Initial dips may reach up to 15°, with a general decrease in grain size away from the framework. The upper part of the slope is a spur and groove (Shinn, 1963) or a groove and buttress system (Figure 14–7), a boulder rampart (Figure 14–8), and the windward reef flat (Figure 14–9). The latter contains various corals which are covered with encrusting calcareous red (coralline) algae with some scattered debris. Marine grasses may be present.

The reef lagoon represents an area of contrast. Included are the cellular and pinnacle reefs, where the reef-building framework and related organisms flourish, and the lagoon floor which is dominated by biogenic debris covered with calcareous green algae (*Halimeda, Penicillus*; see Chapter 2). Depths of lagoons range widely, with some over 100 m deep in Pacific atolls.

Chapter 14   Reefs and the Carbonate Platform System   **489**

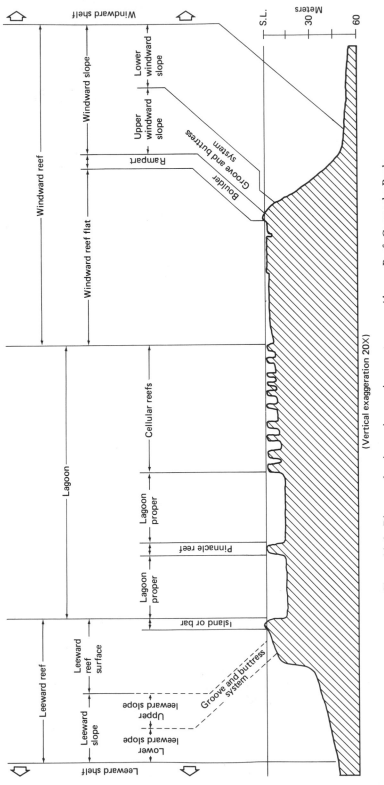

Figure 14-6 Diagram showing various environments across Alacran Reef, Campeche Bank, Mexico. These environments are representative of most closed reefs. (From Hoskin, 1963, Figure 4, p. 19.)

Figure 14-7 Spur and groove system on windward reef near Key West, Florida. (Courtesy of E. A. Shinn.)

Figure 14-8 Boulder rampart on windward side of Heron Reef, Capricorn Group, Great Barrier Reef. This debris accumulates on an algal ridge.

Figure 14-9 Reef flat area at Heron Reef, Australia. Primary framework organism are short branching corals and coralline algae.

The leeward part of the reef is basically a small-scale version of the windward reef. The same general subdivisions are present but are less extensive, due to the slow rate of growth of the leeward reef compared to its windward counterpart.

The zonation of framework organisms and growth forms across the reef is reasonably consistent throughout the world. Typically, however, the Caribbean Sea, Indian Ocean, and Pacific Ocean have some distinctions among them, although those in the latter two waters are quite similar. Generally, there is most species diversity in Indo-Pacific reefs (Milliman, 1974). Caribbean reefs contain luxuriant sea fans, see lillies, and sea whips, whereas they are uncommon on Indo-Pacific reefs. These are not, however, significant sediment producers.

Up the fore-reef slope and across the reef flat there is a general pattern of framework organisms and growth forms, which corresponds to the respective position on the reef and conditions prevailing at that site. The spur and groove or groove and buttress system on the windward slope (Figure 14-7) is the result of growth of corals in response to wave activity. The corals achieve an oriented growth form creating the spurs, whereas the biogenic debris is concentrated in channel-like grooves in between (Shinn, 1963; Maxwell, 1968). The crest is typically an algal ridge constructed by encrusting coralline algae (Figure 14-8). It is commonly smooth, due to the high energy created by breaking waves and related surge, but with much rubble on its landward portion. The reef flat behind this ridge is typically dominated by a variety of branching corals and coralline algae (Figure 14-9).

Reef sediments

One of the most characteristic aspects of reef sediments is their almost exclusively biogenic origin. The framework builders contribute to the unconsolidated reef sediment, but most material is derived from the diverse fauna and flora that live on and near the reef. The bulk of reef sediment is produced by five broad taxa: corals, coralline algae, green algae (*Halimeda*), foraminifera, and mollusks (Milliman, 1974). There is much local variation and some reefs may also contain nonbiogenic particles such as ooids and intraclasts. Pellets may be common and are the only important nonskeletal biogenic constituent.

Reef sediments are mostly sand and gravel, and are moderate to poorly sorted. There is considerable variation within the various niches of the reef system. Reef sediment texture is the result of three major factors: its biogenic origin, wave activity, and breakdown by organisms. Gravel is most common in the boulder rampart area, whereas silt is most abundant in deep, sheltered portions of the lagoon and in small protected cavities and crevasses within the framework. Sand is by far the most abundant grain size and is found throughout the reef.

Biogenic breakdown of reef material is a significant contributor to the volume and texture of reef sediments. Numerous organisms scrape the reef framework to obtain nourishment. Prominent among these is the parrotfish, which feeds almost constantly on corals and produces much sediment in the form of large pellets which readily disaggregate. Other organisms, such as sea urchins, also scrape the framework surface, whereas sea cucumbers ingest unconsolidated sediment. These, plus gastropods, are the primary pellet producers on the reef. Boring organisms, such as pelecypods, bacteria, sponges, and algae, also cause sediment breakdown and contribute to the ultimate texture of reef sediment.

Sediment Piles

Carbonate particles of varying composition may accumulate through physical transport into banks or piles. Included are bars, dunes, and tidal deltas (Wilson, 1975). Among the more extensive modern sediment piles are the ooid shoals, which are common on the Bahama Platform.

The sand-size, spherical grains called ooids (Chapter 1) are formed under rather narrow conditions and commonly occur almost at the exclusion of other carbonate grains. Ooids are presently forming in less than 5 m of water, with some areas intertidal (Ginsburg and James, 1974). They are favored by processes that cause agitation of the grains by currents and are generally found along platform edges or near tidal channels. Tidal currents are the dominant process in the ooid facies. The dynamic substrate typically prohibits grass or algae from becoming established.

Although ooids may occur as scattered grains or as small localized patches on carbonate platforms, they also form extensive shoals. The two most prominent locations are on the Great Bahama Bank and the Trucial Coast of the Persian Gulf. On the Bahama Bank, ooid shoals parallel the edge of the platform margin (Ball, 1967), whereas they are associated with tidal channels on the Trucial Coast (Purser and Evans, 1973; Loreau and Purser, 1973).

Almost continual motion of the ooid sand substrate results in ubiquitous bedforms and well-sorted and rounded particles. Ripples and megaripples typically cover large, low-amplitude sand waves (Figure 14–10). The elongate ooid shoals are characterized by large spillover lobes generated by tidal currents. Large-scale, well-developed, bidirectional cross-stratification is produced by these lobes (Figure

Figure 14–10 Oblique aerial photo of oolite shoals on Great Bahama Bank, showing large-scale bedforms generated by tidal currents.

14–11). Migration of the ooid sand generally prohibits establishment of an extensive benthic infauna and related bioturbation. Typically, these ooid shoals rest over pelleted muds and skeletal sand of the carbonate platforms.

Organic Banks

Detrital particles of biogenic carbonate may be accumulated through a combination of waves and currents with trapping or baffling, especially by benthic organisms. These organic banks may take on a variety of sizes and shapes. Biogenic debris in modern carbonate environments is generally a combination of sand and mud, although gravel may be abundant locally. The composition of these skeletal particles ranges widely depending on the depth of water, local processes, terrigenous influx if any, and other factors. Much of the carbonate sand material currently found on the outer portion of open shelves is relict in nature (Ginsburg and James, 1974). The origin of the skeletal particles is commonly dominated by echinoderms and mollusks, with algae, foraminifera, bryozoans, and coral also present. Planktonic foraminifera typically become relatively abundant on the outer shelf and may be a good indicator of relative depth of water.

Physical processes are generally only modest in rigor, so that sorting, rounding, and overall shape modification are not at a high level. Bioturbation may be extensive and except for local ripples and related cross-strata, physical structures are uncommon.

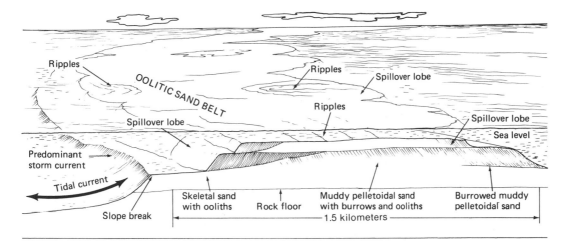

Figure 14-11 Diagram showing aerial and cross-section at view of oolite shoals with tidal-generated, bidirectional cross-stratification. (After Ball, 1967, Figure 9, p. 563.)

Lime Mud Accumulations

Extensive areas of lime mud are present in the modern environment and were probably more widespread in the past. Primary sediment particles are elongate, needle- and lath-shaped aragonite grains which are only a few micrometers in length. They occur in shallow marine areas that are protected from rigorous wave and current attack. The water depth is generally less than 4 m. The substrate in these environments is rather stable, and as a consequence it supports a thick carpet of marine grasses (*Thalassia*) and green algae (*Penicillus, Halimeda*). Various mollusks are also common.

Although the sediment is dominated by fine mud, there is a sizable sand fraction as well, giving it a distinct bimodality. The coarser fraction is primarily mollusks and *Halimeda* plates. Many investigators believe that the mud itself is derived largely from *Penicillus*, which produces needle-shaped aragonite particles. Others suggest that degradation of molluscan debris produces similar particles.

Typically, these sediments are highly bioturbated by infauna and roots of the grasses. The result is a massive and structureless lime mud facies. The combination of cohesion of the mud and the protection afforded by the grass and algae makes transport difficult except under extreme conditions such as storms.

Extensive shallow areas of shelves or platforms are comprised largely of small ovoid fecal pellets. These pellets are composed of unoriented lime mud particles and they generally range from 0.15 to about 2.0 mm in maximum diameter. Most, however, are less than 0.5 mm. Fecal pellets typically contain significant particulate organic matter. Because of some difficulties in determining origin and diagenetic alterations, the term peloid or pelletoid is now in common usage for ancient grains of this type. Many people consider that all peloids are fecal in origin (Bathurst, 1975); however, some ovoid and spherical grains may appear similar to fecal pellets.

Many fecal pellets are quite cohesive due to the nature of the micrite and to mucus that may be secreted by the producing organism. It is common that only one or at most a few taxa in any given environment may produce the pellets. The fact that most pelleted sediment is well sorted may be attributed to their origin rather than to any physical processes operating in the environment of deposition. Typically abundant pelleted mud indicates low-energy conditions which permit accumulation of the lime mud that has been pelletized. In addition, agitation would cause many, perhaps most of the sand size pellets to disaggregate.

Pelleted mud is generally structureless. Some areas contain aggregates of these pellets which are held together by fragile cement. These **grapestones** (Illing, 1954) are particularly widespread in some areas of the Great Bahama Bank.

MODERN CARBONATE SYSTEMS

Even though modern marine environments are rather restricted in the extent of carbonate sediments compared to much of the geologic past, there is a diversity of depositional systems present. The following discussion considers some of the larger and better known shallow carbonate systems in order to cover the spectrum of settings available.

Great Bahama Bank

The Great Bahama Bank and adjacent Little Bahama and Cay Sal Bank comprises one of the most extensive modern carbonate depositional systems. The region is isolated from all terrigenous sediment sources by deep channels and straits with depths in excess of 650 m. Great Bahama Bank itself covers about 100,000 km² and rises abruptly up to 4 km above the surrounding depths. Throughout most of the platform, the water depth is less than 5 m. The general profile resembles a shallow saucer with the edges of the bank exhibiting shallower depths than the center. Ginsburg and James (1974) consider the Great Bahama bank to be a rimmed shelf. Tidal range is only about 0.5 m and is markedly influenced by wind. Tidal currents are greatest over the edge of the bank and become negligible near the center (Ginsburg and Hardie, 1975). Rainfall is high enough to prevent high salinity even though circulation in the center of the bank is poor.

Sediment facies distribution

Considerable effort has been expended in examining the sediments of the Bahama Bank (Illing, 1954; Newell and Rigby, 1957; Cloud, 1962; Enos, 1974b). Essentially all of the major carbonate sediment facies are represented on the bank, including extensive tidal flats (see Chapter 10). A generalized map of the distribution of these lithofacies is presented as Figure 14-12. Although reefs are not widespread, they are well developed along part of the west side of the Tongue of the Ocean (Figure 14-12), a deep reentrant into the bank. Skeletal sands are associated with the reefs and are widespread along the bank margin throughout the northern half. Ooids are associated with relatively high tidal currents which occur along the northwest margin

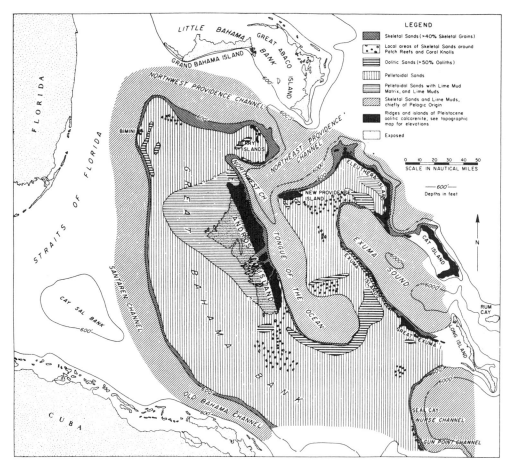

Figure 14-12 Surface sediment facies of Bahama Platform. (Map prepared by R. N. Ginsburg.)

of the bank. The interior of the bank is a relatively low wave energy environment and is dominated by lime mud or some modifications of it (Figure 14-12). Mud itself is in the central area and pelleted mud covers an extensive region west of Andros Island. Probably the most extensive lithofacies is pellet sand and mud. This includes grapestones, the aggregates of pellets, ooids, or other peloid particles. These sediment facies can be depicted by a general cross section across the Bahama Bank (Figure 14-13). Remember that biogenic debris occurs as an important constituent in all these facies, including the ooid shoals, where the nuclei are skeletal in origin.

Persian Gulf

The Persian Gulf is a narrow arm of the northwest part of the Indian Ocean and is separated from the Gulf of Oman by the Straits of Hormuz. It covers about 225,000 km ^{2} and possesses a mean depth of 35 m with a maximum of 100 m (Purser and Seibold, 1973). The northeast side of the elongate basin is bounded by high-relief

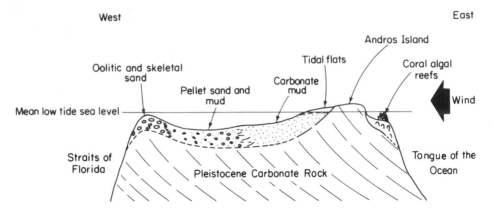

Figure 14-13 Generalized cross section of Bahama Banks from Florida Straits to Tongue of the Ocean. (From Blatt et al., 1980, p. 706.)

mountains and supplies terrigenous sediment to that half of the Gulf. The southwest side is generally a low-lying desert. It is the side of the Gulf, especially the southern portion, called the Trucial Coast, which will be discussed here.

The Trucial Coast and adjacent marine area has been termed an attached platform (Blatt et al., 1980), which contrasts with the Great Bahama Banks. This area is also somewhat shut off from influences of the Indian Ocean by the constriction at the mouth of the Persian Gulf. These phenomena, coupled with the arid climate, result in elevated salinity. Off the Trucial coast, values range typically from 40 to 50 ‰, with higher concentrations adjacent to coastal sabkhas (see Chapter 10). Wind-driven currents are most prominent in this region, but tidal currents are locally important in constricted areas (Purser and Seibold, 1973).

Distribution of sediments and sedimentary environments

Although 14 different sediment types have been delineated from this region, this number of textural types can be simplified into sand, muddy sand, and mud. The typical distribution of these grain sizes is from coarser near the coast to finer in the center of the Gulf (Figure 14-14). There is a wide variation in textures and composition near the coast, but regular and monotonous patterns persist offshore (Wagner and van der Togt, 1973).

Of the 14 sediment types, 12 are carbonate, one is gypsum, and one is quartz. Carbonate types include a variety of skeletal sands, pelletoidal sands, and ooid sands, as well as muds containing skeletal debris. Most of the variety in composition and distribution patterns takes place near and behind Great Pearl Bank, an extensive barrier between the peninsulas at Qatar and Oman (Figure 14-15). Localized current patterns, topography, and benthic communities influence these patterns.

Sedimentation and sediment types along the Trucial Coast can be divided into three parts: western, central, and northeastern (Purser and Evans, 1973). The western region has a diverse suite of lithofacies and a patchy distribution. Various skeletal sands and muddy sands dominate the shallow shelf, with some skeletal-bearing muds in deeper and quieter areas (Figure 14-15). Reefs are present locally

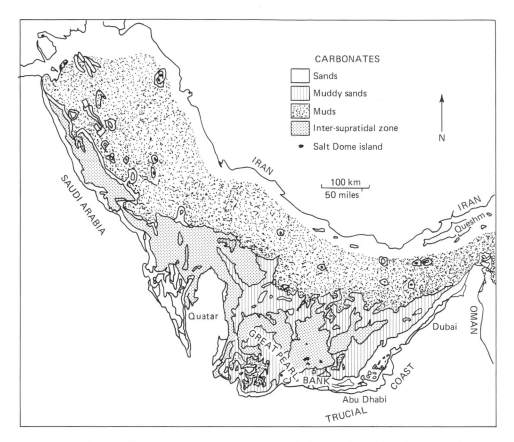

Figure 14-14 Carbonate sediment textures of the Persian Gulf. (From Wagner and van der Togt, 1973.)

and have coral-algal sands associated with them. Shoals and constructions where tidal currents persist are characterized by ooid sands. A few shallow embayed areas are dominated by pelleted sediments.

The central region of the trucial Coast is dominated by the barrier complex of the Great Pearl Bank. This region has been considered previously (see Chapter 11 and Figure 11–28). The northeast region presents a contrast to the other two areas in that sediment distribution is uniform and generally parallels the coast. Local patches of reefs, ooid sands, and pelleted mud are present adjacent to the coastal sabkha. The region is dominated by a belt of skeletal sand with a border belt of muddy skeletal sand on its gulfward side. This belt extends to or near the 40 m contour (Purser and Evans, 1973). Figure 14–16 is a nice general summary diagram based on the area off the Qatar Peninsula which shows the distribution of sediment textures and lithofacies with respect to water depth and distance from shore. Rounded skeletal sands give way to muddy sand and eventually skeletal-bearing lime muds across the shelf. This general distribution is typical of the shelf adjacent to the Trucial Coast and is complicated only by the variety of lithofacies present in the nearshore areas.

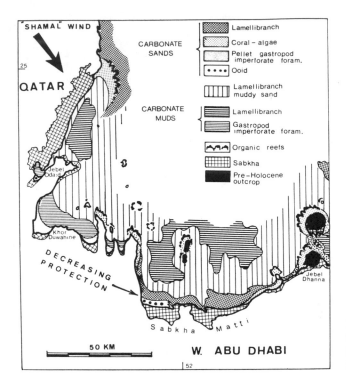

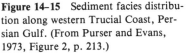

Figure 14-15 Sediment facies distribution along western Trucial Coast, Persian Gulf. (From Purser and Evans, 1973, Figure 2, p. 213.)

South Florida Shelf

This region is at the southern tip of Florida, separating the Atlantic Ocean and the Gulf of Mexico and lying between the Everglades and the Florida Straits. It is a shallow-rimmed shelf which covers several thousand square kilometers. An arcuate string of Pleistocene islands called the Florida Keys separates the open marine, high-

	Near-shore terrace	First offshore terrace	Second offshore terrace	Central Gulf level
Lithofacies	Rounded sand	Skeletal sand / Muddy sand	Very muddy sand / Muddy sand	Sandy mud
Wt. % finer than 53 μ (4.3 ϕ)	100 / 60 / 20	Gravel and sand grades	Silt and mud grades	
Depth m diagrammatic	20 / 60	18	31	53

Figure 14-16 Relationships between bathymetry and lithofacies off the Qatar Peninsula, Persian Gulf. (From Bathurst, 1975, Figure 190, p. 186.)

energy part of this shelf from Florida Bay, a shallow, protected environment which receives some freshwater influx from the adjacent Everglades (Figure 14–17).

The open marine area is dominated by wave-generated processes but is also affected by ocean currents and tidal currents. Salinity is normal and depths are less than 20 m on the forereef area, beyond which they drop to hundreds of meters in the Florida Straits. Most of the open marine shelf is less than 10 m deep and is comprised of the barrier reef system with a lagoon behind. Tidal range is only about 0.5 to 0.7 m. The Florida Keys are Pleistocene carbonates, some of which are reefs similar to the present-day reefs on the shelf.

Some swift tidal currents flow in channels between islands, forming tidal deltas, but in general tidal currents have little effect on sedimentation in front of or behind the keys. Florida Bay exhibits a general trend of increasing salinity, from low values of 10 ‰ to normal marine levels, with lowest concentrations in the northeast corner (Ginsburg, 1956). Runoff does not include a significant amount of terrigenous sediment. The bay is well protected and very shallow. Waves commonly stir up the muddy substrate, but little net transport occurs. The tidal range is less than 30 cm.

Distribution of sediments and sedimentary environments

Among the more comprehensive studies on sediments of the South Florida Shelf are those by Ginsburg (1956), Swinchatt (1965), and Enos and Perkins (1977). The natural subdivision into open and protected environments afforded by the Florida Keys results in two rather distinct depositional systems on this carbonate shelf: the reef tract and Florida Bay. The reef tract includes a forereef zone, reef, and back reef or lagoon. The fore-reef area is characterized by a steep slope with gravitational processes dominating. Coarse skeletal sand and gravel is deposited, with high initial dips typical of all forereef or reef flank deposits. These skeletal particles are

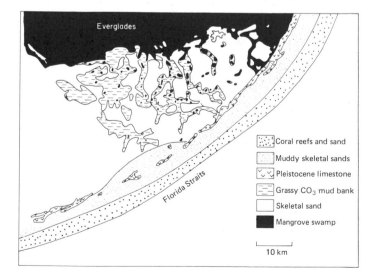

Figure 14–17 Map of Florida Bay and Keys, showing general bathymetry. (After Ginsburg, 1956, p. 2392.)

Chapter 14   Reefs and the Carbonate Platform System

dominated by coral and coralline algae. Foraminifera, *Halimeda*, and carbonate mud become more abundant as depth increases (Ginsburg, 1956).

The reef proper is a combination of framework carbonates and skeletal debris. The latter is dominated by coral with abundant coralline algae and *Halimeda*. This is the highest-energy environment of the shelf and intense wave activity removes fines, resulting in either sand or gravel, which shows better sorting and rounding on the open water side of the reef than on the crest (Swinchatt, 1965). The outer portion of the reef is characterized by well-developed spur and groove structure (Figure 14-7).

The back reef or lagoon area is relatively protected from wave attack and consequently carbonate mud is a common constituent of these skeletal sands, but *Halimeda* dominates. This area has scattered coral knolls on a substrate of rippled sand. Green algae, especially *Halimeda*, is widespread (Figure 14-18). The sediment is coarse sand to gravel, poor to moderately sorted and not well rounded (Swinchatt, 1965).

A summary of the reef tract system shows an increase in grain size, a decrease in sorting and rounding, and an increase in fines as the shore is approached. There is also a distinct trend in composition of the sediment composition, with mud essentially absent on the reef proper, where coral dominates. *Halimeda* increases markedly toward the shore (Figure 14-18).

The Florida Bay environment and its sediments display a marked contrast to the reef tract in nearly all respects. Water depth is less than 2 m and the only relief is due to a complex pattern of so-called mud banks, which form complicated patterns in the bay (Figure 14-17). Sediments display a general homogeneity which reflects conditions in the bay. There are scattered mangrove islands with associated supratidal algal

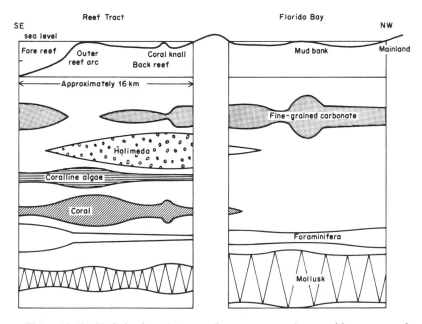

Figure 14-18 Variation in carbonate sediment texture and composition across reef tract and Florida Bay. (From Ginsburg, 1956, Figure 1, p. 2384.)

mat development (Ginsburg, 1964). The soft substrate is typically covered with a carpet of grass (*Thalassia*) and related calcareous green algae such as *Penicillus, Halimeda,* and *Acetabularia*, with *Penicillus* dominant. Benthic mollusks, both in-faunal and epifaunal, are abundant (Figure 14–18).

The bay sediments are essentially lime mud and sandy lime mud with local muddy sands (Figure 14–17). Nearly all sediment is bimodal, with the modes being lime mud and coarse skeletal sand. There is much bioturbation of the sediment by roots and by burrowers. Little or no evidence of sediment transport is present (Bathurst, 1975). The source of the lime mud has been attributed to the *Penicillus,* which produces needle-shaped aragonite grains that are 15 m in length. Sediment budget calculations indicate that all such lime mud can be attributed to this source (Stockman et al., 1967).

Yucatan Shelf

This open carbonate shelf is located to the north and northwest of the Yucatan Penin-sula of Mexico in the southern Gulf of Mexico [Figure 14–19(a)]. As one proceeds down the east side of this peninsula the shelf narrows and some terrigenous influx oc-curs, although reefs are well developed (Purdy et al., 1975). The Yucatan shelf is broad and typically flat, with a gently inclined slope. It covers about 35,000 km² and may be subdivided into an inner and outer shelf based largely on a change in slope near the 60 m contour [Figure 14–19(b)]. The outer shelf limit, which is an abrupt break in slope, ranges widely throughout the region; on the east side of the peninsula it ranges from about 80 to 200 m, whereas it is typically between 190 and 300 m in the north and northwest (Logan et al., 1969). There are three erosional terraces, which apparently developed during the Holocene rise in sea level. They are at 30 to 46 m, 56 to 73 m, and 105 to 158 m.

Major current patterns across the shelf are due to wind directions. The northern shelf is subjected to currents which sweep from east to west throughout the year. On the more westerly side there is a seasonal shift so that currents are southerly in sum-mer but reverse in winter. Tropical storms frequently pass over the Yucatan shelf and cause significant sediment transport, especially adjacent to reef areas (Hoskin, 1963).

Distribution of sediments

Unconsolidated sediments on the Yucatan Shelf range from only a few cen-timeters to a few meters in thickness except near the reefs. They represent sedimenta-tion during the Holocene transgression and overlie older sediments and rock, also of carbonate composition. Much of the sediment on this shelf, especially the outer por-tion, appears to be relict based on radiocarbon dates (Logan et al., 1969).

The inner shelf is dominated by skeletal sand, with molluscan debris the domi-nant constituent. This lithofacies is the most widespread and extends to the reefs near the break in slope within the shelf (Figure 14–20). Numerous shelf reefs (Figure 14–5) are located along this break in slope. These reefs display the typical morphology and biota of Caribbean reefs (Hoskin, 1963). They have been characterized as having four major zones: the reef flat, upper slope, lower scarp, and basal slope. Each is characterized by various communities and related sediment types. The reef flat is in

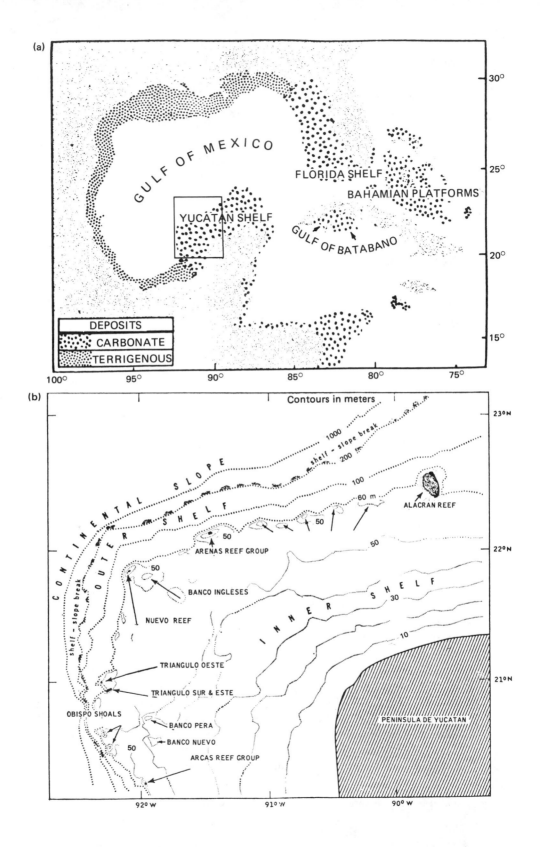

(a)

30°

GULF OF MEXICO

25°

FLORIDA SHELF

BAHAMIAN PLATFORMS

YUCATAN SHELF

GULF OF BATABANO

20°

DEPOSITS

CARBONATE

TERRIGENOUS

15°

100°   95°   90°   85°   80°   75°

(b)

Contours in meters

23°N

1000

shelf - slope break

200

100

60 m

ALACRAN REEF

50

CONTINENTAL SLOPE

OUTER SHELF

50

ARENAS REEF GROUP

22°N

shelf - slope break

50

BANCO INGLESES

50

INNER SHELF

30

NUEVO REEF

10

TRIANGULO OESTE

21°N

TRIANGULO SUR & ESTE

OBISPO SHOALS

BANCO PERA

PENINSULA DE YUCATAN

BANCO NUEVO

50

ARCAS REEF GROUP

92°W   91°W   90°W

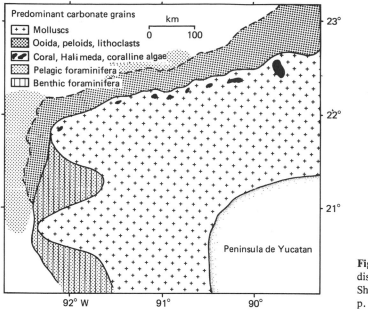

Predominant carbonate grains

km

0   100

+ + Molluscs

Ooida, peloids, lithoclasts

Coral, Halimeda, coralline algae

Pelagic foraminifera

Benthic foraminifera

23°

22°

21°

Peninsula de Yucatan

92° W   91°   90°

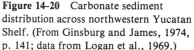

Figure 14-20 Carbonate sediment distribution across northwestern Yucatan Shelf. (From Ginsburg and James, 1974, p. 141; data from Logan et al., 1969.)

the surf zone and contains a diverse assemblage. The upper slope is dominated by *Acropora palmata*. The lower scarp is characterized by massive corals and the lower slope is dominated by coralline algae, which can thrive with little light.

The outer shelf displays a more diverse suite of lithofacies which are primarily relict in nature (Logan et al., 1969). The northern portion is dominated by sand-size sediment comprised of ooids, pellets, and clasts of lime mud (Figure 14-20). Original deposition was in shallow, agitated water near the coast during a glacial advance. The western area is a combination of pelagic and benthic foraminifera. Pelagic forms dominate beyond the shelf. All outer shelf sediments contain a fairly high volume of mud (Logan et al., 1969).

Great Barrier Reef

The shelf and reef complex that lies adjacent to the east coast of Queensland in northeastern Australia is probably the most famous reef complex in the world; it is certainly the largest. This system covers more than 200,000 km² and extends for a distance of 1500 km along the Australian continent. The Great Barrier Reef–Queensland Shelf system differs markedly from those previously discussed in that it contains a significant amount of terrigenous sediment influx. The adjacent mainland is one of moderate relief with a narrow coastal plain. Three significant rivers enter the shelf, all in the southern portion (Maxwell, 1968).

Salinity of the system is typical of open marine conditions (35 ‰). Ocean currents tend to flow in a north-to-south pattern over the shelf. Tidal ranges are high; the

Figure 14-19 Index map of the Gulf of Mexico, showing (a) Yucatan Shelf and (b) bathymetry of northwestern portion of shelf with major features. (From Logan et al., 1969.)

entire shelf is at least mesotidal, with the southern inner shelf in the macrotidal range (Maxwell, 1968). Most of the individual reefs are only a few kilometers in diameter and occur in clusters on the outer portion of the shelf. The entire system is somewhat like the coast of Belize in the southern Gulf of Mexico in that the reefs act as a barrier with a lagoon between them and the mainland. In the case of the Great Barrier Reef, the lagoon is most of the shelf and the real barrier is near the shelf-slope break. There is an abrupt drop into the Queensland Trench in the northern area, but in the south it is a gentle slope to the Coral Sea platform (Ginsburg and James, 1974).

Maxwell (1968) has subdivided this shelf system into the Northern, Central, and Southern Regions, based on shelf depths, width, and reef morphology (Figure 14-21). For example, depths in the Northern Region are less than 40 m, in the Central Region they range from 40 to 70 m, and in the Southern Region the depth is commonly in excess of 70 m (Maxwell and Swinchatt, 1970).

Four physiographic zones are recognized on the shelf, plus a shelf edge and slope zone. These shelf zones are: nearshore zone (to 10 m), inner shelf (10 to 40 m), marginal shelf (40 to 100 m), and the southern shelf embayment, which separates the

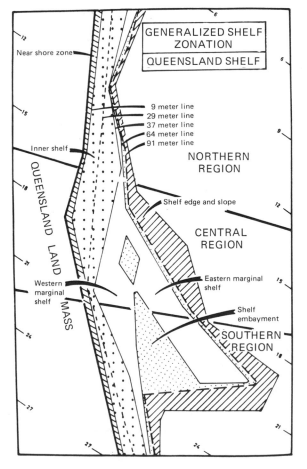

Figure 14-21 Diagram representing shelf system and showing zonation. (From Maxwell and Swinchatt, 1970, Figure 2, p. 698.)

southern marginal shelf into eastern and western portions (Figure 14–21). Terracelike features that occur at about 30, 40, and 60 m are interpreted as old shoreline features (Maxwell and Swinchatt, 1970).

Distribution of sediments and lithofacies

Sediment comprising the present Barrier Reef-Queensland Shelf system is from two sources; part is terrigenous, derived from the adjacent coast range, and part is carbonate, produced on and near the reef complexes. The terrigenous sediments dominate the landward half of the shelf and the carbonates the seaward half. Terrigenous sediments include gravel, sand, and mud. Because modern rivers appear to be supplying only mud to the shelf, the coarser size fractions are interpreted as being relict in nature (Maxwell, 1968).

Although there is a generally similar pattern to the overall sediment distribution on this shelf system, the Southern Region displays the greatest diversity and complexity of sediment types. It is the widest portion of the system and it includes a shelf embayment, the Capricorn Channel (Figure 14–22). Reef groups occur both landward and seaward of this shelf embayment.

Maxwell and Swinchatt (1970) describe seven lithofacies for the Southern Region.

1. Terrigenous sand is dominantly quartz and fairly well sorted; it contains less than 1% mud. It covers most of the nearshore and inner shelf zones to 40 m (Figures 14–21 and 14–22) except in the southernmost area, where reefs of the Capricorn Group produce carbonates which extend across part of the inner shelf.

2. Terrigenous muddy sand and mud dominate the shelf embayment and part of the nearshore zone at the south, where modern river influx is prominent (Figure 14–22).

3. Interreef carbonate sand covers more of the Southern Region than any other lithofacies. Texturally, it is fine to coarse sand with some mud. It is comprised of *Halimeda*, mollusks, coral, and bryozoa in varying amounts.

4. A nonreef carbonate sand is characterized by low percentages of coral and foraminifera. It grades into the terrigenous sand and contains some mud.

5. Carbonate muddy sand is similar to the previously mentioned facies but has more mud and abundant pelagic foraminifera.

6. Carbonate muds rim the interreef facies and probably were derived from the reef complex (Maxwell and Swinchatt, 1970).

7. The reef facies includes the sediments on the reefs themselves and around their margins. In addition to the unconsolidated sediments, the in situ reef textures and fabrics are found here. Sand and gravel particles are skeletal debris from coral, coralline algae, *Halimeda*, and foraminifera.

One of the most studied of the individual shelf reefs is Heron Reef (Maxwell et al., 1961, 1964), located in the Capricorn Group of the Southern Region (Figure

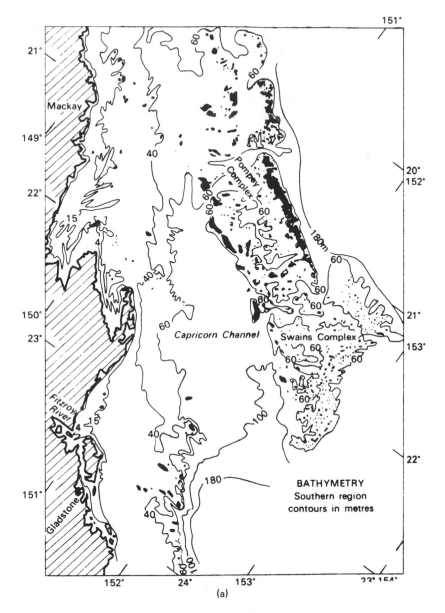

Figure 14-22 Bathymetry (a) and sediment distribution (b) in the Southern Area, Great Barrier Reef, Australia. (From Reading, 1978, p. 285; data from Maxwell, 1968, and Maxwell and Swinchatt, 1970.)

14-22). Classed as a lagoonal platform type (Figure 14-5), it is about 12 km long, has distinct windward and leeward sides, and there is a sand cay at the western end (Figure 14-23). The reef edge extends seaward from the algal ridge and falls steeply with interruptions of flat terraces. Much dense framework dominates the high-energy area, with a near absence of sediment particles due to wave action. The reef flat is

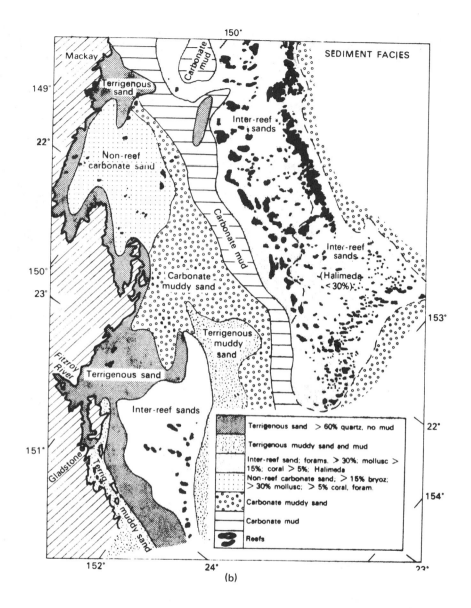

subdivided into the outer coral-algal, the living coral, and the inner sandy zones (Maxwell et al., 1961) (Figure 14–24). This reef flat is better developed and smoother on the windward reef than on the leaward side (Figure 14–23). The shallow lagoon is a broad zone of skeletal sands with few framework builders. This area is less than a meter deep at low tide. The blue lagoon averages a depth of 4 m and contains small luxuriant patch reefs on a skeletal sand substrate [Figure 14–24(c)].

subdivided into the outer coral-algal, the living coral, and the inner sandy zones (Maxwell et al., 1961) (Figure 14–24). This reef flat is better developed and smoother on the windward reef than on the leaward side (Figure 14–23). The shallow lagoon is a broad zone of skeletal sands with few framework builders. This area is less than a meter deep at low tide. The blue lagoon averages a depth of 4 m and contains small luxuriant patch reefs on a skeletal sand substrate [Figure 14–24(c)].

These zones of framework organisms and skeletal sands are fairly typical of shelf reefs throughout this system. In fact, they are comparable to most of the Indo-Pacific coral reefs.

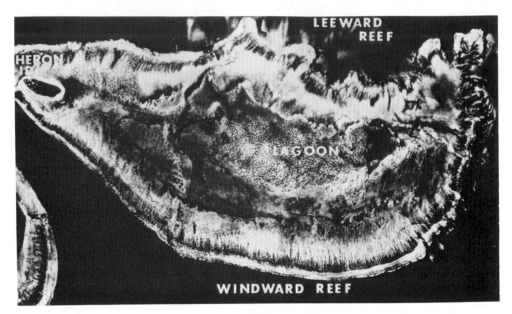

(a)

(b)

Figure 14–23 (a) Vertical and (b) oblique aerial photos of Heron Reef, Capricorn Group, Australia.

(a)

(b)

(c)

Figure 14–24 Series of photographs showing main zones from Heron Reef; (a) near vertical wall (11 m); (b) algal ridge; (c) shallow lagoon.

ANCIENT CARBONATE SHELF AND PLATFORM DEPOSITS

Ancient carbonate deposits are well known back to the Precambrian. Extensive and thick accumulations of ancient carbonates are typically related to abundant organisms and their skeletal remains, as are their modern counterparts. The term **carbonate buildup** has become one used extensively in association with thick carbonates. It carries no necessary connotation of reefal development, although reefs are a common element in such sequences. An excellent and comprehensive summary of these carbonate buildups is presented by Heckel (1974), who defines the term as a carbonate rock unit which (1) differs somewhat from adjacent and overlying strata, (2) is generally thicker than equivalent carbonates, and (3) probably was a topographic high.

Not all of the ancient carbonate shelf and platform deposits contain thick buildups, but most do. The typical ancient carbonate deposit contains facies which represent intertidal, shallow subtidal, deep subtidal, and organic buildups of reefs. The following discussion considers some examples of these systems but stresses the subtidal environment (see Chapter 10 for a discussion of intertidal carbonates).

Before examples are considered, it is pertinent to mention briefly the changes in dominant organisms associated with reefs and other carbonate buildups through geologic time (Heckel, 1974). Blue-green algae (stromatolites) have persisted throughout. Modern reef-type corals have been known only since the Mesozoic. The great reefs of the Late Paleozoic were dominated by a mixed assemblage of invertebrates together with coralline and blue-green algae.

Cretaceous of Texas and Northern Mexico

The Cretaceous has long been known as one of the periods during which carbonate platform and shelf deposits were widespread and well developed. They are widespread throughout the old Tethyian Sea and the Gulf of Mexico (Arthur and Schlanger, 1979). Extensive reefs were associated with the carbonate platforms, and these reefs were dominated by **rudistids** as framework builders. These large, attached bivalves grew into many forms and developed in somewhat the same niche as that of modern reef-forming corals. Although the rudists were dominant, other organisms, such as coralline algae, corals, and stromatoporoids, were also present.

In the area of central Texas and northernmost Mexico, the Edwards Formation of the Lower Cretaceous displays the various facies and buildups of a complex carbonate platform. This unit was deposited on an extensive platform that rose above the Tyler Basin and the ancestral Gulf of Mexico (Figure 14–25). The platform was characterized by carbonate skeletal sand (grainstone) and lime muds, with rudist reefs along the platform margins and the intraplatform lagoons (Fisher and Rodda, 1969).

Several lithofacies are represented in this carbonate platform accumulation. The reef complex is dominantly a rudistid framestone (biolithite) with coarse skeletal sand (grainstone) on the flanks (Figure 14–26). Thick accumulations of pelletoidal sand lies behind the reef complex. The remainder of the carbonate facies is largely skeletal with scattered pellets. An oncolitic facies is present at the northern end.

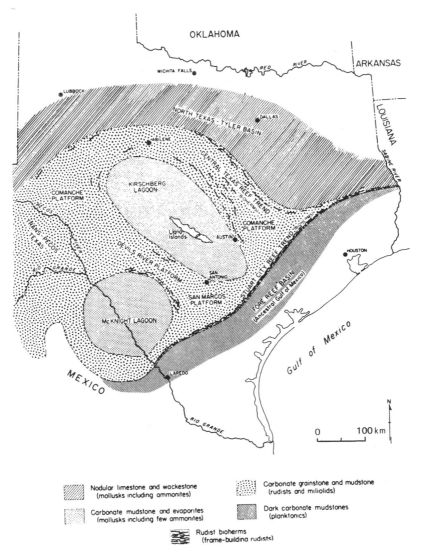

Figure 14-25 Paleogeographic reconstruction of Edwards reef system in central Texas. (From Fisher and Rodda, 1969, p. 56.)

Pelagic lime muds containing skeletal material (wackestone) cover the entire section. A shaly facies is also present across the center of the platform complex (Griffith et al., 1969). This representation of the Edwards–Stuart City system is comparable to that found throughout much of the Texas–Northern Mexico platform and platform-margin deposits (Lozo and Smith, 1964; Enos, 1974a).

The idealized Cretaceous bank and related facies may be characterized as shown in Figure 14-27. The offshore and deepest facies is dominated by pelagic micrite. This grades into a toe-of-slope facies with some slump structures and rhythmites. Both facies have ammonites and planktonic foraminifera. The forereef

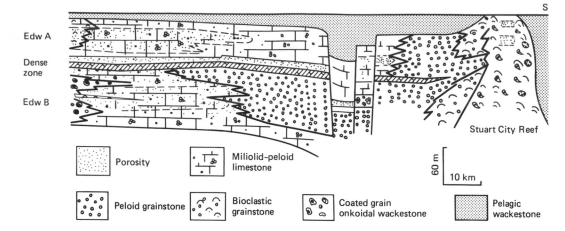

Edw A

Dense
zone

Edw B

S

Stuart City Reef

| | Porosity | | | Miliolid-peloid limestone | | |
| | Peloid grainstone | | | Bioclastic grainstone | | | Coated grain onkoidal wackestone | | | Pelagic wackestone |

60 m
10 km

Figure 14-26 Stratigraphic cross section across the Stuart City reef trend in Texas. (From Griffith et al., 1969, in Wilson, 1975, p. 337.)

facies has a steep initial dip and contains coarse skeletal debris plus lithiclasts. The framework facies is dominated by rudists and rises tens of meters above the deeper carbonate facies (Figure 14-27). Immediately behind the reef facies is an ooid and skeletal sand (grainstone) which intertongues with the reef. The inner bank facies is mostly micrite with some skeletal fragments and foraminifera with local burrowing. It is characteristic of a restricted environment. More-restricted areas are characterized by evaporites and essentially no fossils (Wilson, 1975).

Permian of Guadalupe Mountains (Texas)

The Permian Reef Complex of west Texas is one of the most famous carbonate platform buildups in the rock record. It has been studied and reported on extensively dating back to the initial recognition of reef strata by Lloyd (1929). Classic works on this area include those of King (1942, 1959) and Newall et al. (1953).

The strata under discussion are located in the Guadalupe Mountains astride the New Mexico-Texas line and along the margin of the Delaware Basin (Figure 14-28). During Permian time three distinct basins, Midland, Delaware and Marfa, occupied the west Texas-southeastern New Mexico region and were bordered by relatively shallow shelves and protruding platforms. One of the unique aspects of this Permian complex is the ability of one to view the major stratigraphic facies and the original morphology in the present outcrop (Figure 14-29).

Although the Permian sequences of this area have been examined for a half-century and there is general agreement regarding the stratigraphy of these strata, there is not agreement on the specific interpretations of the depositional environments in which these strata accumulated. Most of the disagreement centers around the nature of the organic buildup and its role in separating the deeper basin from the platform carbonates (Kendall, 1969; Silver and Todd, 1969; Motts, 1972; Dunham, 1972).

Extent	Envir. interpretation	Biota	Microfacies
Inner bank 100 km across		Essentially no biota except miliolids.	Anhydrite facies within Golden Lane bank.
	Salinity increase	Restricted biota, algal stromatolites, biostromes of oysters, forams algae.	Light-colored micrite, thin to medium beds. Minor cycles of miliolid grainstone; bioturbated wackestone to laminated micrite dolomite crusts.
	El Abra Limestone	Rudist debris. (Red alga) (Green alga) (Large forams).	Oolitic-biogenic grainstone.
Shelf margin knolls and patch reefs few km wide	Interreef sands / Inner knolls	Rudistids, forams.	Micrite rudist knolls with creamy, shelly, thick-bedded carbonates.
	Outer knolls	Rudistids, colonial corals, stromatoporoids, encrusting algae, boring clams, benthonic foram.	
Forereef clastics 2 km wide	Tamabra Lst	Mixed biota, mainly debris from upslope.	Coarse intraclastic-biogenic boulders imbedded in micrite.
Toe of slope Tamaulipas 2 km wide	2°–15° slope	Planktonic microfossils: ammonites, globigerinids.	Dark microbioclastic and pelagic micrite, thick to thin rhythmic bedding. Dark chert nodules. Slump structures.
Bathyal-Tamaulipas facies many km wide		Planktonic microfossils: ammonites, globigerinids.	Light to dark, pelagic micrite, fine sand in places.

— Overlapping facies —

Figure 14-27 Facies across Cretaceous buildup in northern Mexico. (From Wilson, 1975, Figure 11-3, p. 323.)

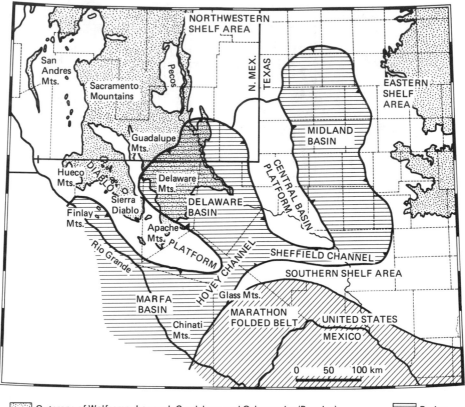

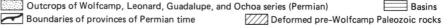 Outcrops of Wolfcamp, Leonard, Guadalupe, and Ochoa series (Permian) ▭ Basins
Boundaries of provinces of Permian time ▨ Deformed pre-Wolfcamp Paleozoic rocks

Figure 14-28 Paleogeographic map of southwestern United States during Permian time. (From King, 1942.)

Regardless of which author one chooses to follow, there is a distinct three-part appearance to these Permian strata. These are (1) relatively deep basin strata containing terrigenous and carbonate lithofacies; (2) organic buildups and (3) platform carbonates, evaporites, and terrigenous strata (Figure 14-30). The basinal facies include a black terrigenous mudstone interpreted as a starved basin and a very fine skeletal wackestone/packstone with planktonic organisms supplying the biogenic debris (Dunham, 1972). The toe-of-slope deposits are poor to moderately sorted skeletal sands with slumping, grading, and other slope-related structures.

The organic buildup includes a very coarse skeletal debris facies with dips up to 30°. The adjacent framework facies consists of sponges, brachiopods, encrusting algae, and crinoids. Numerous large cavities and collapse features are present (Dunham, 1969, 1972). A facies of generally sorted skeletal sands and ooids is present in and around the buildup facies. Behind the organic buildup are three facies which are interpreted as being deposited in a restricted platform environment. Included are a red, terrigenous mudstone; evaporites dominated by gypsum; and mixed carbonate

Figure 14–29 Photograph of El Capitan reef complex in the Permian of west Texas, McKittrick Canyon. (Courtesy of L. C. Pray.)

grain types (Figure 14–30). Included in the latter are stromatolites, ooids, intraclasts, pellets, and some skeletal debris.

The vertical distribution of these lithofacies indicates cyclic patterns to the accumulation of the Permian sequence (Silver and Todd, 1969). A four-stage sequence for each cycle has been proposed:

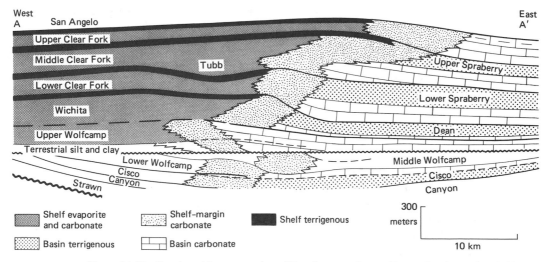

Figure 14–30 Stratigraphic cross section of Permian strata in west Texas, showing major shelf, buildup, and basinal facies. (From Silver and Todd, 1969, Figure 9, p. 2232.)

1. A barrier reef separating a lagoon and related sabkha from a deeper basin
2. Progradation of the reef and concomitant lowering of sea level decreasing the lagoonal area
3. Continued lowering of sea level with an extensive sabkha complex replacing the lagoon and a well-developed barrier island over the reef
4. Continued sea-level drop, resulting in a coastal desert with active dune field (Silver and Todd, 1969).

This depositional model closely parallels the present situation along the Trucial Coast of the Persian Gulf.

Mott (1972) draws a comparison between the Permian carbonate system and the Andros Island–Tongue of the Ocean system. He believes that the organic buildup is similar to those along and south of Andros Island. Tidal shoals are an important aspect of the Permian strata and relate to those described by Ball (1967) from the Bahamas.

Regardless of the details of the depositional model for this system, it does closely parallel some modern systems and also the model proposed by Wilson (1975) for the Cretaceous (Figure 14–27).

Devonian of Alberta, Canada

Carbonate buildups were widespread during the Late Devonian in Belgium, the Moscow Basin, the Canning Basin in northwestern Australia, western Canada, and in New York in the United States. Most of these show a similar biological assemblage. The reefs represent one of the early appearances in the geologic record of essentially continuous shelf-margin buildups (Wilson, 1975). An important facet of these Late Devonian reefs is the rise to dominance of colonial tetracorals, with tabulate corals becoming subsidiary (Heckel, 1974).

The Central Alberta Basin in Canada contains some of the most numerous, diverse, and well developed of all Devonian carbonate buildups. The discovery of oil in 1947 and the subsequent extensive research in this region make it among the best known in the world. The reefs or buildups occur in extensive tracts, each of which may have dozens of distinct individual reefs (Klovan, 1974). Indications are that the various reef tracts were developed on structurally controlled topographic highs (Wilson, 1975). Both linear and broad groupings are present.

The carbonate buildups are typically flat on the bottom with an overall lenticular shape. They intertongue laterally with dark basinal shales. Initial dip of the reef flank deposits is from 1 to 15°. Carbonate buildups may reach hundreds of meters in thickness. There are three distinct communities which produce the reefal accumulation in each of these buildups: (1) the basal coral community, developed in a quiet water environment; (2) a tabular coral community in moderately rough conditions; and (3) a massive coral community in high-energy conditions (Lecompte, 1958). These framework facies may be combined with adjacent talus and debris slope facies to produce a facies distribution diagram for a typical Devonian reef from the Alberta Basin (Figure 14–31). The depositional fabrics represented are those discussed earlier in the chapter (see Figure 14–3).

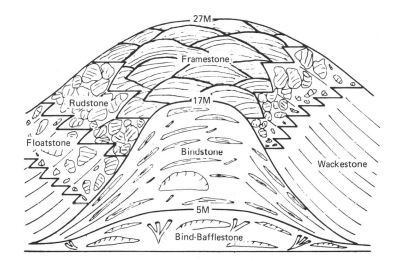

Figure 14-31 Facies in Devonian reefs of Alberta. See Figure 14-2 for rock types. (From Klovan, 1974, Figure 14, p. 796.)

Adjacent and related basinal and platform facies in this depositional system are quite similar to those previously described for the Permian and Cretaceous Periods. The deeper basins were accumulating dark terrigenous mud transported by oceanic currents, whereas shallower platforms were occupied by a variety of benthic organisms and accumulated their skeletal remains. These platform deposits typically served as the locus for reefal development. Skeletal sands and fine carbonate muds were common. Many of these reefs contain evaporite and intertidal deposits capping the buildups or behind them, indicating a shallow and restricted environment (Klovan, 1974). Cyclicity was apparently common, with sea-level fluctuations the major triggering mechanism. The base of each cycle is marked by an unconformity. Above this surface the carbonate sequence exhibits a shoaling-upward trend.

Devonian of New York

The carbonate platform and shelf sequences described above are dominated by reefal buildups. The Devonian carbonates of New York contain some of these buildups, but they are not a major part of this shallow marine system. These sediments accumulated to the west of the present-day Appalachian Mountains. Much of the research on these strata has been conducted by L. F. Laporte and his students. Their work will serve as the basis for discussion.

There are four major facies recognized in these carbonate sequences: tidal flat, shallow subtidal, deep subtidal, and carbonate buildups. The tidal flat facies has been considered in Chapter 10. Emphasis here is in the subtidal shelf system. The shallow subtidal facies represents the zone from low tide to normal wave base. In that respect it is analogous to the nearshore zone; however, it is many kilometers wide. This facies is characterized by a spectrum from skeletal sands (grainstones) to skeletal-bearing muds (wackestones). The former may show rounding, sorting, and cross-stratification, with ooids also in patches. The finer end of the spectrum typically displays bioturbation and contains pellets. The fauna in this facies shows great diversity (Laporte, 1971).

The deep subtidal facies contains terrigenous mud mixed with fine-grained carbonate. Skeletal debris is scattered throughout with much bioturbation. Fossil assemblages are diverse and well preserved; some individuals are articulated. This facies is interpreted as being below wave base in relatively quiet water but with enough current activity to supply food and oxygen to the benthic organisms.

The carbonate buildup facies is characterized by massive framework, carbonate accumulations of various sizes and shapes (Laporte, 1971). Some are in the form of stromatoporoid biostromes, some are stromatoporoid patch reefs, and others are buildups of carbonate sediment which is stabilized by solitary corals.

During Early Devonian time a generally transgressive situation (Laporte, 1969) with tidal flat facies was succeeded by shallow subtidal and deep subtidal facies [Figure 14–32(a)]. A few small carbonate buildups occurred as biostromes. Laporte (1971) interprets these as tidal channel deposits. The Middle Devonian was characterized by subtidal shelf deposits with some buildups at the transition between the shallow and deep subtidal facies [Figure 14–32(b)]. As the Acadian Orogeny influenced this area with terrigenous deposits from the east, some large and thick carbonate buildups developed on the deep subtidal platform [Figure 14–32(c)].

A generalized diagram of the reconstructed carbonate system (Figure 14–33) shows essentially the same three major elements that were discussed in the previous examples. Relief on the system is considerably less and the buildups are less pronounced, but the same relative features and positions with respect to the basin and the landmass are present.

Silurian of the Michigan Basin

Extensive carbonate shelf systems were present during Middle and Late Silurian time in the Gotland Islands region of Sweden, in Great Britain, the mid-continent of North America, and in the New Mexico–west Texas region. One of the best known and most diverse such systems is that associated with the Michigan Basin in the midwestern United States. Although this area includes nearly all of Michigan, some of the adjacent states of Wisconsin, Illinois, Indiana, Ohio, and New York, as well as part of Canada, contain strata deposited in this system (Figure 14–34). The Michigan Basin is the prominent element in this Silurian complex, but the Illinois Basin, Kankakee Arch, and Findlay Arch are also involved.

The general lithofacies distribution shows carbonate dominance throughout with some being limestone and some dolomite. Barrier reefs and pinnacle reefs abound around the periphery of the Michigan Basin, with a total of 160 known shelf buildups and 70 pinnacle reefs (Wilson, 1975). Some of these carbonates contain a variety of terrigenous particles, with a general increasing trend to the south in the Illinois Basin implying a source area in that direction (Lowenstam, 1957). The total stratigraphic section is less than 200 m thick. Reef buildups range widely in thickness from a few meters to nearly 100 m. Depth of water is probably about equal to buildup thickness (Wilson, 1975). Faunal diversity is extreme, with about 400 species reported

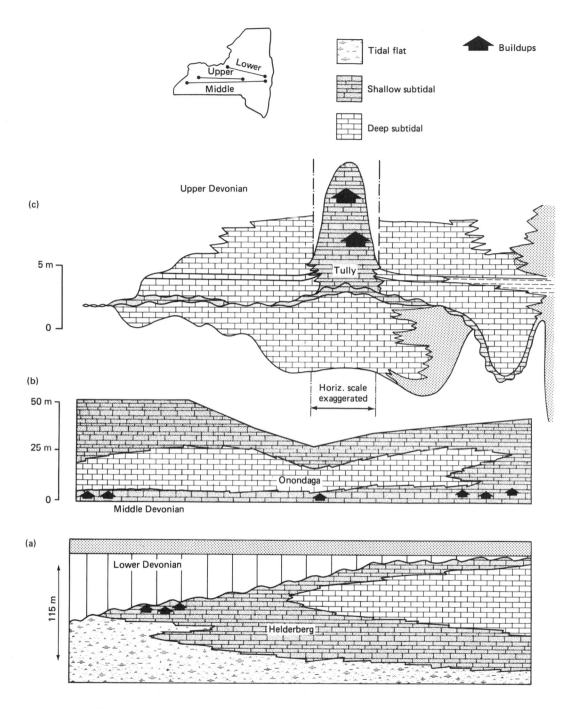

Figure 14-32 Stratigraphic cross sections across the Devonian of New York for each series, showing major depositional environment. (Modified from Laporte, 1971).

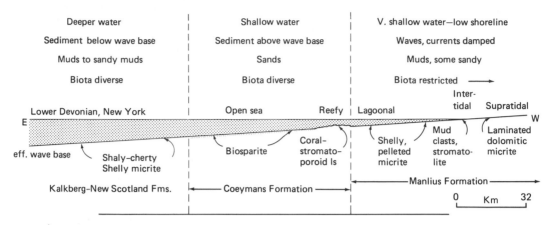

Figure 14-33 Sedimentation model for carbonate shelf during Devonian time in New York. (From Heckel, 1972, Figure 11, p. 258.)

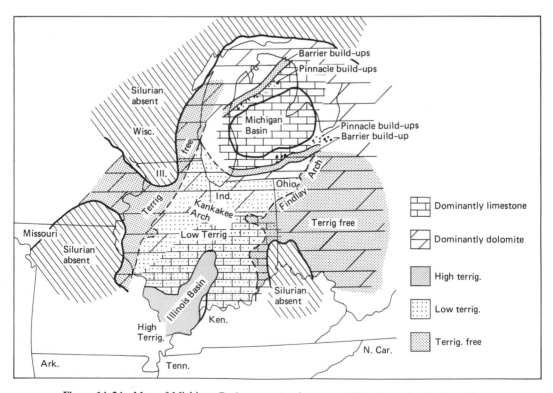

Figure 14-34 Map of Michigan Basin area, showing general lithofacies distribution. (From Sellwood, 1978, Figure 10-47A, p. 310.)

522

(Shaver, 1974). Primarily from outcrop data, Wilson (1975) describes five distinct reef types, based largely on variations in environmental parameters: (1) stromatolitic mounds, (2) mud mounds with bryozoa, (3) mud mounds with crinoids and tabulate corals, (4) wave-resistant framework reefs, and (5) pinnacle reefs. This group includes those defined by dominant taxa and those by environment. The framework or barrier reefs and the pinnacle reefs may be comprised of the same dominant organisms as are present in the other three types. A reconstruction of a typical profile section across this carbonate shelf shows the relative location of the types listed (Figure 14–35). Trends in biota composition as the carbonate shelf matures through a reef development sequence show that initially stromatactis dominates in the quiet water phase but then gives way to a diverse assemblage of reef formers as rougher conditions and more mature reefs persist.

Extensive three-dimensional data, largely from the subsurface of Michigan, show that there is a cyclicity to these Silurian carbonates and that three basic sequences occur (Mesolella et al., 1974): nonreef, barrier reef, and pinnacle reef. The nonreef sequence consists of cycles of alternating carbonates and evaporites (Figure 14–36). The carbonates include three facies: a biosparite (skeletal grainstone), a micritic crinoidal biosparite, and a micritic facies which displays a nodular habit. The evaporites consist of a relatively thin anhydrite overlain by halite.

The barrier reef sequence shows large stromatoporoid buildups which typically have debris on the flanks and intertongue with the nonreefal carbonates (Figure 14–36). These barriers form a large belt which nearly surrounds the basin (Figure 14–34). The smaller pinnacle reefs may have similar reef-building organisms as the barrier type, but they commonly are dominated by algal framework also. These pinnacles develop on the carbonate surface and show intertonguing with the evaporite facies (Figure 14–36).

Mesolella et al. (1974) developed three models for Silurian reef development. The preferred one invokes sea-level changes to bring about the cyclic nature of carbonates and reef deposition with the evaporite sequences. The fauna do show a general decrease in water depth or a shoaling-upward trend, terminating in algal stromatolite accumulations at the reef crests.

Summary

The examples presented above cover a fairly broad range of the geologic time scale and include carbonate buildups of diverse communities of reef builders. Nonetheless, there is a common theme to both the overall nature of the system and the sequence that accumulates in the buildups or reefs themselves. The three-part scheme mentioned earlier is applicable to each of the examples: a relatively deep fore-reef area; the buildup zone, which is typically narrow; and the back reef, which is commonly a shallow and restricted environment.

The development that begins the buildup sequence tends to stabilize the area. This is followed by a colonization by framework builders, succeeded by a great diversity of these organisms. The final stage may be a return to low diversity as one taxon dominates.

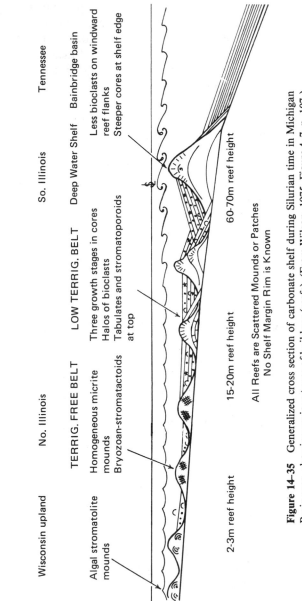

Wisconsin upland   No. Illinois   So. Illinois   Tennessee

TERRIG. FREE BELT   LOW TERRIG. BELT   Deep Water Shelf   Bainbridge basin

Algal stromatolite mounds

Homogeneous micrite mounds
Bryozoan-stromatactoids

Three growth stages in cores
Halos of bioclasts
Tabulates and stromatoporoids at top

Less bioclasts on windward reef flanks
Steeper cores at shelf edge

2-3m reef height

15-20m reef height

60-70m reef height

All Reefs are Scattered Mounds or Patches
No Shelf Margin Rim is Known

Figure 14-35   Generalized cross section of carbonate shelf during Silurian time in Michigan Basin area, showing various types of buildups (reefs). (From Wilson, 1975, Figure 4-7, p. 107.)

524

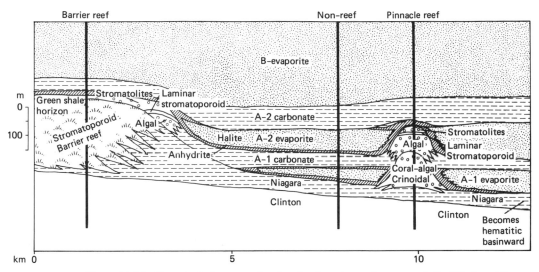

Figure 14-36 Hypothetical cross section showing three types of sequences in the Silurian of the Michigan Basin. (From Mesolella et al., 1974, Figure 6, p. 40.)

SUMMARY OF CHARACTERISTICS OF SEDIMENT/SEDIMENTARY ROCK BODIES REPRESENTING THE REEF AND CARBONATE PLATFORM SYSTEM

The shallow marine carbonate system is a diverse and widespread portion of the stratigraphic record. Because this is the only depositional system that generates its own sediment, it has a unique set of characteristics compared to marine systems dominated by terrigenous sediments.

Tectonic setting. Shallow carbonate systems require an absence of significant terrigenous influx. As a consequence, a stable tectonic setting with little relief facilitates their development.

Shape. A variety of shapes may be displayed by various components of the carbonate platform; however, the entire system is tabular in configuration. Various types of buildups present small and local or linear accumulations such as occur on reefs, ooid shoals, or sediment piles.

Size. The carbonate platform system may extend over many thousands of square kilometers and may range from hundreds to thousands of meters in thickness. Specific environments of deposition may produce sediment bodies which range from small pinnacle reefs covering hundreds of square meters with thick accumulations to more extensive but thin carbonate mud accumulations only a few to ten meters or so thick.

Textural trends. Broad textural trends of the type that are rather common in terrigenous systems are absent in carbonate systems. The major factor in this characteristic is the fact that carbonate sediment is made within the basin of deposition rather than being transported to it from a distant source. Consequently, in transgressive or progradational situations, conditions do not lend themselves to either fining- or coarsening-upward conditions. In many situations a rising sea level permits maintenance of existing depositional conditions for extended periods of time at a particular site, resulting in thick vertical accumulations.

Lithology. The mineralogy of carbonate platform systems is not complex; aragonite, calcite, and dolomite account for virtually all of the sediment that accumulates. Although most of the sediment is biogenic, carbonate mud and ooids may also account for significant percentages of the total carbonate material. Evaporite minerals and quartz sand from adjacent terrigenous environments comprise the non-carbonate portion of the strata.

Sedimentary structures. A wide variety of structures may be present. Physical structures such as cross-strata are common in oolite units or skeletal grainstones. Desiccation features, birds-eye structures, and fenestrate fabric are abundant in shallow wackestones. Bioturbation may occur throughout any of the environments but is uncommon in ooid grainstones. In reef boundstones, bioturbation commonly occurs as varieties of boring rather than burrowing.

Paleontology. The fossil content of carbonate platform strata is abundant and diverse. Reefs contain a complex but indicative fauna composed of both framework builders and other related taxa. Various other types of organisms may be useful in reconstructing depositional environments as well as in determining the origin of the strata.

Associations. Different facies within the carbonate platform system may display a variety of lateral and stratigraphic associations with one another. Almost any combination is possible. There are also associations between carbonate strata and adjacent terrigenous strata. Examples include carbonate reef and platform deposits, which accumulate on the outer continental shelf and grade laterally into terrigenous inner shelf and coastal deposits, such as is the situation in Australia adjacent to the Great Barrier Reef. Another possibility is the stratigraphic association of carbonate platform deposits and deeper seaward deposits, which may be a combination of fine terrigenous and pelagic muds.

ADDITIONAL READING

BATHURST, R. G. C., 1975. *Carbonate Sediments and Their Diagenesis,* 2nd ed. Elsevier, New York, 660 p. This is an excellent book on all aspects of carbonate sediments and environments. There is little on ancient examples.

FRIEDMAN, G. M. (ED.), 1969. *Depositional Environments in Carbonate Rocks.* Soc. Econ. Paleontologists and Mineralogists, Spec. Publ. No. 14, Tulsa, Okla., 209 p. This volume contains several excellent papers on ancient carbonate deposits with detailed interpretations of their depositional environments.

FROST, S. H., WEISS, M. P., AND SAUNDERS, J. B. (EDS.), 1977. *Reefs and Related Carbonates—Ecology and Sedimentology.* Amer. Assoc. Petroleum Geologists, Studies in Geol. No. 4, Tulsa, Okla., 421 p. An extensive collection of papers on reefs and related environments. Both modern and ancient examples are included. Emphasis is on the Caribbean and on reef biota.

LAPORTE, L. F. (ED.), 1974. *Reefs in Time and Space.* Soc. Econ. Paleontologists and Mineralogists, Spec. Publ. No. 18, Tulsa, Okla., 256 p. Several fine papers on modern and ancient reef deposits.

RÜTZLER, K., AND MACINTYRE, I. G., (EDS.) 1982, *The Atlantic Barrier Reef Ecosystem at Carrie Bow Cay, Belize,* I: *Structure and Communities,* Smithsonian Contrb. Mar. Sciences No. 12, Washington, D.C., 539 p. Although it is concerned with one small reef, this is the most comprehensive single volume on the reef environment.

WILSON, J. L. 1975. *Carbonate Facies in Geologic History.* Springer-Verlag, New York, 471 p. This is the best single book available on ancient carbonates. It is comprehensive, well written, and well illustrated. Many case histories are summarized.

15 The Continental Slope and Rise System

The outer portion of the continental margin is characterized by truly deep water sedimentation. The setting for this depositional system lies seaward of the shelf-slope break and extends to the abyssal plain. The processes and sediments on the slope and rise have become better known and understood in the past few decades as increased interest in offshore petroleum, plate tectonics, and overall history of ocean basins have fostered much research. These areas are generally beyond the influence of waves, tidal currents, and wind-driven currents on sediment transport, although these processes may occur locally. They are the sites of the greatest accumulations of sediments in the world ocean; most of these sediments are terrigenous.

CHARACTERISTICS AND FEATURES
OF THE OUTER CONTINENTAL MARGIN

The slope-rise portion of the continental margin represents a stratigraphically and structurally complex portion of the earth's crust. It is essentially the transition between continental influence and ocean basin influence, with most of the sediments being terrigenous in nature. Great thickness of sediment accumulates in this region, with the average about 2 km (Emery, 1970). The fact that this part of the crust is along a very dynamic boundary between oceanic and continental crust causes a variety of structurally complex internal configurations (Bouma, 1979). This characteristic also creates problems of both structural and petrologic nature when trying to recognize and interpret depositional environments from study of the rock record.

General Morphology and Structure

The profile of the continental margin is commonly depicted as containing a broad, gently sloping shelf adjacent to a coastal plain. Seaward of this is a relatively steep slope and then a less steep rise which terminates at the abyssal plain (Figure 15–1). There is great vertical exaggeration in all such profile diagrams; the mean inclination of the shelf is 0°07' and that of the slope is only a little more than 4° (Shepard, 1973). Whereas this shelf-rise type of margin is morphologically simple and is typical of trailing-edge margins, there are other profile schemes.

Curray (1969) recognized four other types of margin morphology, one of which is the bordering depression type of margin [Figure 15–2(b)] which is typical of the

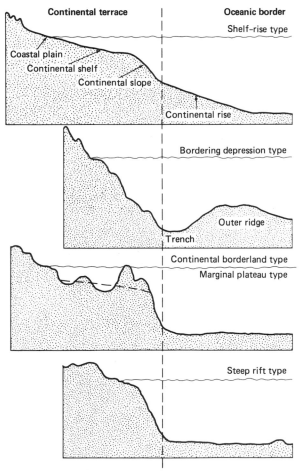

Figure 15–1 Generalized profile across the continental margin to the abyssal plain. Vertical exaggeration is about × 200. (After Curray, 1969.)

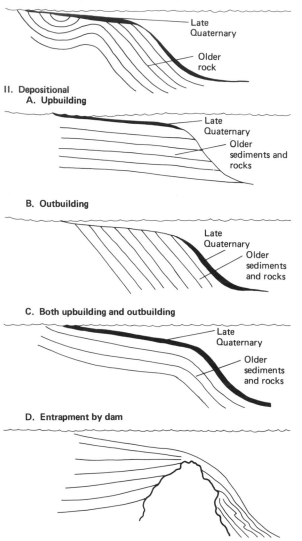

I. Tectonic or erosional

Late Quaternary

Older rock

II. Depositional
  A. Upbuilding

Late Quaternary

Older sediments and rocks

B. Outbuilding

Late Quaternary

Older sediments and rocks

C. Both upbuilding and outbuilding

Late Quaternary

Older sediments and rocks

D. Entrapment by dam

Figure 15-2 Generalized profiles of various types of continental margins. (From Reineck and Singh, 1980; after Curray, 1969.)

Pacific Ocean area, where leading-edge margins are present. Trenches generally are present adjacent to the continental terrace. The continental borderland type of margin is also characteristic of tectonically active areas but is without a trench. The area off the southern California coast is an example. Marginal plateaus are rather smooth platformlike features on the slope, such as the Blake Plateau. Steep rift margins are where no rise has developed as yet, such as off the west coast of the Iberian peninsula.

Prior to the development of seismic reflection profile equipment, it was only possible to speculate about the internal structure of the continental margin. As the result of extensive geophysical studies all over the world, it is now possible to demonstrate that there is great variety in the origin and composition of the margins. Using the same simple surface profile, Curray (1969) has shown five different styles of continental margin, including both erosional and depositional mechanisms. It has also been shown that the type formed by damming (II-D) may utilize a variety of features as the dam. It may be basement rock, coral reef, salt domes, or volcanoes (Shepard, 1973). Actual simplified seismic traces illustrate the nature of the margin stratigraphy (Figure 15-2). Fault blocks giving a horst and graben pattern superimposed on an anticlinorium develop basins of sediment accumulation in the California borderland [Figure 15-2(a)]. A large Cretaceous reef acted like a dam in developing the west Florida shelf [Figure 15-2(b)], and much progradation produced the gently sloping margins of Nova Scotia and northeastern Florida [Figure 15-2(c) and (d)].

Using the various styles of continental margins which have been recognized and placing them in a tectonic setting, Emery (1980) has formulated a classification based on stages of development. After an initial stage of formation (Figure 15-3) the margin proceeds through youthful, mature, and old age (destruction) stages (using the geomorphic terminology of William Morris Davis).

The initial stage shows tensional block faulting with considerable relief and little or no sediment throughout (Figure 15-3). During the youthful stage thick sediment fills the topographic lows on the shelf, but little or no rise is present (Figure

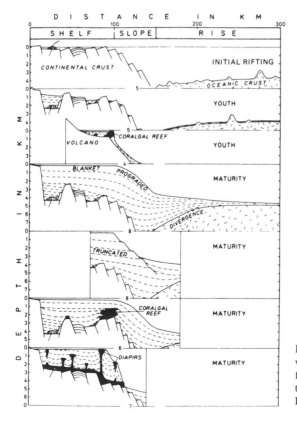

Figure 15-3 Generalized diagrams of various stages of development and configurations of continental margins. Vertical exaggeration is about 10×. (From Emery, 1980, p. 301.)

15-3). This is similar to type II-D of Curray (Figure 15-2). Such youthful margins are present on the west coasts of the United States, Europe, and Australia.

At maturity the margins contain thick sediment cover with a large and thick rise (Figure 15-3). Prograding, truncation, and intrusion of diapiric structures may characterize the margin (see Figures 15-2 and 15-3). These mature, trailing-edge margins are common adjacent to broad coastal plains such as the eastern United States, eastern South America, and most of western Africa (Emery, 1980).

As the margin continues to age, a trench develops as the result of convergence. Shearing, folding, and stacking of underthrust sheets eventually results in the destruction of the margin (Figure 15-3), such as along eastern South America and the Aleutian Islands of Alaska.

Relief Elements of the Outer Margin

Throughout its worldwide extent the continental margin exhibits a fairly broad range of relief; some is extreme and widespread, whereas some is local and only on a small scale. Although this chapter is concerned with the outer continental margin, that is, the slope and rise, it is necessary to extend beyond in both landward and seaward directions to include continuing and related features.

Shelf-slope break

A sedimentologically important relief feature is the shelf-slope break. A glance at the margin profiles illustrates the fairly abrupt change in gradient from the outer shelf to the upper slope. This narrow zone is a first-order topographic feature which serves as a boundary between two of the largest physiographic features on the earth's surface. It has been recognized as important because of its role in the transport of sediment from the shelf to the deeper ocean (Southard and Stanley, 1976). This break is at an average distance of 75 km from shore at an average depth of 132 m (Shepard, 1973). Remember that these figures are averages; there is a standard deviation (Table 15-1). Obviously, the origin of the margin at a particular location will have a great affect on the nature and location of the shelf-slope break.

Submarine valleys

Most of the detailed bathymetry of the world ocean has been completed since World War II. These data have shown that there are numerous types of submarine valleys on the ocean floor (Shepard and Dill, 1966), some of which are found on the outer continental margin.

TABLE 15-1 VALUES FOR SHELF–SLOPE BREAK, CONTINENTAL UNITED STATES

Region	Shelf width (km)	Depth at break (m)	Standard deviation on depth at break (m)
Western United States	20	142	± 50
Eastern United States	126	120	± 100
Gulf of Mexico	118	81	± 30

SOURCE: Data from Blankenship, 1978.

Without any doubt, **submarine canyons** are the most impressive of the submarine valleys. They are steep-walled, typically with a V-shaped cross section. They branch and are sinuous in their course. These features of up to thousands of meters of relief occur worldwide on the slope but extend both landward to the shelf and seaward to the rise. **Fan valleys** comprise the other common valley type located on the outer margin. The valleys typically begin at the base of the submarine canyon and cross the fans that form the rise. They may be V- or U-shaped with varying steepness. The relief rarely exceeds 180 m. The winding valley commonly displays natural levees and has many distributaries (Shepard and Dill, 1966). The submarine canyon–fan valley system forms the most important sediment dispersal path on the outer continental margin.

OUTER MARGIN ENVIRONMENTS

The outer margin may be characterized by two general types of processes, which in turn interact with sediment. One is a relatively slow process which persists through time, such as oceanic currents and accumulation of pelagic sediment. The other represents more intense and short-lived phenomena which may result in much sediment transport during each event. Included would be severe storms, earthquakes, or other phenomena that might generate turbidity currents. Often the latter type of phenomenon tends to overwhelm and obliterate effects of the first. The following discussion treats the various major elements of the outer margin from the process point of view. Some of the elements are affected largely by the slow, continual processes, whereas others are dominated by the more catastrophic events.

Shelf-Slope Break

There are data to support the viewpoint that the shelf-slope break is a region of relatively high turbulence and current energy in comparison to the areas immediately landward and seaward. A dearth of the fine fraction in the bottom sediment supports this generalization (Southard and Stanley, 1976). A variety of bottom currents persists in this zone, including those generated by surface waves, tides, internal waves, barometric pressure, and turbidity currents. The latter are typically related to submarine canyons and will be considered under that heading.

Surface waves do not act as a major transporter of sediment in the shelf-slope break area; however, data indicate that those with a long period do interact with the bottom. Oscillation ripples are formed at depths up to 200 m (Komar et al., 1972) and waves show refraction at the shelf edge (Ewing, 1973). Internal waves are known to be present on the shelf and slope, but they are ephemeral in nature and do not transport substantial sediment (Southard and Stanley, 1976).

Tidal currents show a marked increase at the shelf-slope break based on known relationships, but supportive data are lacking. If the tidal wave is parallel to the coast and the earth's rotation is neglected, it is possible to demonstrate that the maximum tidal velocity is proportional to the distance from shore divided by water depth (Southard and Stanley, 1976). Greatest effects would therefore be on shallow, gently sloping margins. Many data have been collected on tidal currents within the submarine canyons. These currents show distinct alternation of flow up and down the canyon axes at speeds of less than 50 cm/sec (Shepard, 1979).

One of the most important sedimentation phenomena is the transfer of sediment from the shelf edge down to the slope. This has been called **shelf-edge bypassing** by Stanley et al. (1972). After passing over the shelf edge, sediment is commonly transported downslope by gravity processes such as slides, slumps, turbidity currents, grain flow, or debris flow (see Chapter 2). Often this bypassing is caused by incisement of a submarine canyon into the shelf or by an offshelf-moving sand stream being terminated at the shelf-slope break (Southard and Stanley, 1976). This happens on both wide and narrow margins but is more common on narrow ones, such as off southern California. Bedforms with seaward orientations have been observed and indicate offshelf transport direction. In summary, this shelf-slope break zone experiences sediment transport via submarine canyons or on the open areas, with this sediment causing either progradation of the shelf or deposition at the base of the slope due to gravitational processes (McCave, 1972; Southard and Stanley, 1976).

Outer Continental Margin

This extensive area includes four related but somewhat different environments: the open slope, submarine canyons, deep-sea fans at the base of the slope, and the rise apron, which is the area between fans. Although the same processes operate throughout the outer margin, their intensity and degree of influence may vary among these environments.

Four major categories of processes operate on the outer margin (Kelling and Stanley, 1976): (1) atmospheric and tidal, (2) thermohaline, (3) gravitational, and (4) biological. The first two are closely related in that both involve clear-water motion which is capable of moving sediment. Included are such phenomena as wave motion, tidal currents, and particularly the strong oceanic currents. Most of the sediment being transported is clay size and is carried in suspension. The thermohaline type is essentially restricted to **geostrophic currents**, whereas atmospheric and tidal processes include a broad spectrum of phenomena (Kelling and Stanley, 1976). Actual accumulation of this fine sediment may take place in multiple fashions. The idea of pelagic sedimentation with particles settling vertically through the water column is much too simplistic and is not compatible with phenomena such as density stratification or the alternating textures found in layers of this sediment. **Contour currents**, which flow parallel to bathymetry, play an important role in outer margin sedimentation.

Gravity processes are probably the most important of the outer margin sedimentation phenomena in that they transport the greatest volume of sediment in this region. Although turbidity currents were long considered the prime agent of sediment transport, it is now known that, in fact, a spectrum of gravitational processes operate on the outer margin (Walker, 1973; Middleton and Hampton, 1973; Lowe, 1979). These range from dilute turbidity flow to sliding and slumping, which involves coherent movement of cohesive material. Included between are grain flows and debris flows (Figure 15–4).

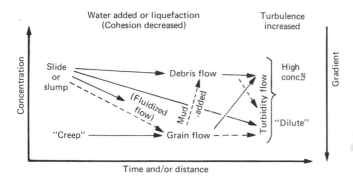

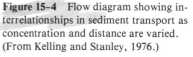

Figure 15-4 Flow diagram showing interrelationships in sediment transport as concentration and distance are varied. (From Kelling and Stanley, 1976.)

Although biological activity on the outer margin is low relative to the inner margin or even to abyssal areas, there are significant interactions that should be mentioned. Bioturbation is present, with holothurians, crustaceans, and some fish being the chief agents. There are fecal pellet producers, with worms being dominant. The basic contribution of this bioturbation activity is the disturbance of the existing sediments and alteration of their textures (Kelling and Stanley, 1976). Biologic erosion (see Chapter 2) is fairly common along the walls of submarine canyons, where boring organisms have been observed (Warme et al., 1971).

Open slope

The open portion of the continental slope, away from submarine canyons, is largely affected by gravity-generated phenomena such as slumping, sliding, debris flow, and some turbidity currents, coupled with relatively slow currents called contour currents. The transport of sediment by gravity processes is downslope in rather rapid pulses of material, whereas contour currents move at rates of generally less than 20 cm/sec but in a continuous fashion.

The gravity processes, especially turbidity currents and debris flows, excavate small gullies, and scars of large slumps create some relief (Kelling and Stanley, 1976). Seismic profiles of slope regions show discontinuities and distortions in stratification due to these processes.

The recognition of contour currents as an important process in the deep marine environment has answered many questions. Thermohaline circulation is caused by density gradients in the oceanic water masses. Typically, the currents generated by such gradients are slow, but numerous data demonstrate that they not only exist but that they may entrain and transport sediment. In actual fact the greatest influence of these contour currents is on the rise, but they are also present on the slope (Hollister and Elder, 1969; Heezen and Hollister, 1971). Numerous deep-water photographs show directional phenomena caused by currents. These include deflection of stalked benthic organisms, scouring adjacent to organisms or pebbles, lineations, and even small bedforms. Rather systematic and detailed data off the northeastern United States describe and define the Western Boundary Undercurrent, which flows counter to the Gulf Stream along the outer continental margin (Figure 15-5). Such currents

Chapter 15   The Continental Slope and Rise System

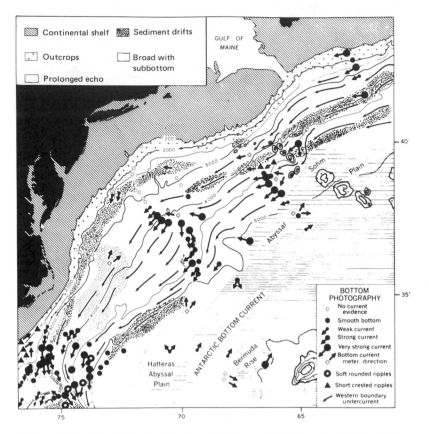

Figure 15-5 Map of Western Boundary Undercurrent off the eastern United States, showing flow patterns and general sediment distribution. (From Heezen and Hollister, 1971.)

are typical of continental margins on the west side of ocean basins. Notice the distinct parallelism between directional features and contours—thus the name for these currents and their sediments (contourites). The typical pattern displayed by such current systems is characterized by relatively strong currents below 3000 to 3500 m and weak currents or their total absence in shallower areas.

Submarine canyons

Submarine canyons experience mass movement in the upper canyon, high sediment concentration flows in the lower canyon, and turbidity currents in the fan area. Turbidity currents are probably not as prevalent now as they were during low stands of sea level when the shelf was narrow or absent. The intermittent and almost catastrophic nature of turbidity current development creates a rather diagnostic signature in the sedimentologic record.

One of the best types of documentation of larger-scale and rapid sediment movement in canyons is that associated with transoceanic cable breaks. The initial event of this type that was recognized as such was on the Grand Banks off Newfoundland in 1929 and was triggered by an earthquake (Heezen et al., 1954; Heezen and Drake, 1964). Subsequent similar phenomena have been recorded off the Congo and Magdalena Rivers, and New Guinea (Heezen and Hollister, 1971). The confining nature of the canyons, coupled with the steep slope, unstable substrate, and triggering mechanisms, results in dramatic sediment transport in this environment. Current ripples have been observed at depths of 4000 m (Keller and Shepard, 1978).

Base-of-slope environments

Great quantites of sediment accumulate at the base of the continental slope in the form of the continental rise. The rise is a system of numerous coalescing **submarine fans** or cones, each of which typically contains **fan valleys**, distributary channels, and **suprafans** (Figure 15-6). The fan valley typically displays levees and is a basal continuation of a submarine canyon. As the turbidity currents and other gravity-flow phenomena exit the canyon across the fan apex, there is a loss of con-

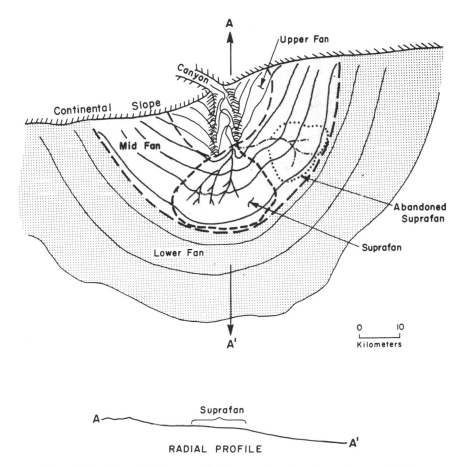

Figure 15–6 Schematic diagram of submarine fan, showing major morphologic elements including a radial profile. (From Normark, 1978.)

Chapter 15 The Continental Slope and Rise System

finement and distributaries are formed in much the same fashion as on alluvial fans (Chapter 6) or riverine deltas (Chapter 9).

The fan complex may be subdivided into three divisions based on dominant sedimentary processes: (1) upper or proximal fan, (2) middle fan, and (3) lower or distal fan (Normark, 1970, 1978). The upper fan is the area of the leveed fan valley (Figure 15-6). The middle fan contains distributary channels and has a convex-upward profile due to deposition. This bulge is called the suprafan and probably results from rapid deposition at the end of the leveed valley (Normark, 1970). The lower fan is usually not channelized and is characterized by sheetflow (Normark, 1974). The distal margin of the fan lacks definition because it merges with similar sediments of the abyssal plain.

Carbonate slope-basin

Carbonate platforms rise high above the ocean floor and present a variety of "outer margin" type of system. Under existing conditions the Bahama Platform rises above the Atlantic and provides opportunity to examine a platform-basin system. Such a system was apparently widespread in geologic time (Cook and Taylor, 1977; Cook, 1979).

The eastern portion of the Bahama Platform, including the Tongue of the Ocean, serves as an excellent example of the slope-to-basin carbonate system. A wide variety of carbonate sediment is produced on the platform and is shed to the basin by turbidity currents, debris flow, and grain flow (Schlager and Chermak, 1979). Sediment moves down the slope via many small gullies rather than in large, deeply incised canyons. Three distinct facies belts are present: (1) the gullied slope, mostly a bypass area with some mud and rubble; (2) the basin margin, with fan-shaped accumulations of proximal, relatively coarse turbidites; and (3) the basin interior, with distal turbidites and ooze (Schlager and Chermak, 1979).

SUBMARINE CANYON–FAN SYSTEM

The discussion above considered the general characteristics and dominant processes of the canyon and fan environments separately. To understand the sedimentology and stratigraphy of the accumulated materials in these environments, it is necessary to consider the system as a whole.

General Morphology

Whereas there is great morphologic variability and range in size of the canyon-fan systems of the world, there are also some generalizations which persist (Shepard and Dill, 1966). Numerous modern systems have been studied but hundreds remain unknown in terms of their detail. The length of the canyon ranges widely but is generally related to the type of margin present. Those that cut the California borderland are less than 20 km, whereas on broad trailing-edge margins canyons may be a few hundred kilometers long. Fan size is typically directly related to canyon length but there are exceptions (Nelson et al., 1970). There is also a general relationship between the canyon's gradient and length; short canyons are steep, and vice versa. The same relationship tends to also be applicable to fans. No widely applicable generalizations can be made about the depths of canyons or fans because they begin and terminate at varying locations on the outer margin.

The Astoria Canyon-Fan system adjacent to the coast of Oregon provides a rather typical example of this depositional system. Its dimensions and general characteristics fall near the middle of the spectrum for such systems, but it does contain a large number of distributary channels (Nelson et al., 1970; Nelson and Kulm, 1973). The canyon heads on the shelf and cuts across the slope in a fairly straight path. It terminates at the slope base, where the Astoria Channel fan valley begins (Figure 15–7). The upper canyon on the shelf has a U-shaped transverse profile which becomes V-shaped on the slope. This variation may be the result of the rate of infilling, with the U-shaped portion reflecting more rapid sediment accumulation. The longitudinal profile shows a general decrease in gradient downslope, yielding a slightly concave-upward profile. There is no abrupt change in gradient from the canyon to Astoria Channel on the fan (Nelson et al., 1970). The primary criteria for

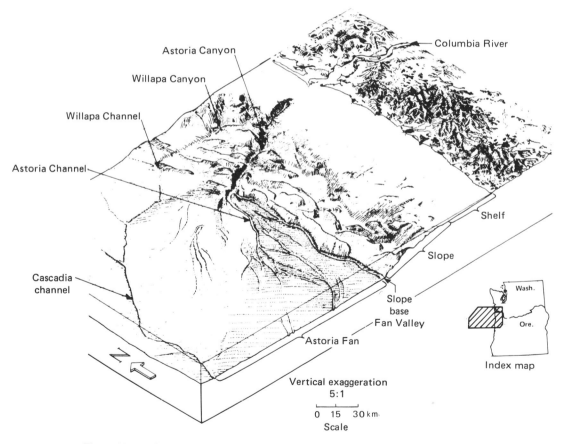

Figure 15–7 Block diagram of Astoria Canyon and adjacent fan system of the west coast of Washington at the Columbia River. (From Nelson et al., 1970, p. 260.)

Chapter 15   The Continental Slope and Rise System   **539** determining the boundary between the canyon and the fan valley are (1) an abrupt change in relief and (2) the presence of levees on the fan valley.

The Astoria Fan itself exhibits three physiographic divisions. These are well displayed in Figure 15-8, which includes three radiating longitudinal profiles across the fan. The upper fan is steep and generally concave upward overall, but it contains small relief elements, especially the natural levees. The middle fan shows the characteristic convex bulge and the lower fan is gently sloping and smooth. Several older and inactive distributary channels are present, yielding a spectrum of gradients and profile shapes (Nelson et al., 1970).

Sediments

The canyon-fan system is a complex one, with several different environments present. The process-response conditions in each of these environments are such that sediments are accumulated with rather unique characteristics. It is generally possible to relate a particular type of sequence to a specific environment.

Recent papers summarizing sediments on submarine fans have described five types of sediment accumulation in this system (Walker and Mutti, 1973; Walker, 1978, 1979): (1) turbidites (see Chapter 2), (2) massive sandstones, (3) pebbly sandstones, (4) clast-supported conglomerates, and (5) matrix-supported beds, including debris flow deposits, pebbly mudstones, and slump deposits. In all but the last type, fluid turbulence is important in transport.

The **turbidite sequence,** or **Bouma sequence** (Bouma, 1962) as it is commonly called, represents the standard or typical sediment accumulation from turbidity current deposition. It is characterized by fine components which occur in a predictable sequence (Figure 15-9) and each sequence represents deposition by a single turbidity current. There is great variation in the total thickness of the Bouma sequence and it

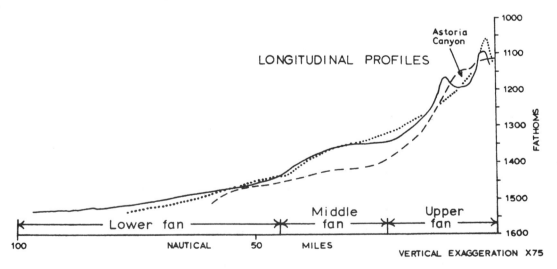

Figure 15-8 Longitudinal profiles across the Astoria fan. Note the typical midfan bulge. (From Nelson et al., 1970.)

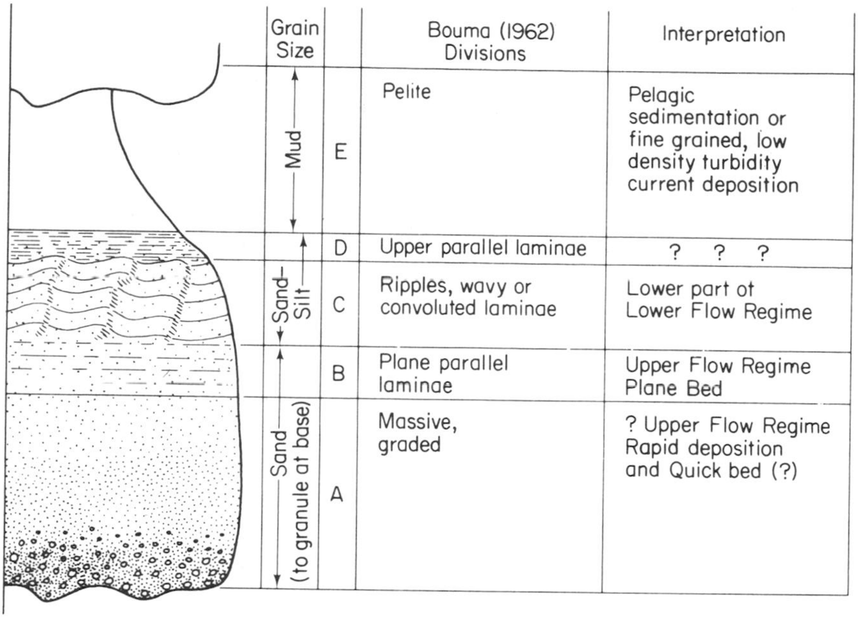

Grain Size		Bouma (1962) Divisions	Interpretation
Mud	E	Pelite	Pelagic sedimentation or fine grained, low density turbidity current deposition
Sand–Silt	D	Upper parallel laminae	? ? ?
Sand–Silt	C	Ripples, wavy or convoluted laminae	Lower part of Lower Flow Regime
Sand–Silt	B	Plane parallel laminae	Upper Flow Regime Plane Bed
Sand (to granule at base)	A	Massive, graded	? Upper Flow Regime Rapid deposition and Quick bed (?)

Figure 15-9 Bouma turbidite sequence, showing trends of sedimentary structures, grain size, and depositional conditions. (After Bouma, 1962).

may occur with one or more of the five elements absent. This can result from erosion of the upper portion by subsequent turbidity currents or from lack of deposition due to available sediment or conditions within the turbidity current.

The Bouma sequence contains a basal element (A) which is massive and graded (Figure 15-9). The coarsest grains within a given sequence are in this element and its base may display various bottom marks. It is typically the thickest unit in the turbidite sequence. The basal unit is interpreted as representing rapid deposition in the upper flow regime.

The overlying element (B) is relatively thin, finer grained than A, and is comprised of thin, parallel laminae (Figure 15-9). Deposition is in the plane bed mode of the upper flow regime. Element C displays ripple cross-stratification representing the lower flow regime. It is composed of fine sand or mud and is generally thicker than the underlying and overlying portions of the sequence (B and D). The thin, laminated muds of element D have an undocumented origin and cannot be distinguished from the uppermost portion of the sequence in most cases (Walker, 1979). The top portion of the Bouma sequence (E) is fine mud which results from the waning stages of the turbidity current and from pelagic accumulation (Figure 15-9). This unit is commonly referred to as interturbidite.

Some authors have termed the Bouma sequence as the "classic" turbidite (Walker, 1978), and they tend to typify the middle to distal portions of the fan. These are equivalent to facies C of Mutti and Ricci Lucchi (1975). Beyond these areas are finer-grained and thinner-bedded units called thin-bedded turbidites, the D and E facies of Mutti and Ricci Lucchi (1975).

Massive sandstones are gradational from these turbidites. They display little mud, coarse grain size, are massive to irregularly bedded (facies A of Mutti and Ricci Lucchi, 1975), and commonly display dish structure (see Figure 3–34). All indications are that these units were deposited rapidly (Lowe, 1976).

Pebbly sandstones are generally devoid of mud, display grading, contain cross-strata, and have bottom marks at their base. Imbrication of pebbles may be present. The units are commonly lenticular in shape. Deposition was probably similar to that for massive sandstones, with some traction causing stratification and imbrication.

The increase in clasts provides a gradation from the pebbly sandstone facies into clast-supported conglomerates. These conglomerates display a variety of features which is important to their relative position in the depositional system. The upcurrent position may display totally disorganized beds (Figure 15–10). These grade downcurrent into a succession of inverse to normally graded beds, graded beds, and graded-stratified beds. The latter three types exhibit imbrication and show increasing organization and decreasing grain size in the downcurrent direction.

The matrix-supported units owe their origin primarily to debris flow. Their internal makeup is chaotic, with great range in composition and texture (Hampton, 1972). Generally, sedimentary structures are not discernible.

Canyons

Whereas studies of canyon morphology and origin have been numerous and sediment transport processes in the canyon have also received much attention, there is much less data on sediments that accumulate in the canyons. Most information on

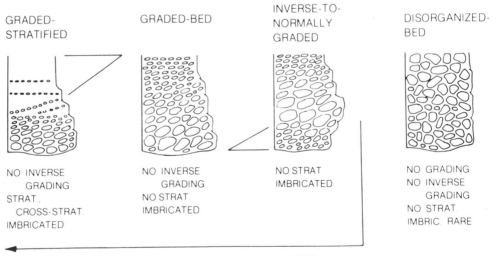

THESE THREE MODELS SHOWN IN SUGGESTED
RELATIVE POSITIONS DOWNCURRENT

Figure 15-10 Four varieties of resedimented conglomerates. The arrangement relative to downcurrent position is based on theoretical grounds only. (From Walker, 1978, p. 942.)

canyon accumulations is from underwater photography or short box cores (Shepard et al., 1969). Although data are somewhat limited, there is a general pattern to canyon sediments.

The grain size of canyon sediments ranges widely depending on the source. Some canyons contain coarse sand and gravel (Bouma, 1965), whereas others contain fine sand or coarse silt as their coarsest particles (Scott and Birdsall, 1978). The combination of sediment texture and structures permits three canyon facies to be differentiated: (1) canyon axis, which contains the coarsest sediment and is dominated by physical structures; (2) the canyon floor, where bioturbation is widespread; and (3) the canyon wall, which is characterized by laminated muddy sand.

Sediments in the canyon axis display ripple cross-stratification and truncation of strata. Thin graded units, < 10 cm in thickness, suggest gravity flow or a decreasing current velocity (Scott and Birdsall, 1978). Away from the axis on the canyon floor there is much bioturbation, but some remnants of physical structures may be preserved. The style of the biogenic structures indicates that they were formed by a spectrum of organisms, including polychaetes and sea urchins. Fecal pellets are common in canyon floor sediments. Apparently, bioturbation is able to keep up with the rate of accumulation. Cohesive muddy sands which display relatively uniform stratification persist on the canyon walls. Slumping and convolute bedding is common, due to the steepness of the depositional surface (Shepard et al., 1969; Scott and Birdsall, 1978).

Upper fan

Much diversity and complexity typifies the sediments that accumulate in the proximal portion of a submarine fan. Within this area is the fan valley, which includes a meandering channel, terraces, and levees (Figure 15-11). The channel itself and adjacent terraces receive coarse conglomeratic deposits of the types described above. Theoretically, one would expect that these coarse sediment types would display the down-gradient distribution shown in Figure 15-10. Actual relationships between channel facies and terrace facies are not known because of limited core data from this environment (Walker, 1978). Upcurrent feeder channels which are the lower part of the submarine canyon are likely to contain matrix-supported sediments and disorganized beds (Figure 15-11).

Thin-bedded turbidites dominate the natural levees and interchannel areas of the upper fan (Nelson et al., 1978). These thin-bedded turbidites represent the upper portion of a Bouma sequence, with ripple cross-strata and overlying mud being the sediment accumulated. There have been data which show that interchannel sequences increase in thickness down the fan gradient from about 2 cm in the upper fan to 10 cm near the lower fan (Nelson and Kulm, 1973). As the thickening occurs there is a parallel increase in the completeness of the Bouma sequence.

Mid-fan

The dominant features of the midfan region are the distributaries, with their associated suprafan lobes. Massive and pebbly sandstones are dominant in the distributary channels (Figure 15-11). In general, the pebbly sandstone would be ex-

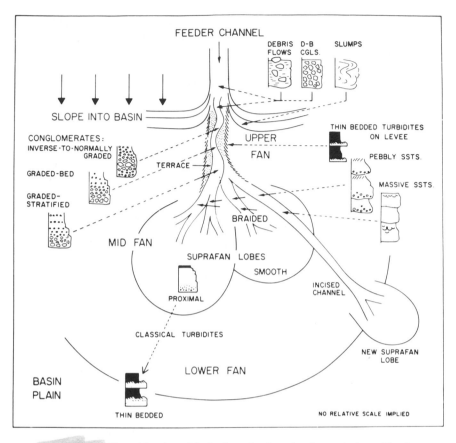

Figure 15–11 Depositional model of submarine fan, showing typical stratification sequences for each of the major elements. (From Walker, 1976.)

pected to accumulate upstream from the massive sandstone. The overall nature of the suprafan region is much like the braided stream system, with lenticular sand bodies formed as channels shift and fine sediments are scoured away (Walker, 1978). Some classical turbidites with complete Bouma sequences would be expected on the smooth outer part of the midfan beyond the well-developed channels. Note that as the fan progrades a new suprafan may be established on the lower fan, thus shifting these relatively coarse facies to that location.

Lower fan

In the typical fan, the lower fan is generally lacking channels and therefore relief. It is dominated by thin-bedded turbidite units in the distal end, grading toward more "classic" Bouma sequences upslope (Figure 15–11). Bedding tends to be regular and fairly continuous. Individual units are generally less than 10 cm thick (Nelson et al., 1978). There are some fans which are connected to the abyssal plain and have channels extending across the outer fan.

STRATIGRAPHIC SEQUENCES

Outer margin sediments may accumulate to great thickness and may be incorporated in the rock record. Because the thick submarine fan and related sediments are located at a potentially dynamic part of the earth's crust, it is possible that much structural deformation may occur as these sediments are incorporated into the rock record. It is therefore extremely important that details of the sedimentologic and stratigraphic characteristics of these sequences be understood in order to be able to interpret ancient sequences properly. Among the most important features to be considered are paleocurrent data, sediment texture and fabric, and stratigraphic sequence.

The outer margin accumulates a number of sequences, but the most widespread and volumetrically significant is one accumulated in the submarine fan and related environments. The simplified geometry is that of a thick clastic wedge which thins and fines in a seaward direction. The coalescing of adjacent fans creates a nearly continuous belt of these sediments, with some geographic variation due to valley locations (see Figure 15–12).

The dominant theme during development of a submarine fan system is progradation as more and more sediment is supplied to the base of the slope via submarine canyons. Using the general depositional model of Walker (Figure 15–11), it is a relatively straightforward sequence with the lower fan being successively covered by midfan and upper fan deposits. Such a sequence would show an overall coarsening-upward and thickening-upward trend (Figure 15–13). Within this sequence there would be both coarsening-upward and fining-upward units, depending on the specific conditions represented.

The basal part of the sequence consists of thin, classic turbidite units, with those at the bottom being of the thin-bedded type and representing the most distal part of the fan. Progressing up the section the Bouma sequences become thicker, with coarser grains and less mud. Each Bouma turbidite sequence represents the smooth outer portion of a suprafan lobe (Figure 15–13), and some are capped by a massive or pebbly sandstone (Walker, 1978). As the higher portion of the midfan is deposited there is more channel development, and consequently massive and pebbly sandstones which themselves show normal grading (Figure 15–13). These channel-fill deposits have similarities with analogous fluvial and deltaic deposits. The upper fan deposits are dominated by various types of channel fill in the form of resedimented conglomerates (Figure 15–10). Within this sequence there would therefore be both coarsening-upward and fining-upward units, depending on the specific conditions represented. The fining-upward scheme would occur due to channel abandonment such as avulsion, which takes place in a similar fashion as it does in the fluvial system. Commonly, the complete sequence is capped by matrix-supported slump and debris flow (Figure 15–13), representing the feeder channel at the base of the submarine canyon.

The entire sequence could be tens or hundreds of meters thick. The model sequence developed by Walker (1978) is a generalized and simplified situation. It is similar to a model developed by Mutti and Ricci Lucchi (1975) based on studies in the

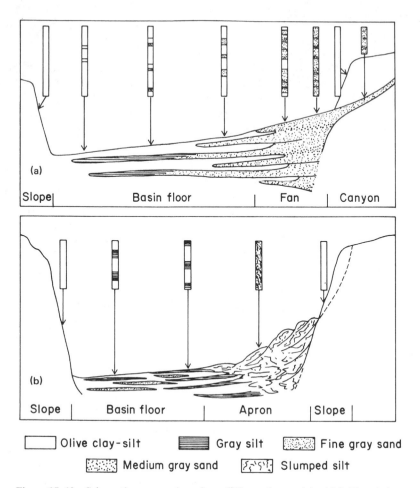

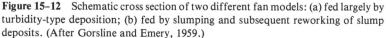

| Olive clay-silt | Gray silt | Fine gray sand |
| Medium gray sand | Slumped silt | |

Figure 15-12 Schematic cross section of two different fan models: (a) fed largely by turbidity-type deposition; (b) fed by slumping and subsequent reworking of slump deposits. (After Gorsline and Emery, 1959.)

Italian Apennines. In reality the abandonment of channels and canyons, large-scale slumping, or mass wasting and other phenomena can cause pronounced changes in the sequence. The simplified model does serve as an excellent tool to aid in interpreting the rock record.

ANCIENT OUTER MARGIN SEQUENCES

The recognition of the outer margin depositional system as incorporated in the rock record and the reconstruction of the environments represented has become a rather comprehensive and sophisticated exercise in geology. The first demonstrable connec-

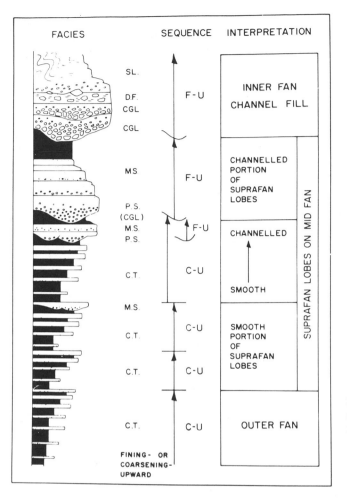

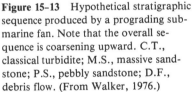

Figure 15–13 Hypothetical stratigraphic sequence produced by a prograding submarine fan. Note that the overall sequence is coarsening upward. C.T., classical turbidite; M.S., massive sandstone; P.S., pebbly sandstone; D.F., debris flow. (From Walker, 1976.)

tion between graded beds and turbidity currents was made by Kuenen and Migliorini (1950). The definition of the turbidite sequence of primary structures by Bouma (1962) followed, as did other significant developments in turbidite sedimentology (Walker, 1973). The latter include an atlas of turbidite structures (Dzulynski and Walton, 1965) and the use of turbidite sequences in basin analysis (Potter and Pettijohn, 1977). During the first two decades, the emphasis in turbidite research seemed to be on paleocurrents, whereas during the decade of the 1970s, emphasis was on the construction of turbidite depositional models and reconstruction of details within the system (Mutti and Ricci Lucchi, 1975; Walker, 1978).

The preceding section emphasized the stratigraphic sequences and overall facies relationships in the outer margin. Actually, recognition of the various sedimentary facies must include numerous criteria, such as (1) grain size, (2) bed thickness and sand/shale ratios, (3) nature of the stratification, (4) bottom marks, (5) paleoecological indicators, and (6) internal structures and textures (Walker and Mutti, 1973). The latter include grading, conglomerate fabric, massive sandstones, and variations in the Bouma sequences preserved. Sediment body geometry is also important.

Ancient canyon-fill deposits are best recognized by their comparison with modern characteristics, such as size, geometry, location with respect to known shallow- and deep-water deposits, nature of infilling sediments, and a fauna of mixed ages and water depths (Whitaker, 1974).

Paleocurrent patterns of the outer margin may be complex. Canyons display downslope directions of sediment dispersal, fans show a somewhat radial pattern in their channel deposits, but overbank sediments display a range of directions. The entire basin may show paleocurrent directions which are aligned essentially parallel to the outer edge of the slope or axial to the turbidite basin. Early authors (McBride, 1962; Enos, 1969a) explained this apparently anomalous paleocurrent pattern by the loss of basin margin deposits (fans) in the process of deformation during mountain building (Walker, 1979). Present knowledge of contour currents provides a better explanation for this axial paleocurrent pattern.

Generally, ancient fan deposits display a decrease in grain size and sand/shale ratio toward the center of the basin and an overall coarsening-upward trend in the sequence. Considering the three prominent subdivisions of the fan, the upper fan is dominated by coarse sandstones and conglomerates which represent channels, with some mudstones which were deposited as levees. The middle fan is characterized by a combination of coarse channel deposits and Bouma turbidite sequences, with the latter most common in the distal half of the midfan. The outer fan deposits are relatively continuous and are dominated by thin turbidite sequences.

A good stratigraphic model of this three-part fan sequence was presented by Rupke (1977), who was able to recognize successive development of fan lobes (Figure 15–14). His model shows the lowest portion to be dominantly fine and thin turbidites which are produced by low-density turbidity flow and represent the inactive part of the fan (Figure 15–15). Some settling of fine pelagic sediment also contributes to this portion of the sequence.

Avulsion results in generation of a new active fan which progrades over the fines deposited during the previous period of inactivity for a particular area (Figure 15–15). This active fan is represented in the rock record by turbidite sequences with some channels. Sandstone dominates, with some mudstone interbeds. In Rupke's (1977) model there are two parts to this sandstone portion of the sequence. The thicker and coarser lower portion is produced by a combination of grain flow, fluidized sediment flow, and high-density turbidity flow, whereas the overlying and thinner unit is produced by turbidity flow and generally represents interchannel and levee deposition (Figure 15–14). The coarsest portion of the sequence represents channel fill, with grain flow and fluidized sediment flow dominating over high-density turbidity flow.

This fan sequence may contain **olisthostrome** deposits above or below it, representing slumps or debris flows along the slope (Figure 15–15).

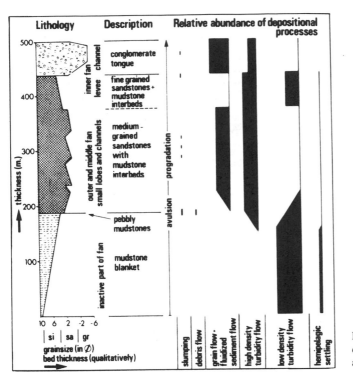

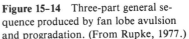

Figure 15–14 Three-part general sequence produced by fan lobe avulsion and progradation. (From Rupke, 1977.)

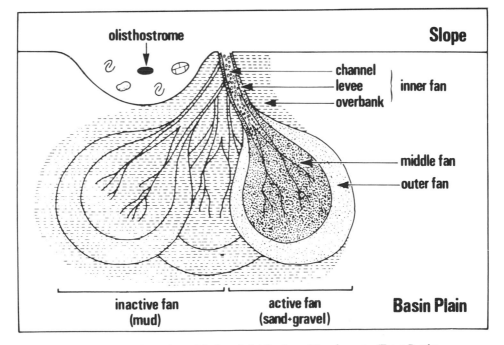

Figure 15–15 Schematic model of areal distribution of fan elements. (From Rupke, 1977.)

Tertiary of California

The Tertiary section of southern California includes several thousand meters of varied lithologies whose depositional environments are interpreted to represent the entire spectrum of conditions from the deep sea to the continents. Turbidites represent much of the Eocene-Oligocene in the Santa Ynez Mountains and the Pliocene adjacent to the south in the Ventura Basin (van de Kamp et al., 1974; Hsu, 1977). Emphasis in the following discussion will be on the Eocene and Oligocene sequence.

The Tertiary sequence in the Santa Ynez Mountains unconformably overlies Cretaceous rocks. There is a marked thickening of the section from west to east across about 100 km (Figure 15–16). Five mappable lithofacies have been recognized from this region, one of which is continental in origin. The marine facies include turbidite and marine mudstones (shown as two units in Figure 15–17), proximal turbidites, shallow marine, and coastal facies. Only the turbidite facies will be discussed here.

The turbidite and mudstone (lutite) facies is dominated by silt- and clay-size particles, with sandstones comprising up to 30% locally (van de Kamp et al., 1974). The sandstone shows grading, whereas the mudstones are typically structureless or display thin, plane stratification. The bases of the coarser units have sharp and generally erosional contacts. Many types of bottom marks are present and organic material is widespread. The relatively coarse units display features of the typical Bouma turbidite sequence, but van de Kamp et al. (1974) recognize only the basal (A and B) part (Figure 15–17). Dish structures (Figure 3–34) are common in the upper part of the basal unit, whereas the thinly stratified B unit may display convolutions at the top.

The proximal turbidite facies is dominated by sandstone, with minor amounts

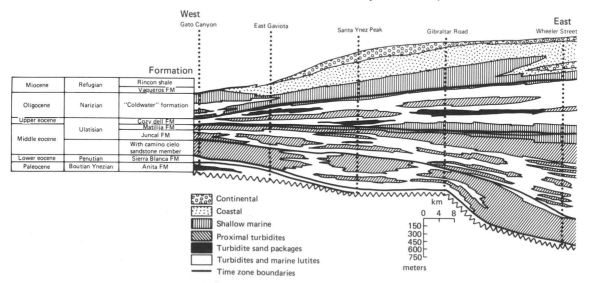

Figure 15–16 General lithofacies section of Tertiary strata in Santa Ynez Mountains area of California. (From van de Kamp et al., 1974.)

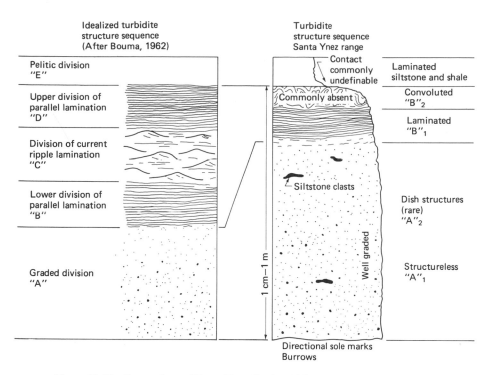

Figure 15-17 Comparison of Santa Ynez distal turbidite sequence to the classic sequence of Bouma. (From van de Kamp et al., 1974.)

of mudstone and conglomerate interbedded. Thickness of beds ranges from 30 cm to 6 m. Basal contacts are sharp, with bottom marks abundant (Stauffer, 1967). The conglomerates may contain clasts up to 60 cm in diameter, but imbrication is not apparent. Convolutions may be present in the upper, fine-grained portion of the units. Two distinct sequences have been recognized, with the greatest difference between the two occurring in the sandstone portion of the sequence (van de Kamp et al., 1974). The sequence typical of the Camino Cielo Formation contains abundant dish structures, whereas the Matilija Formation has massive, poorly graded sandstone strata in the units.

Using the models of Walker (Figure 15-13) and Rupke (Figure 15-14) it is possible to interpret the environments of deposition for these lithofacies from the California Tertiary. The so-called turbidite and marine lutite facies is analogous to the thin turbidites of Walker and represents deposition on the lower fan, probably with some pelagic sediments incorporated as well (Figure 15-14). This type of environment characterizes the Oligocene and some of the Eocene (Figure 15-16). The so-called proximal turbidites represent the middle fan environment, where some channeling occurs. The two distinct types of units probably can be interpreted as representing dominance by distinct processes. They fall into the category of massive sandstones rather than classical turbidites. The type characterized by dish structures (Camino Cielo) probably was dominated by fluidized flow, whereas the Matilija, which is massive, was produced by mixed sediment gravity flow processes (van de Kamp, et al, 1974). This is somewhat in disagreement with Stauffer (1967), who considered the Matilija to represent grain flow. By contrast, Link (1975) interprets the same unit as representing a prograding delta. This demonstrates well the need for careful and systematic observation of the rocks and interpretation of their depositional environments; even then disagreements arise. The net result is the need for more careful scrutiny and, hopefully, resolution of the disagreements.

Upper Carboniferous of England

One of the early comprehensive studies of an outer margin sequence from the rock record is that by Walker (1966a, 1966b, 1976, 1978) of the Upper Carboniferous in north-central England. The rocks in question have an aggregate thickness of 800 to 900 m. Walker's efforts were concentrated on the Shale Grit and Grindslow Shales. He recognized 10 lithofacies in these two units, six of which are restricted to the Shale Grit (Walker, 1966b).

Emphasis here will be given to the Shale Grit, which displays the typical sequence that would be expected to represent a submarine fan system. The six lithofacies present are (1) turbidite, (2) sandstone, (3) thick sandstone, (4) mudstone, (5) pebbly mudstone, and (6) very thinly stratified mudstone. The turbidite lithofacies contain interbedded sandstones and mudstones which are characterized by several varieties of Bouma sequences (Walker, 1966b). Various types of bottom marks are present at the base of these units. Each unit ranges up to about 60 cm in thickness. Mean grain size is in the very fine to fine sand range, with the maximum being very coarse sand. These units have all the attributes of the classic turbidites.

Walker (1966b) designated a sandstone facies which is similar to the turbidite facies except that the sandstone unit is 60 cm thick; it is in fact transitional with the thick sandstone facies, in which individual units reach 3 m. The sandstone facies contains some small pebbles, but these do not show imbrication and only a small portion of the thick sandstone units are graded. In fact, the sandstone and thick sandstone facies of Walker (1966b) are equivalent to the massive and pebbly sandstones in the submarine fan model (Figure 15–11).

One important aspect of the turbidite and massive sandstone facies is the formation of amalgamated units (Walker, 1966b). As one existing Bouma turbidite sequence is passed over by a turbidity current and subsequently buried by a succeeding and similar sequence there is commonly at least partial erosion of the existing sequence by the succeeding event. The recognition of such phenomena is important to the description and interpretation of the resulting stratigraphic sequence. Four different characteristic relationships may exist. These typically involve removal of the B–E layers of the Bouma sequence, so that the A layer of one unit is immediately overlain by the A layer of another. Recognition is essentially like that of recognizing an unconformity; it may include "truncation," basal conglomerates, or may be invisible to the eye. Typically, tracing the units laterally will reveal a more complete sequence because such erosion in quite local.

The mudstone facies is generally well stratified and contains carbonaceous material locally. This facies is interpreted as representing thin turbidites (Figure

15-11). Pebbly mudstones are massive or display slumping. Clasts up to 1 cm in diameter are scattered throughout with no apparent organization. This facies may be wedge shaped or tabular. It apparently represents deposition under debris flow conditions (Walker, 1966b).

The thinly stratified mudstones have no sedimentary structures or textural patterns other than some alternation of carbonaceous supply. Deposition was very slow and under quiescent conditions, away from the direct influence of gravity-flow processes.

Reconstruction of depositional system

Using the general lithostratigraphy and the details observed in the Shale Grit (Figure 15-18), Walker (1966b, 1978) reconstructed the Late Carboniferous outer margin in which these strata accumulated. The basal Edale Shales are comprised of black mudstones which were deposited beyond the margin in the deep basin. The Mam Tor Sandstones are dominated by classic and thin turbidites and represent outer fan deposition.

The Shale Grit represents a series of suprafans with the massive sandstones being distributary deposits and turbidite units representing interchannel deposition. The general coarsening-upward trend points to a progradation of the channelized part of the fan over the smooth, distal portion (Figure 15-19).

The overlying Grindslow Shales display a combination of coarse debris flow deposits from channels and fine mudstones which accumulated on the slope between feeder channels (Figure 15-19). The Kinderscout Grit caps the sequence and has been interpreted as a steep and rapidly prograding deltaic complex (Collinson, 1969).

The general stratigraphy, sedimentary structures, textural trends, and radial paleocurrent pattern (Walker, 1966b), have led to this sequence being reconstructed into a depositional system similar to the model described previously (compare Figures 15-11 and 15-19). Using Walther's Law, the reconstruction shows a prograding outer margin, with the basinal Edale Shales being covered by lower fan deposits (Mam Tor), and so on up the section.

Silurian, England

Detailed mapping and stratigraphic studies of an area in Herefordshire, southwestern England, have documented several preserved submarine canyon-fill sequences (Whitaker, 1962). These are among the earliest studied and best documented canyon-fill sequences and their relationships to older strata are apparent.

A total of six parallel channel trends shows a deepening toward the depositional basin toward the southwest. These canyons cut up to 200 m into the section of the Wenlockian and Ludovian Series of Silurian age. Although faulting has obscured some of the relationships, the nature of the strata is apparent.

The unconformable relationships between the fill and canyon walls are readily apparent and mapping has delineated the size and geometry of the canyons. Characteristics of the fill include concave-upward bedding, slump structures, boulder beds, cross-strata, and various bottom marks, as well as an unusual mixture of fossils (Figure 15-20). The fauna preserved in these canyon-fill deposits includes

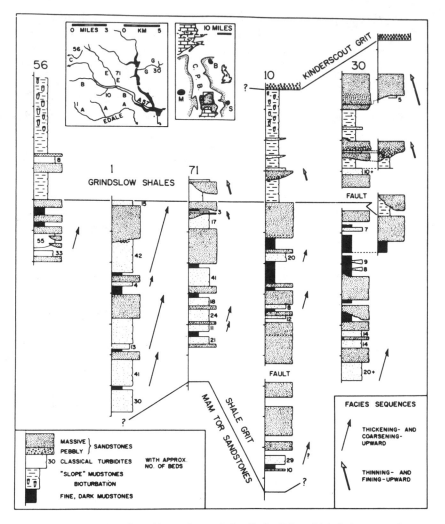

Figure 15-18 Stratigraphic sections of the Shale Grit, which is interpreted as representing active lobes of an ancient submarine fan. (From Walker, 1978, p. 960.)

some organisms that are indigenous to the canyon floor, some that are transported to the canyon with sediments from the adjacent shelf, and fossils that are reworked from the older beds into which the channel is cut (Whitaker, 1962, 1974). Some of the coarse particles display imbrication, and elongate fossils may be oriented by currents parallel to the canyon axis.

Ordovician, Gaspé Peninsula, Canada

Emphasis in the discussion of both the modern and ancient outer margin has been on the submarine fan type of depositional system. This is due in part to the numerous recent studies on such fans and to the interest in such ancient deposits for petroleum ex-

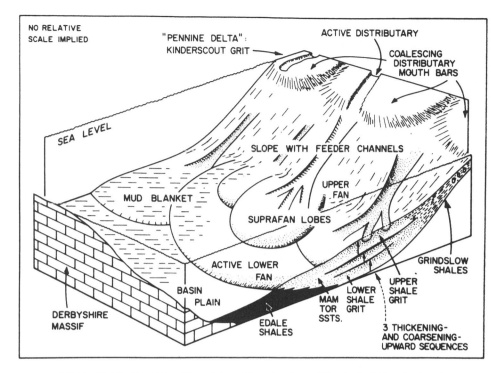

Figure 15–19 Reconstruction of depositional system of the Shale Grit and related units, showing a prograding submarine fan system. (From Walker, 1978.)

ploration. It has been noted that these submarine fan systems may have numerous, thick, coarse channel deposits. Some ancient turbidite sequences lack the coarse channel deposits and apparently do not represent the submarine fan type of depositional system. Examples of thick sections containing hundreds or thousands of Bouma sequences occur especially along the Appalachian and Ouachita fold belts in Paleozoic rocks (e.g., McBride, 1962; Enos, 1969a; Morris, 1974). These somewhat monotonous accumulations of Bouma sequences have commonly been referred to as **flysch**.

Cloridorme formation

The Cloridorme Formation (Middle Ordovician?) is exposed along the northern Gaspé Peninsula (Quebec) of Canada. It has been studied in detail (Enos, 1969a, 1969b; Skipper and Middleton, 1975) and is essentially equivalent to the Martinsburg Formation (McBride, 1962) of the central Appalachians. The unit is estimated to be a few thousand meters thick. Its lithology is dominated by dark gray mudstone with lesser amounts of lithic arenites (15%) and calcisiltites (20%), all of which are interpreted as turbidites (Enos, 1969a).

The overall interpretation of a turbidity current origin for the Cloridorme Formation is based on three types of data: (1) thin beds of terrigenous and carbonate detritus are interbedded with hemipelagic muds; (2) the presence of extensive sedimentary structures, including various bottom marks, convolute bedding,

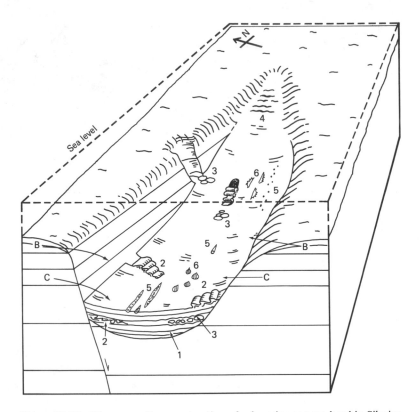

Figure 15–20 Diagrammatic reconstruction of submarine canyon-head in Silurian time. Letters B and C refer to assemblages of fossils (not to scale): (1) concave-upward beds, (2) slump structures, (3) boulder beds with imbrication, (4) bedforms, (5) bottom marks, (6) oriented fossils. (After Whitaker, 1962, 1974.)

grading, and ripple cross-stratification; and (3) pieces of benthic organisms are in the coarse sediments and pelagic tests are restricted to the fine mud (Enos, 1969b).

The geometry and thickness of beds within the Cloridorme reveals some tabular units (Figure 15–21) which may be traceable for 3 km, but most of the beds are discontinuous over the exposure area, which consists of coastal cliffs that are kilometers in extent. There is a range in bed thickness and it is related to grain size; mudstone beds may be up to 20 cm thick but most are 0.5 to 4.0 cm and the lithic arenite beds may exceed a meter in thickness, with the mean about 15 cm (Enos, 1969b).

Thousands of measurements were made of directional structures in the Cloridorme Formation. Various types of bottom marks, including both scour marks and tool marks, are widespread and reveal a fairly consistent east-to-west paleocurrent direction (Enos, 1969b, Figure 3). Other directional features in this unit include cross-stratification and oriented fossils (graptolites and orthocone cephalopods). Graded beds are present in nearly all the units containing sand-size particles. There is a noticeable decrease in grading down the paleocurrent direction. Burrows, slump structures, and pull-aparts are fairly common.

Figure 15-21 Photograph of Cloridorme Formation with tabular turbidite units. (Courtesy of D. Beedon.)

Vertical sequences of texture and structures are repeated many times, as is typically the case in flysch-type strata. Some appear to resemble the Bouma sequence, but some are also different (Skipper and Middleton, 1975). The latter sequences have three distinct divisions or units, consisting of (1) a basal coarse litharenite with abundant bottom marks, (2) a relatively thick mudstone with nodular calcareous masses, and (3) an overlying laminated mudstone (Figure 15-22).

Depositional environments. The initial assumption made regarding the origin of the Cloridorme Formation is that the strata are turbidites. This is well substantiated by the thin tabular nature of the beds, the nature of the sedimentary structures, and the monotonous repetition of the units in the section. It is apparent from the general appearance of these strata that they differ from the submarine fan deposits described above. Cloridorme strata are somewhat like the thin turbidites or

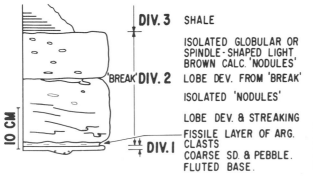

Figure 15-22 Diagram showing typical turbidite sequence in the Cloridorme Formation, Quebec, Canada. (From Skipper and Middleton, 1975, p. 1939.)

Chapter 15   The Continental Slope and Rise System

lower fan deposits, but they also contain sequences more than a meter thick, which is not characteristic of the lower fan environment.

A variety of data indicates that these strata were deposited by large turbidity currents over a very gently sloping ocean floor (Skipper and Middleton, 1975). Coarsest particles settled first but continued to move by traction and produced ripple cross-stratification and plane beds (division 1). The turbid mass of sediment-laden water resulted in rapid accumulation of mud, which was still subjected to flow, and shearing resulted (division 2). The finest particles settled from the tail of the turbidity current and yielded parallel stratified mud (division 3). Commonly pelagic sediments containing fossils are present between turbidite sequences (Skipper and Middleton, 1975).

The large-scale depositional setting of the Cloridorme Formation as well as similar units in the Appalachian and Ouachita foldbelts is that of an outer deep margin adjacent to a foldbelt. This tectonic situation produces a large sediment supply which eminates from the entire extent of the source area. The result is much downslope transport in the form of turbidity currents, but apparently either no significant channelization develops or this part of the system is continually destroyed by tectonic activity (Walker, 1978). The major paleocurrent flow is actually parallel to the foldbelt along tectonic strike (Enos, 1969b, Fig. 3). It is reasonable to assume that contour currents may have played a role in the deposition of these sequences.

Tourelle Formation

The Lower Ordovician Tourelle Formation is exposed along the south side of the St. Lawrence River on the Gaspé Peninsula of Canada. This thick unit is characterized by thick-bedded sandstones which are coarse and massive. Their composition ranges from subarkose to sublitharenite. Rock fragments include shale, chert, carbonates, and some of volcanic origin. They are the equivalent of Walker's (1978) pebble sandstone or massive sandstones (Figure 15–11) and they accumulated in channels on the surfaces of coalescing submarine fans (Hiscott, 1978; Hiscott and Middleton, 1979).

The source of the terrigenous particles that formed the Tourelle was primarily a plutonic and metamorphic continental block. Deposition took place on a submarine fan complex which developed along a foredeep mud basin as two continental blocks converged. Uplift was high at the zone of convergence, with one of the blocks foundering to form the foredeep basin (Hiscott, 1978).

Lower Ordovician, Nevada

Carbonate sediments may also accumulate in the deep, outer margin environment. These deep-water carbonates have received much attention in both modern and ancient depositional environments (Cook and Enos, 1977). One excellent stratigraphic sequence in the Upper Cambrian and Lower Ordovician in Nevada illustrates this type of deep carbonate environment. The sequence consists of about 600 m of shallow water carbonates, including algal stromatolites, grainstones, and fenestral fabrics which are laterally equivalent, to about 150 m of dark micrites and interbedded coarse-grained gravity-flow and slumped deposits (Cook and Taylor, 1977). It is the latter sequence that has been interpreted as a deep-water slope deposit.

The Holes Limestone consists of dark micrite near its base with few fossils, an absence of bioturbation, and scarce terrigenous particles. This lithofacies is overlain by and interbedded with different varieties of allochthonous carbonates, including thinly bedded and deformed limestone, conglomeratic limestone with clasts of micrite, and well-sorted grainstones containing abundant shallow-water algae (Cook and Taylor, 1977). Many large slump blocks, narrow channels filled with clasts over a meter in diameter, and various indications of debris flow are present. Many of the conglomeratic units contain large disk-shaped clasts of dark micrite derived from the slope, mixed with lighter-colored clasts which have a shallow-water origin (Figure 15–23).

Cook and Taylor (1977) have reconstructed their interpretation of the outer margin system at the time these depositional sequences accumulated (Figure 15–24). A broad, shallow marine shelf and adjacent tidal flat environments were accumulating shallow water, biogenic, and micritic carbonates. The combination of shelf currents and extensive slope failure resulted in extensive sediment gravity flows with some channels being excavated. These materials accumulated in channel-fill sequences and in irregular masses at the base of the slope, where they were modified by contour currents moving along the axis of the basin (Figure 15–24).

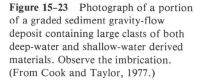

Figure 15–23 Photograph of a portion of a graded sediment gravity-flow deposit containing large clasts of both deep-water and shallow-water derived materials. Observe the imbrication. (From Cook and Taylor, 1977.)

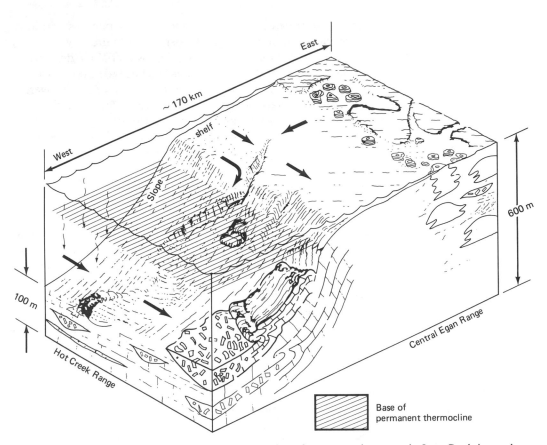

Figure 15-24 Diagrammatic reconstruction of outer margin system in Late Cambrian and earliest Ordovician time in Nevada. (From Cook and Taylor, 1977.)

SUMMARY OF CHARACTERISTICS
OF SEDIMENT/SEDIMENTARY ROCK BODIES
REPRESENTING THE CONTINENTAL SLOPE AND RISE SYSTEM

Many of the very thick sedimentary sequences in the stratigraphic record represent the outer continental margin environments. Although this has been recognized for some time, it is only since the late 1960s and early 1970s that detailed depositional models for these environments have been developed.

Tectonic setting. This system accumulates in a relatively active tectonic setting compared to most depositional systems. The outer margin displays much relief near the juncture between the continental block and oceanic crust. In addition, the slope is disected by submarine canyons.

Shape. The overall configuration of the strata in this system is one of a wedge or thick lens. The system as preserved in the stratigraphic record may display an elongate configuration parallel to the bathymetry. There are some elongate channel deposits within the system, and thin blankets of pelagic material may be associated with the deep-sea fan deposits.

Size. The thickness of the outer margin strata may reach thousands of meters. These strata may extend thousands of kilometers along depositional strike and have a width of tens to hundreds of kilometers. Some depositional situations such as would exist on a continental borderland might produce localized submarine fan accumulations.

Textural trends. Some significant textural trends are exhibited by rise sediments, especially submarine fan sequences. The overall sequence of a prograding fan is one that coarsens upward. The large-scale trend tends to mask trends within component sequences; some of these fine upward and others coarsen upward. Individual channel, levee, and turbidite sequences show a fining-upward trend. The intermediate-scale trend is a coarsening upward trend of several turbidite sequences which collectively represent a suprafan lobe.

Lithology. The mineralogy of the outer margin system is probably the most diverse of all marine deposits except perhaps glacial marine sediments. Because of a range of grain size, a broad potential source area, and rapid accumulation, these strata tend to accumulate anything that is available. Quartz, rock fragments, and clay minerals are all abundant, with feldspars probably somewhat less common. Biogenic debris is rarely abundant, although it is present in most deposits. Composition of these strata has been used to determine source location and transport pathways.

Sedimentary structures. Deep-sea deposits of the outer margin display the broadest spectrum of sedimentary structures. Prevalent varieties include graded bedding, various cross-strata, and many types of bottom marks. The Bouma sequence contains these structures in a predictable order and they may be of great value in paleocurrent studies.

Paleontology. There are two different types of fossils preserved in outer margin sequences. Accumulation of pelagic organisms, especially foraminifera, accounts for one. The other is a displaced fauna derived from shallower waters by the downslope transport of sediment. The result is that skeletal remains of organisms typically living on the shelf or slope may be found in deep-sea fan deposits. Fossils are typically not abundant in these strata.

Associations. Outer margin deposits may be vertically adjacent to pelagic deposits of the deep sea, especially as the result of progradation of the deep-sea fan over the abyssal plain. When the supply of sediment to the fan is cut off, pelagic deposits will accumulate, but the resulting strata are thin. Terrigenous shelf deposits may be preserved in the stratigraphic record landward of the outer margin. Such large-scale relationships are common in folded mountain belts such as the Appalachian Mountains.

ADDITIONAL READING

BOUMA, A. H., 1962. *Sedimentology of Some Flysch Deposits—A Graphic Approach to Facies Interpretation.* Elsevier, New York, 168 p. A classic work on turbidites which spawned the now standard Bouma sequence, which has proven to be one of the truly important concepts in sedimentology.

DOTT, R. H., JR., AND SHAVER, R. H. (EDS.), 1974. *Modern and Ancient Geosynclinal Sedimentation*, Soc. Econ. Paleontologists and Mineralogists, Spec. Publ. No. 19, Tulsa, Okla., 380 p. An excellent and comprehensive compilation of papers on the topic with several on outer margin sedimentation.

DOYLE, L., J., AND PILKEY, O. H., JR. (EDS.), 1979. *Geology of Continental Slopes*. Soc. Econ. Paleontologists and Mineralogists, Spec. Publ. No. 27, Tulsa, Okla., 374 p. A superb volume of excellent papers on a subject that is receiving great attention. Both modern and ancient examples plus overviews are included.

DZULYNSKI, S., AND WALTON, E. K., 1965. *Sedimentary Features of Flysch and Greywackes*. Elsevier, New York, 274 p. Another classic book on turbidite deposits. Fantastic illustrations and overall coverage of sedimentary structures associated with these sequences.

HEEZEN B. C., AND HOLLISTER, C. D., 1971. *The Face of the Deep*. Oxford University Press., New York, 659 p. The best assemblage of deep-sea photographs of the outer margin and deep ocean basins, with an adequate text to cover the subject without distracting from the photos.

LAJOIE, J. (ED.), 1970. *Flysch Sedimentology in North America*. Geol. Assoc. Canada, Spec. Paper No. 7, 242 p. This is probably the best single volume on the topic until its publication date, with papers contributed by most of the prominent researchers in the field. It is now out of date because it precedes the fan models now available.

MIDDLETON, G. V., AND BOUMA, A. H. (EDS.), 1973. *Turbidites and Deep Water Sedimentation*. Soc. Econ. Paleontologists and Mineralogists, Pacific Section Short Course, Anaheim, Calif., 157 p. A great collection of well-written papers on turbidity current processes and resulting accumulations. The best place to start reading on the subject.

16 The Deep Ocean System

This chapter is devoted to a discussion of the modern deep ocean environments and their counterparts as interpreted from the rock record. A glance at a global physiographic map shows that this includes nearly 60% of the earth's surface. Why the obvious imbalance in treatment? There are actually several reasons; if this text had been written 25 years ago there would have been a greater imbalance and more reasons to justify it. Probably the most important of these reasons is that existing data and overall level of knowledge for the deep ocean are much less than for the systems described previously. This deficiency is gradually being reduced as extensive research is conducted. It is safe to state that more has been learned about the geology of the deep ocean since 1965 than in all previous time.

Although the factors discussed above are important, they form only part of the reason for the relative coverage presented here. By comparison with the other environments considered, the deep ocean seems to be more homogeneous. Such a statement tends to raise the ire of marine geologists, but if one considers the relative variation per unit area, it appears true. In other words, more variation exists in sediments and processes in many environments than have been observed so far in the deep ocean. Part of this reasoning is that deep ocean sediments are not readily preserved in the rock record and subsequently brought to or near the surface, where they can be examined. They tend to be preserved under the ocean floor or are commonly consumed and rendered unrecognizable during subduction. The net result is that less is known about ancient deep-sea deposits relative to other depositional systems.

Please remember that the statements above are presented in relation to other environments. Not too many years ago the deep-sea environment was thought to be a flat, rather featureless desert with a "layer cake" type of sediment accumulation (i.e., no significant lateral variation). The extensive efforts of researchers during the 1960s and 1970s have demonstrated that this is an oversimplification. There is in fact much variation. Considerable overall research activity is going on in this remote part of the earth's surface (Warme, et al., 1981), but it is of much less intensity per unit area than on the present continental margin.

Physiography

The ocean floor contains numerous and diverse large-scale physiographic features and many small ones as well. Previous authors (Heezen et al., 1959) subdivide the oceans into three major physiographic divisions, including the continental margin (Figure 16-1). The ocean basin proper in Figure 16-1 includes the ocean basin floor and the oceanic ridge. The ocean basin floor consists largely of abyssal hills and abyssal plains, whereas the oceanic ridge is a broad, mountaneous belt of relatively high relief. Although this simplified subdivision is convenient, it belies the complex nature of submarine topography (Figure 16-2). Numerous features are present, many of which are as large as major physiographic elements of the subaerial crust. It is also important to realize that these relief features may exert much influence on sedimentation in the deep ocean environment.

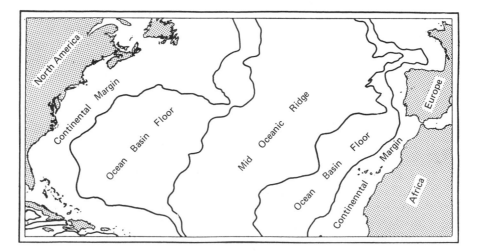

Figure 16-1 North Atlantic Ocean, showing generalized outline of major physiographic provinces. (From Heezen et al., 1959.)

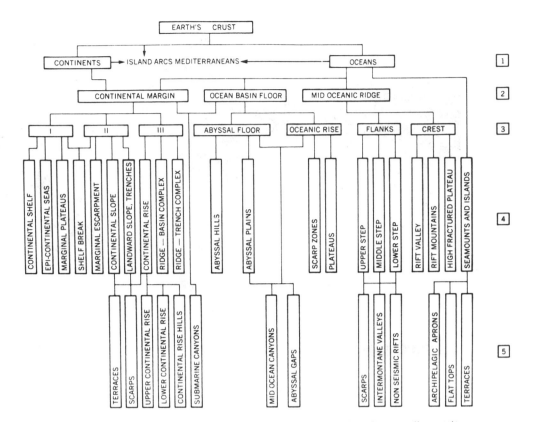

Figure 16-2 Diagram of submarine features, showing hierarchy according to size with five levels. (From Heezen and Hollister, 1971, p. 10.)

Large-scale features

Without question the oceanic ridge system is the largest single feature of the earth's surface. It is rather continuous, and extends for about 60,000 km, and reaches a width of a few thousand kilometers. Relief is in thousands of meters, with a general increase toward the central rift valley. Related to the oceanic ridge system are numerous fracture zones which cut almost perpendicular to the trend of the ridge system. These large faults not only cause lateral displacement along the ridge but also display much relief, over 1000 m in the case of some in the eastern Pacific off the California coast (Krause et al., 1964).

Deep-sea trenches are equally striking in their long arcuate extent and high relief. These features are associated with certain types of continental margins (Figure 16-2) but also occur in other plate-boundary situations (Figure 16-3). The trenches are asymmetrical in profile and provide a unique type of sedimentary environment.

Oceanic rises are large, low-relief, generally nonlinear, areas of the oceanic ridge, which rise above the adjacent abyssal plain. They have been described as large

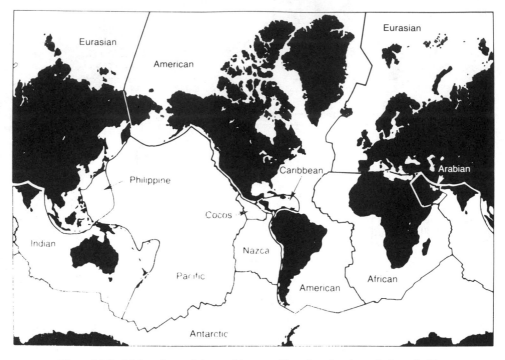

Figure 16–3 Major plates of the earth's crust. Note that plate boundaries coincide with seismically active zone. (From Ross, 1982, Figure 6–11).

blisterlike features on the oceanic crust. The Bermuda Rise and the Blake Plateau are examples.

Small-scale features

The most abundant and widespread small physiographic features of the ocean floor are of volcanic origin. There are thousands of volcanoes scattered throughout the world ocean and they are most common in the Pacific Ocean. Some rise above sea level to form islands, many of which are associated with coral reef development. Some never reach sea level, whereas others have subsided below sea level. The latter are called **guyots** and are characterized as being flat-topped **seamounts**. All are high-relief features and as such they cause local effects on circulation and sedimentation.

Physical Processes

The physical processes of the deep ocean are somewhat limited in nature, extensive in distribution, and generally quite slow in terms of their rate. Their basis is almost exclusively due to gravity. The most significant is the deep circulation of different water masses in the world ocean and their role in sediment distribution. Local processes include downslope transport via turbidity currents and downcurrent transport by bottom currents.

Deep ocean circulation is largely **thermohaline** in that it results from density gradients caused by a variation in some combination of temperature and salinity. Density differences of only 0.001 g/cm³ are sufficient to produce a density gradient that generates water motion. For the most part these currents flow from the high latitudes toward the equator (Figure 16–4), due to the presence of cold, higher-density water throughout the water column in high latitudes. Within the water column are multiple water masses moving in opposite directions to one another, causing transfer of water from one water mass to overlying and underlying water masses. Geostrophic effects of the earth's rotation and tidal influences also affect deep-water circulation.

The rates at which deep-sea currents flow cover a fairly broad spectrum. Some are so slow that it takes several years to move from the Arctic to the low latitudes. Others are much faster and there is evidence that some deep-sea bottom currents affect the substrate. Deep-sea photographs show deflected, stalked organisms, lineations on the sediment surface, and sediment shadows associated with large particles or benthic organisms. Some ripples have been observed.

Most deep-sea sediments are in the silt and clay size range. Except for the relatively coarse biogenic particles, deep-sea sediment can be entrained by currents of less than 5 cm/sec (Figure 16–5). Numerous deep-sea photographs have been taken showing suspended sediment clouds which show lineation indicating current move-

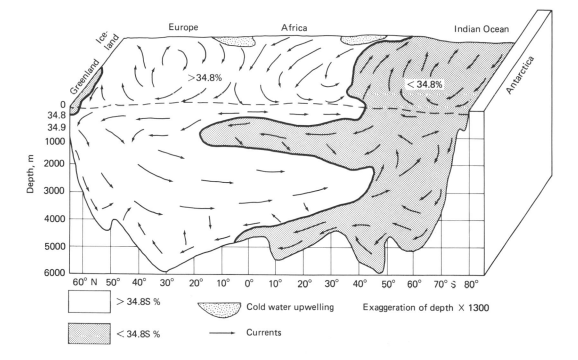

Figure 16–4 Circulation in the deep ocean of the Atlantic. (From Heezen and Hollister, 1971, p. 357.)

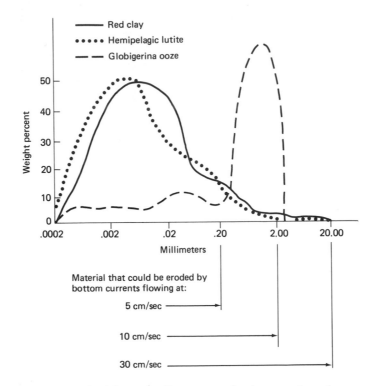

Figure 16–5 Graph of deep-sea sediment types, showing current speeds necessary for erosion of various grain sizes. (From Heezen and Hollister, 1971, p. 355.)

ment (Heezen and Hollister, 1971). Systematic plotting of directional features shows (1) that much of the sea-floor sediment exhibits evidence of currents, and (2) that there is temporal and spatial variation in the grand scheme of deep-sea circulation, as shown in Figure 16–4.

One of the most important oceanic current systems is present in the Southern Ocean, which circumscribes Antarctica. The surface component is the Antarctic Circumpolar Current and the Antarctic Bottom Water is the deep portion; both are high-velocity oceanic currents and have marked effect on sediments. The ocean floor in this region shows widespread manganese nodules, ripples, and lineations (Heezen and Hollister, 1971).

An excellent example of transport of deep-sea sediment was provided by a natural tracer study of a red glacial sediment body that was deposited near the mouth of the St. Lawrence River on the Grand Banks. The distinct color of this unit makes the sediment recognizable that is eroded from it and transported across the North Atlantic. The extent of this red clay includes nearly all of the western North Atlantic, more than 3000 km from its origin (Figure 16–6). Detailed studies of the mineralogy and paleontology have confirmed that its origin is the Pleistocene red clay. (Conolly et al., 1969; Needham et al., 1969). This is good testimony to the extensive effects of deep-sea currents.

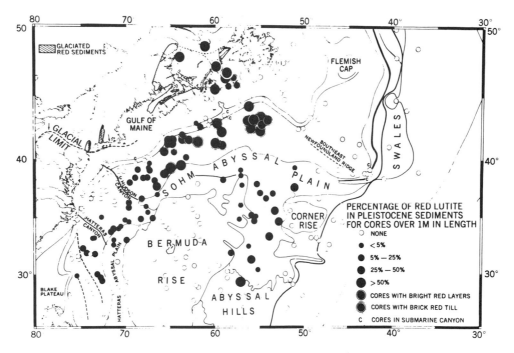

Figure 16-6 Distribution of red glacial sediment over the northwest Atlantic. Such widespread distribution relative to the source area is due to deep circulation in the Western Boundary Current. (From Heezen and Hollister, 1971, p. 397.)

DEEP-SEA SEDIMENTS

The deep-sea depositional system differs from virtually all others on the earth's surface. This is due in part to the fact that much of the sediment comes to rest by other than bed load transport, nor is it the result of in situ chemical precipitation. Second, deep-sea sediments involve various chemical cycles to a far greater degree than other environments; also, organisms play a greater role insofar as geographic extent is concerned. Finally, sediments of the deep sea display much greater geographic homogeneity than is found in other environments.

Sources

The origins of deep-sea sediments are not unlike those in other marine environments, but some of the lesser contributing sources become more significant in deep-sea sediments because of the fact that most land-derived sediment is deposited before it reaches the deep sea. As a consequence, the small amounts from various sources are not as diluted in the deep sea.

According to Shepard (1973), there are four main sources for deep-sea sediments, which produce lithogenous, biogenous, hydrogenous, and cosmogenous particles (Figure 16-7). Lithogenous sediment particles are produced from weathering of previously existing rocks or from volcanic activity; skeletal material produced

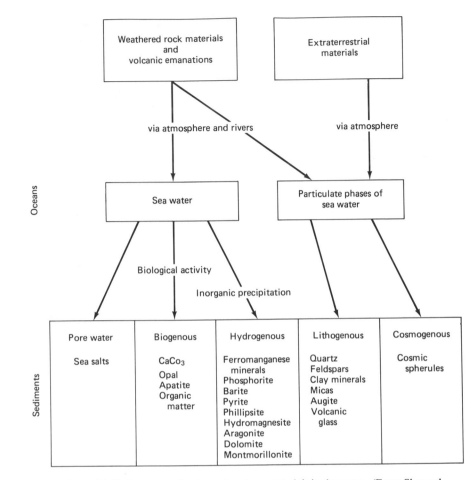

Figure 16-7 Sources and nature of various materials in the ocean. (From Shepard, 1973, p. 406.)

by organisms is biogenous; inorganic precipitates on the sea floor are hydrogenous; and any extraterrestrial material is cosmogenous. Some sea salts may precipitate from pore waters and are not included in the other four categories (Figure 16-7).

Lithogenous sediments include a broad spectrum of mineral species derived from numerous sources but with essentially no change to the solid phase during its residence in seawater (Goldberg, 1964; Riley and Chester, 1971). Included are mineral grains and rock fragments from land as well as clay minerals which are weathering products. Volcanic glass and other pyroclastic material is also placed in this category.

Hydrogenous material is produced by inorganic reactions on the sea floor. It may be formed directly from seawater or by alteration of previously existing minerals (Chester and Hughes, 1967). The phosphates, sulfates, oxides, and hydroxides all are included (Figure 16-7). Especially important are Fe-Mn nodules. Some silicates, such as cherts, feldspar overgrowths, and palygorskite, have also been reported. The altered hydrogenous sediment is comprised of zeolites and some clay minerals, especially smectite, which forms from volcanic material on the sea floor (Riley and Chester, 1971).

Manganese nodules are hydrated oxides of iron and manganese which also contain copper, nickel, and cobalt. They are most common as spheres ranging from about 1 to 30 cm in diameter. The ferromanganese oxides may coat or encrust most any particles or material in the deep sea (Ross, 1980). The nodules may occur throughout the ocean basins but are very abundant in some areas. Mero (1972) estimated that about 25% of the ocean floor is covered. The presence of abundant nodules is coincident with strong bottom currents where scour or at least nondeposition prevails.

Carbonate and silicate skeletal material is the most abundant and widespread of the biogenous sediments. Also present are phosphates in the form of skeletal apatite, some organic matter, and the sulfate barite, which precipitates as the result of the activity of some plankton (Riley and Chester, 1971).

The least abundant category is the cosmogenous sediment, which consists of small, widely scattered magnetic spherules which have compositions comparable to iron and stony meteorites (Hunter and Parkin, 1960).

Classification

Not only were the first studies of deep-sea sediments conducted from material collected on the cruise of H.M.S. *Challenger* (1872–76), but the classification that resulted (Murray and Renard, 1891) persisted until the 1960s. The great wealth of data collected as the result of the Deep Sea Drilling Project (DSDP) has lead to a new classification (Table 16–1).

Regardless of which of the classifications is utilized, the same two major subdivisions are made: pelagic, and terrigenous or clastic. Although these subdivisions are used widely, there is not universal agreement regarding their definitions. Pelagic is commonly designated to indicate deep-sea deposits far from land influence. The definition used here is similar to that of Shepard (1973) and Berger (1974). Pelagic sediment is the material that falls out of suspension slowly through the water column. It characterizes deep-sea deposits but also occurs on the continental margin. The rate of accumulation is very slow, generally from 2 to 50 mm every 1000 years (Berger, 1974), with the rate significantly dependent on the location and specific type of sediment. This rate is so slow that although it is present on the margin, pelagic sediment is masked by the rapid rates of accumulation from land-derived and resedimented terrigenous material.

The other major category has been referred to as terrigenous by most authors (e.g., Shepard,1973), but some authors use the term clastic to categorize essentially the same spectrum (Hay, 1974). In fact, the nonpelagic portion is not really either type of sediment unless the terms are used in the broad sense. "Terrigenous" means derived from land, which would include some of the fine clays typically considered as pelagic, and it excludes some of the carbonate particles that originate on reefs but eventually accumulate in the deep sea. On the other hand, "clastic" implies any previously existing particle that is not currently in situ regardless of its origin. Ter-

TABLE 16-1   CLASSIFICATION OF DEEP-SEA SEDIMENTS

I.   (Eu-) pelagic deposits (oozes and clays)
     $< 25\%$ of fraction $> 5\ \mu m$ is of terrigenic, volcanogenic, and/or neritic origin.
     Median grain size $< 5\ \mu m$ (excepting authigenic minerals and pelagic organisms).
     A.   Pelagic clays. $CaCO_3$ and siliceous fossils $< 30\%$.
          (1) $CaCO_3$ 1–10 %. (Slightly) calcareous clay.
          (2) $CaCO_3$ 10–30 %. Very calcareous (or marl) clay.
          (3)   Siliceous fossils 1–10 %. (Slightly) siliceous clay.
          (4)   Siliceous fossils 10–30 %. Very siliceous clay.
     B.   Oozes. $CaCO_3$ or siliceous fossils $> 30\%$.
          (1) $CaCO_3 > 30\%$. $< \frac{2}{3}\ CaCO_3$: marl ooze. $> \frac{2}{3}\ CaCO_3$: chalk ooze.
          (2) $CaCO_3 < 30\%$. $> 30\%$ siliceous fossils: diatom or radiolarian ooze.

II. Hemipelagic deposits (muds)
     $> 25\%$ of fraction $> 5\ \mu m$ is of terrigenic, volcanogenic, and/or neritic origin.
     Median grain size $> 5\ \mu m$ (excepting authigenic minerals and pelagic organisms).
     A.   Calcareous muds. $CaCO_3 > 30\%$.
          (1) $< \frac{2}{3}\ CaCO_3$: marl mud. $> \frac{2}{3}\ CaCO_3$: chalk mud.
          (2)   Skeletal $CaCO_3 > 30\%$: foram~, nanno~, coquina~.
     B.   Terrigenous muds. $CaCO_3 < 30\%$. Quartz, feldspar, mica dominant.
          Prefixes: quartzose, arkosic, micaceous.
     C.   Volcanogenic muds. $CaCO_3 < 30\%$. Ash, palagonite, etc., dominant.

III. Pelagic and/or hemipelagic deposits
          (1)   Dolomite-sapropelite cycles.
          (2)   Black (carbonaceous) clay and mud: sapropelites.
          (3)   Silicified claystones and mudstones: chert.
          (4)   Limestone.

SOURCE: Berger, 1974, p. 214.

rigenous will be used in this discussion primarily because of its prevalence in the literature.

Pelagic sediments

There is a great diversity within pelagic sediments, far more than was realized during the early work of Murray and Renard (1891).

Biogenic sediments. Tests of small organisms comprise the majority of pelagic sediments in much of the world ocean. These are almost exclusively the tests of microplankton and include both photosynthisizers and animals with tests of carbonate or silica. Carbonate tests are produced by pteropods; cone-shaped planktonic gastropods; foraminifera, especially *Globigerina*; and coccolithophores, submicroscopic photosynthetic organisms (Figure 16–8). All these organisms are found in near-surface waters of the low latitudes. Siliceous tests are produced by radiolaria; radially symmetrical single-celled organisms; and by diatoms, which are photosynthetic and serve as the base of the open ocean food chain (Figure 16–9).

Biogenic pelagic sediments are commonly referred to as oozes, meaning that 30% or more of the sediment is comprised of planktonic tests (Table 16–1). In most

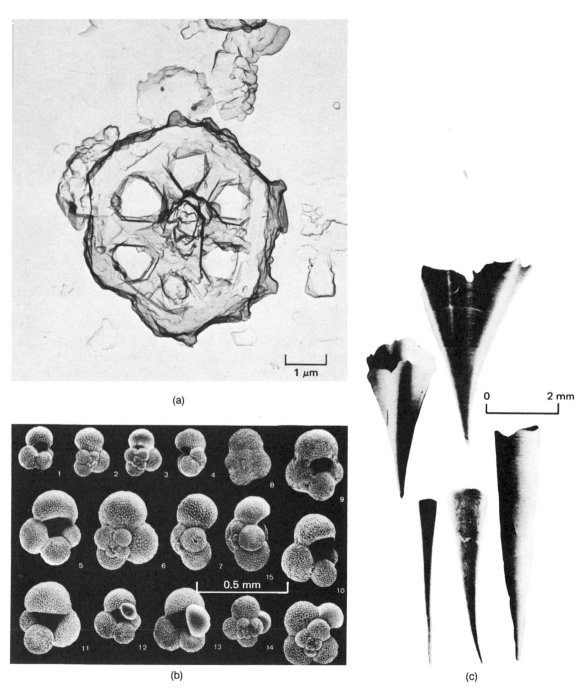

Figure 16–8 Tests of calcareous pelagic organisms; (a) coccolithophore; (b) *Globigerina* foraminifera; (c) pteropod.

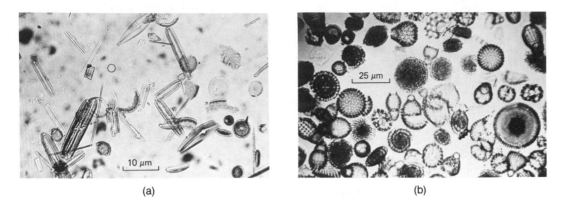

(a)   (b)

Figure 16–9   Tests of siliceous pelagic organisms: (a) diatoms; (b) radiolaria.

of the world ocean where ooze is the dominant sediment type, one type of test is also dominant.

Fish bones and teeth may also contribute to biogenic pelagic sediments. These phosphatic particles are not a common constituent of the deep sea.

Clays and muds.   Previous classifications of deep-sea sediments considered nonbiogenic pelagic deposits as being just clay; however, Berger's (1974) classification subdivides this group into two categories: pelagic clay, in which $< 25\%$ of the > 5 μm fraction is nonbiogenic and median grain size is > 5 μm, and hemipelagic mud, in which $> 25\%$ of the > 5 μm fraction is nonbiogenic and median grain size is > 5 μm (Table 16–1). The latter category is not widespread and is restricted to areas adjacent to the continental margin.

A particularly important pelagic clay is variously called red clay or brown clay. Although its appearance is more brown than red, it is widely termed "red clay." Mineralogically, it is primarily the clay mineral illite, which is generally ascribed to a land origin (Griffin et al., 1968).

Other pelagic sediments.   Also included in this diverse category of deep-sea sediments are volcanic ash, eolian dust, extraterrestrial dust and meteorites, and glacial marine sediment (Figure 16–10). The diagram illustrates well that although these and previously discussed types have various origins, they all reach the sea floor through a common means, that of settling through the long column of water in the deep sea.

The origin and nature of volcanic, cosmic, and eolian sediment particles is rather straightforward, but the glacial marine sediment requires some discussion. Recall the general discussion of glacial marine sediment in Chapter 7. The nature of pelagic sediment of a glacial origin is more specific, but it is quite different from other pelagic sediments. Sediment-laden icebergs move out over the continental margin and the deep sea. Melting causes the sediment to fall to the ocean floor in a true pelagic manner except that some particles are large and fall rapidly. It is evident, therefore, that some glacial marine sediment is pelagic due to ice-rafting, and some is not due to its direct deposition on the continental margin from the glacier. Some

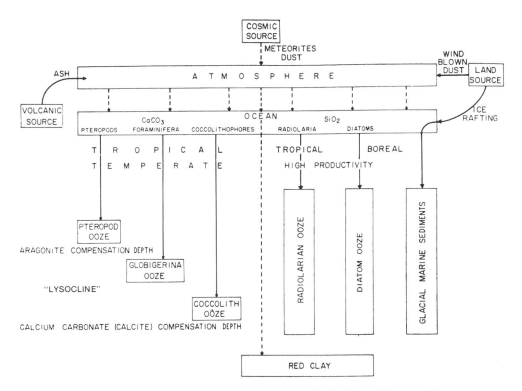

Figure 16-10 Flow diagram of pelagic sedimentation in the world ocean. (From Hay, 1974, p. 3.)

authors (e.g., Shepard, 1973) include glacial marine deep-sea sediments as terrigenous.

Terrigenous sediments

The vast volume of terrigenous deep-sea sediments owes its origin to turbidity currents or other types of sediment gravity flow. Slumps, slides, and other gravity phenomena carry sediment from relatively shallow depths to the deep sea. Tongues of fine-grained turbidites extend across the abyssal plain from the outer continental margin and interfinger with deep-sea pelagic deposits. Islands also serve as a source for similar sediments and phenomena. Although not generally considered as terrigenous, reef debris and piles of volcanic deposits can be moved in this fashion and thereby reach the deep-sea floor.

Near the bottom of large areas of the world ocean there is a thick layer of sediment-laden, dense water called the nepheloid layer. Thickness may exceed 2 km and sediment concentrations range widely but are typically less than 0.3 mg/liter. This sediment reaches the nepheloid layer through fluvial discharge, turbidity currents, resuspension by bottom currents, and pelagic contributions (Pierce, 1976). It may be transported by deep ocean currents. The sediment particles tend to aggregate, commonly facilitated by organic material. Rates of accumulation vary greatly but are generally very slow; residence time for a given particle in the nepheloid layer would be up to a few years.

Distribution

The nature of the deep-sea environment, with its extremely low temperatures and high pressures, causes some pertubations on accumulation of sediments. Such factors as currents, depth, biologic activity, and physicochemical cycles are important considerations in characterizing sediment distribution in the deep sea.

Carbonate compensation depth

Calcium carbonate is carried in solution via rivers to the world ocean. A variety of types of precipitation of this compound occurs, with a significant portion being in the form of tests of planktonic organisms such as coccolithophores, foraminifera, and pteropods. These organisms die and the tests settle through the water column, where increased pressure and decreasing temperature cause solution of the calcium carbonate due to undersaturation of the deep ocean waters. The dissolution of planktonic carbonate tests has been known since the early work of Murray and Renard (1891), who recognized that the deep ocean water is undersaturated with respect to both calcium carbonate and silica.

The carbonate compensation depth (CCD) is the depth at a given place in the ocean where there is no net accumulation of carbonate due to the balance between the rate of supply and the rate of dissolution (Bramlette, 1961). This depth has a range of more than a kilometer throughout the world ocean (Figure 16–11) except in Antarctica, where it is only a few hundred meters (Anderson, 1975). A related aspect of the deep sea is the lysocline, which is the depth at which the effects of $CaCO_3$ dissolution

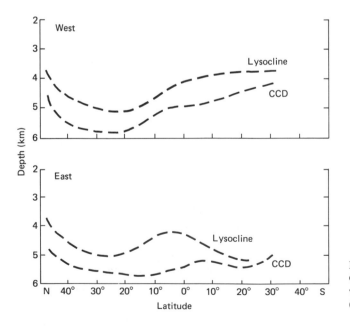

Figure 16–11 Depth distributions of the CCD and the lysocline on the east and west sides of the Mid-Atlantic Ridge. (From Berger, 1974, p. 223.)

are evident but not totally effective at removing carbonate (Figure 16–11). The lysocline and the CCD are not parallel within a given ocean basin. This is due to rates of dissolution and is largely controlled by the temperature of bottom waters (Berger, 1974). In general, the CCD is deeper in the center of the ocean basin and decreases toward the margin.

Silica precipitation

Three primary pelagic contributors of silica to the deep ocean floor are diatoms, dinoflagellates, and radiolarians, with sponge spicules also important. Both diatoms and radiolaria are favored by high fertility and their greatest production and abundance is adjacent to the continents where upwelling occurs. Oceanic circulation then distributes the organisms into "silica belts" around the ocean basins (Lisitzin, 1972).

In contrast to carbonate, silica experiences greatest corrosion in the upper waters of the ocean. This then produces contrasting trends:

1. Silica is favored by high fertility, whereas carbonate is not.
2. Silica corrosion is greatest in undersaturated upper water, whereas carbonate corrosion is greatest at great depth.
3. This results in silica accumulation at great depth, with carbonate restricted to shallower ocean floors (Berger, 1974).

Geographic distribution

It is possible to present the distribution of sediments in the world ocean using only five types in the deep sea in addition to the continental margin sediments (Figure 16–12). The global patterns show glacial marine sediments circumscribing the Antarctic and adjacent to Greenland; belts of siliceous sediment occur in the high latitudes and near the equator. Terrigenous sediment is patchy and adjacent to the continental margins. The remaining majority is about evenly divided between deep-sea clay and calcareous oozes (Figure 16–12), with depth a primary factor in controlling the distribution. An excellent summary of sediment distribution by Davies and Gorsline (1976) provides data on regional distribution within the world ocean.

The North Atlantic probably has more data on sediment distribution currently available than does any other part of the world ocean. Much of the sediment in the northernmost part of this region is terrigenous sediment that has been transported great distances (Figure 16–12). A complexity of source areas of all rock types and ages contributes lithogenous sediment, which is then mixed with the biogenic pelagic deposits, resulting in an extremely complex stratigraphy (Davies and Gorsline, 1976). The rate of pelagic carbonate ooze accumulation in this region is 35 to 60 mm every 10^3 years (Berger, 1974), among the highest in the world, yet much of the area is characterized by terrigenous sediment. This indicates that the terrigenous rate must be high enough to mask the highest rate of calcareous ooze accumulation.

A marked contrast is shown between the North Atlantic and the Central Pacific, where there is considerably less terrigenous influx and where accumulation of biogenic ooze is less than 20 mm every 10^3 years (Berger, 1974). The absence of ter-

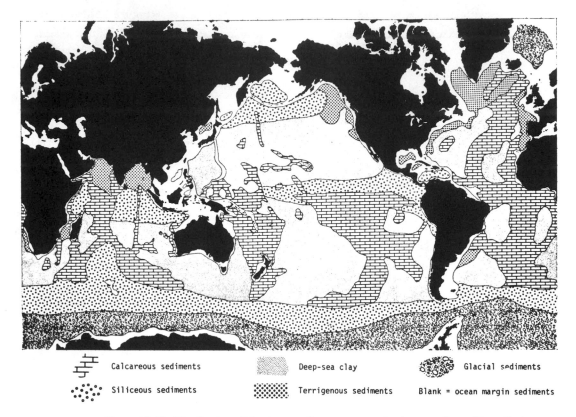

⊞	Calcareous sediments	⣿	Deep-sea clay	⣿	Glacial sediments
∴	Siliceous sediments	⣿	Terrigenous sediments		Blank = ocean margin sediments

Figure 16–12 Distribution of dominant sediment types on the deep-sea floor. (From Davies and Gorsline, 1976, p. 26.)

rigenous influx is in part due to the Pacific being mainly surrounded by mountainous coasts with short rivers that transport less clays than do long rivers, to the distance from landmasses, and also to trapping of sediment by trenches which nearly circumscribe the Pacific. Here the biogenic contributions far exceed terrigenous or pelagic clay, which accumulates at about 2 mm every 10^3 years. Major oceanic circulation patterns have a marked influence on distribution patterns in this region (Davies and Gorsline, 1976).

Although much smaller, the Indian Ocean shows a complex pattern of sediment distribution. Some large areas of rapid terrigenous influx are deep-sea fans related to riverine input, especially the Bengal Fan served by the Ganges-Brahmaputra River complex (Figure 16–12). Topography is important as calcareous ooze accumulates above the CCD along the edges of the basin and the north-south trending ridge that nearly bisects the ocean. Siliceous ooze is concentrated near the equator as it is in the Pacific, and pelagic clay characterizes the deep midlatitude areas (Figure 16–12).

A few general summary statements can be used to characterize deep-sea sediment distribution. Areas of upwelling typically are dominated by terrigenous mud and diatomaceous oozes regardless of depth. Below the CCD, radiolarian ooze prevails in regions of oceanic divergence and red clay is prevalent at the subtropical convergence (Berger, 1974). Calcareous ooze is abundant throughout these regions above the CCD. Distribution of clay minerals also tends to conform to some general patterns. Kaolinite is abundant in the low latitudes, reflecting the general intensity of weathering in these regions. Illite is more common in the higher latitudes. Smectite is more abundant in the Pacific Ocean than in the Atlantic, due to the relative abundance of volcanic activity in the Pacific over riverine runoff in the Atlantic.

Models for Oceanic Sedimentation

Sedimentation in ocean basins in the result of (1) source, (2) transportation, and (3) accumulation, which together with diagenesis, provide the distribution of sediment in the ocean with respect to both space and time. Using the present conditions in the world ocean, it is possible to construct a static model assuming a two-layered ocean (Davies and Gorsline, 1976). The surface layer responds to atmospheric circulation and produces gyres controlled by the interaction of the atmosphere, the upper layer of the ocean, and the landmasses. The pattern shown in Figure 16–13(a) is similar to that presented in Chapter 13 (see Figure 13–4). Upwelling and downwelling occur on eastern sides and western sides of oceans, respectively.

Deep ocean circulation is thermohaline in nature but is also affected by landmasses and to some extent by the oceanic ridge which is shown bisecting the model ocean *F*igure 16–13(b). The sediment distribution that emerges from these patterns of circulation is complex but predictable Figure 16–13(c).

The continental margin has been discussed in detail in Chapters 13 and 15, but it is worth observing the sediment types and comparing these with Figures 13–4 and 13–5. The ocean basin itself contains some glacial marine sediments in the high latitudes. Siliceous sediments are concentrated at locations of convergence and divergence of surface current circulation, where downwelling and upwelling is present and where biogenic productivity is high Figure 16–13(c). The oceanic ridge contains volcanic sediments flanked by calcareous ooze which persists because of the relatively shallow depths. This model shows quite extensive deposits of deep-sea clay, but calcareous ooze may occupy some of these areas if depths are above the CCD. Contourites are present along the outer continental margin where strong deep water currents occur *F*igure 16–13(c). Although this model applies theoretically to any major ocean, it strongly resembles the North Atlantic.

This model is applicable to present-day conditions of the globe. If earlier periods of geologic time are considered, numerous significant changes would have to be accommodated. Certainly, one of the most important is the location of landmasses on the globe and the resulting shape of oceanic bodies. Such features as bathymetry, currents, and climate would be altered and would thereby alter deep ocean sedimentation. Climatic changes such as extensive ice caps (Pleistocene) or evaporite deposition (Jurassic) would also be important changes in the model for oceanic sedimentation as proposed by Davies and Gorsline (1976).

Specific sedimentologic conditions superimposed on the conditions noted above tend to produce differing deep ocean sequences. Using the terminology of Gorsline (1980), ocean basins will be considered in two categories: (1) oceanic, which have active margins and little terrigenous sediment reaching the deep ocean, such as

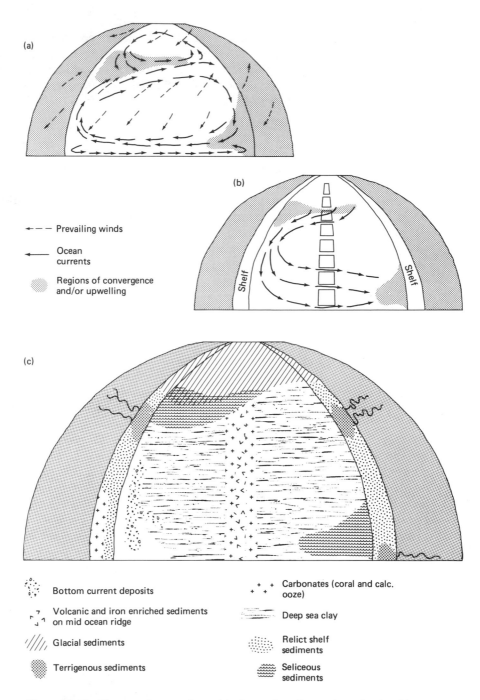

(a)

(b)

— ← - - Prevailing winds

←——— Ocean
currents

Regions of convergence
and/or upwelling

Shelf

Shelf

(c)

Bottom current deposits

Carbonates (coral and calc.
ooze)

Volcanic and iron enriched sediments
on mid ocean ridge

Deep sea clay

Glacial sediments

Relict shelf
sediments

Terrigenous sediments

Seliceous
sediments

Figure 16-13 Diagrams for a static model of oceanic sedimentation, showing (a) surface circulation, (b) deep circulation, and (c) predicted pattern of sediment distribution. (From Davies and Gorsline, 1976, p. 64.)

the Pacific; and (2) margin-affected, which have passive margins with much land-derived sediment reaching the basin, such as the Atlantic. In oceanic basins sedimentation is considered continuous because it is dominated by discontinuous processes such as turbidity currents.

The margin-affected basins may have three distinct sequences. One is characterized by the canyon–fan–abyssal plain sequences as the margin progrades basinward. The second is the slope and base-of-slope sequence, which is characterized by mass movement on steep gradients where conditions are unstable (Booth, 1979). The third type occurs as particles settle from the nepheloid layer or the highly productive surface waters (Gorsline, 1980). These sequences occur in decreasing relative abundance as listed.

Stratigraphic model

The concepts of sea-floor spreading and continental drift are now widely accepted. If one thinks for a moment about the impact of the mobility of the earth's crust on sediment distribution in both space and time, it is apparent that profound changes have been and are taking place. The relatively young age of the oceanic crust, the spreading rates of the sea floor, and the slow accumulation of pelagic sediments interact to produce a rather predictable stratigraphy in the deep sea.

Most students of geology are aware that the relative ages of basal deep-sea sediments increases away from the spreading centers or oceanic ridges. Good examples are present in both the Atlantic and Pacific Oceans, where Jurassic strata are found near the continents away from the ridges and very young sediments are found at or near the ridges (Figure 16–14).

These types of data, coupled with the data provided by sediment distribution on the ocean floor and the static model presented above (Figure 16–13), permit construction of a generalized model of sedimentation on the sea floor through time. The static model was empirically generated. The model was generalized from a wealth of data on sediment distribution and circulation phenomena. The stratigraphic model presented in Figure 16–15 is based largely on theory; data are still being collected that will determine its validity.

Assuming that the CCD has not changed markedly throughout the Cenozoic, perhaps even the Mesozoic, a particular stratigraphy is expected across an ocean from the ridge to the distal part of the basin. As the spreading center is fed by hydrothermal activity, iron- and manganese-rich products of volcanism will accumulate at the ridge crest. The crest of the ridge is well above the CCD and calcareous ooze accumulates. Through the spreading process the lithosphere also subsides, such that at some distance from the ridge crest, the CCD is exceeded and pelagic clay accumulates [Figure 16–15(a)]. Now the pelagic clay protects the underlying and older calcareous ooze, so that by drilling away from the ridge system one should encounter a stratigraphy such as that at the edge of the model. In fact, cores taken by the DSDP have supported this hypothesis (Broecker, 1972). Actually, the structure of this region is complex, due to the widespread faulting on the ridge. The actual cross section shows this stratigraphy, but it is superimposed on numerous fault blocks with many interruptions in the continuity of the pelagic sediments [Figure 16–15(b)].

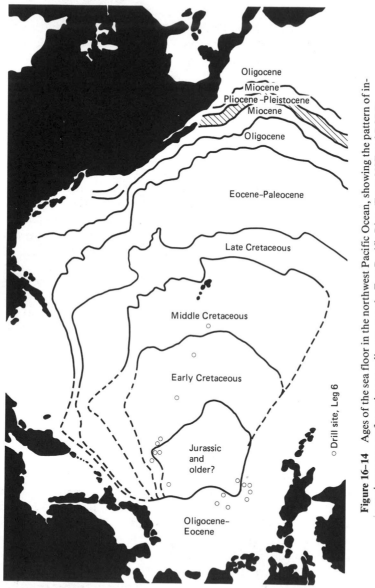

Figure 16-14 Ages of the sea floor in the northwest Pacific Ocean, showing the pattern of increasing age away from the spreading center, the East Pacific Rise. (From NSF and DSDP news release, 1970.)

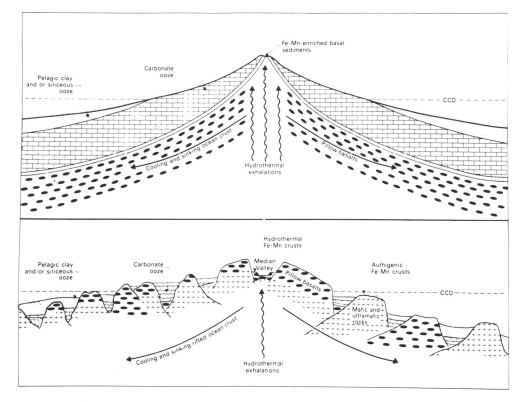

Figure 16-15 Stratigraphy and structure on a fast-spreading ridge, showing (a) general stratigraphic model and (b) structural complexities caused by fault block development. (From Jenkyns, 1978, p. 324.)

A perturbation on this theme occurs when such spreading is not along latitudinal lines but develops a northward or southward component, especially if equatorial or arctic waters are crossed. The best example is in the east-central Pacific Ocean, where spreading away from the East Pacific Rise is to the northwest and crosses the equator.

Siliceous ooze occurs as a belt at the equator (Figures 16-12 and 16-13). Spreading should therefore produce a section of iron- and manganese-rich material overlain by calcareous ooze on the ridge flank. Pelagic clay is added below the CCD, and at the equatorial zone, siliceous ooze accumulates. Once beyond the equatorial zone, pelagic clay is again dominant (Figure 16-16). These data are based on findings of the Deep Sea Drilling Project (Broecker, 1972). This is another excellent example of Walther's Law.

DEEP-SEA STRATIGRAPHY

It is pertinent here to reiterate the role of the Deep Sea Drilling Project in providing data on the history of the ocean basins. Since the first cruise in the Gulf of Mexico in 1968, several hundred sites have been drilled throughout the world, with hundreds of

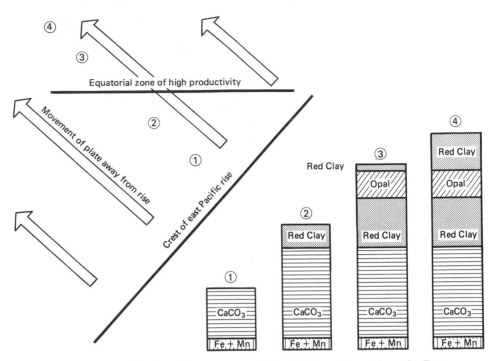

Figure 16-16 Stratigraphic sequences as plate migrates northwest from the East Pacific Rise across the equator. (From Broecker, 1972, p. 53.)

scientists from many nations participating. Although this project has not received as much publicity as the space program, it has at least matched it for contributions to our knowledge. The combination of this drilling project together with the hundreds of thousands of kilometers of seismic reflection records of the ocean floor has provided a rather good knowledge of the stratigraphy of this remote part of the earth's crust (Warme, et al., 1981).

Data show that there have been major climatic changes, rates of spreading have varied during the past, and there have been long periods when extensive areas were not accumulating sediment. The CCD ranged from near 3000 m to almost 5000 m during Cenozoic time.

In addition to major climatic changes there have also been important changes in oceanic currents through geologic time. Such changes may involve different patterns or velocities. A particularly important example of the latter is in the Southern Ocean, where Antarctic Bottom Water has shown a marked increase in velocity during the past few million years (Kennett and Watkins, 1976). Associated with the erosion pavement produced by these swift currents is an extensive pavement of manganese nodules, which covers more than 1 million km^2 in the southern Indian Ocean.

Major shifts in oceanic currents during the geologic past appear to have been caused by plate movement and separation of land masses. These shifts and coincident increases in current velocity have caused widespread scour on the ocean floor. An excellent example is the widespread Oligocene unconformity in the southwest Pacific and Southern Ocean. This unconformity extends over basins and ridges and is characterized by a distinct faunal break but no lithologic change (Kennett et al., 1972). More recent unconformities in the South Pacific display a horizon of manganese nodules (Kennett and Watkins, 1975).

Central Pacific Ocean

An excellent synthesis of the history of sedimentation as interpreted by the DSDP is presented by van Andel et al. (1975). Although four lithospheric plates are present in the region discussed, the summary here involves only the largest, the Pacific Plate. Some discrepencies exist between spreading rates as determined by magnetic anomalies (3.6 to 4.9 cm per year) compared to DSDP ages (7.1 to 9.5 cm per year) (van Andel and Bukry, 1973). In either case it provides an idea of the mobility of the crust in this region.

An east-west stratigraphic cross section from the East Pacific Rise to the center of the ocean (Figure 16–17) demonstrates well the types of accumulation shown in the models discussed previously (Figures 16–13 to 16–16). The basal iron- and manganese-rich zone is overlain virtually everywhere by calcareous ooze with siliceous ooze on the top. Much of the Oligocene through present is absent in the western part of the section. This demonstrates lack of accumulation or erosion of these regions over long periods of time.

A contrasting cross section is present at the equator, 10° south of the one shown in Figure 16–17. Here the Cenozoic section is fairly complete and little of the expected section is missing. This section includes the crest of the East Pacific Rise and demonstrates well the absence of deep-sea sediments on the crest, with thickening and subsidence away from it (van Andel and Bukry, 1973).

Rates of accumulation show great variation in time and space. They range from as little as a meter or so per million years to more than 50 m every 10^6 years. A composite graph showing rates through time as determined from all drill holes in the Central Pacific demonstrates the great variability (Figure 16–18). Even though this variation does exist, there are some obvious periods of widespread slow rates (e.g., Eocene-Oligocene boundary) and high rates (e.g., Late Oligocene and Late Miocene time).

ANCIENT DEEP-SEA DEPOSITS ON LAND

The nature of lithospheric dynamics is such that preservation and uplift of deep-sea sediments to a position above sea level is not commonplace. Subduction and the collapse of the outer continental margin commonly transforms deep-sea sediments into unrecognizable rock bodies. There are, however, some large-scale uplifts whereby deep-sea deposits are brought to the surface during subduction as plates collide. They may be well preserved, thus enabling their recognition and interpretation.

Often the terms pelagic and deep-sea sediments are used interchangeably. Such usage is incorrect and confusing. Pelagic sediments accumulate on the continental margin including the shelf, as well as in the deep sea. In situations where no terrigenous influx is present, pelagic accumulation may dominate as it apparently has in

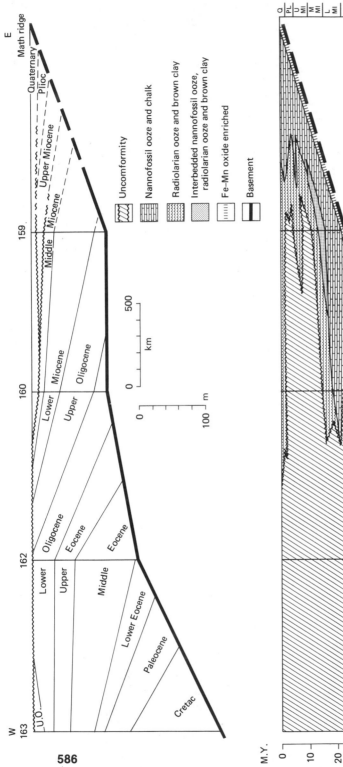

Figure 16-17 Biostratigraphic-lithostratigraphic section north of the equator in the central Pacific. (From van Andel et al., 1975, p. 24.)

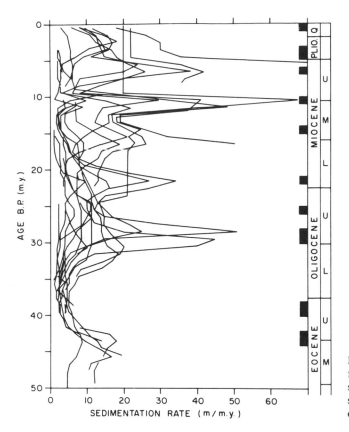

Figure 16–18 Sedimentation rates from many cores of the Central Pacific showing great variation in both space and time. (From van Andel et al., 1975, p. 59.)

the geologic past, such as the Cretaceous chalks of the British Isles. Deep-sea sediments are those deposited on the ocean floor and are those to which this discussion is devoted.

Recognition of Ancient Deep-Sea Deposits

Many criteria have been used to support an interpretation of a deep-sea environment of deposition for various stratigraphic sequences. The dominance of pelagic organisms has been used widely, especially when incorporated in sequences of thin and regularly bedded deposits. Other deep-sea criteria include manganese nodules and association with features of the oceanic crust such as **ophiolite** assemblages and pillow lavas.

An excellent, but perhaps extreme example of the difficulty associated with interpretation of deep-sea deposits is the conflict of opinion on the depositional environment of the Caballos Formation, a Paleozoic unit in west Texas. R. L. Folk and E. F. McBride have studied the Caballos and have markedly different interpretations of its origin (Folk and McBride, 1976; McBride and Folk, 1977). McBride takes the more traditional stand that the Caballos is a deep-sea deposit based on the radiolaria, manganese nodules, similarity of red shale and chert interbeds with modern deep-sea deposits, and its stratigraphic position within other supposed deep-sea deposits. On the other side, R. L. Folk interprets the Caballos as representing shallow-water deposition based on possible algal structures, geopetal structures, tidal channels, karst features, fossil logs, and paleosols. The object here is not to pass judgment but to emphasize the difficulty of making interpretations of deep-sea deposits; as mentioned above, this is an extreme example.

Troodos Massif (Cretaceous-Tertiary) of Cyprus

The island of Cyprus is of great interest geologically since the formation of the modern concepts of plate tectonics. The Troodos Massif in the southwestern part of the island is regarded as a piece of the ocean floor. A thorough study of the upper Troodos pillow lavas and their overlying pelagic sediments has been presented by Robertson and Hudson (1974). The following discussion is primarily from their work.

A composite section of the pillow lavas and the younger sedimentary strata shows various lithologies with a total of up to almost 800 m (Figure 16–19). Pillow lavas at the base show all the typical features of subaqueous flows. No evidence of emergence is present.

Immediately above the pillow basalts are fine-grained sediments which are brown to black in color with a chemical composition dominated by iron and manganese oxides. The thickness is generally only a few meters but may reach 30 m. X-ray diffraction patterns show this material to be mostly amorphous. The term umbers has been applied to these strata (Robertson and Hudson, 1974). The umbers are interbedded with light-colored tuffs. This material is very similar to ferromanganese sediments currently being formed on the East Pacific Rise. Umbers of the Troodos are interpreted as having formed as volcanism subsided, with thermal springs being active.

At most locations the umbers are overlain by thin radiolarian mudstone and radiolarites. These pink or pale gray strata are well bedded, with differing clay content being the cause of the layering. Thickness ranges up to 35 m but is typically much less. Like the umbers, these radiolarian strata tend to fill in the topographic lows in the pillow basalts below (Robertson and Hudson, 1974). The radiolarians are abundant and are well preserved. The siliceous cement is thought by some authors (e.g., Gibson and Towe, 1971) to result from devitrification of volcanic glass. In the Troodos section its presence above the umbers suggests that the silica may have been provided by the thermal springs that produce the umbers.

Scattered patches of illite and smectite clay overlie the radiolarian strata. They also tend to fill the upper part of topographic lows, which contain the umbers and radiolarites. Locally, the clay may be interbedded with sandstone composed of volcanic debris.

The earliest deposition of carbonates in this region was during the late Cretaceous and consisted of somewhat discontinuous accumulations of chert-free chalk (Figure 16–19). Coccoliths are numerous and poorly preserved planktonic foraminifera are also present. This unit is interpreted as having been deposited close

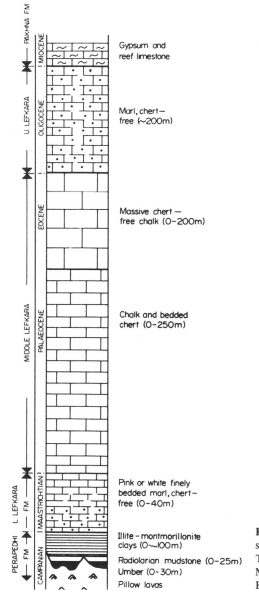

Figure 16-19 Composite stratigraphic section of Upper Cretaceous and Lower Tertiary strata from the Troodos Massif, Cyprus. (From Robertson and Hudson, 1974, p. 406.)

to the CCD, due to the presence of large coccoliths and the poor condition of the foraminifera (Robertson and Hudson, 1974).

Thick carbonate sequences reaching as much as 65 m accumulated in the Troodos region during Tertiary time. The Paleocene and Lower Eocene chalks contain interbedded cherts, some of which are nodular. They are interpreted as having a replacement origin; petrographic examination reveals foraminiferal ghosts in the rocks. Succeeding Eocene and Miocene carbonates are dominated by planktonic organisms. Decreasing depth of water is indicated by the taxa present and the section is capped by reef limestones with evaporites and evidence of subaerial emergence. This marked decrease in depth through the period of sediment accumulation is due to the extreme tectonic events in the area of the Tethys Ocean (Moores and Vine, 1971; Robertson and Hudson, 1974).

Ligurian Apennines (Jurassic) of Italy

Another intensely studied sequence of strata attributed by most to a deep-sea origin occurs in the Mesozoic and Cenozoic of the Ligurian Apennines of Italy. These rocks crop out in northwestern Italy adjacent to the Mediterranean Sea (Figure 16-20). Like the example discussed previously, these rocks owe their preservation on land to the intense tectonic activity in the Tethyian Sea area during the Cenozoic.

Rocks representing oceanic crust are typically used as strong evidence in support of a deep-sea origin for overlying sedimentary strata. In Liguria the ophiolitic sequence of the Mesozoic ocean floor consists of serpentine, gabbro, pillow, and massive basalts together with breccia derived from these facies (Barrett and Spooner, 1977). The breccias range in composition and are interpreted to represent talus deposits at the base of submarine faults scarps.

The general nature of the Mesozoic section in Liguria from the base upward is ophiolites, radiolarite, pelagic limestone, and prethrust flysch (Folk and McBride, 1978). This sequence is not according to the model presented in Figure 16-15(a). A postulated rationale for this discrepancy will be presented.

The radiolarites range widely in thickness, with the maximum near 200 m. At some localities these units thin from a few meters to a featheredge. The typical

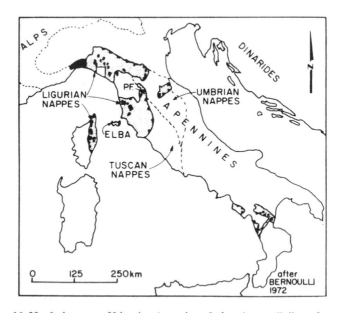

Figure 16-20 Index map of Ligurian Apennines, Italy, where radiolites of supposed deep-sea origin are exposed. (From Folk and McBride, 1978, p. 1070.)

radiolarite lithology is a red, interbedded shale and chert which has been called "ribbon-bedded" (Figure 16–21). The shale strata are dominated by clay minerals with < 5% radiolarians and the chert strata are > 50% radiolarians (Folk and McBride, 1978). These radiolarites contain a wide variety of sedimentary structures, including some ripples, grading, pull-aparts, and other types of soft-sediment deformation. Trace fossils are common, especially grazing structures on bedding plane surfaces. All these features have been observed on the present ocean floor (Heezen and Hollister, 1971).

Deposition of these radiolites is generally considered to have taken place on the ocean floor below the CCD. This was probably about 2500 m during Jurassic time (Hsu, 1976), which is the average depth of modern spreading ridges. However, R. L. Folk interprets this radiolite sequence as representing shallow water (Folk and McBride, 1978).

The radiolites are succeeded in sequence by carbonate strata of the Calpionella Limestone (Folk and McBride, 1978), which is Late Jurassic in age. This white carbonate lithofacies accumulated through Miocene time. Initial studies and cursory examination indicate a thick, monotonous coccolith ooze which is micritic in outcrop. Close inspection reveals flute casts and other sole marks which are defined by thin shale breaks. Grading is present but is very subtle, due to the overall uniformity in grain size of the coccoliths. According to McBride's interpretation, they are resedimented biogenic particles analogous to thin turbidites (see Chapter 15).

This sequence is capped by Miocene-to-Pliocene flysch sequences similar to those described by Ricci-Lucchi (1975). These turbidites were deposited on deep-sea

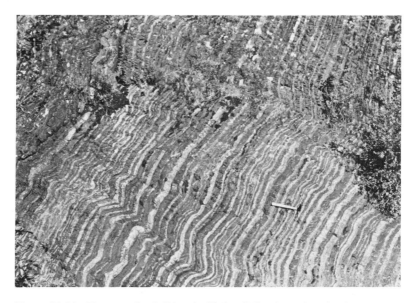

Figure 16–21 Photograph of ribbon-bedded radiolite from the Ligurian Appennines, Italy. (Courtesy of R. L. Folk.)

fans and contain the classic types of sequences associated with this type of environment (see Chapter 15).

The complete Ligurian sequence consists of a shallowing-upward type of section going from oceanic crust below the CCD to a coccolith-accumulating environment with a marginal deep-sea fan system on top. Obviously, tectonic activity was an important influence and spreading played a role. It seems that the collision of the spreading ocean floor with the European continent resulted in some shallowing and uplift, which produced the flysch accumulation. The reversal of the stratigraphic sequence compared to modern spreading centers such as the East Pacific Rise [Figure 16–15(a)] is due at least in part to the relatively small size of the basin and the tectonics associated with the collision of the spreading oceanic crust with the continental crust (Folk and McBride, 1978).

Paleozoic of the Cordilleran Foldbelt

Strata of the pre-Mesozoic portion of the rock record present more difficulty insofar as deep-sea depositional environments are concerned. Planktonic foraminifera and radiolarians are not common in Paleozoic strata and coccoliths do not occur in this interval. Both siliceous and calcareous plankton became abundant in the Jurassic (Schopf, 1980). The most commonly used criteria for identification of Paleozoic deep-sea deposits are dark shales and abundant graptolites.

Good examples of supposed deep-sea accumulations occur in the Cambrian through Devonian strata of the Cordilleran foldbelt in North America (Figure 16–22). A synthesis of these sections by Churkin (1974) provides a good summary of the stratigraphy and the interpreted environments of deposition. In his study, Churkin considered three broad systems; only the graptolitic shale and chert belt are discussed here.

The complex structural relationships in the regions of these graptolitic shales and chert demonstrate that most were thrust over various other Palezoic rocks; however, there are a few locations, especially in Nevada (Figure 16–22), where pillow basalts are interbedded with the shale and chert (Churkin and Kay, 1967) near the base of the sections. These data suggest an association with oceanic crust.

The graptolitic sequences are quite condensed in thickness relative to their temporal equivalent carbonates and quartzites to the east (Figure 16–22). These are thin carbonate and quartizite layers interbedded with the graptolitic units (Figure 16–23). Rates of accumulation of up to 1 mm/10^3 yr have been estimated (Churkin, 1974). Silica content is 75 to 90% SiO_2, which is high even for radiolarites. It is postulated that ash from explosive volcanoes enriched the shales; some relict pieces of pumice and glass shards have been found (Churkin, 1974).

Only chitinophosphatic and siliceous fossils such as linguloid brachiopods, radiolarians, and sponge spincules are found with the graptolites. The absence of calcareous varieties suggest a depth of water below the CCD. A somewhat restricted circulation is envisioned, with volcanic sources to the west. The basin was narrow but deep (Churkin, 1974).

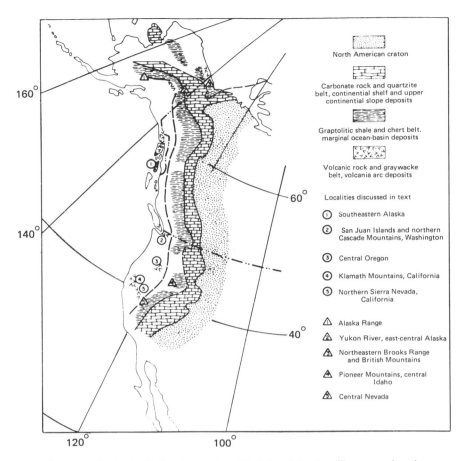

Figure 16-22 Major Paleozoic stratigraphic belts of the Cordilleran province in western North America. (From Churkin, 1974, p. 175.)

SUMMARY OF CHARACTERISTICS
OF SEDIMENT/SEDIMENTARY ROCK BODIES
REPRESENTING THE DEEP OCEAN DEPOSITIONAL SYSTEM

Stratigraphic models of the deep ocean system have been produced recently, largely due to the extensive contributions of the Deep Sea Drilling Project (DSDP). Although geologists who study the stratigraphic record of this system are still seeking a thorough understanding of its overall nature, much progress has been made.

Tectonic setting. The deep ocean system is so extensive that it actually encompasses a broad spectrum of tectonic conditions. These range from stable abyssal floors to the active spreading centers of the oceanic ridges. A stable ocean floor with little relief is most conducive to the accumulation of typical deep ocean sequences.

Shape. Deep ocean sediments accumulate in extensive tabular or blanket-

Chapter 16 The Deep Ocean System

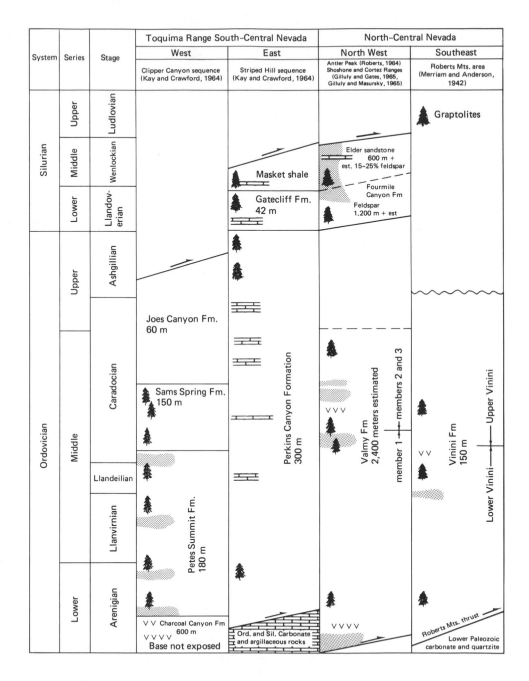

Figure 16–23 Generalized stratigraphic sections of grapholitic shale and chert belt in the Lower Paleozoic in Nevada. (From Churkin, 1974, p. 178.)

shaped deposits. There is little variation on this theme except that these tabular deposits may be discontinuous in high-relief areas within the oceanic ridge province.

Size. This is the most widespread of all depositional systems; however, when viewed in the rock record, the strata tend to possess limited extent. The reasons include their consumption during subduction and the extensive faulting that was necessary to bring the strata above sea level where they can be viewed. The rate of accumulation of sediment is extremely low; thus sequences are thin compared to other systems which represent a comparable amount of time.

Textural trends. No significant textural trends are displayed by deep ocean strata. Virtually everything that accumulates is very fine grained and the variation in particle size is controlled primarily by the origin of the particle.

Lithology. The composition of deep ocean strata is rather simple. These strata consist primarily of clay minerals such as illite, kaolinite, and smectite; aragonite in the form of coccoliths and foraminifera; and silica produced by diatoms and radiolarians. Other minerals are uncommon except for iron-manganese nodules which may be locally abundant.

Sedimentary structures. Horizontal, fine laminations are nearly ubiquitous and are the only common physical features of deep ocean strata. Rarely, bioturbation features are present.

Paleontology. The fossils contained in deep ocean sequences represent one of the most diagnostic characteristics of this depositional system. Extensive accumulations of diatoms, radiolarians, planktonic foraminifera, and coccoliths are not only indicative of the deep ocean environment, but the relative abundance may be used in interpreting water depth. Most interpretations of apparent deep ocean deposits rest heavily on the contained fossils.

Associations. Some common stratigraphic associations are prevalent in deep ocean and related strata. The sedimentary sequence typically overlies an oceanic crust sequence which is generally capped by pillow lavas. A thin zone enriched in iron and manganese oxides separates the crustal basement from overlying deep ocean sedimentary strata. The sedimentary sequence may consist of nearly any arrangement of composition types, depending on water depth and other factors.

ADDITIONAL READING

HAY, W. W., (ED.), 1974. *Studies in Paleooceanography.* Soc. Econ. Paleontologists and Mineralogists, Spec. Publ. No. 20, Tulsa, Okla., 218 p. A collection of papers summarizing the state of knowledge at the time on various topics concerned with deep-sea sedimentation, such as carbonate dissolution, history of circulation patterns, and chemical cycles.

HEEZEN, B. C., AND HOLLISTER, C. D., 1971. *The Face of the Deep.* Oxford University Press, New York, 659 p. A great compilation of photographs on the deep-sea floor with a brief but good text.

HSU, K. J., AND JENKYNS, H. C. (EDS.), 1974. *Pelagic Sediments on Land and under the Sea.* Internat. Assoc. Sedimentologists, Spec. Publ. No. 1., Blackwell, London, 447 p. A fine collection of papers on various aspects of deep-sea sedimentation with a good balance between modern and ancient deposits.

LISITZIN, A. P., 1972. *Sedimentation in the World Ocean*. Soc. Econ. Paleontologists and Mineralogists, Spec. Publ. No. 17, Tulsa, Okla., 218 p. An excellent summary of the distribution of surface sediments throughout the world. Much Soviet literature and data are synthesized, together with good coverage of all available data.

RILEY, J. P., AND CHESTER, R. L. (EDS.), 1976. *Chemical Oceanography*, v. 5, 2nd ed. Academic Press, New York, 401 p. The volume is somewhat misnamed in that it is devoted entirely to various aspects of sedimentation in the deep-sea environment.

WARME, J. E., DOUGLAS, R. G., AND WINTERER, E. L. (EDS.), 1981. The Deep Sea Drilling Project: A Decade of Progress. Soc. Econ. Paleontologists and Mineralogists, Spec. Publ. No. 32, Tulsa, Okla., 564 p. A tremendous summary of the first 10 years of the DSDP, organized on the topical basis. A must for anyone interested in the geology of the ocean basins.

Glossary

Allochem Sediment grains of calcium carbonate which are formed by physico-chemical or biochemical precipitation within a depositional basin. Included are intraclasts, ooids, biogenic grains, and pellets.

Allochthonous shelf A continental shelf system where the sediment being accumulated comes from outside the shelf.

Amictic A type of lake that displays no seasonal cycles.

Amphidromic point A point of no tide on a chart of cotidal lines from which the cotidal lines radiate.

Antidunes A type of bedform that moves progressively upcurrent.

Atoll A ring-shaped reef with a central lagoon; the reef rises from the ocean floor.

Autochthonous shelf A continental shelf system where the sediment being accumulated is derived from the shelf itself.

Autosuspension The suspension of sediment caused by fluid turbulence, which itself is generated by the motion of the current. The current itself is, in turn, impelled by the contrast in weight between the suspension and the overlying fluid.

Backshore That part of the beach that is usually dry and is reached only by the highest tides and storm waves; approximately equal to the supratidal zone.

Bafflestone Rock containing in situ stalked or branching fossils that trapped sediment by baffling (e.g., bryozoans or crinoids).

Bajada A broad, sloping depositional surface at the base of a mountain range and extending into an adjacent basin. It is formed by the coalescing of alluvial fans.

Barrier reef A reef that is separated from a landmass by a lagoon in which normal reef coral does not grow.

Bedform A morphologic feature having various systematic patterns of relief, and created by the conditions of flow at the dynamic interface between a body of cohesionless sediment particles and a fluid.

Bedload The sediment load that moves by traction along the bed as the result of shearing at the boundary of the flow.

Bindstone Tabular, in situ fossils that bound sediment together during deposition (e.g., stromatoporoids).

Bioerosion Removal of a lithic or consolidated mineral substrate by the direct action of organisms.

Biofacies Assemblages of animals or plants formed at the same time or which display lateral variation in a stratigraphic unit.

Biogenic sediment Sediment that is derived from biologically generated particles.

Bioturbation The reworking of sediment by organisms.

Bird's-eye structure A pattern of small voids in carbonate sediments or rocks. These voids may be open or filled with cement. They range widely in shape.

Bog A saturated, spongy groundmass comprised largely of decaying vegetation, which may develop into peat.

Bore The advancing edge or front of a tidal wave as it ascends a river or an estuary.

Boring An excavation in a rock or solid substrate by an organism.

Bottom marks (sole marks) Sedimentary structures on the bottom surfaces of beds generally composed of cohesionless particles and overlying cohesive material.

Boundary layer The layer at the boundary where a fluid flows over sediment or over a solid surface.

Burrow An excavation, generally cylindrical in shape, made by an organism in unconsolidated sediment.

Calcrete Term used for surficial cementation of sand and gravel by calcium carbonate.

Capacity The quantity of material transported by a fluid past a point in a given time.

Carbonate buildup A general term for any carbonate accumulation which is thicker than laterally equivalent deposits and was morphologically higher than adjacent sediment during deposition.

Carbonate lake A lake that is accumulating large percentages of carbonate material typically in the form of marl.

Chalk A soft, somewhat friable, very fine grained limestone comprised almost totally of planktonic organisms, largely coccoliths or foraminifera.

Channel-fill deposits Stream deposits that accumulate within the banks of a channel.

Cheniers A beach ridge built on swamp or marsh deposits.

Chute A rather narrow channel through which water flows rapidly during or near overbank conditions. They typically are developed on the point bar or the inside of a meander.

Chute bar   Lobate or parabola-shaped accumulation of sediment at the end of a chute in the point-bar environment.

Clastic dike   A tabular body of clastic material that transects the strata of sedimentary units by invasion from below.

Climbing-ripple cross-stratification   Cross-stratification formed by the combination of ripple migration and additional sediment.

Coarse-grained meander belt   Coarse-grained sediment accumulation by meandering streams. It is characterized by higher gradients and more flashy discharge than those of fine-grained meander belts.

Coastal bay   Enclosed or semienclosed water bodies adjacent to the open marine environment. Lagoons and estuaries are included.

Competence   The maximum particle size of a given density that a fluid can transport under a given velocity.

Concretions   A nodular or irregular concentration of authigenic constituents developed by localized precipitation of material, generally around a nucleus.

Contour current   Oceanic currents that flow parallel to bathymetric contours and are typically found along the outer continental margin.

Crevasse-splay   A fan-shaped sediment body formed by overbank breeching of a levee. Sediment-laden water moves through a crevasse and spreads in a thin accumulation.

Debris flow   A type of sediment gravity flow in which fine sediment and fluids support large particles.

Delta front   The front sloping portion of a delta just seaward of the delta plain. It is characterized by sand and grades seaward into deeper prodelta muds.

Density current   A gravity-induced current caused by a density gradient between adjacent fluids. This density range may be due to temperature, salinity, or suspended sediment.

Diagenesis   All physical and chemical changes in sediment after deposition and up to metamorphism.

Diamictite   An unsorted, terrigenous sediment that is comprised of a mixture of mud with coarser particles. It is a nongenetic term for till.

Dimictic   A type of lake that experiences two seasonal cycles, each of which contains a long period of thermal stratification followed by a short period during which there is no stratification.

Distal bar   The outer bar of the delta front province just landward of prodelta. It is commonly a mixture of mud and sand with both physical and biogenic structures.

Distributary   An outflowing branch that flows away from the main stream and does not rejoin it, such as in a delta.

Dreikanter   A pebble shaped by the wind into three faces with sharp angles. A type of ventifact.

Drumstick barrier   A barrier island characterized by one narrow end and one wide end which contains numerous accretion ridges.

Ebb delta   A tidal delta formed by ebbing tidal currents and modified in shape by waves.

Ebb dominated   The condition of a tidal water body having a greater ebb tidal prism than flood tidal prism.

Ebb slack   That portion of the ebb tidal cycle during which no tidal currents are flowing. It generally has a duration of only a few minutes.

Eolianites   A lithified dune deposit. The term is most commonly used for Pleistocene age deposits comprised largely of calcium carbonate.

Epilimnion   The upper layer in a stratified lake.

Epsilon cross-stratification   A large-scale stratification produced by successive accretion units commonly found on point bars. It is gently sloping toward the thalweg of the channel.

Equivalent diameter   Sediment particle size for various sizes, shapes, and densities which fall through water at an equal velocity.

Erg   A sea of sand containing extensive dune complexes, such as the Sahara Desert.

Esker   A long, narrow, and typically sinuous accumulation of sand and gravel deposited by an englacial or subglacial stream.

Estuary   A coastal bay that receives significant freshwater runoff and has open tidal exchange with the open marine environment.

Eutrophic   A water body that contains an abundance of nutrients.

Fabric   The arrangement of sediment particles in an accumulation.

Facies   The total lithologic and biologic characteristics of a sedimentary deposit resulting from accumulation in a depositional environment.

Fan valley   A valley in a deep-sea fan. They generally are arranged in a distributary fashion.

Fining-upward sequence   A stratigraphic sequence that displays a decrease in grain size (fining) from bottom to top.

Flaser bedding   A type of stratification which consists of alternations of ripple cross-strata and discontinuous mud lenses.

Flood delta   A tidal delta that is generated by flood currents and is located on the landward side of an inlet.

Flood dominated   A tidal body that is subjected to a flood tidal prism that is larger than the ebb prism.

Flood stack   That portion of the flood tidal cycle when no tidal currents are flowing. It generally lasts for only a few minutes.

Flow regime   The relationships that exist between the various flow conditions and bedforms at the sediment/water interface and the water surface.

Fluidized sediment flow   A type of sediment gravity flow in which the sediment is supported by upward-flowing fluid as grains settle.

Flute cast   A cast formed by a parabolic- or spatulate-shaped scour formed by a turbulent current moving over cohesive sediment.

Flysch   A general term used to describe sequences of interstratified mudstones and sandstones which form thick sequences in folded mountain belts.

Forced waves   Waves generated and maintained by a continuous force such as tidal waves.

Foredunes   Dune ridge developed along the shoreward portion of a coastal dune complex.

Foreshore   The portion of the beach between normal high tide and normal low tide. It is relatively steep in comparison with the backshore. Also called beach face.

Framestone Massive, in situ fossils that form the framework of a reef.

Fringing reef A reef that is contiguous with its adjacent landmass.

Froude number A dimensionless number that expresses the relationship between inertial and gravity forces in a fluid. It is typically applied to flow regime and generation of bedforms.

Geostrophic current A current that is defined by the assumption that an exact balance exists between the horizontal pressure gradient and the Coriolis effect.

Glacial drift A general term used to describe all sediment carried and deposited by or associated with glaciers.

Glacial marine Refers to marine terrigenous sediment which was derived from a glacier.

Glaciofluvial Refers to streams that flow from glaciers or the sediment deposited by these streams.

Glauconite A green phyllosilicate mineral which is essentially a hydrous potassium, iron silicate. It is restricted to sedimentary rocks of marine origin.

Graded bedding A type of stratification that displays a gradation in grain size from bottom to top. It is typically coarsest at the bottom, but the reverse situation may also occur.

Grain flow A type of sediment gravity flow in which the sediment is supported by direct grain-to-grain interactions.

Gravity waves A wave whose velocity of propagation is controlled primarily by gravity.

Half-tide level The condition or time of the tide which is at the level midway between any given high tide and the following or preceding low tide.

Heavy minerals Detrital minerals which have a specific gravity greater than 2.85 and are a common but minor constituent of most terrigenous sediments.

Homopyncnal A condition in which the water has uniform density.

Humid fan A sediment fan accumulation which develops in a humid climate, commonly associated with glacial melting. Humid fans are much less steep than arid fans.

Hummocky cross-stratification A gently undulating type of cross-stratification having sets which may be concave-up or convex-up. Sets intersect one another at angles of about 3 to 6°.

Hypersaline Pertains to water that is typically above normal marine concentration due to lack of circulation and a high rate of evaporation.

Hypolimnion The lower layer in a stratified lake.

Hypopycnal Refers to a type of flow such that the inflowing water is less dense than the body of water it enters (e.g., a river entering the marine environment).

Imbrication The condition whereby sediment particles having at least one long axis are inclined with the current. The particles dip upstream.

Internal waves A wave developed on a density surface in a well-stratified water column; typically the heights, periods, and wavelengths are large.

Intraclasts Aggregates of semi-lithified lime mud which are eroded from their site of accumulation and transported within the basin of deposition.

Inverse grading The type of textural grading in strata which shows an increase in grain size upward.

Iron formation A rock type in the Precambrian consisting of alternating strata of iron-rich material and siliceous material.

Isolith map A portrayal of variations in aggregate thickness of a given lithologic facies perpendicular to bedding at selected points.

Kame An elongate accumulation of poorly stratified sand and gravel deposited by a subglacial stream at or near the terminus of a glacier.

Kame terrace A terracelike accumulation of sand and gravel deposited at or near the terminus of a glacier by a meltwater stream.

Kurtosis The peakedness of a frequency distribution curve expressed as a percent.

Ladderback ripples Two sets of superimposed ripples oriented essentially at right angles to each other.

Lagoon A coastal bay that lacks significant runoff and appreciable tidal circulation. Salinity is typically elevated with respect to normal marine levels.

Laminar flow Fluid flow in which there is sliding within the fluid along surfaces that conform to the boundary of the fluids.

Late-stage emergence Refers to the emergence or exposure of tidal flats near the end of the ebb cycle. Commonly, water is flowing due to gravity only at this stage and bedforms may be generated at large angles to those generated by forced tidal currents.

Lateral moraine Accumulations of glacial drift along valley sides after the glacier melts.

Lebenspurren Features in sediment made by interaction of organisms with the sediment; trace fossils.

Lenticular bedding A type of stratification that consists of discontinuous lenses of ripple-bedded sand within mud.

Lime mud Calcium carbonate material which is less than 4 μm in diameter; micrite.

Limnology The study of lakes.

Linguoid bar Term used for sediment bodies with a spoon-shaped, lobate outline that are common in braided streams.

Liquified sediment flow A type of sediment gravity flow in which sediment particles are supported by the upward flow of fluid escaping between grains as the grains settle out by gravity.

Lithiclast (lithoclast) A carbonate rock fragment, generally of gravel size, which was derived mechanically from a previously existing carbonate rock.

Lithofacies Term commonly used to represent the various lithologic attributes of a sediment or sedimentary rock body.

Loess Homogeneous, massive silt deposited by wind and typically derived from glaciofluvial deposits.

Log-normal distribution A frequency distribution whose logarithms follow a normal distribution.

Longitudinal bar Bar-shaped accumulation of sediment which is elongate parallel to flow and typical of braided streams.

Longshore current A coastal current generated by waves approaching the shore at an oblique angle and refracting in shallow water. These currents are most prominent in the zone of breaking waves.

Lower flat bed   Refers to conditions in the flow regime characterized by low Froude numbers with minimal bedload sediment transport producing horizontal strata.

Marl   Mixture of calcium carbonate and clay, typically fine grained.

Marsh   An area of low, wet land which is covered with grasses. Coastal marshes are high in the intertidal zone.

Mean   The arithmetic average value in a group of data.

Median   The 50th percentile in a population; the middle.

Megaripples   Bedforms with a wave length or spacing between 0.5 and 5.0 m.

Micrite   Fine-grained calcium carbonate with particle size less than 4 μm.

Mode   The most abundant class in a population. The highest point on a frequency curve.

Molasse   Terrigenous shallow marine strata of various textures and compositions which combine to form a thick sequence. They are not as rhythmic as flysch deposits and represent continental or shallow marine deposition.

Moments   Statistical measurements on a population which are based on all values in the population. Included are the mean, standard deviation, skewness, and kurtosis.

Monomictic   Refers to lakes that have one cycle of stratification and overturn per year.

Mud (mudstone)   Sediment or sedimentary rock which is composed of a combination of silt- and clay-size particles without regard to composition.

Natural levee   A low ridge of sediment accumulation bordering the channel of a meandering stream. An overbank deposit generally of fine sand and mud.

Neap tide   Tide having the lowest range in the lunar cycle. Caused by the sun and moon being at right angles to the earth. It occurs every two weeks.

Nodules   Small, somewhat spherical bodies which commonly have some type of nucleus. They are commonly the result of local chemical conditions.

Normal distribution   A population which may be expressed by an equation that when plotted displays a symmetrical curve which rises from zero to a maximum and then declines to zero.

Oligotrophic   Pertaining to water bodies that have a low concentration of nutrients.

Olistostrome   A sedimentary deposit composed of a complex mass of heterogeneous materials that accumulated as a result of gravity sliding or slumping of unconsolidated sediments.

Oncolites   A somewhat spherical mass formed by encrustations of algae around a nucleus. Also called algal pisolites.

Ooids   Sand-size spherical or elliptical bodies of calcium carbonate with a central nucleus surrounded by concentric rings.

Ophiolites   General assemblage of basic igneous rocks, including serpentine, chlorite, epidote, and albite.

Orthochem   The material that acts as cement or matrix to allochems in carbonate rocks. Included are sparry cement and micrite.

Orthoconglomerates   Conglomerate that is texturally mature and deposited by normal but turbulent aqueous conditions. The opposite of paraconglomerate.

Orthogonal   Line on a refraction diagram that is everywhere perpendicular to the wave crest.

Outwash Sediment deposited by glacial meltwater streams.

Overbank deposits Sediment that accumulates beyond the channel in the fluvial system. Included are levees, crevasse splays, and floodplains.

Overgrowths Secondary deposit on a host grain in optical continuity. Common in sedimentary rocks, especially on quartz grains.

Packing The spatial density of particles in a sediment accumulation.

Palimpsest sediment Relict sediments that have been reworked by the environment which they now occupy.

Paraconglomerate The type of conglomerate that is deposited by sediment gravity flow or glacial ice rather than to normal aqueous flow. They are without internal organization.

Parting lineation A small but distinct sedimentary structure consisting of near-parallel ridges and grooves only a few millimeters wide or in relief. They are aligned parallel to current flow and are characteristic of plane bedded sandstones.

Pebbly mudstone A rock comprised of mud matrix with dispersed clasts. Clasts may display crude bedding. No origin is implied, but the term is commonly applied to rocks of suspected glacial origin.

Pelagic The division of the marine environment which includes the entire mass of water in the neritic and oceanic provinces.

Pellets A sand-size aggregate of fine particles which is excreted by an organism, generally in elliptical or cylindrical form.

Permeability The ability of a porous material to transmit fluids.

Planar cross-stratification The type of cross-stratification that is only two-dimensional. It is developed by migration of linear bedforms.

Plane beds The stratification which is also a bedform in the upper flow regime. The layering is planar or two-dimensional.

Playa lake An ephemeral lake in a desert system. It is typically characterized by thin accumulations containing evaporites.

Plunge point Location of the most landward-breaking wave at the shore; typically, the base of the foreshore beach. Sometimes called the plunge step.

Plunging breaker A breaking wave that tends to curl over and crash as it breaks.

Polyhaline bays Coastal bays that experience wide variation in salinity, generally on a seasonal basis. They may be below normal marine salinity in the wet season and above normal in the dry season.

Polymictic conglomerate A conglomerate that contains particles of many rock types.

Porosity The ratio of void space to the total volume of rock expressed in percent.

Provenance Refers to the origin of sedimentary particles. Determination of location and composition of source rocks from which sediments are derived.

Pyroclastic sediment A term covering all particles exploded from a volcano; tephra.

Reactivation surface A broad, curving surface that truncates cross-stratification and is caused by changes in currents which impart a change in bedform profile.

Red beds General term applied to red-colored sedimentary rocks, usually shale and sandstone. The red color is due to oxidized iron disbursed throughout the rocks.

Reef A wave-resistant, bioconstructed marine structure.

Regression The gradual emergence of land as sea level falls.

Relict sediment   Sediment that was deposited in an environment other than the one it now occupies; it is out of equilibrium with its present environment.

Reynolds number   A dimensionless number that is proportional to the ratio between inertial and viscous forces in a fluid.

Rhythmites   Rhythmic stratification such as varves.

Ridge and runnel   The most landward wave-built sand bar and associated trough. It is typically intertidal and ephemeral.

River-dominated delta   A river delta that develops a digitate and irregular outer margin due to extreme river influence and low wave and tidal energy.

Roundness   The sharpness or curvature of the edges and corners of a particle. It may be expressed as the ratio of the average radius of the curvature of several corners or edges to that of the maximum inscribed circle.

Rudistid   A sessile bivalve mollusk of the Mesozoic Era. These oysterlike organisms have one large valve attached to the substrate and another smaller one attached like a lid on a can.

Sabkha   A deflation surface in an arid environment where the sediment particles are removed to the groundwater level.

Saline lake   Any lake that has a total dissolved solid concentration of at least 5 ‰.

Saltation   The bouncing mode of particle transport in a fluid.

Scroll bar   A crescent-shaped, constructed feature built on the inside of a meander in a stream. They give the undulating surface to the point-bar accumulation.

Sea waves   Waves that are under the direct influence of wind.

Sediment pile   An accumulation of sediment in various shapes which originated largely through mechanical piling of particles. Examples are dunes, longshore bars, spits, and ooid shoals.

Sedimentary structures   Internal features of sediment or sedimentary rock formed contemporaneously with or after deposition.

Sedimentation   The study of sedimentary processes.

Sedimentation unit   An accumulation of sediment deposited under essentially constant conditions.

Sedimentology   The study of sediments and sedimentary rocks.

Set-up   The increase in water level along or near a coast due to an onshore wind. Off-shore winds would cause the opposite effect; set-down.

Shear stress   The component of stress that acts tangential to a plane through any given point.

Shelf-edge bypassing   The transport of silt and sand across the shelf edge onto the upper continental slope.

Shoal retreat massif   A body of sand on the continental shelf which was left behind due to submergence as the sea level rose.

Shoreface   The inner portion of the continental shelf, extending to a depth of storm wave base.

Skewness   Deviation of a population's distribution from the symmetry of a normal distribution.

Slump   A type of mass movement resulting from gravity and characterized by a rotational movement of a mass of rock or sediment.

Sorting   Statistically, this refers to the distribution of particles about the mean;

essentially the standard deviation. It is also used to express the selection of particles according to physical attributes during transport.

Sphericity The tendency for a particle to become equidimensional for all radii; approaching a sphere.

Spilling breaker A wave breaking in shallow water that breaks over a considerable distance.

Spring tide Tide of the highest range in a lunar cycle, caused by the sun and moon reinforcing each other. It occurs twice during a lunar month.

Standard deviation The deviation from the mean of a normal distribution equal to 34% of the population.

Steady flow Hydraulic conditions in which flow conditions of magnitude and direction remain inchanged.

Stokes' law A relationship that expresses the rate of settling of particles in a fluid.

Storm tide (storm surge) The rise in water level that accompanies a storm and is caused by the action of wind stress on the water surface.

Stromatolites A thinly stratified deposit of detrital particles which were accumulated by blue-green algae and rapidly lithified. They are generally indicative of the intertidal environment.

Submarine canyon A large, sinuous, V-shaped valley on the outer continental margin. Depth may be thousands of meters and length reaches hundreds of kilometers.

Submarine fan A fan-shaped sediment accumulation at the mouth of a submarine canyon.

Suprafan A fan-shaped lobe of sediment deposited at the mouth of a distributary channel on the midfan area of a submarine fan.

Suspended load Sediment particles dispersed within a fluid, transported in the main body of flow, and having essentially no contact with the substrate below.

Swale General term used for small topographic depression.

Swamp A low-lying, wet environment with trees as a primary type of vegetation. Coastal swamps are common in low latitudes where mangroves are abundant.

Swash bar An intertidal, wave-constructed, ephemeral sand bar along the open coast. Equivalent to ridge.

Swell waves Relatively long and low waves with a sinusoidal profile that are no longer under the direct influence of the wind.

Syneresis A dewatering in sediment during hardening that commonly results in large cracks. The cracks may fill with mineral precipitate and are then known as septaria.

Tabular cross-stratification The type of cross-stratification that displays sets which are broad in lateral dimension compared to set thickness. They are commonly planar but may be undulating.

Tephra A general term for all volcanic ejecta.

Terrigenous sediments Sediment composed of particles derived from previously existing rocks on land.

Textural inversion The term used to describe sediments that do not fit into Folk's textural maturity classification. It usually results from a dual source and the sediment is commonly bimodal.

Textural maturity A concept that describes the interaction of environmental processes with sediment. It is based on mud content, sorting, and roundness.

Thalweg The lowest or deepest line along a stream channel.

Thermocline The narrow layer in a water body where there is maximum change in temperature with depth.

Threshold velocity The minimum velocity at which a fluid will move a sediment particle of a specific size, shape, and density.

Tidal bedding A sequence of strata that shows alternation of horizontal layers of sand and mud corresponding to the ebb and flood tidal cycle.

Tidal delta A generally fan-shaped sediment body, typically associated with inlets and formed by tidal currents.

Tidalites Accumulations of sediment that are the direct result of transport by tidal currents; they may be intertidal or subtidal.

Tidal prism The water budget of an inlet, estuary, or other tidal body during a tidal cycle.

Tide-dominated delta A river delta that displays elongate sand bodies oriented essentially normal to the shoreline trend due to extreme tidal currents.

Till Unsorted and unstratified glacial drift.

Tillite Consolidated or indurated glacial till. Generally applied to all pre-Pleistocene glacial till.

Time-velocity asymmetry The phenomenon of maximum flood- and ebb-tidal currents occurring at other than midtide. It is the rule rather than an exception.

Traction The various processes that move sediment in the bed load.

Transgression The advance of water over the adjacent land due to sea-level rise or land subsidence.

Trough cross-stratification The cross-stratification which is three-dimensional in a trough shape. It results from migration of crescent- or linguoid-shaped bedforms.

Turbidite sequence **(Bouma sequence)** A sedimentation unit produced by deposition from a single turbidity current. It has a regular and predictable sequence of texture and sedimentary structures.

Turbidity current A density current that is a type of sediment gravity flow in which fluid disturbance causes sediment to become suspended, thereby creating the density gradient.

Turbulent flow Fluid flow that is characterized by vortices and eddies.

Uniform flow Flow of a current in a fluid where there is no convergence or divergence; that is, the width and depth of the channel or pipe remain unchanged.

Upper flat bed Refers to conditions at the transition between lower and upper flow regime producing horizontal (plane) strata.

Upwelling The process by which water rises from lower to higher depths, usually as a result of surface currents diverging from the coast.

Varves The cyclic and thin strata which show seasonal changes. They are commonly associated with lake deposits.

Wadi A dry channel of an intermittent stream.

Walther's Law The relationship between modern depositional environments and the stratigraphic record whereby adjacent modern environments become stratigraphically adjacent in vertical arrangements.

Wash load The suspended sediment in transport by a fluid; suspension load.

Washover fan A fan-shaped sediment accumulation on the landward side of a barrier or beach formed by waves overtopping the island and carrying sediment with them.

Wave-dominated delta A river delta that displays a rather smooth, accurate outer margin due to high wave energy.

Wave-drift currents Slow currents moving in the direction of wave propagation and caused by the net transport of water as it moves in orbital motion.

Wavy bedding A type of stratification in which continuous ripple-stratified sand alternates with continuous and undulating mud layers.

Wind tidal flat Extensive low lying supratidal areas that are inundated irregularly due to storm tides or wind tides.

X-radiograph An x-ray picture of a sediment or rock section which shows internal structures that are commonly too subtle to see with the unaided eye. Commonly used in the analysis of sediment cores.

References

ADIE, R. J., 1975. Permo-Carboniferous glaciation of the southern hemisphere, *in* Wright, A. E., and Moseley, F. (eds.), *Ice Ages: Ancient and Modern*. Seel House Press, Liverpool, England, 287–300.

ALLEN, E. A., 1977. *Petrology and Stratigraphy of Holocene Coastal-Marsh Deposits along the Western Shore of Delaware Bay*. University of Delaware Sea Grant Rept. 20–77, Newark, Del., 287 p.

ALLEN, J. R. L., 1962. Petrology, origin and deposition of the highest Lower Old Red Sandstone of Shropshire, England. Jour. Sed. Petrology, 30:193–208.

ALLEN, J. R. L., 1963. Sedimentation in the modern delta of the River Niger, West Africa, *in* Van Straaten, L. M. J. U. (ed.), *Deltaic and Shallow Marine Deposits*. Elsevier, New York, 26–34.

ALLEN, J. R. L., 1964a. The Nigerian continental margin: bottom sediments, submarine morphology and geological evolution. Mar. Geol. 1:289–332.

ALLEN, J. R. L., 1964b. Studies in fluviatile sedimentation: six cyclothems from the Lower Old Red Sandstone, Anglo-Welsh Basin. Sedimentology, 3:163–198.

ALLEN, J. R. L., 1965a. Late Quaternary Niger Delta, and adjacent areas: sedimentary environments and lithofacies. Amer. Assoc. Petroleum Geologists Bull., 49:547–600.

ALLEN, J. R. L., 1965b. A review of the origin and characteristics of recent alluvial sediments. Sedimentology, 5:89–191.

ALLEN, J. R. L., 1965c. The sedimentation and paleogeography of the Old Red Sandstone of Anglesey, North Wales. Yorkshire Geol. Soc. Proc., 35:139–185.

ALLEN, J. R. L., 1965d. Fining upward cycles in alluvial succession. Liverpool, Manchester Geol. Jour., 4:229–246.

ALLEN, J. R. L., 1968. *Current Ripples, Their Relation to Patterns of Water and Sediment Motion.* North-Holland, Amsterdam, 433 p.

ALLEN, J. R. L., 1970a. *Physical Processes of Sedimentation: An Introduction.* Allen & Unwin, London, 248 p.

ALLEN, J. R. L., 1970b. Sediments of the modern Niger Delta: a summary and review, *in* Morgan, J. R., and Shaver, R. H. (eds.), *Deltaic Sedimentation.* Soc. Econ. Paleontologists and Mineralogists, Spec. Publ. 15, Tulsa, Okla., 138–151.

ALLEN, J. R. L., 1974. Studies in fluviatile sedimentation: implications of pedogenic carbonate units, Lower Old Red Sandstone, Anglo-Welsh outcrops. Geol. Jour., 9:181–208.

AMERICAN COMMISSION ON STRATIGRAPHIC NOMENCLATURE, 1961. Code of stratigraphic nomenclature. Amer. Assoc. Petroleum Geologists Bull., 45:645–660.

ANDERSON, F. E., BLACK, L., WATLING, L. E., MOOK, W., AND MAYER, L. M., 1981. A temporal and spatial study of mudflat erosion and deposition. Jour. Sed. Petrology, 51:729–736.

ANDERSON, J. B., 1975. Factors controlling $CaCO_3$ dissolution in the Weddell Sea from foraminiferal distribution patterns. Mar. Geol., 19:315–332.

ANDERTON, R., 1976. Tidal shelf sedimentation: an example from the Scottish Dalradian. Sedimentology, 23:429–458.

ANDREWS, P. B., 1970. *Facies and Genesis of a Hurricane Generated Washover Fan, St. Joseph Island, Central Texas Coast.* University of Texas, Bur. Econ. Geol. Rept. Inv. No. 67, Austin, Tex., 147 p.

ANDREWS, S., 1981. Sedimentology of great sand dunes, Colorado, *in* Ethridge, F. G., and Flores, R. M. (eds.), *Recent and Ancient Nonmarine Depositional Environments: Models for Exploration.* Soc. Econ. Paleontologists and Mineralogists, Spec. Publ. 31, Tulsa, Okla., 279–291.

ARTHUR, M. A., AND SCHLANGER, S. O., 1979. Cretaceous "oceanic anoxic event" as causal factors in development of reef-reservoired giant oil fields. Amer. Assoc. Petroleum Geologists Bull., 63:870–885.

ASQUITH, D. O., 1970. Sedimentary models, cycles and delta, Upper Cretaceous, Wyoming. Amer. Assoc. Petroleum Geologists Bull., 58:2274–2283.

BAGNOLD, R. A., 1941. *The Physics of Blown Sand and Desert Dunes.* Metheun, London, 265 p. (reprinted 1954).

BAGNOLD, R. A., 1954. Experiments on a gravity-free dispersion of large solid spheres in a Newtonian fluid under shear. Proc. Roy. Soc. London, Ser. A, 225:49–63.

BAGNOLD, R. A., 1962. Auto-suspension of transported sediments; turbidity currents. Proc. Roy. Soc. London, Ser. A, 265:315–319.

BALL, M. M., 1967. Carbonate sand bodies of Florida and the Bahamas. Jour. Sed. Petrology, 37:556–591.

BARRELL, J., 1912. Criteria for the recognition of ancient delta deposits. Geol. Soc. Amer. Bull., 23:377–446.

BARRELL, J., 1913. The Upper Devonian delta of the Appalachian geosyncline, Part I. The delta and its relations to the interior sea. Amer. Jour. Sci., 4th ser., 36:429–472.

BARRELL, J., 1914. The Upper Devonian delta of the Appalachian geosyncline, Part II. Amer. Jour. Sci., 4th ser. 37:87–104, 225–253.

BARRETT, T. J., AND SPOONER, E. T. C., 1977. Ophiolitic breccias associated with allochthonous oceanic crustal rocks in the East Ligurian Apennines, Italy—a comparison with observations from rifted ocean ridges. Earth & Planetary Sci. Lett., 35:79–91.

BARWIS, J. H., 1976. Internal geometry of Kiawah Island beach ridges, in Hayes, M. O., and Kana, T. W. (eds.), Terrigenous Clastic Depositional Environments. Tech. Rept. No. 11-CRD, Coastal Res. Div., University of South Carolina, Columbia, S. C., II-115 to II-125.

BARWIS, J. H., AND MAKURATH, J. H., 1978. Recognition of ancient tidal inlet sequences: an example from the Upper Silurian Keyser Limestone in Virginia. Sedimentology, 25:61–82.

BASAN, P. B., AND FREY, R. W., 1977. Actual-palaeontology and neoichnology of salt marshes near Sapelo Island, Georgia, in Crimes, T. P., and Harper, J. C. (eds.), Trace Fossils 2. Geol. Jour., Spec. Issue 9. Seel House Press, Liverpool, England, 41–70.

BASS, N. W., 1934, Origin of Bartlesville shoestring sands, Greenwood and Butler Counties, Kansas. Amer. Assoc. Petroluem Geologists Bull., 18:1313–1345.

BATES, C. C., 1953. Rational theory of delta formation. Amer. Assoc. Petroleum Geologists Bull., 37:2119–2161.

BATHURST, R. G. C., 1975. Carbonate Sediments and Their Diagenesis, 2nd ed. Elsevier, Amsterdam, 620 p.

BEADLE, L. C., 1974. The Inland Waters of Tropical Africa. Longman, London, 365 p.

BEHRENS, E. W., 1969. Hurricane effects on a hypersaline bay, in Castaneres, A. A., and Phleger, F. B. (eds.), Coastal Lagoons—A Symposium. Universidad Nacional Autónoma de Mexico/Unesco, Mexico City, 301–312.

BEHRENS E. W. AND LAND, L. S., 1972, Subtidal Holocene dolomite, Baffin Bay, Texas. Jour. Sed. Petrology, 42:155–161.

BELL, W. C., FENIAK, O. W., AND KURTZ, V. E., 1952. Trilobites of the Franconia Formation, southeast Minnesota. Jour. Paleontol. 26:175–198.

BELT, E. S., 1975. Scottish Carboniferous cyclothem patterns and their paleoenvironmental significance, *in* Broussard, M. L. (ed.), *Deltas Models for Exploration.* Houston Geological Society, Houston, Tex., 427–449.

BERG, R. R., 1954. Franconia Formation of Minnesota and Wisconsin. Geol. Soc. Amer. Bull., 65:857–881.

BERG, R. R., 1975. Depositional environment of Upper Cretaceous Sussex Sandstone, House Creek Field, Wyoming. Amer. Assoc. Petroleum Geologists Bull., 59:2099–2110.

BERG, R. R., AND DAVIES, D. K., 1968. Origin of the Lower Cretaceous Muddy Sandstone at Bell Creek Field, Montana. Amer. Assoc. Petroleum Geologists Bull., 52:1888–1898.

BERGER, W. H., 1974. Deep-sea sedimentation, *in* Burk, C. A., and Drake, C. L. (eds.), *The Geology of Continental Margins.* Springer-Verlag, New York, 213–241.

BERNARD, H. A., AND LeBLANC, R. J., 1965. Résumé of the Quaternary geology of the northeastern Gulf of Mexico Province, *in* Wright, H. E., Jr., and Frey, D. G. (eds.), *The Quaternary of the United States.* Princeton University Press, Princeton, N.J., 137–185.

BERNARD, H. A., LeBLANC, R. J., AND MAJOR, C. F., 1962. *Recent and Pleistocene Geology of Southeast Texas.* Geol. Gulf Coast and Central Texas and guidebook of excursion, Houston Geological Society, Houston, Tex., 225 p.

BIGGS, R. B., 1978. Coastal bays, *in* Davis, R. A., (ed.), *Coastal Sedimentary Environments.* Springer-Verlag, New York, 69–99.

BJØRLYKKE, K., 1974. Glacial striations on clasts from the Moelo Tillite of the Late Precambrian of southern Norway. Amer. Jour. Sci., 274:443–448.

BLANKENSHIP, J. B., 1978. A comparison of shelf width and depth to shelf break, continental United States. Unpubl. rept., Univ. Southern California, Geol. Sci. Dept., 23 p.

BLATT, H., 1967. Original characteristics of clastic quartz grains. Jour. Sed. Petrology, 37:401–424.

BLATT, H., AND CHRISTIE, J. M., 1963. Undulatory extinction in quartz of igneous and metamorphic rocks and its significance in provenance studies of sedimentary rocks. Jour. Sed. Petrology, 33:559–579.

BLATT, H., MIDDLETON, G. V., AND MURRAY, R. C., 1980. *Origin of Sedimentary Rock,* 2nd ed. Prentice-Hall, Englewood Cliffs, N.J., 782 p.

BLISSENBACH, E., 1954. Geology of alluvial fans in semi-arid regions. Geol. Soc. Amer. Bull., 65:175–190.

BLOOM, A. L., 1978. *Geomorphology.* Prentice-Hall, Englewood Cliffs, N.J., 510 p.

BLUCK, B. J., 1967. Deposition of some Upper Old Red Sandstone conglomerates in the Clyde area; a study in the significance of bedding. Scottish Jour. Geol., 3:139–167.

BOERSMA, J. R., 1967. Remarkable types of mega cross-stratification in the fluviatile sequence of a subrecent distributary of the Rhine Amerongen, the Netherlands. Geol. Mijnbouw, 46:217–235.

BOERSMA, J. R., 1970, Distinguishing features of wave-ripple cross-stratification and morphology. Doctoral Thesis, Univ. Utrecht, The Netherlands, 65 p.

BOOTH J. S., 1979, Recent history of mass wasting on the upper continental slopes, northern Gulf of Mexico, as interpreted from the consolidation state of the sediment, *in* Doyle, L. J., and Pilkey, O. H. (eds.), *Geology of Continental Slopes*. Tulsa, Soc. Econ. Paleontologists and Mineralogists, Spec. Publ., 27:153–164.

BOOTHROYD, J. C., 1972. *Coarse-Grained Sedimentation on a Graded Outwash Fan, Northeast Gulf of Alaska*. Tech. Rept. No. 6-CRD, Coastal Res. Div., University of South Carolina, Columbia, 127 p.

BOOTHROYD, J. C., 1978. Mesotidal inlets and estuaries, *in* Davis, R. A. (ed.), *Coastal Sedimentary Environments*. Springer-Verlag, New York, 287–360.

BOOTHROYD, J. C., AND ASHLEY, G. M., 1975. Processes, bar morphology and sedimentary structures on braided outwash fans, northeastern Gulf of Alaska, *in* Jopling, A. V., and McDonald, B. C. (eds.), *Glaciofluvial and Glaciolacustrine Sedimentation*. Soc. Econ. Paleontologists and Mineralogists, Spec. Publ. No. 23, Tulsa, Okla., 193–222.

BOOTHROYD, J. C., AND HUBBARD, D. K., 1975. Genesis of bedforms in mesotidal estuaries, *in* Cronin, L. E. (ed.), *Estuarine Research,* v. 2: *Geology and Engineering*. Academic Press, New York, 217–234.

BOOTHROYD, J. C., AND NUMMEDAL, D., 1978. Proglacial braided outwash: a model for humid alluvial-fan deposits, *in* Miall, A. D. (ed.), *Fluvial Sedimentology*. Can. Soc. Petroleum Geologists, Mem. No. 5, Calgary, Alberta, 641–668.

BOSENCE, D. W. J., 1973. Facies relationships in a tidally influenced environment, a study from the Eocene of the London Basin. Geol. Mijnbouw, 52:63–67.

BOULTON, G. S., 1975. Processes and patterns of subglacial sedimentation: a theoretical approach, *in* Wright, A. E., and Moseley, F. (eds.), *Ice Ages; Ancient and Modern*. Seel House Press, Liverpool, England, 7–42.

BOUMA, A. H., 1962. *Sedimentology of Some Flysch Deposits: A Graphic Approach to Facies Interpretation*. Elsevier, Amsterdam, 168 p.

BOUMA, A. H., 1963. A graphic presentation of facies model of salt marsh deposits. Sedimentology, 2:122–129.

BOUMA, A. H., 1965. Sedimentary characteristics of some samples collected from submarine canyons. Mar. Geol., 3:291–320.

BOUMA, A. H., 1972. Distribution of sediments and sedimentary structures in the Gulf of Mexico, *in* Rezak, R., and Henry, V. J. (eds.), *Contributions on the Geological and Geophysical Oceanography of the Gulf of Mexico*. Texas A&M Univ. Ocean. Studies, 3:35–66.

BOUMA, A. H., 1979. Continental slopes, *in* Doyle, L. J., and Pilkey, O. H. (eds.), *Geology of Continental Slopes*. Soc. Econ. Paleontologists and Mineralogists, Spec. Publ. No. 27, Tulsa, Okla., 1–16.

BOURGEOIS, J., 1980. A transgressive shelf sequence exhibiting hummocky stratification, the Cape Sebastian (Upper Cretaceous), southwestern Oregon. Jour. Sed. Petrology, 50: 681–702.

BOWEN, R. L., 1959. Late Paleozoic glaciation of eastern Australia. Ph.D. thesis, Melbourne University, 285 p.

BRADLEY, W. H., 1930, The varves and climates of the Green River epoch. U. S. Geol. Survey, Prof. Paper 158-E:87–110.

BRAMLETTE, M. N., 1946. *The Monterey Formation of California and The Origin of Its Siliceous Rocks.* U. S. Geol. Survey, Prof. Paper No. 212, 57 p.

BRAMLETTE, M. N., 1961. Pelagic sediments, *in* Sears, M. (ed.), *Oceanography*, Amer. Assoc. Adv. Sci., Publ. No. 67, Washington, D.C., 345–366.

BRENCHLEY, P. J., AND NEWALL, G., 1970. Flume experiments on the orientation and transport of models and shell valves. Palaeogeog. Palaeoclimatol. Palaeoecol., 7:185–220.

BRENNER, R. L., 1978. Sussex sandstone of Wyoming–example of Cretaceous offshore sedimentation. Amer. Assoc. Petroleum Geologists Bull., 62:181–200.

BRENNER, R. L., 1980. Construction of process-response models for ancient epicontinental seaway depositional systems using partial analogs. Amer. Assoc. Petroleum Geologists Bull., 64:1223–1244.

BRENNER, R. L., AND DAVIES, D. K., 1973. Storm-generated coquinoid sandstone: genesis of high-energy marine sediments from the Upper Jurassic of Wyoming and Montana. Geol. Soc. Amer. Bull., 84:1685–1698.

BRENNER, R. L., AND DAVIES, D. K. 1974. Oxfordian sedimentation in western interior United States. Amer. Assoc. Petroleum Geologists Bull., 58:407–428.

BRIDGES, P. H., 1976. Lower Silurian transgressive barrier islands, southwest Wales. Sedimentology, 23:347–362.

BROECKER, W. S., 1972. *Chemical Oceanography*. Harcourt Brace Jovanovich, New York, 214 p.

BROOKFIELD, M. E., 1977. The origin of bounding surfaces in ancient aeolian sandstones. Sedimentology, 24:303–332.

BROUSSARD, M. L. (ED.), 1975. *Deltas, Models for Exploration*. Houston Geological Society, Houston, Tex., 555 p.

BROWN, L. F., JR., 1969a. Geometry and distribution of fluvial and deltaic sandstones (Pennsylvanian and Permian), north-central Texas. Gulf Coast Assoc. Geol. Soc. Trans., 19:23–47.

BROWN, L. F., JR., 1969b. Late Pennsylvanian poralic sediments, *in Guidebook to the late Pennsylvanian Shelf Sediments, North-Central Texas.* Dallas Geological Society, Dallas, Tex., 21–33.

BROWN, L. F., JR., CLEAVES, A. W., AND ENXLEBEN, A. W., 1973. *Pennsylvanian Depositional Systems in North-Central Texas: A Guide for Interpreting Terrigenous Clastic Facies in a Cratonic Basin.* Guidebook No. 14, University of Texas, Bur. Econ. Geol., Austin, Tex., 122 p.

BRUNSKILL, G. J., 1969. Fayetteville Green Lake, New York: II. Precipitation and sedimentation of calcite in a meromictic lake with laminated sediments. Limnol. Oceanogr., 14:830–847.

BULL, W. B., 1964. Geomorphology of segmented alluvial fans in western Fresno County, California. U.S. Geol. Survey, Prof. 352-E, 89–129.

BULL, W. B., 1968. Alluvial fan, *in* Fairbridge, R. W. (ed.), *Encyclopedia of Geomorphology*, Reinhold, New York, 7–10.

BULL, W. B., 1972. Recognition of alluvial-fan deposits in the stratigraphic record, *in* Rigby, K. J., and Hamblin, W. K. (eds.), *Recognition of Ancient Sedimentary Environments*. Soc. Econ. Paleontologists and Mineralogists, Spec. Publ. No. 16, Tulsa, Okla., 68–83.

BURKE, K. C. A., 1972, Longshore drift, submarine canyons, and submarine fans in development of Niger Delta. Amer. Assoc. Petroleum Geologists Bull., 56:1975–1983.

BURKE, K. C. A., AND DEWEY, J. F., 1973. Plume-generated triple junctions: key indicators in applying plate tectonics to old rocks. Jour. Geol., 81:406–433.

BYRNE, J. V., LE ROY, D. O., AND RILEY, C. M., 1959. The Chenier plain and its stratigraphy, southwestern Louisiana. Gulf Coast Assoc. Geol. Soc. Trans., 9:237–260.

CACCHIONE D., AND SOUTHARD, J. R., 1974. Incipient sediment movement by shoaling internal gravity waves. Jour. Geophys. Res., 79:2237–2242.

CAMPBELL, C. V., AND OAKS, R. Q., 1973. Estuarine sandstone filling tidal scours, Lower Cretaceous Fall River Formation, Wyoming. Jour. Sed. Petrology, 43:765–767.

CAMPBELL, D. H., 1963. Percussion marks on quartz grains. Jour. Sed. Petrology, 33:855–859.

CANT, D. J., 1978. South Saskatchewan River and the Battery Point Formation, *in* Miall, A. D. (ed.), *Fluvial Sedimentology*. Can. Soc. Petroleum Geologists, Mem. No. 5, Calgary, Alberta, 627–640.

CANT, D. J., AND WALKER, R. G., 1976. Development of a braided fluvial facies model for the Devonian Battery Point Sandstone, Quebec. Can. Jour. Earth Sci., 13:102–119.

CANT, D. J., AND WALKER, R. G., 1978. Fluvial processes and facies sequences in the sandy, braided South Saskatchewan River. Sedimentology, 25:625–648.

CARLSON, P. A., AND MOLNIA, B. F., 1977. Submarine faults and slides on the continental shelf, Northern Gulf of Alaska. Mar. Geotech., 2:275–290.

CAROZZI, A. V., 1962. Observations of algal biostromes in the Great Salt Lake, Utah. Jour. Geol., 70:246–252.

CARTER, C. W., 1975. Miocene-Pliocene beach and tidal flat deposits, southern New Jersey, *in* Ginsburg, R. N. (ed.), *Tidal Deposits*. Springer-Verlag, New York, 109–116.

CARTER, C. W., 1978. A regressive barrier and barrier-protected deposit: depositional environments and geographic setting of the Late Tertiary Cohansey Sand. Jour. Sed. Petrology, 48:933–950.

CASTANARES, A. A., AND PHLEGER, F. B. (EDS.), 1969. *Coastal Lagoons—A Sym-*

posium. Universidad Nacional Autónoma de México/UNESCO, Mexico City, 686 p.

CASTER, K. E., 1934. The stratigraphy and paleontology of northwest Pennsylvania. Bull. Amer. Paleontol., 21:1–185.

CHAPMAN, V. J., 1976. *Coastal Vegetation.* Pergamon Press, New York, 292 p.

CHESTER, R., AND HUGHES, 1967. A chemical technique for the separation of ferro-manganese minerals, carbonate minerals and adsorbed trace elements from pelagic sediments. Chem. Geol., 2:249–262.

CHURKIN, M., JR., 1974. Paleozoic marginal ocean-basin-volcanic arc systems in the Cordilleran Foldbelt, *in* Dott, R. H., Jr., and Shaver, R. H. (eds.), *Modern and Ancient Geosynclinal Sedimentation.* Soc. Econ. Paleontologists and Mineralogists, Spec. Publ. No. 19, Tulsa, Okla., 174–192.

CHURKIN, M., JR., AND KAY, G. M., 1967. Graptolite bearing Ordovician siliceous and volcanic rocks, northern Independence Range, Nevada. Geol. Soc. Amer. Bull., 78:651–668.

CLIFTON, H. E., 1969. Beach lamination: nature and origin. Mar. Geol., 7:553–559.

CLIFTON, H. E., 1971. Orientation of empty pelecypod shells and shell fragments in quiet water. Jour. Sed. Petrology, 41:671–682.

CLIFTON, H. E., 1976. Wave-formed sedimentary structures; a conceptual model, *in* Davis, R. A., and Ethington, R. L. (eds.), *Beach and Nearshore Sedimentation.* Soc Econ. Paleontologists and Mineralogists, Spec. Publ. No. 24, Tulsa, Okla., 126–148.

CLIFTON, H. E., 1977. Rain impact ripples. Jour. Sed. Petrology, 47:678–79.

CLIFTON, H. E., HUNTER, R. E., AND PHILLIPS, R. L., 1971. Depositional structures and processes in the non-barred high-energy nearshore. Jour. Sed. Petrology, 41:651–670.

CLOUD, P. E., JR., 1959. Geology of Saipan Mariana Islands: Part 4. Submarine topography and shoal water ecology. U. S. Geol. Survey, Prof. Paper 280-K, 361–445.

CLOUD, P. E., JR., 1962. *Environment of Calcium Carbonate Deposition West of Andros Island, Bahamas.* U.S. Geol. Survey, Prof. Paper 350, 138 p.

COHEN, A. D., 1973. Petrology of some Holocene peat sediments from Okefenokee swamp-marsh complex of south Georgia. Geol. Soc. Amer. Bull., 84:3867–3878.

COHEN, A. D., AND SPACKMAN, W., 1974. The petrology of peats from the Everglades and coastal swamps of southern Florida, *in* Gleason, P. J. (ed.), *Environment of South Florida: Past and Present.* Miami Geol. Soc., Mem. No. 2, 233–255.

COLE, G. A., 1975. *Textbook of Limnology.* C. V. Mosby, St. Louis, Mo., 283 p.

COLEMAN, J. M., 1969. Brahmaputra River; channel processes and sedimentation. Sed. Geol. (special issue), 3:122–239.

COLEMAN, J. M., 1976. *Deltas: Processes of Deposition and Models for Exploration.* Continuing Education Publication Co., Champaign, Ill., 102 p.

COLEMAN, J. M., AND GAGLIANO, S. M., 1964. Cyclic sedimentation in the Mississippi River deltaic plain. Gulf Coast Assoc. Geol. Soc. Trans., 14:67–80.

COLEMAN, J. M., AND PRIOR, D. B., 1981, Subaqueous sediment in stabilities in the offshore Mississippi River delta, *in Offshore Geologic Hazards.* Tulsa, Amer. Assoc. Petroleum Geologists, Short Course Notes No. 18:5.1–5.53.

COLEMAN, J. M., AND WRIGHT, L. D., 1971. *Analysis of Major River Systems and Their Deltas; Procedures and Rationale, with Two Examples.* Tech. Rept. No. 95, Louisiana State University, Coastal Studies Inst., Baton Rouge, La., 125 p.

COLEMAN, J. M., AND WRIGHT, L. D., 1973. Variability of modern river deltas. Gulf Coast Assoc. Geol. Soc. Trans., 23:33–36.

COLEMAN, J. M., AND WRIGHT, L. D., 1975. Modern river deltas: variability of processes and sand bodies, *in* Broussard, M. L. (ed.), 1975, *Deltas, Models for Exploration.* Houston Geological Society, Houston, Tex., 99–149.

COLEMAN, J. M., GAGLIANO, S. M., AND WEBB, J. E., 1964. Minor sedimentary structures in a prograding distributary. Mar. Geol., 1:240–258.

COLEMAN, J. M., SUHEYADA, J. N., WHELAN, T., AND WRIGHT, L. D., 1974. Mass movements of Mississippi River delta sediments. Gulf Coast Assoc. Geol. Soc. Trans., 24:49–68.

COLLINSON, J. D., 1968. Deltaic sedimentation units in the Upper Carboniferous of northern England. Sedimentology, 10:223–254.

COLLINSON, J. D., 1969. The sedimentology of the Grindslow Shales and the Kinderscourt Grit: a deltaic complex in the Namurian of northern Ireland. Jour. Sed. Petrology, 39:194–221.

COLLINSON, J. D., 1970. Bedforms in the Tana River, Norway. Geog. Ann., Ser. A, 52:31–56.

COLLINSON, J. D., 1978a. Fluvial system, *in* Reading, H. G. (ed.), *Sedimentary Environments and Facies.* Elsevier, New York, 15–79.

COLLINSON, J. D., 1978b. Lakes, *in* Reading, H. G. (ed.), *Sedimentary Environments and Facies.* Elsevier, New York, 61–79.

COLLINSON, J. D., 1978c. Deserts, *in* Reading, H. G. (ed.), *Sedimentary Environments and Facies*, Elsevier, New York, 80–96.

COLLINSON, J. D., AND THOMPSON, D. B., 1982. *Sedimentary Structures.* George Allen & Unwin, London, 194p.

CONOLLY, J. R., NEEDHAM, N. D., AND HEEZEN, B. C., 1969. Late Pleistocene and Holocene sedimentation in the Laurentian Channel. Jour. Geol., 77:131–147.

CONYBEARE, C. E. B., AND CROOK, K. A. W., 1968. *Manual of Sedimentary Structures.* Bur. Min. Resources, Geol. and Geophys., Canberra A.C.T., Australia, Bull. No. 102, 327 p.

COOK, D. O., 1970. The occurrence and geologic work of rip currents off southern California. Mar. Geol., 13:31–55.

Cook, H. E., 1979. Ancient continental slope sequences and their value in understanding modern slope development, *in* Doyle, L. R., and Pilkey, O. H., Jr. (eds.), *Geology of Continental Slopes*. Soc. Econ. Paleontologists and Mineralogists, Spec. Publ. No. 27, Tulsa, Okla., 287–305.

Cook, H. E., and Enos, P. (eds.), 1977. *Deep-Water Carbonate Environments*. Soc. Econ. Paleontologists and Mineralogists, Spec. Publ. No. 25, Tulsa, Okla., 336 p.

Cook, H. E., and Taylor, M. E., 1977. Comparison of continental slope and shelf environment in the Upper Cambrian and lowermost Ordovician of Nevada, *in* Cook, H. E., and Enos, P. (eds.), *Deep-Water Carbonate Environments*. Soc. Econ. Paleontologists and Mineralogists, Spec. Publ. No. 25, Tulsa, Okla., 51–81.

Cooke, R. U., and Warren, A., 1973. *Geomorphology in Deserts*. University of California Press, Berkeley, Calif., 374 p.

Cotter, E., 1978. The evolution of fluvial style, with special reference to the central Appalachian Paleozoic, *in* Miall, A. D. (ed.), *Fluvial Sedimentology*. Can. Soc. Petroleum Geologists, Mem. No. 5, Calgary, Alberta, 361–384.

Cram, J. M., 1979. The influence of continental shelf width on tidal range: paleoceanographic implications. Jour. Geol., 87:441–447.

Crimes, T. P., 1975. The stratigraphical significance of trace fossils, *in* Frey, R. W. (ed.), *The Study of Trace Fossils*. Springer-Verlag, New York, 109–130.

Crimes, T. P., and Harper, J. C. (eds.), 1970, *Trace Fossils*. Liverpool, Seel House Press, 547 p.

Cronin, L. E. (ed.), 1975. *Estuarine Research*, v. 2: *Geology and Engineering*. Academic Press, New York, 587 p.

Crowell, J. C., 1954, Geology of the Ridge Basin area, Los Angeles and Ventura Counties, California. Calif. Div. Mines, Bull. 170 p. with map.

Crowell, J. C., 1955. Directional-current structures from the pre-alpine flysch, Switzerland. Geol. Soc. Amer. Bull., 66:1351–1384.

Crowell, J. C., 1957. Origin of pebbly mudstones. Geol. Soc. Amer. Bull., 68:993–1009.

Crowell, J. C., and Frakes, L. A., 1971. Late Paleozoic glaciation: Part IV. Australia. Geol. Soc. Amer. Bull., 82:2515–2540.

Crowell, J. C., and Frakes, L. A., 1972. Late Paleozoic glaciation: Part V. Karoo Basin, South Africa. Geol. Soc. Amer. Bull., 83:2887–2912.

Culbertson, J. A., 1940. Downdip Wilcox (Eocene) of coastal Texas and Louisiana. Amer. Assoc. Petroleum Geologists Bull., 24:1891–1922.

Curray, J. R., 1960. Sediments and history of the Holocene transgression, continental shelf, Gulf of Mexico, *in* Shepard, F. P., Phleger, F. B., and Van Andel, T. J. (eds.), *Recent Sediments of the Northwest Gulf of Mexico*. Amer. Assoc. Petroleum Geologists, Tulsa, Okla., 221–266.

Curray, J. R. 1964. Transgressions and regressions, *in* Miller, R. L. (ed.), *Papers in Marine-Geology, Shepard Commemorative Volume*. Macmillan, New York, 175–203.

Curray, J. R., 1965. Late Quaternary history, continental shelves of the

United States, *in* Wright, H. E., Jr., and Frey, D. G. (eds.), *The Quaternary of the United States.* Princeton University Press, Princeton, N.J., 723–735.

CURRAY, J. R., 1969. History of continental shelves, *in* Stanley, D. J. (ed.), *New Concepts of Continental Margin Sedimentation.* Amer. Geol. Inst., Washington, D. C., JC-6-1 to JC-6-7.

DALZIEL, I. W. D., AND DOTT, R. H., JR., 1970. *Geology of the Baraboo district, Wisconsin.* Wisconsin Geol. and Nat. Hist. Surv. Inf. Circ. No. 14, 164 p.

DAVIDSON-ARNOTT, R. G. D., AND GREENWOOD, B., 1974. Bedforms and structures associated with bar topography in the shallow-water wave environment, Kouchibouguac Bay, New Brunswick, Canada. Jour. Sed. Petrology, 44:698–704.

DAVIDSON-ARNOTT, R. G. D., AND GREENWOOD, B., 1976. Facies relationships on a barred coast, Kouchibouguac Bay, New Brunswick, Canada, *in* Davis, R. A., Jr., and Ethington, R. L. (eds.), *Beach and Nearshore Sedimentation.* Soc. Econ. Paleontologists and Mineralogists, Spec. Publ. No. 24, Tulsa, Okla., 149–168.

DAVIES, D, K., ETHRIDGE, F. G., AND BERG, R. R., 1971. Recognition of barrier environments. Amer. Assoc. Petroleum Geologists Bull., 55:550–565.

DAVIES, J. L., 1964. A morphogenic approach to world shorelines. Z. Geomorphol., 8:27–42.

DAVIES, J. L., 1980, *Geographical Variation in Coastal Development.* 2nd edition, Longman, New York, 212 p.

DAVIES, T. A. AND GORSLINE, D. S., 1976. Oceanic sediments and sedimentary processes, *in* Riley, J. P., and Chester, R. (eds.), *Chemical Oceanography,* v. 5, 2nd ed. Academic Press, New York, 1–80.

DAVIS, R. A., JR., 1965. Underwater study of ripples, southeastern Lake Michigan. Jour. Sed. Petrology, 35:857–866.

DAVIS, R. A., JR., 1978. Beach and nearshore zone, *in* Davis, R. A., Jr. (ed.), *Coastal Sedimentary Environments.* Springer-Verlag, New York, 237–285.

DAVIS, R. A., JR., AND FOX, W. T., 1972. Coastal processes and nearshore sandbars. Jour. Sed. Petrology, 42:401–412.

DAVIS, R. A., JR., AND FOX, W. T., 1981. Interaction of beach and tide generated processes in a microtidal inlet: Matanzas Inlet, Florida. Mar. Geol., 40:49–68.

DAVIS, R. A., JR., AND MCGEARY, D. F. R., 1965. Stability in nearshore bottom topography and sediment distribution. Proc. 8th Conf. Great Lakes Res., 222–231.

DAVIS, R. A., JR., AND MALLETT, C. W., 1981, Sedimentation in a Permian subglacial channel. Jour. Sed. Petrology, 51:185–190.

DAVIS, R. A., JR., BURKE, R. B., AND BRAME, J. W., 1979, Origin and development of barrier islands on west-central peninsula of Florida. (abs.) Amer. Assoc. Petroleum Geologists, Ann. Meeting, Houston, 74–75.

DAVIS, R. A., JR., FOX, W. T., HAYES, M. O., AND BOOTHROYD, J. C., 1972. Comparison of ridge and runnel systems in tidal and non-tidal environments. Jour. Sed. Petrology, 42:413–421.

References

DAVIS, R. A., JR., FINGLETON, W. G., ALLEN, G. R., CREALESE, C. D., JOHANNS, W. M., O'SULLIVAN, J. A., REIVE, C. L., SCHEETZ, S. J., AND STRANALY, G. L., 1973. Corpus Christi Pass: a hurricane modified inlet on Mustang Island, Texas. Contr. Mar. Sci., 17:123–131.

DE, BEAUMONT, L. E., 1845. *Leçons de geologie practique.* Septième leçon. P. Bertrand, Paris, 221–252.

DENNY, C. S., 1967. Fans and pediments. Amer. Jour. Sci., 265:81–105.

DEWEY, J. F., AND BIRD, J. M., 1970. Mountain belts and the new global tectonics. Jour. Geophys. Res. 75:2625–2647.

DEWEY, J. F., AND BURKE, K. C. A., 1974. Hot spots and continental breakup: implications for collisional orogeny. Geology, 2:57–60.

DICKINSON, K. A., 1971. Grain-size distribution and the depositional history of northern Padre Island, Texas. U.S. Geol. Survey, Prof. Paper 750-C, p. 1C–6C.

DICKINSON, K. A., BERRYHILL, H. L., AND HOLMES, C. W., 1972. Criteria for recognizing ancient barrier coastlines, *in* Rigby, J. K., and Hamblin, W. K. (eds.), *Recognition of Ancient Sedimentary Environments.* Soc. Econ. Paleontologists and Mineralogists, Spec. Publ. No. 16, Tulsa, Okla., 192–214.

DICKINSON, W. R., 1970. Global tectonics. Science, 168:1250–1256.

DICKINSON, W. R., 1974a. Plate tectonics and sedimentation, *in* Dickinson, W. R. (ed.), *Tectonics and Sedimentation.* Soc. Econ. Paleontologists and Mineralogists, Spec. Publ. No. 22, Tulsa, Okla., 1–27.

DICKINSON, W. R., 1974b. Sedimentation within and beside ancient and modern magmatic arcs, *in* Dott, R. H., Jr., and Shaver, R. H. (eds.), *Modern and Ancient Geosynclinal Sedimentation.* Soc. Econ. Paleontologists and Mineralogists, Spec. Publ. No. 19, Tulsa, Okla., 230–239.

DICKINSON, W. R., AND SUCZEK, C. A., 1979. Plate tectonics and sandstone composition. Amer. Assoc. Petroleum Geologists Bull., 65:2164–2182.

DIETZ, R. S., AND HOLDEN, J. C., 1966. Miogeoclines (miogeosynclines) in space and time. Jour. Geol., 74:566–583.

DOEGLAS, D. J., 1962. The structure of sedimentary deposits of braided rivers. Sedimentology, 1:167–190.

DONALDSON, A. C., MARTIN, R. H., AND KANES, W. H., 1970, Holocene Guadalupe Delta of Texas Gulf coast, *in* Morgan, J. P., and Shaver, R. H. (eds.), *Deltaic Sedimentation, Modern and Ancient.* Tulsa, Soc. Econ. Paleontologists and Mineralogists, Spec. Publ. No. 15:107–137.

DOTT, R. E., JR., 1961. Squantum "tillite" Massachusetts—evidence of glaciation or subaqueous mass movements? Geol. Soc. Amer. Bull., 72:1289–1304.

DOTT, R. H., JR., 1966. Eocene deltaic sedimentation at Coos Bay, Oregon. Jour. Geol., 74:373–420.

DOTT, R. H., JR., 1974. Cambrian tropical storm waves in Wisconsin. Geology, 2:243–246.

DOTT, R. H., JR., AND BATTEN, R. L., 1981. *Evolution of the Earth*, 2nd ed. McGraw Hill, New York, 649 p.

DOTT, R. H., JR., AND HOWARD, J. K., 1962. Convolute lamination in nongraded sequences. Jour. Geol., 70:114–121.

DRAKE, D. E., 1976. Suspended sediment transport and mud deposition on continental shelves, *in* Stanley, D. J., and Swift, D. J. P. (eds.), *Marine Sediment Transport and Environmental Management*. John Wiley, New York, 127–158.

DRAKE, D. E., AND GORSLINE, D. S., 1973. Distribution and transport of suspended particulate matter in Hueneme, Redondo, Newport, and La Jolla submarine canyons. Geol. Soc. Amer. Bull., 84:3949–3968.

DREIMANIS, A., 1976. Tills, their origin and properties, *in* Legget, R. F. (ed.), *Glacial Till, an Interdisciplinary Study*. Roy. Soc. Canada, Spec. Publ. No. 12, 11–49.

DUNBAR, C. O., AND RODGERS, J., 1957. *Principles of Stratigraphy*. John Wiley, New York, 356 p.

DUNHAM, R. J., 1962. Classification of carbonate rocks according to texture, *in* Ham, W. E. (ed.), *Classification of Carbonate Rocks*. Amer. Assoc. Petroleum Geologists, Mem. No. 1, Tulsa, Okla., 279 p. 108–121.

DUNHAM, R. J., 1969. Vadose pisolite in the Capitan Reef (Permian), New Mexico and Texas, *in* Friedman, G. M. (ed.), *Depositional Environments in Carbonate Rocks*. Soc. Econ. Paleontologists and Mineralogists, Spec. Publ. No. 14, Tulsa, Okla., 182–191.

DUNHAM, R. J., 1972. Capitan Reef, New Mexico and Texas: facts and questions to aid interpretation and group discussion. Soc. Econ. Paleontologists and Mineralogists, Permian Basin Sec., Publ. 72-14, 270.

DYER, K. R., 1973. *Estuaries: A Physical Introduction*. John Wiley, New York, 140 p.

DZULYNSKI, S., AND SANDERS, J. E., 1962. Current marks on firm mud bottoms. Conn. Acad. Arts and Sci., Trans., 42:57–97.

DZULYNSKI, S., AND WALTON, E. K., 1965. *Sedimentary Features of Flysch and Greywackes*. Elsevier, Amsterdam, 274 p.

EARDLEY, A. J., 1938. Sediments of Great Salt Lake, Utah. Amer. Assoc. Petroleum Geologists Bull., 22:1305–1411.

ECHOLS, D. J., AND MALKIN, D. S., 1948. Wilcox (Eocene) stratigraphy, a key to production. Amer. Assoc. Petroleum Geologists Bull., 32:11–33.

EDWARDS, J. M., AND FREY, R. W., 1977. Substrate characteristics within a Holocene salt marsh, Sapelo Island, Georgia. Senckenbergiana Maritima, 9:215–259.

EHRLICH, R., AND WEINBERG, B., 1970. An exact method for characterization of grain shape. Jour. Sed. Petrology, 40:205–212.

EMBLETON, C. AND KING, C. A. M., 1975, *Periglacial Geomorphology,* 2nd edition, Halsted Press, New York, 203 p.

EMBRY, A. F., AND KLOVAN, J. E., 1972. Absolute water depth limits of Late Devonian paleoecologic zones. Geol. Rundschau, 61:672–686.

EMERY, K. O., 1952. Continental shelf sediments of southern California. Geol. Soc. Amer. Bull., 63:1105–1108.

EMERY, K. O., 1968. Relict sediments on continental shelves of the world. Amer. Assoc. Petroleum Geologists Bull., 52:445–464.

EMERY, K. O., 1969. Continental shelves. Sci. Amer., 221:106–122.

EMERY, K. O., 1970. Continental margins of the world. Geol. east Atlantic continental margin. I. Gen. Econ. Papers, ICSU/SCOR Working Party 31 Symp. Cambridge, Rept. No. 70/13, 7–29.

EMERY, K. O., 1980. Continental margins—classification and petroleum prospects. Amer. Assoc. Petroleum Geologists Bull., 64:297–315.

EMERY, K. O., AND GALE, J. F., 1951. Swash and swash mark. Amer. Geophys. Union Trans., 32:31–36.

EMRICH, G. O., 1966. *Ironton and Galesville (Cambrian) Sandstones in Illinois and Adjacent Areas*. Illinois Geol. Survey, Circ. 403, 55 p.

ENOS, P., 1969a. Anatomy of a flysch. Jour. Sed. Petrology, 39:680–723.

ENOS, P., 1969b. *Chloridorme Formation, Middle Ordovician Flysch, Northern Gaspe Peninsula, Quebec*. Geol. Soc. Amer., Spec. Paper No. 117, 66 p.

ENOS, P., 1974a. Reefs, platforms, and basins of middle Cretaceous in northeast Mexico. Amer. Assoc. Petroleum Geologists Bull., 58:800–809.

ENOS, P., 1974b. Surface sediment facies map of the Florida-Bahamas Plateau. Geol. Soc. Amer., 5 p.

ENOS, P., AND PERKINS, R. D., 1977. *Quaternary Sedimentation in South Florida*. Geol. Soc. Amer., Mem. No. 147, 198 p.

ETHRIDGE, F. G., GOPINATH, T. R., AND DAVIES, D.K., 1975. Recognition of deltaic environments from small samples, *in* Broussard, M. L. (ed.), *Deltas, Models for Exploration*. Houston Geological Society, Houston, Tex., 151–164.

EUGSTER, H. P., 1967. Hydrous sodium silicates from Lake Magadi, Kenya: precursors of bedded chert. Science, 157:1177–1180.

EUGSTER, H. P., 1969. Inorganic bedded cherts from the Magadi area, Kenya. Contr. Mineral. and Petrol., 22:1–31.

EUGSTER, H. P., AND HARDIE, L. A., 1975. Sedimentation in an ancient playa-lake complex: The Wilkins Peak Member of the Green River Formation of Wyoming. Geol. Soc. Amer. Bull., 86:319–334.

EUGSTER, H. P., AND HARDIE, L. A., 1978. Saline lakes, *in* Lerman, A. (ed.), *Lakes: Chemistry, Geology, Physics*. Springer-Verlag, New York, 237–294.

EVANS, G., 1965. Intertidal flat sediments and their environments of deposition in the Wash. Geol. Soc. London, Quart. Jour., 121:209–245.

EVANS, G., 1975. Intertidal flat deposits of the Wash, western margin of the North Sea, *in* Ginsburg, R. N. (ed.) *Tidal Deposits*, Springer-Verlag, Inc., New York, 13–20.

EVANS, G., AND BUSH, P. R., 1969. Some oceanographical and sedimentological observations on a Persian Gulf lagoon, *in* Castanares, A. A., and Phleger, F. B. (eds.), *Coastal Lagoons—A Symposium*. Universidad Nacional Autónoma de México/UNESCO, Mexico City, 155–170.

EVANS, G., SCHMIDT, V., BUSH, P., AND NELSON, H., 1969. Stratigraphy and geologic history of the sebkha, Abu Dhabi, Persian Gulf. Sedimentology, 12:145–159.

EWING, J. A., 1973. Waved-induced bottom currents on the outer shelf. Mar. Geol., 15:M31–M36.

FARKAS, S. E., 1960. Cross-lamination analysis in the Upper Cambrian Franconia Formation of Wisconsin. Jour. Sed. Petrology, 30:447–458.

FERM, J. C., 1970. Allegheny deltaic deposits, *in* Morgan, J. P., and Shaver, R. H. (eds.), *Deltaic Sedimentation*. Soc. Econ. Paleontologists and Mineralogists, Spec. Publ. No. 15, Tulsa, Okla., 246–255.

FERM, J. C., 1974. Carboniferous environmental models in eastern United States and their significance, *in* Briggs, G. (ed.), *Carboniferous of the Southeastern United States*. Geol. Soc. Amer., Spec. Paper No. 148, 79–95.

FERM, J. C., AND CARAVOC, V. V., JR., 1968. A nonmarine sedimentary model for the Allegheny rocks of West Virginia, *in* Klein, G. D. (ed.), *Late Paleozoic and Mesozoic Continental Sedimentation, Northeastern North America—A Symposium*. Geol. Soc. Amer., Spec. Paper No. 106, 1–19.

FERM, J. C., AND WILLIAMS, E. G., 1965. Characteristics of a carboniferous marine invasion in western Pennsylvania. Jour. Sed. Petrology, 35:319–330.

FERM, J. C., MILICI, R. C., AND EASON, J. E., 1972, Carboniferous depositional environments in the Cumberland Plateau of southern Tennessee and northern Alabama. Tenn., Div. Geol., Rep. Inv. No. 33, 32 p.

FETH, J. H., 1964. Review and annotated bibliography of ancient lake deposits (Precambrian to Pleistocene) in the western states. U.S. Geol. Survey Bull. 1080, 119 p.

FIELD, M. E., 1980. Sand bodies on coastal plain shelves: Holocene record of the U.S. Atlantic inner shelf off Maryland. Jour. Sed. Petrology, 50:505–528.

FIELD, M. E., AND DUANE, D. B., 1976, Post-Pleistocene history of the United States inner continental shelf: significance to origin of barrier islands. Geol. Soc. Amer. Bull., 87:691–702.

FISCHER, A. G., 1961. Stratigraphic record of transgressing seas in light of sedimentation on Atlantic coast of New Jersey. Amer. Assoc. Petroleum Geologists Bull., 45:1656–1666.

FISHER, J. J., 1968. Barrier island formation: discussion. Geol. Soc. Amer. Bull., 79:1421–1426.

FISHER, W. L., 1969. Facies characterization of Gulf Coast basin systems, with some Holocene analogies. Gulf Coast Assoc. Geol. Soc. Trans., 19:239–261.

FISHER, W. L., AND BROWN, L. F., 1972. *Clastic Depositional Systems—A Genetic Approach to Facies Analysis*. University of Texas, Bur. Econ. Geol., Austin, Tex., 203 p.

FISHER, W. L., AND MCGOWEN, J. H., 1967. Depositional systems in the Wilcox Group of Texas and their relationship to occurrence of oil and gas. Gulf Coast Assoc. Geol. Soc. Trans., 17:105–125.

FISHER, W. L., AND MCGOWEN, J. H., 1969. Depositional systems in the Wilcox Group (Eocene) of Texas and their relation to occurrence of oil and gas. Amer. Assoc. Petroleum Geologists Bull., 53:30–54.

FISHER, W. L., AND RODDA, P. U., 1969. Edwards Formation (Lower Cretaceous), Texas: dolomitization in a carbonate platform system. Amer. Assoc. Petroleum Geologists Bull., 53:55–72.

FISHER, W. L., BROWN, L. F., JR., SCOTT, A. J., AND MCGOWEN, J. H., 1969. *Delta Systems in the Exploration for Oil and Gas—A Research Colloquium*. University of Texas, Bur. Econ. Geol., Austin, Tex., 102 p. and 168 illus.

FISK, H. N., 1944. *Geological Investigations of the Alluvial Valley of the Lower Mississippi River*. Mississippi River Commission. Vicksburg, Miss., 78 p.

FISK, H. N., 1947. *Fine-Grained Alluvial Deposits and Their Effects on Mississippi River Activity*. Mississippi River Commission. Vicksburg, Miss., 82 p.

FISK, H. N. 1955. Sand facies of recent Mississippi delta deposits. World Petroleum Cong. 4, Rome, Proc., Sec. 1–C, 377–398.

FISK, H. N., 1960. Recent Mississippi River sedimentation and peat accumulation. 4th Internat. Cong. Carboniferous Strat. and Geol., Comptes Rendus, 187–199.

FISK, H. N., 1961. Bar-finger sands of Mississippi Delta, *in* Peterson, J. A., and Osmond, J. C. (eds.), *Geometry of Sandstone Bodies—A Symposium*. Amer. Assoc. Petroleum Geologists, Tulsa, Okla., 29–52.

FISK, H. N., MCFARLAN, E., JR., KOLB, C. R., AND WILBERT, L. J., JR., 1954. Sedimentary framework of the modern Mississippi Delta. Jour. Sed. Petrology, 24:76–99.

FLINT, R. F., 1961. Geological evidence of cold climate, *in* Nairn, A. E. M. (ed.), *Descriptive Paleoclimatology*. Wiley-Interscience, New York, 140–155.

FLINT, R. F., 1971. *Glacial and Quaternary Geology*. John Wiley, New York, 892 p.

FOLK, R. L., 1951. Stages of textural maturity in sedimentary rocks. Jour. Sed. Petrology, 21:127–130.

FOLK, R. L., 1954. The distinction between grain size and mineral composition in sedimentary rock nomenclature. Jour. Geol. 62:344–359.

FOLK, R. L., 1955. Student operator error in the determination of roundness, sphericity, and grain size. Jour. Sed. Petrology, 25:297–301.

FOLK, R. L., 1956. The rock of texture and composition in sandstone classification. Jour. Sed. Petrology, 26:166–171.

FOLK, R. L., 1959. Practical petrographic classification of limestones. Amer. Assoc. Petroleum Geologists Bull., 43:1–38.

FOLK, R. L., 1960. Petrography and origin of the Tuscarora, Rose Hill and Keefer Formation, Lower and Middle Silurian of eastern West Virginia. Jour. Sed. Petrology, 30:1–58.

FOLK, R. L., 1962. Spectral subdivision of limestone types, *in* Ham, W. E. (ed.), *Classification of Carbonate Rocks*. Amer. Assoc. Petroleum Geologists, Mem. No. 1, Tulsa, Okla., 62–84.

FOLK, R. L., 1974. *Petrology of Sedimentary Rocks*. Hemphills, Austin, Tex., 170 p.

FOLK, R. L., AND MCBRIDE, E. F., 1976. The Caballos Novaculite revisited: Part I. Origin of novaculite members. Jour. Sed. Petrology, 46:659–669.

FOLK, R. L., AND MCBRIDE, E. F., 1978. Radiolites and their relations to subjacent "ocean crust" in Liguria, Italy. Jour. Sed. Petrology, 48:1069–1102.

FOLK, R. L., AND WARD, W. C., 1957. Brazos River bar: a study in the significance of grain size parameters. Jour. Sed. Petrology, 27:3–26.

FOLK, R. L., AND WEAVER, C. E., 1952. A study of the texture and composition of chert. Amer. Jour. Sci., 250:498–510.

FOREL, F. A., 1885. Les ravins savo-lacustres des fleuves glaciaires. Acad. Sci. Paris, Comptes Rendus, 101:725–728.

FOX, W. T., LADD, J. W., AND MARTIN, M. K., 1966. A profile of the four moment measures perpendicular to a shore line, South Haven, Michigan. Jour. Sed. Petrology, 36:1126–1130.

FRAKES, L. A., AND CROWELL, J. C., 1970. Late Paleozoic glaciation: Part II. Africa exclusive of the Karoo Basin. Geol. Soc. Amer. Bull., 81:2261–2286.

FRAZIER, D. E., AND OSANIK, A., 1969. Recent peat deposits—Louisiana coastal plain, in Dapples, E. C., and Hopkins, M. E. (eds.), Environments of Coal Deposition. Geol. Soc. Amer., Spec. Paper No. 114, 63–84.

FREEMAN, T. J., 1962. Quiet water oolites from Laguna Madre, Texas. Jour. Sed. Petrology, 32:475–483.

FREEMAN, T. J., 1968a. "Petrographic" unconformities in the Ordovician of northern Arkansas. Okla. Geol. Notes, 26:21–28.

FREEMAN, T. J., 1968b. Physical evidence of erosion at the Louisville-Jeffersonville contact (the "type" paraconformity), Louisville, Kentucky (abs.). Geol. Soc. Amer., Spec. Paper No. 115, 72.

FREEMAN, T. J., 1969. Cement-composition discontinuity in the Cambrian of Texas and its stratigraphic implications. Geol. Soc. Amer. Bull., 80:2095–2096.

FREY, R. W. (ED.), 1975. The Study of Trace Fossils. Springer-Verlag, New York, 562 p.

FREY, R. W., AND BASAN, P. B., 1978. Coastal salt marshes, in Davis, R. A., JR. (ed.), Coastal Sedimentary Environments. Springer-Verlag, New York, 101–169.

FRIEDMAN, G. M., AND JOHNSON, K. G., 1966. The Devonian Catskill deltaic complex of New York, type example of a "tectonic delta complex," in Shirley, M. L. (ed.), Deltas in Their Geologic Framework. Houston Geological Society, Houston, Tex., 171–188.

FRIEDMAN, G. M., AND SANDERS, J. E., 1978. Principles of Sedimentology. John Wiley, New York, 792 p.

GADD, P. E., LAVELLE, J. W., AND SWIFT, D. J. P., 1978. Estimates of sand transport on the New York shelf near-bottom current meter observations. Jour. Sed. Petrology, 48:239–252.

GALLAGHER, J. L., 1977. Zonation of westlands vegetation, in Clark, J. R. (ed.), Coastal Ecosystem Management. John Wiley, New York, 752–758.

GALLOWAY, W. E., 1975. Process framework for describing the morphologic and stratigraphic evolution of deltaic depositional systems, *in* Broussard, M. L. (ed.), *Deltas, Models for Exploration*. Houston Geological Society, Houston, Tex., 87–98.

GALLOWAY, W. E., 1977. *Catahoula Formation of the Texas Coastal Plain: Depositional Systems, Composition, Structural Development, Groundwater Flow, History, and Uranium Distribution*. University of Texas, Bur. Econ. Geol., Rept. Inv. No. 87, Austin, Tex., 59 p.

GARY, M., MCAFEE, R., JR. AND WOLF, C. L. (eds.), 1972, *Glossary of Geology*. Amer. Geol. Inst., Washington, D. C., 805 p.

GIBSON, T. G., AND TOWE, K. M., 1971. Eocene volcanism and the origin of Horizon A. Science, 172:152–154.

GILBERT, G. K., 1885. The topographic features of lake shores. U. S. Geol. Survey, 5th Ann. Rept., 69–123.

GILBERT, G. K., 1890. *Lake Bonneville*. U.S. Geol. Survey, Monograph 1, 438 p.

GINSBURG, R. N., 1956. Environmental relationships of grain size and constituent particles in some Florida carbonate sediments. Amer. Assoc. Petroleum Geologists Bull., 40:2384–2427.

GINSBURG, R. N., 1964. South Florida carbonate sediments. Guidebook for field trip no. 1. Geol. Soc. Amer. Ann. Mtg., Miami, Fla., 72 p.

GINSBURG, R. N. (ED.), 1975. *Tidal Deposits*. Springer-Verlag, New York, 428 p.

GINSBURG, R. N., AND HARDIE, L. W., 1975. Tidal and storm deposits, northeastern Andros Island, Bahamas, *in* Ginsburg, R. N. (ed.), *Tidal Deposits*. Springer-Verlag, New York, 201–208.

GINSBURG, R. N., AND JAMES, N. P., 1974. Holocene carbonate sediments of continental shelves, *in* Burk, C. A., and Drake, C. L. (eds.), *The Geology of Continental Margins*. Springer-Verlag, Berlin, 137–155.

GLAESER, J. D., 1978, Global distribution of barrier islands in terms of tectonic setting. Jour. Geol., 86:283–297.

GLENNIE, K. W., 1970. *Desert Sedimentary Environments*. Developments in Sedimentology, v. 14. Elsevier, New York, 222 p.

GLENNIE, K. W., 1972. Permian Rotliegendes of northwest Europe interpreted in light of modern desert sedimentation studies. Amer. Assoc. Petroleum Geologists Bull., 56:1048–1071.

GODFREY, P. J., AND GODFREY, M. M., 1973. Comparison of ecological and geomorphic interactions between altered and unaltered barrier island systems in North Carolina, *in* Coates, D. R. (ed.), *Coastal Geomorphology*. State University of New York, Binghamton, N. Y., 239–258.

GOLDBERG, E. D., 1964. The ocean as a chemical system. Trans. New York Acad. Sci., Ser. 2, 27:7–20.

GOLDSMITH, V., 1973. Internal geometry and origin of vegetated coastal sand dunes. Jour. Sed. Petrology, 43:1128–1142.

GOLDSMITH, V., 1978. Coastal dunes, *in* Davis, R. A. (ed.), *Coastal Sedimentary Environments*. Springer-Verlag, New York, 171–236.

GOLDTHWAIT, R. P., (ED.), 1971. *Till: A Symposium*. Ohio State University Press, Columbus, Ohio, 402 p.

GORSLINE, D. S., 1980, Deep-water sedimentologic conditions and models, Mar. Geol., 38:1-21.

GORSLINE, D. S., AND EMERY, K. O., 1959. Turbidity-current deposits in San Pedro and Santa Monica basins off southern California. Geol. Soc. Amer. Bull., 70:279-290.

GORSLINE, D. S., AND SWIFT, D. J. P. (EDS.), 1977. Shelf sediment dynamics: a national overview. Proc. of Workshop, University of Southern California, Los Angeles, 134 p.

GOULD, H. R., AND MCFARLAN, E., JR., 1959. Geologic history of the chenier plain, southwestern Louisiana. Gulf Coast Assoc. Geol. Soc. Trans., 9:261-270.

GOVETT, G. J. S., 1966. Origin of banded iron formations. Geol. Soc. Amer. Bull., 77:1191-1212.

GRAF, W. H., 1971. *Hydraulics of Sediment Transport*. McGraw-Hill, New York, 513 p.

GRAHAM, S. A., DICKINSON, W. R., AND INGERSOLL, R. V., 1975. Himalayan-Bengal model for flysch dispersal in the Appalachian-Ouachita system. Geol. Soc. Amer. Bull., 86:273-286.

GREENSMITH, J. T., 1966. Carboniferous deltaic sedimentation in eastern Scotland: a review and reappraisal, *in* Shirley, M. L. (ed.), *Deltas in Their Geologic Framework*. Houston Geological Society, Houston, Tex., 189-211.

GREENWOOD, B., AND DAVIDSON-ARNOTT, R. G. D., 1976. Textural variation in subenvironments of the shallow-water wave zone, Kouchibouguac Bay, New Brunswick. Can. Jour. Earth Sci., 9:679-688.

GREER, S. A., 1975. Sandbody geometry and sedimentary facies at the estuary-marine transition zone. Ossabaw Sound, Ga.: a stratigraphic model. Senckenbergiana Maritima, 7:105-136.

GRESSLY, A., 1838. Observation géologique sur le Jura Soleurais. Neue Denkschr. Allg. Schweizerische Gesell. ges. Naturw., 2:1-112.

GRIFFIN, J. J., WINDOM, H., AND GOLDBERG, E. D., 1968. The distribution of clay minerals in the world ocean. Deep-Sea Res., 15:433-459.

GRIFFITH, L. S., PITCHER, M. G., AND RICE, G. W., 1969. Quantitative analysis of a Lower Cretaceous reef complex, *in* Friedman, G. M. (ed.), *Depositional Environments in Carbonate Rocks*. Soc. Econ. Paleontologists and Mineralogists, Spec. Publ. No. 14, Tulsa, Okla., 120-138.

GUILCHER, A., 1967, Origin of sediments in estuaries, *in* Lauff, G. A. (ed.), *Estuaries*. Amer. Assoc. Adv. Sci., Washington, D. C., 149-157.

GUSTAVSON, T. C., 1974. Sedimentation on gravel outwash fans, Malaspina Glacier Foreland, Alaska. Jour. Sed. Petrology, 44:374-389.

GUSTAVSON, T. C., 1975a. Bathymetry and sediment distribution in proglacial Malaspina Lake, Alaska. Jour. Sed. Petrology, 45:450-461.

GUSTAVSON, T. C., 1975b. Sedimentation and physical limnology in proglacial

Malaspina Lake, southeastern Alaska, *in* Jopling, A. V., and McDonald, B. C. (eds.), *Glaciofluvial and Glaciolacustrine Sedimentation*. Soc. Econ. Paleontologists and Mineralogists, Spec. Publ. No. 23, Tulsa, Okla., 249–263.

HAM, W. E. (ED.), 1962. *Classification of Carbonate Rocks*. Amer. Assoc. Petroleum Geologists, Mem. No. 1, Tulsa, Okla., 279 p.

HAMBLIN, A. P., AND WALKER, R. G., 1979. Storm-dominated shallow marine deposits: the Fernie-Kootenary (Jurassic) transition, southern Rocky Mountains. Can. Jour. Earth Sci., 16:1673–1690.

HAMBLIN, W. K., 1961. Paleogeographic evolution of the Lake Superior region from Lake Keewenawan to Late Cambrian time. Geol. Soc. Amer. Bull., 72:1–18.

HAMILTON, W., AND KRINSLEY, D. H., 1967. Upper Paleozoic deposits of South Africa and South Australia. Geol. Soc. Amer. Bull., 78:783–800.

HAMPTON, M. A., 1972. The role of subaqueous debris flow in generating turbidity currents. Jour. Sed. Petrology, 42:775–793.

HAMPTON, M. A., AND BOUMA, A. H., 1977. Shelf instability near the shelf break, western Gulf of Alaska. Mar. Geotech., 2:309–331.

HAND, B. M., WESSEL, J. M., AND HAYES, M. O., 1969. Antidunes in the Mount Toby Conglomerate (Triassic), Massachusetts. Jour. Sed. Petrology, 39: 1310–1316.

HAPP, S. C., 1940, Significance of texture and density of alluvium deposits in the middle Rio Grande Valley. Jour. Sed. Petrology, 10:3–19.

HARDIE, L. A., SMOOT, J. P., AND EUGSTER, H. P., 1978. Saline lakes and their deposits: a sedimentological approach, *in* Matter, A., and Tucker, M. E. (eds.), *Modern and Ancient Lake Sediments*. Internat. Assoc. Sedimentologists, Spec. Publ. No. 2. Blackwell, London, 7–42.

HARLAND, W. B., 1964a. Critical evidence for a great Infra-Cambrian glaciation. Geol. Rundschau, 54:45–51.

HARLAND, W. B., 1964b. Evidence of Late Precambrian glaciation and its significance, *in* Nairn, A. E. M. (ed.), *Problems in Paleoclimatology*. Wiley-Interscience, New York, 119–149.

HARMS, J. C., 1969. Hydraulic significance of some sand ripples. Geol. Soc. Amer. Bull., 80:363–396.

HARMS, J. C., AND FAHNESTOCK, R. N., 1965. Stratification, bed form, and flow phenomena (with an example from the Rio Grande), *in* Middleton, G. V. (ed.), *Primary Sedimentary Structures and Their Hydrodynamics Interpretation-A Symposium,* Soc. Econ. Paleontologists and Mineralogists, Spec. Publ. No. 12, 84–115.

HARMS, J. C., SOUTHARD, J. B., SPEARING, D. R., AND WALKER, R. G., 1975. *Depositional Environments As Interpreted from Primary Sedimentary Structures and Stratification Sequences*. Soc. Econ. Paleontologists and Mineralogists, Short Course No. 2, Lecture Notes, Tulsa, Okla., 161 p.

HATCH, F. H., RASTALL, R. H., AND GREENSMITH, J. T., 1965. *Petrology of Sedimentary Rocks*, 4th ed. Thomas Murby, London, 408 p.

HATCHER, P. G., AND SEGAR, D. A.,1976. Chemistry and continental sedimentation, *in* Stanley, D. J., and Swift, D. J. P. (eds.), *Marine Sediment Transport and Environmental Management*. John Wiley, New York, 461–477.

HAVEN, D. S., AND MORALES-ALAMO, R., 1972. Biodeposition as a factor in sedimentation of fine suspended solids in estuaries, *in* Nelson, B. W. (ed.), *Environmental Framework of Coastal Plain Estuaries*. Geol. Soc. Amer., Mem. No. 133, 121–130.

HAY, R. L., 1966. *Zeolites and Zeolitic Reactions in Sedimentary Rocks*. Geol. Soc. Amer., Spec. Paper 85, 130 p.

HAY, W. W., 1974. Introduction, *in* Hay, W. W. (ed.), *Paleooceanography*. Soc. Econ. Paleontologists and Mineralogists, Spec. Publ. No. 20, Tulsa, Okla., 1–5.

HAYES, M. O., 1967a. *Hurricanes as Geological Agents: Case Studies of Hurricane Carla, 1961 and Cindy, 1963*. University of Texas, Bur. Econ. Geol., Rept. Inv. No. 61, Austin, Tex., 56 p.

HAYES, M. O., 1967b. Relationship between coastal climate and bottom sediment type on the inner continental shelf. Mar. Geol., 5:111–132.

HAYES, M. O. (ED.), 1969. Coastal environments in northeastern Massachusetts and New Hampshire. Soc. Econ. Paleontologists and Mineralogists, Eastern Section Guidebook, University of Massachusetts, Amherst, Mass., 462 p.

HAYES, M. O., 1972. Forms of sediment accumulation in the beach zone, *in* Meyer, R. E. (ed.), *Waves on Beaches*. Academic Press, New York, 297–356.

HAYES, M. O., 1975. Morphology of sand accumulations in estuaries, *in* Cronin, L. E. (ed.), *Estuarine Research*, v. 2: *Geology and Engineering*. Academic Press, New York, 2:3–22.

HAYES, M. O., AND KANA, T. W., (EDS.), 1976. *Terrigenous Clastic Depositional Environments*. Tech. Rept. No. 11–CRC, Coastal Res. Div., University of South Carolina, Columbia, S.C., 302 p.

HECKEL, P. H., 1972. Recognition of ancient shallow marine environments, *in* Rigby, J. K., and Hamblin, W. K. (eds.), *Recognition of Ancient Sedimentary Environments*. Soc. Econ. Paleontologists and Mineralogists, Spec. Publ. No. 16, Tulsa, Okla., 226–296.

HECKEL, P. H., 1974. Carbonate buildups in the geologic record, *in* Laporte, L. F. (ed.), *Reefs in Time and Space*. Soc. Econ. Paleontologists and Mineralogists, Spec. Publ. No. 18, Tulsa, Okla., 90–154.

HEDBERG, H. (ED.), 1976. *International Stratigraphy Guide, a Guide to Stratigraphic Classification, Terminology and Procedure*. John Wiley, New York, 200 p.

HEEZEN, B. C., AND DRAKE, C. L., 1964, Grand Banks slump. Amer. Assoc. Petroleum Geologists Bull., 48:221–225.

HEEZEN, B. C., AND EWING, M., 1952, Turbidity currents and submarine slumps and the 1929 Grand Banks earthquake. Amer. Jour. Sci., 250:849–878.

HEEZEN, B. C., AND HOLLISTER, C. D., 1971. *The Face of the Deep*. Oxford University Press, New York, 659 p.

HEEZEN, B. C., ERICSON, D. B., AND EWING, M., 1954, Further evidence for a turbidity current following the 1929 Grand Banks earthquake. Deep-Sea Research, 1:193–202.

HEEZEN, B. C., THARP, M., AND EWING, M., 1959. *The Floors of the Ocean*. Geol. Soc. Amer., Spec. Paper No. 65, 122 p.

HEWARD, A. P., 1981. A review of wave-dominated clastic shoreline deposits. Ear. Sci. Reviews, 17:223–276.

HIGH, L. R., JR., AND PICARD, M. D., 1965. Sedimentary petrology and origin of analcime-rich Popo Agie Member, Chugwater (Triassic) Formation, west-central Wyoming. Jour. Sed. Petrology, 35:49–70.

HIGH, L. R., JR., AND PICARD, M. D., 1969. Stratigraphic relations within Upper Chugwater Group (Triassic), Wyoming. Amer. Assoc. Petroleum Geologists Bull., 53: 1091–1104.

HILL, G. W., AND HUNTER, R. E., 1976. Interaction of biological and geological processes in the beach and nearshore, northern Padre Island, Texas, *in* Davis, R. A. JR., and Ethington, R. L. (eds.), *Beach and Nearshore Sedimentation*. Soc. Econ. Paleontologists and Mineralogists, Spec. Publ. No. 24, Tulsa, Okla., 169–187.

HISCOTT, R. N., 1978. Provenance of Ordovician deep-water sandstone, Tourelle Formation, Quebec, and implications for initiation of the Taconic orogeny. Can. Jour. Earth Sci., 15:1579–1597.

HISCOTT, R. N., AND MIDDLETON, G. V., 1979. Depositional mechanics of thick-bedded sandstones at the base of a submarine slope, Tourelle Formation (Lower Ordovician), Quebec, Canada, *in* Doyle, L. J., and Pilkey, O. H., Jr. (eds.), *Geology of Continental Slopes*. Soc. Econ. Paleontologists and Mineralogists, Spec. Publ. No. 27, Tulsa, Okla., 307–326.

HOBDAY, D. K., 1974. Beach- and barrier-island facies in the Upper Carboniferous of Northern Alabama, *in* Briggs, G. (ed.), *Carboniferous of the Southeastern U.S.* Geol. Soc. Amer., Spec. Paper No. 148, 209–223.

HOLLISTER, C. D., AND ELDER, R. B., 1969. Contour currents in the Weddell Sea. Deep Sea Res., 16:99–101.

HOLLISTER, C. D., HEEZEN, B. C., AND NAFE, K. E., 1975, Animal traces on the deep-sea floor, *in* Frey, R. W. (ed.), *The Study of Trace Fossils*, New York, Springer-Verlag, Inc., 493–510.

HOLMES, A., 1965. *Principles of Physical Geology*, 2nd ed., Thomas Nelson, London, 1288 p.

HOOKE, R. LeB., 1967. Processes on arid-region alluvial fans. Jour. Geol., 75:438–460.

HORNE, J. C., FERM, J. C., CARUCCIO, F. T., AND BAGANZ, D. P., 1978. Depositional models in coal exploration and mine planning in Appalachian region. Amer. Assoc. Petroleum Geologists Bull., 62:2379–2411.

HOROWITZ, D. H., 1966. Evidence for deltaic origin of an upper Ordovician sequence in the central Appalachians, *in* Shirley, M. L. (ed.), *Deltas in Their Geologic Framework*. Houston Geologic Society, Houston, Tex., 159–169.

HOSIER, P. E., AND CLEARY, W. J., 1977. Cyclic geomorphic patterns of washover on a barrier island in southeastern North Carolina. Environ. Geol., 2:23–31.

HOSKIN, C. M., 1963. *Recent Carbonate Sediments on Alacran Reef*. Natl. Acad. Sci.–Natl. Res. Counc., Publ. 1089, Washington, D.C., 160 p.

HOUBOLT, J. J. H. C., 1968. Recent sediments in the southern bight of the North Sea. Geol. Mijnbouw, 47:245–273.

HOUBOLT, J. J. H. C., AND JONKER, J. B. M., 1968. Recent sediments in the eastern part of the Lake Geneva (Lac Léman). Geol. Mijnbouw, 47:131–148.

HOUGH, J. L., 1958a. Fresh-water environment of deposition of Precambrian banded iron formation. Jour. Sed. Petrology, 28:414–430.

HOUGH, J. L., 1958b. *Geology of the Great Lakes*. University of Illinois Press, Urbana, Ill., 313 p.

HOWARD, J. D., AND FREY, R. W., 1975. Estuaries of the Georgia coast, U.S.A.: Sedimentology and biology: II. Regional animal-sediment characteristics of Georgia estuaries. Senckenbergiana Maritima, 7:33–103.

HOWARD, J. D., AND REINECK, H. E., 1972a. Georgia coastal region, Sapelo Island, U.S.A.: Sedimentology and biology: IV. Physical and biogenic structures of nearshore shelf. Senckenbergiana Maritima, 4:81–123.

HOWARD, J. D., AND REINECK, H. E., 1972b. Georgia coastal region, Sapelo Island, U.S.A.: Sedimentology and biology: VIII. Conclusions. Senckenbergiana Maritima, 4:217–222.

HOWARD, J. D., ELDERS, C. A., AND HEINBOKEL, J. F., 1975. Estuaries of the Georgia coast, U.S.A.: Sedimentology and biology: V. Animal-sediment relationships in estuarine point bar deposits, Ageechee River–Ossabaw Sound, Georgia, Senckenbergiana Maritima, 7:181–203.

HOYT, J. H., 1967. Barrier island formation. Geol. Soc. Amer. Bull., 78:1125–1135.

HOYT, J. H., 1968. Barrier island formation. Reply. Geol. Soc. Amer. Bull., 79:947.

HOYT, J. H., 1969. Chenier versus barrier, genetic and stratigraphic distinction. Amer. Assoc. Petroleum Geologists Bull., 53:299–306.

HSU, K. J., 1960. Paleocurrent structures and paleogeography of the ultra-shelvetic flysch basins, Switzerland. Geol. Soc. Amer. Bull., 71:577–610.

HSU, K. J., 1973. Origin of saline giants: a critical review of the discovery of the Mediterranean evaporite. Earth Sci. Rev., 8:371–396.

HSU, K. J., 1976. *Paleoceanography of the Mesozoic Alpine Tethys*. Geol. Soc. Amer., Spec. Paper No. 170, 44 p.

HSU, K. J., 1977. Studies of the Ventura Field, California: I. Facies geometry and genesis of Lower Pliocene turbidites. Amer. Assoc. Petroleum Geologists Bull., 61:137–168.

HUBBARD, D. K., AND BARWIS, J. H., 1976. Discussion of tidal inlet sand deposits: examples from the South Carolina coast, *in* Hayes, M. O., and Kana, T. W. (eds.), *Terrigenous Clastic Depositional Environments*. Tech. Rept. No. 11-CRD, Coastal Res. Div., University of South Carolina, Columbia, S.C., II–128 to II–142.

HUBER, N. K., 1975. The geologic story of Isle Royale National Park. U.S. Geol. Surv., Bull., 1309, 66 p.

HUBERT, J. F., 1966. Sedimentary history of Upper Ordovician geosynclinal rocks, Girvan, Scotland. Jour. Sed. Petrology, 36:677–699.

HUBERT, J. F., 1972. Sedimentology of Upper Cretaceous Cody-Parkman delta, southwestern Powder River Basin, Wyoming. Geol. Soc. Amer. Bull., 83: 1649–1670.

HUBERT, J. F., REED, A. A., AND CAREY, P. J., 1976. Paleogeography of the East Berlin Formation, Newark Group, Connecticut Valley. Amer. Jour. Sci., 276:1183–1207.

HUFFMAN, G. G., AND PRICE, W. A., 1949. Clay dune formation near Corpus Christi, Texas. Jour. Sed. Petrology, 19:118–127.

HUMPHREYS, M., AND FRIEDMAN, G. M., 1975. Late Devonian Catskill deltaic complex in north-central Pennsylvania, *in* Shirley, M. L. (ed.), *Deltas, Models for Exploration*. Houston Geological Society, Houston, Tex., 369–379.

HUNT, C. B., AND WASHBURN, A. L., 1960. Salt features that simulate ground patterns formed in cold climates. U.S. Geol. Survey, Prof. Paper 40–B, B403.

HUNT, C. B., ROBINSON, T. W., BOWLES, W. A., AND WASHBURN, A. L., 1966, Hydrologic Basin. Death Valley, California. U.S. Geol. Survey, Prof. Paper 494–B:1–138.

HUNTER, R. E., 1977a. Basic types of eolian and subaqueous sand-flow cross-strata. Sedimentology, 24:361–388.

HUNTER, R. E., 1977b. Terminology of cross-stratified sedimentary layers and climbing-ripple structures. Jour. Sed. Petrology, 47:697–706.

HUNTER, R. E., 1981. Stratification styles in eolian sandstones: some Pennsylvanian to Jurassic examples from the western interior U.S.A., *in* Ethridge, F. G., and Flores, R. M. (eds.), *Recent and Ancient Nonmarine Depositional Environments: Models for Exploration*. Soc. Econ. Paleontologists and Mineralogists, Spec. Publ. No. 31, Tulsa, Okla., 315–329.

HUNTER, R. E., CLIFTON, H. E., AND PHILLIPS, R. L., 1979. Depositional processes, sedimentary structures and predicted vertical sequences in barred nearshore systems, southern Oregon coast. Jour. Sed. Petrology, 49:711–726.

HUNTER, W., AND PARKIN, D. W., 1960. Cosmic dust in recent deep-sea sediments. Proc. Roy. Soc. London, Ser. A, 225:382–397.

HUTCHINSON, R. W., AND ENGELS, G. G., 1972. Tectonic evolution in the southern Red Sea and its possible significance to older rifted continental margins. Geol. Soc. Amer. Bull., 83:2989–3002.

ILLING, L. V., 1954. Bahamian calcareous sands. Amer. Assoc. Petroleum Geologists Bull., 38:1–95.

INGLE, J. C., JR., 1966, *The Movement of Beach Sand—An Analysis Using Fluorescent Grains*. Dev. in Sedimentology, New York, Elsevier Publ. Co., 221 p.

INMAN, D. L., 1953, Areal and seasonal variations in beach and nearshore sediments at La Jolla, California. U.S. Army, Corps of Engineers, Beach Erosion Board, Tech. Memo. 34, 82 p.

INMAN, D. L., 1957. *Wave-Generated Ripples in Nearshore Sands*. U.S. Army, Corps of Engineers, Beach Erosion Board, Tech. Mem. No. 100, 42 p.

INMAN, D. L., AND NORDSTROM, C. E., 1971. On the tectonic and morphologic classification of coasts. Jour. Geol., 79:1–21.

ISACKS, B., OLIVER, J., AND SYKES, L. R., 1968. Seismology and the new global tectonics. Jour. Geophys. Res., 73:5855–5899.

JACKSON, R. G., 1976. Largescale ripples of the Wabash River. Sedimentology, 23:593–624.

JACKSON, R. G., 1978. Preliminary evaluation of lithofacies models for meandering alluvial streams, *in* Miall, A. D. (ed.), *Fluvial Sedimentology*. Can. Soc. Petroleum Geologists, Mem. No. 5, Calgary, Alberta, 543–576.

JAHNS, R. H., 1969. California's ground-moving weather. Eng. and Sci. Mag., California Institute of Technology, Pasadena, Calif.

JAMIESON, T. F., 1860. On the drift and rolled gravel of the north of Scotland. Geol. Soc. London, Quart. Jour., 16:347–371.

JAMES, H. L., 1954. Sedimentary facies of iron formation. Econ. Geol., 49:235–293.

JAMES, N. P., 1977. Shallowing-upward sequences in carbonates. Geoscience Canada, 4:126–136.

JAMES, N. P., 1979a. Introduction to carbonate facies models, *in* Walker, R. G. (ed.), *Facies Models*. Geoscience Canada, Reprint Ser. 1, 105–108.

JAMES, N. P., 1979b. Reefs, *in* Walker, R. G. (ed.), *Facies Models*. Geoscience Canada, Reprint Ser. 1, 121–132.

JENKYNS, H. C., 1978, Pelagic environments, in Reading, H. G. (ed.), *Sedimentary Environments and Facies*, Elsevier, New York, 314–371.

JOHANNSON, C. E., 1963. Orientation of pebbles in running water: a laboratory study. Geog. Annl., 45:85–111.

JOHNSON, D. W., 1919. *Shoreline Processes and Shoreline Development*. John Wiley, New York, 584 p.

JOHNSON, H. D., 1979. Shallow siliciclastic seas, *in* Reading, H. G. (ed.), *Sedimentary Environments and Facies*. Elsevier, New York, 207–258.

JOPLING, A. V., 1963. Hydraulic studies on the origin of bedding. Sedimentology, 2:115–121.

JOPLING, A. V., 1965. Hydraulic factors and the shape of laminae. Jour. Sed. Petrology, 35:777–791.

JOPLING, A. V., 1967. Origin of laminae deposited by the movement of ripples along a stream bed: laboratory study. Jour. Geol., 75:287–305.

JOPLING, A. V., AND McDONALD, B. C., (EDS.), 1975. *Glaciofluvial and Glaciolacustrine Sedimentation*. Soc. Econ. Paleontologists and Mineralogists, Spec. Publ. No. 23, Tulsa, Okla., 320 p.

JOPLING, A. V., AND WALKER, R. G., 1968. Morphology and origin of ripple-drift cross lamination, with examples from the Pleistocene of Massachusetts. Jour. Sed. Petrology, 38:971–984.

KAHLE, C. F., 1974. Ooids from Great Salt Lake, Utah, as an analogue for the genesis and diagenesis of ooids in marine limestone. Jour. Sed. Petrology, 44:30–39.

KANES, W. H., 1970, Facies and development of the Colorado River delta in Texas, *in* Morgan, J. P., and Shaver, R. H. (eds.) *Deltaic Sedimentation, Modern and Ancient*. Tulsa, Soc. Econ. Paleontologists and Mineralogists, Spec. Publ. No. 15:78–106.

KARCZ, I., 1972. Sedimentary structures formed by flash floods in southern Israel. Sed. Geol., 7:161–182.

KAY, G. M., 1951. *North American Geosynclines*. Geol. Soc. Amer., Mem. No. 48, 143 p.

KELLER, G. H., AND SHEPARD, F. P., 1978. Currents and sedimentary processes in submarine canyons off the northeast United States, *in* Stanley, D. J., and Kelling, G. (eds.), *Sedimentation in Submarine Canyons, Fans and Trenches*. Dowden, Hutchinson & Ross, Stroudsburg, Pa., 15–32.

KELLER, W. D., 1952. Analcime in the Popo Agie Member of the Chugwater Formation. Jour. Sed. Petrology, 22:70–82.

KELLING, G., AND STANLEY, D. J., 1976. Sedimentation in canyon, slope and base-of-slope environments, *in* Stanley, D. J., and Swift, D. J. P. (eds.), *Marine Sediment Transport and Environmental Management*. John Wiley, New York, 379–435.

KELLING, G., AND WALTON, E. K., 1957. Load-cast structures: their relationships to upper-surface structures and their mode of formation. Geol. Mag., 49:481–490.

KELLING, G., AND WILLIAMS, P. F., 1967. Flume studies of the reorientation of pebbles and shells. Jour. Geol., 75:243–267.

KELTS, K., AND HSU, K. J., 1978. Calcium carbonate sedimentation in freshwater lakes and the formation of non-glacial varves in Lake Zurich, *in* Lerman, A. (ed.), *Lakes: Chemistry, Geology, Physics*. Springer-Verlag, New York, 295–324.

KENDALL, A. C., 1979. Continental and supratidal (sabkha) evaporites. Geoscience Canada, 5:66–78.

KENDALL, C. G. ST. C., 1969. An environmental reinterpretation of the Permian evaporite/carbonate shelf sediments of the Guadalupe Mountains. Geol. Soc. Amer. Bull., 80:2503–2526.

KENDALL, C. G. ST. C., AND SKIPWITH, P. A. D'E., 1968. Recent algal mats of a Persian Gulf lagoon. Jour. Sed. Petrology, 38:1040–1058.

KENNETT, J. P., AND WATKINS, N. D., 1975. Deep-sea erosion and manganese nodules development in the southeast Indian Ocean. Science, 188:1011–1013.

KENNETT, J. P., AND WATKINS, N. D., 1976. Regional deep-sea dynamic processes recorded by Late Cenozoic sediments of the south-east Indian Ocean. Geol. Soc. Amer. Bull., 87:321–339.

KENNETT, J. P., BURNS, R. E., ANDREWS, J. E., CHURKIN, M., DAVIES, T. A., DUNITRICA, P., EDWARDS, A. R., GALEHOUSE, J. S., PACKHAM, G. H., AND VANDER LINGEN, G. J., 1972. Australian-Antarctic continental drift, paleocirculation changes, and Oligocene deep-sea erosion. Nature, Phys. Sci., 239:51–55.

KENYON, N. H., 1970. Sand ribbons of European tidal seas. Mar. Geol., 9:25–39.

KENYON, N. H., AND STRIDE, A. H., 1970. The tideswept continental shelf sediments between the Shetland Isles and France. Sedimentology, 14:159–173.

KING, C. A. M., AND WILLIAMS, M. W., 1949. The formation and movement of sand bars by wave action. Geog. Jour., 113:70–84.

KING, P. B., 1942. Permian of West Texas and southeastern New Mexico. Amer. Assoc. Petroleum Geologists Bull., 26:535–563.

KING, P. B., 1959. *The Evolution of North America*. Princeton University Press, Princeton, N.J., 190 p.

KINSMAN, D. J. J., 1964. The recent carbonate sediments near Holot el Baharani, Trucial coast, Persian Gulf, *in* Van Straaten, L. M. J. U. (ed.), *Deltaic and Shallow Marine Deposits*. Developments in Sedimentology. Elsevier, New York, 1:189–192.

KINSMAN, D. J. J., 1969. Modes of formation, sedimentary associations and diagnostic features of shallow-water supratidal evaporites. Amer. Assoc. Petroleum Geologists Bull., 53:830–840.

KINSMAN, D. J. J., AND PARK, R. K., 1976. Algal belt and coast sabkha evolution, Trucial Coast, Persian Gulf, *in* Walther, M. R. (ed.), *Interpreting Stromatolites*. Elsevier, Amsterdam, 421–433.

KLEIN, G. DEV., 1963a. Analysis and review of sandstone classifications in the North American literature, 1940–1960. Geol. Soc. Amer. Bull., 74:555–576.

KLEIN, G. DEV., 1963b. Bay of Fundy intertidal zone sediments. Jour. Sed. Petrology, 33:844–854.

KLEIN, G. DEV. 1963c. Boulder surface markings in Quaco Formation (Upper Triassic), St. Martins, New Brunswick, Canada. Jour. Sed. Petrology, 33:49–52.

KLEIN, G. DEV., 1965. Diverse origins of graded bedding. Geol. Soc. Amer., Spec. Paper No. 82, 1–112.

KLEIN, G. DEV., 1967, Paleocurrent analysis in relation to modern marine sediment dispersal patterns: Amer. Assoc. Petroleum Geologists, Bull., 51:366–382.

KLEIN, G. DEV., 1970. Depositional and dispersal dynamics of intertidal sand bars. Jour. Sed. Petrology, 40:1095–1127.

KLEIN, G. DEV., 1971. A sedimentary model for determining paleotidal range. Geol. Soc. Amer. Bull., 82:2585–2592.

KLEIN, G. DEV., 1972a. Determination of paleotidal range in clastic sedimentary rocks. Internat. Geol. Cong., 24th Montreal 1972, Comptes Rendus, Sec. 6, 397–405.

KLEIN, G. DEV., 1972b. A sedimentary model for determining paleotidal range. Reply. Geol. Soc. Amer. Bull., 83:539–546.

KLEIN, G. DEV., 1975. Tidalites in the Eureka Quartzite (Ordovician) eastern California and Nevada, *in* Ginsburg, R. N. (ed.), *Tidal Deposits*. Springer-Verlag, New York, 145–151.

KLEIN, G. DEV., 1977a. *Clastic Tidal Facies*. Continuing Education Publication Co., Champaign, Ill., 149 p.

KLEIN, G. DEV., 1977b. Tidal circulation model for deposition of clastic sediment in epeiric and mioclinal shelf seas. Sed. Geol., 18:1–12.

KLEIN, G. DEV., 1980. *Sandstone Depositional Models for Exploration for Fossil Fuels*, 2nd ed. CEPCO Division, Burgess, Minneapolis, Minn., 149 p.

KLEIN, G. DEV., AND RYER, T. A., 1978. Tidal circulation patterns in Precambrian, Paleozoic, and Cretaceous epeiric and mioclinal seas. Geol. Soc. Amer. Bull., 89:1050–1058.

KLOVAN, J. E., 1974. Development of western Canadian Devonian reefs and comparison with Holocene analogues. Amer. Assoc. Petroleum Geologists Bull., 58:787–799.

KNIGHT, R. J., AND DALRYMPLE, R. W., 1975. Intertidal sediments from the south shore of Cobequid Bay, Bay of Fundy, Nova Scotia, Canada, *in* Ginsburg, R. N. (ed.), *Tidal Deposits*. Springer-Verlag, New York, 47–55.

KOLB, C. R., AND VAN LOPIK, J. R., 1966. Depositional environments of the Mississippi River deltaic plain-southeastern Louisiana, *in* Shirley, M. L. (ed.), *Deltas in Their Geologic Framework*. Houston Geological Society, Houston, Tex., 17–61.

KOMAR, P. D., 1971. Nearshore cell circulation and the formation of giant cusps. Geol. Soc. Amer. Bull., 82:2643–2650.

KOMAR, P. D., 1976. *Beach Processes and Sedimentation*. Prentice-Hall, Englewood Cliffs, N.J., 429 p.

KOMAR, P. D., NEUDECK, R. H., AND KULM, L. D., 1972. Observations and significance of deep water oscillatory ripple marks on the Oregon continental shelf, *in* Swift, D. J. P., Duane, D. B., and Pilkey, O. H. (eds.), *Shelf Sediment Transport: Process and Pattern*. Dowden, Hutchinson & Ross, Stroudsburg, Pa., 601–619.

KRAFT, J. C., 1971. Sedimentary facies patterns and geologic history of a Holocene marine transgression. Geol. Soc. Amer. Bull., 82:2131–2158.

KRAFT, J. C., 1978. Coastal stratigraphic sequences, *in* Davis, R. A., JR. (ed.), *Coastal Sedimentary Environments*. Springer-Verlag, New York, 361–383.

KRAEUTER, J. N., 1976. Biodeposition by salt-marsh invertebrates. Mar. Biol., 35:215–223.

KRAUSE, P. C., MENARD, H. W., AND SMITH, S. M., 1964. Topography and lithology of the Mendicino Ridge. Jour. Mar. Res., 22:236–249.

KRINSLEY, D. H., AND DOORNKAMP, J. C., 1973. *Atlas of Quartz Sand Surface Textures*. Cambridge University Press, Cambridge, 91 p.

KRUMBEIN, W. C., 1934. Size frequency distribution of sediments. Jour. Sed. Petrology, 4:65–77.

KRUMBEIN, W. C., 1940. Flood gravels of San Gabriel Canyon, California. Geol. Soc. Amer. Bull., 51:639–676.

KRUMBEIN, W. C., AND SLOSS, L. L., 1963. *Stratigraphy and Sedimentation*, 2nd ed. W. H. Freeman, San Francisco, 497 p.

KRYNINE, P. D., 1948. The megascopic study and field classification of sedimentary rocks. Jour. Geol., 56:130–165.

KRYNINE, P. D., 1949. Origin of red beds. New York Acad. Sci., 2:60–68.

KRYNINE, P. D., 1950. *Petrology, Stratigraphy and Origin of Triassic Sedimentary Rocks of Connecticut.* Connecticut Geol. Survey Bull., 73, 247 p.

KRYNINE, P. D., 1951. A critique of geotectonic elements. Amer. Geophys. Union Trans., 32:743–748.

KUENEN, P. H., 1953. Significant features of graded bedding. Amer. Assoc. Petroleum Geologists Bull., 37:1044–1066.

KUENEN, P. H., 1958. Experiments in geology. Geol. Soc. Glasgow, Trans., 23:1–28.

KUENEN, P. H., 1965. Value of experiments in geology. Geol. Mijnbouw, 44: 22–36.

KUENEN, P. H., AND MIGLIORINI, C. I., 1950. Turbidity currents as a cause of graded bedding. Jour. Geol., 58:91–127.

KUENEN, P. H., AND PERDOK, W. G., 1962. Experimental abrasion: 5. Frosting and defrosting of quartz grains. Jour. Geol., 70:648–658.

KUENZI, W. D., HORST, O., AND McGEHEE, R. V., 1979. Effect of volcanic activity on fluvial-deltaic sedimentation in a modern arc-trench gap, southwestern Guatemala. Geol. Soc. Amer. Bull., 90:827–838.

KUIJPERS, E. P., 1971, Transition from fluvial to tidal marine sediments in the Upper Devonian of Sevenheads Peninsula, South County Cork, Ireland. Geol. en Mijnb., 50:443–450.

KULM, L. D., ROUSCH, R. C., HARTLETT, J. C., NEUDECK, R. H., CHAMBERS, D. M., AND RUNGE, E. J., 1975. Oregon continental shelf sedimentation: interrelationships of facies distribution and sedimentary processes. Jour. Geol., 83: 145–176.

KUMAR, N., AND SANDERS, J. E., 1974. Characteristics of nearshore storm deposits: examples from modern and ancient sediments (abs.). Amer. Assoc. Petroleum Geologists, Ann. Mtg., San Antonio, 1:55.

KUMAR, N., AND SANDERS, J. E., 1975. Inlet sequence formed by the migration of Fire Island inlet, Long Island, New York, *in* Ginsburg, R. N. (ed.), *Tidal Deposits* Springer-Verlag, New York, 75–83.

LAMING, D. J. C., 1966. Imbrication, paleocurrents and other sedimentary features in the lower New Red Sandstone, Devonshire, England. Jour. Sed. Petrology, 36:940–959.

LAND, L. S., 1964. Eolian cross-bedding in the beach dune environment, Sapelo Island, Georgia, Jour. Sed. Petrology, 34:389–394.

LAPORTE, L. F., 1967. Carbonate deposition near mean sea level and resultant facies mosaic, Manlius Formation (Lower Devonian) of New York state. Amer. Assoc. Petroleum Geologists Bull., 51:73–101.

LAPORTE, L. F., 1969. Recognition of a transgressive carbonate sequence within an epeiric sea: Helderberg Group (Lower Devonian) of New York state, *in* Friedman,

G.M. (ed.), *Depositional Environments in Carbonate Rocks*. Soc. Econ. Paleontologists and Mineralogists, Spec. Publ. No. 14, Tulsa, Okla., 98–118.

LAPORTE, L. F., 1971. Paleozoic carbonate facies of the Central Appalachian shelf. Jour. Sed. Petrology, 41:724–740.

LAPORTE, L. F., 1975. Carbonate tidal flat deposits of the Early Devonian Manlius Formation of New York state, *in* Ginsburg, R. N. (ed.), *Tidal Deposits*. Springer-Verlag, New York, 243–250.

LARSONNEUR, C., 1975. Tidal deposits, Mont Saint-Michel Bay, France, *in* Ginsburg, R. N. (ed.), *Tidal Deposits*. Springer-Verlag, New York, 21–30.

LAUFF, G. H. (ED.), 1967. Estuaries. Amer. Assoc. Adv. Sci., Spec. Publ. No. 83, Washington, D.C., 757 p.

LAVELLE, J. W., ET AL., 1978. Fair weather and storm sand transport on the Long Island, New York, inner shelf. Sedimentology 25:823–842.

LeBLANC, R. J., 1972. Geometry of sandstone reservoir bodies, *in* Cook, T. D. (ed.), *Underground Waste Management and Environmental Implications*. Amer. Assoc. Petroleum Geologists, Mem. No. 18, Tulsa, Okla., 133–189.

LeBLANC, R. J., 1975. Significant studies of modern and ancient deltaic sediments, *in* Broussard, M. L. (ed.), *Deltas, Models for Exploration*. Houston Geological Society, Houston, Tex., 13–85.

LECOMPTE, M., 1958. Les récifs paléozoïques en Belgique. Geol. Rundschau, 47: 384–401.

LEES, A., AND BUTLER, A. T., 1972. Modern temperate water and warm water shelf carbonate sediments contrasted. Mar. Geol., 13:1767–1773.

LEOPOLD, L. B., AND WOLMAN, M. G., 1957. *River Channel Patterns; Braided, Meandering, and Straight*. U.S. Geol. Survey, Prof. Paper 282-B: 39–85.

LEOPOLD, L. B., AND WOLMAN, M. G., 1960. River meanders. Geol. Soc. Amer. Bull., 71:769–794.

LEOPOLD, L. B., WOLMAN, M. G., AND MILLER, J. P., 1964. *Fluvial Processes in Geomorphology*. W. H. Freeman, San Francisco, 522 p.

LERMAN, A. (ED.), 1978. *Lakes: Chemistry, Geology, Physics*. Springer-Verlag, New York, 359 p.

LINDSEY, D. A., 1969. Glacial sedimentology of the Precambrian Gowganda Formation, Ontario, Canada, Geol. Soc. Amer. Bull., 80:1685–1702.

LINDSEY, D. A., 1972. Sedimentary petrology and paleocurrents of the Harebell Formation, Pinyon Conglomerate, and associated coarse clastic deposits, northwestern Wyoming. U. S. Geol. Survey, Prof. Paper 734-B, 68 p.

LINEBACK, J. A., 1971. Pebble orientation and ice movement in southcentral Illinois, *in* Goldthwait, R. P. (ed.), *Till: A Symposium*. Ohio State University Press, Columbus, Ohio, 328–334.

LINK, M. H., 1975. Matilija Sandstone: a transition from deep-water turbidite to shallow-marine deposition in the Eocene of California. Jour. Sed. Petrology, 45:63–78.

LINK, M. H., AND OSBORNE, R. H., 1978. Lacustrine facies in the Pliocene Ridge

Basin Group: Ridge Basin, California, *in* Matter, A., and Tucker, M. E. (eds.), *Modern and Ancient Lake Sediments*. Internat. Assoc. Sedimentologists, Spec. Publ. No. 2. Blackwell, London, 169–186.

LISITZIN, A. P., 1972. *Sedimentation in the World Ocean*. Soc. Econ. Paleontologists and Mineralogists, Spec. Publ. No. 17, Tulsa, Okla., 218 p.

LLOYD, E. R., 1929. Capitan limestone and associated formations of New Mexico and Texas. Amer. Assoc. Petroleum Geologists Bull., 13, 645–658.

LOGAN, B. W., AND CEBULSKI, D. E., 1970. Sedimentary environments of Shark Bay, Western Australia, *in* Logan, B. W., et al. (eds.), *Carbonate Sedimentation and Environments, Shark Bay, Western Australia*. Amer. Assoc. Petroleum Geologists, Mem. No. 13, Tulsa, Okla., 1–37.

LOGAN, B. W., REZAK, R., AND GINSBURG, R. N., 1964. Classification and environmental significance of algal stromatolites. Jour. Geol., 72:68–83.

LOGAN, B. W., HARDING, J. L., AHR, W. M., WILLIAMS, J. D., AND SNEAD, R. G., 1969. *Carbonate Sediments and Reefs, Yucatan Shelf, Mexico*. Amer. Assoc. Petroleum Geologists, Mem. No. 11, Tulsa, Okla., 198 p.

LOGAN, B. W., READ, J. F., AND DAVIES, G. R., 1970. History of carbonate sedimentation, Quaternary Epoch, Shark Bay, Western Australia, *in* Logan, B. W., et al. (eds.), *Carbonate Sedimentation and Environments: Shark Bay, Western Australia*. Amer. Assoc. Petroleum Geologists, Mem. No. 13, Tulsa, Okla., 38–84.

LOGAN, B. W., HOFFMAN, P., AND GEBELEIN, C. D., 1974a. Algal mats, cryptalgal fabrics and structures, Hamelin Pool, Western Australia, *in* Logan, B. W., et al. (eds.), *Evolution and Diagenesis of Quaternary Carbonate Sequences, Shark Bay, Western Australia*. Amer. Assoc. Petroleum Geologists, Mem. No. 22, Tulsa, Okla., 140–194.

LOGAN, B. W., READ, J. F., HAGAN, G. M., HOFFMAN, P., BROWN, R. G., WOODS, P. J., AND GEBELIEN, C. D., 1974b. *Evolution and Diagenesis of Quaternary Carbonate Sequences, Shark Bay, Western Australia*. Amer. Assoc. Petroleum Geologists, Mem. No. 22, Tulsa, Okla., 358 p.

LONG, D. G. F., 1978. Proterozoic stream deposits: some problems of recognition and interpretation of ancient sandy fluvial systems, *in* Miall, A. D.(ed.), *Fluvial Sedimentology*. Can. Soc. Petroleum Geologists, Mem. No. 5, Calgary, Alberta, 313–342.

LONGWELL, C. R., 1949. *Sedimentary Facies in Geologic History*. Geol. Soc. Amer., Mem. No. 39, 171 p.

LOREAU, J. P., AND PURSER, B. H., 1973, Distribution and ultrastructure of Holocene ooids in the Persian Gulf, *in* Purser, B. H. (ed.), *The Persian Gulf*, New York, Springer-Verlag Inc., 279–328.

LOWE, D. R., 1975. Water escape structures in coarse-grained sediments. Sedimentology, 22:157–204.

LOWE, D. R., 1976. Subaqueous liquified and fluidized sediment flows and their deposits. Sedimentology, 23:285–308.

Lowe, D. R., 1979. Sediment gravity flows: their classification and some problems of application to natural flows and deposits, *in* Doyle, L. J., and Pilkey, O. H., Jr. (eds.), *Geology of Continental Slopes*. Soc. Econ. Paleontologists and Mineralogists, Spec. Publ. No. 27, Tulsa, Okla., 75–82.

Lowe, D. R., and LoPiccolo, R. D., 1974, The characteristics and origins of dish and pillar structures. Jour. Sed. Petrology, 44:484–501.

Lowenstam, H. A., 1957. Niagaran reefs of the Great Lakes area, *in* Ladd, H. S. (ed.), *Treatise on Marine Ecology and Paleoecology*. Geol. Soc. Amer., Mem. No. 67, 2:214–248.

Lozo, F. E., and Smith, C. I., 1964. Revision of Comanche Cretaceous stratigraphic nomenclature southern Edwards Plateau, southwest Texas. Gulf Coast Assoc. Geol. Soc. Trans., 14:285–306.

Ludwick, J. C., 1975. Tidal currents, sediment transport and sand banks in Chesapeake Bay entrance, Virginia, *in* Cronin, L. E. (ed.), *Estuarine Research*, v. 2: *Geology and Engineering*. Academic Press, New York, 365–380.

Lynch-Blosse, M. A., and Davis, R. A. jr., 1977. Stability of Dunedin and Hurricane passes, Florida. Coastal Sediments '77, 5th Symp., ASCE, Charleston, S.C., 744–789.

Lustig, L. K., 1965. *Clastic Sedimentation in Deep Springs Valley, California*. U.S. Geol. Survey, Prof. Paper 352-F: 131–192.

MacDonald, K. B., 1977, Coastal salt marsh, *in* Barbour, M. G., and Major, J. (eds.), *Terrestrial Vegetation of California*. New York, John Wiley and Sons, 263–294.

Mandelbaum, H., 1966. Sedimentation in the St. Clair River delta. Great Lakes Res. Div., University of Michigan, Publ. No. 15, 192–202.

Mason, C. C., and Folk, R. L., 1958, Differentiation of beach, dune and aeolian flat environments by size analysis, Mustang Island, Texas. Jour. Sed. Petrology, 28:211–226.

Matter, A., and Tucker, M. E. (eds.), 1978. *Modern and Ancient Lake Sediments*. Internat. Assoc. Sedimentologists, Spec. Publ. No. 2, Blackwell, London, 290 p.

Matthews, R. K., 1974. *Dynamic Stratigraphy*, Englewood Cliffs, New Jersey, Prentice-Hall, Inc., 370 p.

Maxwell, W. G. H., 1968. *Atlas of the Great Barrier Reef*. Elsevier, New York, 268 p.

Maxwell, W. G. H., and Swinchatt, J. P., 1970. Great Barrier Reef: variation in a terrigenous carbonate province. Geol. Soc. Amer. Bull., 81:691–724.

Maxwell, W. G. H., Day, R. W., and Fleming, P. J. G., 1961, Carbonate sedimentation on the Heron Island Reef. Jour. Sed. Petrology, 31:215–230.

Maxwell, W. G. H., Jell, J. S., and McKellar, R. G., 1964. Differentiation of carbonate sediments in the Heron Island reef, Great Barrier Reef. Jour. Sed. Petrology, 31:215–230.

Mazullo, S. J., and Friedman, G. M., 1975. Developing algal mounds in competitive restricting environments of an Ordovician shelf. Amer. Assoc. Petroleum Geologists Bull., 59:2123–2141.

McBRIDE, E. F., 1962. Flysch and associated beds of the Martinsburg Formation (Ordovician), central Appalachians. Jour. Sed. Petrology, 32:39–91.

McBRIDE, E. F., 1963. A classification of common sandstones. Jour. Sed. Petrology, 33:664–669.

McBRIDE, E. F., 1971. Mathematical treatment of size distribution data, *in* Carver, R. E. (ed.), *Procedures in Sedimentary Petrology*. John Wiley, New York, 109–127.

McBRIDE, E. F., AND FOLK, R. L., 1977. The Caballos Novaculite revisited: Part II. Chert and shale members and synthesis. Jour. Sed. Petrology, 47:1261–1276.

McBRIDE, E. F., AND HAYES, M. O., 1962. Dune cross-bedding on Mustang Island, Texas. Amer. Assoc. Petroleum Geologists Bull., 46:546–551.

McCAVE, I. N., 1972. Transport and escape of fine-grained sediment from shelf areas, *in* Swift, D. J. P., et al. (eds.), *Shelf Sediment Transport: Process and Pattern*. Dowden, Hutchinson & Ross, Stroudsburg, Pa., 225–248.

McCAVE, I. N., AND GEISER, A. C., 1979. Megaripples, ridges and runnels on intertidal flats of The Wash, England. Sedimentology, 26:353–369.

McCLENNEN, C. E., 1973. New Jersey continental shelf near bottom current meter records and recent sediment activity. Jour. Sed. Petrology, 43:371–380.

McDONALD, B. C., AND BANERJEE, I., 1971. Sediment and bedforms on a braided outwash plain. Can. Jour. Earth Sci., 8:1282–1301.

McGEE, W. J., 1890. Encroachments of the sea. Forum, 9:437–449.

McGOWEN, J. H., AND GARNER, L. H., 1970. Physiographic features and stratification types of coarse-grained point bars; modern and ancient examples. Sedimentology, 14:77–112.

McGOWEN, J. H., AND GROAT, C. G., 1971. *Van Horn Sandstone, West Texas: An Alluvial Fan Model for Mineral Exploration*. University of Texas, Bur. Econ. Geol., Rept. Invest. No. 72, Austin, Tex., 57 p.

McGREGOR, A. A., AND BIGGS, C. A., 1968. Bell Creek Field, Montana: a rich stratigraphic trap. Amer. Assoc. Petroleum Geologists Bull., 52:1869–1887.

McKEE, E. D., 1965. Experiments on ripple lamination, *in* Middleton, G. V. (ed.), Primary sedimentary structures and their hydrodynamic interpretation. Soc. Econ. Paleontologists and Mineralogists, Spec. Publ. No. 12, Tulsa, Okla., 66–83.

McKEE, E. D., 1966. Structures of dunes at White Sands National Monument, New Mexico (and comparison with structures of dunes from other selected areas). Sedimentology, 7:1–69.

McKEE, E. D., (ED.), 1979. *A Study of Global Sand Seas*. U.S. Geol. Survey, Prof. Paper 1052, 429 p.

McKEE, E. D., AND MOIOLA, R. J., 1975. Geometry and growth of the White Sands dune field, New Mexico. U.S. Geol. Survey, Jour. Res. 3:59–66.

McKee, E. D., Crosby, E. J., and Berryhill, H. L., 1967. Flood deposits, Bijou Creek, Colorado. Jour. Sed. Petrology, 37:829–851.

Meade, R. H., 1972. Transport and deposition of sediments in estuaries, *in* Nelson, B. W. (ed.), *Environmental Framework of Coastal Plain Estuaries*. Geol. Soc. Amer., Mem. No. 133, 91–210.

Meckel, L. D., 1967. Origin of Pottsville conglomerates (Pennsylvanian) in the central Appalachians. Geol. Soc. Amer. Bull., 78:223–258.

Mero, J. L., 1972, Potential economic value of ocean-floor manganese nodule deposits (with discussion) *in* papers from a conference on ferromanganese deposits on the ocean floor, National Sci. Found., Washington, D. C., 191–203.

Mesolella, K. J., Robinson, J. D., McCormick, L. M., and Ormistan, A. R., 1974. Cyclic deposition of Silurian carbonates and evaporites in Michigan Basin. Amer. Assoc. Petroleum Geologists Bull., 58:34–62.

Miall, A. D., 1977. A review of the braided river depositional environment. Earth Sci. Rev., 13:1–62.

Miall, A. D. (ed.), 1978. *Fluvial Sedimentology*. Can. Soc. Petroleum Geologists, Mem. No. 5, Calgary, Alberta, 859 p.

Michelson, P. C., and Dott, R. H., jr., 1973. Orientation analysis of trough cross stratification in Upper Cambrian sandstones of western Wisconsin. Jour. Sed. Petrology, 43:784–794.

Middleton, G. V. (ed.), 1965. *Primary Sedimentary Structures and Their Hydrodynamic Interpretation—A Symposium*. Soc. Econ. Paleontologists and Mineralogists, Spec. Publ. No. 12, Tulsa, Okla., 265 p.

Middleton, G. V., 1969. Turbidity currents, *in* Stanley, D. J. (ed.), *The New Concepts of Continental Margin Sedimentation*. Amer. Geol. Inst., Washington, D.C., GM-A-1 to GM-A-20.

Middleton, G. V., 1970. Experimental studies related to problems of flysch sedimentation, *in* Lajoie, J. (ed.), *Flysch Sedimentology in North America*. Geol. Assoc. Canada, Spec. Paper No. 7, 253–272.

Middleton, G. V., 1973. Johannes Walther's law of correlation of facies. Geol. Soc. Amer. Bull., 84:979–988.

Middleton, G. V., and Hampton, M. A., 1973. Sediment gravity flows: mechanics of flow and deposition: Part I, *in* Middleton, G. V., and Bouma, A. H. (eds.), *Turbidites and Deep Water Sedimentation*. Soc. Econ. Paleontologists and Mineralogists, Pacific Section, Short Course Lecture Notes, Los Angeles, 1–38.

Middleton, G. V., and Southard, J. B., 1977. *Mechanics of Sediment Movement*. Soc. Econ. Paleontologists and Mineralogists, Eastern Section, Short Course No. 3, Lecture Notes, Tulsa, Okla.

Miller, J. A., 1975. Facies characteristics of Laguna Madre wind-tidal flats, *in* Ginsburg, R. A. (ed.), *Tidal Deposits*. Springer-Verlag, New York, Inc., 67–73.

Miller, R. L., and Zeigler, J. M., 1964. A study of sediment distribution in the zone of shoaling waves over complicated bottom topography, *in* Miller, R. L.

(ed.), *Papers in Marine Geology—Shepard Commemorative Volume*. Macmillan, New York, 133–153.

MILLIMAN, J. D., 1974. *Marine Carbonates*. Springer-Verlag, Berlin, 375 p.

MOERS, C. N. K., 1976. Wind-driven currents on the continental margin, *in* Stanley, D. J., and Swift, D. J. P. (eds.), *Marine Sediment Transport and Environmental Management*. John Wiley, New York, 29–52.

MOIOLA, R. J., AND SPENCER, A. B., 1973. Sedimentary structures and grain size distribution, Mustang Island, Texas. Gulf Coast Assoc. Geol. Soc. Trans., 23:324–332.

MOODY-STUART, N., 1966. High and low sinuosity stream deposits, with examples from the Devonian of Spitsbergen. Jour. Sed. Petrology, 36:1102–1117.

MOORE, D. G., 1977. Submarine slides, *in* Voigt, B. (ed.), *Rockslides and Avalanches*. Developments in Geotechnical Engineering, 14a. Elsevier, Amsterdam, 1:563–604.

MOORES, F. M., AND VINE, F. J., 1971. The Troodos Massif, Cyprus and other ophiolites as oceanic crust: evolution and implications. Philos. Trans. Roy. Soc. Ser. A, 268:443–466.

MORGAN, J. P., 1951. Mudlumps at the mouths of the Mississippi River. Proc. First Coastal Eng. Conf., Houston, Tex., 130–144.

MORGAN, J. P., 1970, Deltas, a resumé. Jour. Geol. Education, 18:107–117.

MORGAN, J. P., AND SHAVER, R. H. (EDS.), 1970. *Deltaic Sedimentation, Modern and Ancient*. Soc. Econ. Paleontologists and Mineralogists, Spec. Publ. No. 15, Tulsa, Okla., 312 p.

MORRIS, R. C., 1974. Sedimentary and tectonic history of the Ouachita Mountains, *in* Dickinson, W. R. (ed.), *Tectonics and Sedimentation*. Soc. Econ. Paleontologists and Mineralogists, Spec. Publ. No. 22, Tulsa, Okla., 120–142.

MOTTS, W. S., 1972, Geology and Paleoenvironments of the northern segment, Captain shelf, New Mexico and West Texas. Geol. Soc. Amer., Bull. 83:701–722.

MULLER, G., AND WAGNER, F., 1978. Holocene carbonate evolution in Lake Balaton (Hungary): a response to climate and impact of man, *in* Matter, A., and Tucker, M. E. (eds.), *Modern and Ancient Lake Sediments*. Internat. Assoc. Sedimentologists, Spec. Publ. No. 2, Blackwell, London, 57–81.

MULLER, G., IRION, G., AND FORSTNER, U., 1972. Formation and diagenesis of inorganic Ca-Mg-carbonates in the lacustrine environment. Naturwissenschaften, 59:158–164.

MURPHY, M. A., AND SCHLANGER, S. O., 1962. Sedimentary structures in Ilhas and Sao Sebastiao Formations (Cretaceous) Reconcavo Basin, Brazil. Amer. Assoc. Petroleum Geologists Bull., 46:457–477.

MURRAY, J., AND RENARD, A. F., 1891. *Report on Deep-Sea Deposits Based on Specimens Collected during the Voyage of H.M.S. Challenger in the Years 1872–1876*. Challenger Reports, H.M.S.O., Edinburgh, 525 p.

MUTTI, E., AND RICCI LUCCHI, F., 1975. Turbidites of the northern Apennines: introduction to facies analysis. Internat. Geol. Rev., 20:125–166. (Translated from 1972 article in Mem. de Soc. Geol. Ital., 161–199.)

NAIRN, A. E. M., 1961. *Descriptive Paleoclimatology*. Wiley-Interscience, New York, 380 p.

NAIRN, A. E. M., 1964. *Problems in Paleoclimatology*. Wiley-Interscience, New York, 705 p.

NAMI, M., 1976. An exhumed Jurassic meander belt from Yorkshire, England. Geol. Mag., 113:47–52.

NAMI, M., AND LEEDER, M. R., 1978. Changing channel morphology and magnitude in the Scalby Formation (M. Jurassic) of Yorkshire, England, *in* Miall, A. D. (ed.), *Fluvial Sedimentology*. Can. Soc. Petroleum Geologists, Mem. No. 5, Calgary, Alberta, 431–440.

NEEDHAM, H. D., HABIB, D., AND HEEZEN, B. C., 1969. Upper Carboniferous palynomorphs as a tracer of red sediment dispersal patterns in the northwest Atlantic. Jour. Geol., 77:113–120.

NEEV, D., AND EMERY, K. O., 1967. *The Dead Sea—Depositional Processes and Environments of Evaporites*. Israel Geol. Survey Bull. No. 41, 147 p.

NELSON, B. W. (ED.), 1972. *Environmental Framework of Coastal Plain Estuaries*. Geol. Soc. Amer., Mem. No. 133, 619 p.

NELSON, C. H., AND KULM, L. D., 1973. Submarine fans and channels, *in* Turbidites and deep water sedimentation. Soc. Econ. Paleontologists and Mineralogists, Pacific Section, Short Course Notes, 39–78.

NELSON, C. H., CARLSON, P. R., BYRNE, J. V., AND ALPHA, T. R., 1970. Development of the Astoria canyon-fan physiography and comparison with similar systems. Mar. Geol., 8:259–291.

NELSON, C. H., NORMARK, W. R., BOUMA, A. H., AND CARLSON, P. R., 1978. Thin-bedded turbidites in modern submarine canyon and fans, *in* Stanley, D. J., and Kelling, G. (eds.), *Sedimentation in Submarine Canyons, Fans, and Trenches*. Dowden, Hutchinson & Ross, Stroudsburg, Pa., 177–189.

NEUMANN, A. C., 1966. Observations on coastal erosion in Bermuda and measurements of the boring rate of sponge, *Cliona lampa*. Limnol. Oceanogr., 11:92–108.

NEWELL, N. D., AND RIGBY, J. K., 1957, Geological studies on the Great Bahama Bank, *in* LeBlanc, R. J., and Breeding, J. G. (eds.), *Regional Aspects of Carbonate Deposition*, Soc. Econ. Paleontologists and Mineralogists, Tulsa, Okla., Spec. Publ. No. 5:15–72.

NEWELL, N. D., RIGBY, J. K., FISCHER, A. G., WHITEMAN, A. J., HICKOX, J. E., AND BRADLEY, J. S., 1953. *The Permian Reef Complex of the Guadalupe Mountains Region; A Study in Paleoecology*. W. H. Freeman, San Francisco, 236 p.

NEWTON, R. S., 1968. Internal structure of waveformed ripple marks in the near-shore zone. Sedimentology, 11:275–292.

NICHOLS, M. M., 1972. Sediments of the James River estuary, Virginia, *in* Nelson, B.

W. (ed.), *Environmental Framework of Coastal Plain Estuaries.* Geol. Soc. Amer., Mem. No. 133, 169–212.

NILSEN, T. H., 1969, Old Red sedimentation in the Buelandet-Vaerlandet Devonian district, western Norway. Sed. Geology, 3:35–57.

NIO, S. D., 1976, Marine transgression as a factor in the formation of sand wave complexes. Geol. in Mijnbouw, 55:18–40.

NORMARK, W. R., 1970. Growth patterns of deep-sea fans. Amer. Assoc. Petroleum Geologists Bull., 54:2170–2195.

NORMARK, W. R., 1974. Submarine canyons and fan valleys: factors affecting growth patterns of deep-sea fans, *in* Dott, R. H., Jr., and Shaver, R. H. (eds.), *Modern and Ancient Geosynclinal Sedimentation.* Soc. Econ. Paleontologists and Mineralogists, Spec. Publ. No. 19, Tulsa, Okla., 56–68.

NORMARK, W. R., 1978, Fan valleys, channels and depositional lobes on modern submarine fans; characters for recognition of sandy turbidite environments. Amer. Assoc. Petroleum Geologists, Bull., 62:912–931.

ODUM, I. E., 1975. Feldspar-grain size relations in Cambrian arenites, upper Mississippi valley. Jour. Sed. Petrology, 45:636–650.

OERTEL, G. F., 1975, Ebb-tidal deltas of Georgia estuaries, *in* Cronin, L. E. (ed.), *Estuarine Research*, Academic Press, Inc., New York, 2:267–276.

OFF, T., 1963. Rhythmic linear sand bodies caused by tidal currents. Amer. Assoc. Petroleum Geologists Bull., 47:324–341.

OOMKENS, E., AND TERWINDT, J. H. J., 1960. Inshore estuarine sediments in the Haringvliet (Netherlands). Geol. Mijnbouw, 22:701–710.

OOSTDAM, B. L., AND JORDAN, R. R., 1972. Suspended sediment transport in Delaware Bay, *in* Nelson, B. W. (ed.), *Environmental Framework of Coastal Plain Estuaries.* Geol. Soc. Amer., Mem. No. 133, 143–150.

ORE, H. T., 1964. Some criteria for recognition of braided stream deposits. Wyoming Contr. Geol., 3:1–14.

OTSUKI, A., AND WETZEL, R. G., 1974. Calcium and total alkalinity budgets and calcium carbonate precipitation of a small hard-water lake. Arch. Hydrobiol., 73:14–30.

OTTO, G. H., 1938. The sedimentation unit and its use in field sampling. Jour. Geol., 46:569–582.

OTVOS, E. G., 1970. Development and migration of barrier islands, northern Gulf of Mexico. Geol. Soc. Amer. Bull., 81:241–246.

OXBURGH, E. R., 1974. The plain man's guide to plate tectonics. Geologists' Assoc. (London), Proc., 85:299–357.

PADGETT, G. V., AND EHRLICH, R., 1976. Paleohydrologic analysis of a late Carboniferous fluvial system, southern Morocco. Geol. Soc. Amer. Bull., 87:1101–1104.

PATERSON, W. S. B., 1969. *The Physics of Glaciers.* Pergamon Press, Oxford, 250 p.

PELLETIER, B. R., 1958. Pocono paleocurrents in Pennsylvania and Maryland. Geol. Soc. Amer. Bull., 69:1033–1064.

PETTIJOHN, F. J., 1975. *Sedimentary Rocks,* 3rd ed. Harper & Row, New York, 628 p.

PETTIJOHN, F. J., AND POTTER, P. E., 1964. *Atlas and Glossary of Sedimentary Structures.* Springer-Verlag, New York, 370 p.

PETTIJOHN, F. J., POTTER, P. E., AND SIEVER, R., 1972. *Sand and Sandstone.* Springer-Verlag, New York, 618 p.

PHLEGER, F. B., 1969. Some general characteristics of coastal lagoons, *in* Castanares, A. A., and Phleger, F. B. (eds.), *Coastal Lagoons—A Symposium.* Universidad Nacional Autónoma de México/UNESCO, Mexico City, 5–26.

PICARD, M. D., AND HIGH, L. R., 1972. Criteria for recognizing lacustrine rocks, *in* Rigby, J. K., and Hamblin, W. K. (eds.), *Recognition of Ancient Sedimentary Environments.* Soc. Econ. Paleontologists and Mineralogists, Spec. Publ. No. 16, Tulsa, Okla., 108–145.

PICARD, M. D., AND HIGH, L. R., JR., 1973. *Sedimentary Structures of Ephemeral Streams.* Developments in Sedimentology. Elsevier, Amsterdam, 17, 223 p.

PICARD, M. D., AND HIGH, L. R., JR., 1981. Physical stratigraphy of ancient lacustrine deposits, *in* Ethridge, F. G., and Flores, R. M. (eds.), *Recent and Ancient Nonmarine Depositional Environments: Models for Exploration.* Soc. Econ. Paleontologists and Mineralogists, Spec. Publ. No. 31, Tulsa, Okla., 233–259.

PIERCE, J. W., 1976. Suspended sediment transport at the shelf break and over the outer margin, *in* Stanley, D. J., and Swift, D. J. P. (eds.), *Marine Sediment Transport and Environmental Management.* John Wiley, New York, 437–458.

POSTMA, H., 1961. Transport and accumulation of suspended matter in the Dutch Wadden Sea. Netherlands Jour. Sea Res., 1:148–190.

POSTMA, H., 1967. Sediment transport and sedimentation in the estuarine environment, *in* Lauff, G. H. (ed.), *Estuaries.* Amer. Assoc. Adv. Sci., Spec. Publ. No. 83, Washington, D.C., 158–179.

POTTER, P. E., 1955. The petrology and origin of the Lafayette gravel: Part I. Mineralogy and petrology. Jour. Geol., 63:1–38.

POTTER, P. E., AND PETTIJOHN, F. J., 1977. *Paleocurrents and Basin Analysis,* 2nd ed. Springer-Verlag, New York, 423 p. and plates.

POTTER, P. E., MAYNARD, J. B., AND PRYOR, W. A., 1980. *Sedimentology of Shale.* Springer-Verlag, New York, 303 p.

POWERS, M. C., 1953. A new roundness scale for sedimentary particles. Jour. Sed. Petrology, 23:117–119.

PRAY, L. C., AND MURRAY, R. C., (EDS.), 1965. *Dolomitization and Limestone Diagenesis.* Soc. Econ. Paleontologists and Mineralogists, Spec. Publ. No. 13, Tulsa, Okla., 180 p.

PRICE, W. A., 1958. Sedimentology and quaternary geomorphology of south Texas. Gulf Coast Assoc. Geol. Soc. Trans., 8:41–75.

PRICE, W. A., 1963. Patterns of flow and channeling in tidal inlets. Jour. Sed. Petrology, 33:279–290.

PRINGLE, I. R., 1973. Rb-Sr age determinations on shales associated with the Varanger Ice Age. Geol. Mag., 109:465–560.

PRIOR, D. B., AND COLEMAN, J. M., 1978. Disintegrating retrogressive landslide on very low angle subaqueous slopes, Mississippi Delta. Mar. Geotech., 3:37–60.

PRITCHARD, D. W., 1955. Estuarine circulation patterns. Amer. Soc. Civil Eng. Proc., 81:717/1–717/11.

PRITCHARD, D. W., 1967. What is an estuary? Physical viewpoint, *in* Lauff, G. H. (ed.), *Estuaries*. Amer. Assoc. Adv. Sci., Spec. Publ. No. 83, Washington, D.C., 3–5.

PRYOR, W. A., 1960. Cretaceous sedimentation in upper Mississippi embayment. Amer. Assoc. Petroleum Geologists Bull., 44:1473–1504.

PRYOR, W. A., 1961. Sand trends and paleoslope in Illinois basin and Mississippi embayment, *in* Peterson, J. A., and Osmond, J. C. (eds.), *Geometry of Sandstone Bodies*. Amer. Assoc. Petroleum Geologists, Tulsa, Okla., 119–133.

PRYOR, W. A., 1975. Biogenic sedimentation and alteration of argillaceous sediments in shallow marine environments. Geol. Soc. Amer. Bull., 86:1244–1254.

PRYOR, W. A., AND AMARAL E. J., 1971. Large-scale cross-stratification in the St. Peter sandstone. Geol. Soc. Amer. Bull., 82:239–244.

PRYOR, W. A., AND SABLE, E. G., 1974. Carboniferous of the eastern interior basin, *in* Briggs, G. (ed.), *Carboniferous of the Southeastern United States*. Geol. Soc. Amer., Spec. Paper No. 148, 281–313.

PUIGDEFABREGAS, C., 1973. Miocene point-bar in the Ebro Basin, northern Spain. Sedimentology, 20:133–144.

PUIGDEFABREGAS, C. AND VAN VLIET, A., 1978. Meandering stream deposits from the Tertiary of the southern Pyrenees, *in* Miall, A. D. (ed.), *Fluvial Sedimentology*. Can. Soc. Petroleum Geologists, Mem. No. 5, Calgary, Alberta, 469–486.

PURDY, E. G., 1974. Reef configurations: cause and effect, *in* Laporte, L. F. (ed.), *Reefs in Time and Space*. Soc. Econ. Paleontologists and Mineralogists, Spec. Publ. No. 18, Tulsa, Okla., 9–76.

PURDY, E. G., PUSEY, W. C., AND WANTLAND, K. F., 1975, Continental shelf of Belize-regional shelf attributes *in* Wantland, K. F. and Pusey, W. C. (eds.), *Belize shelf-Carbonate Sediments, Clastic Sediments, and Ecology*. Tulsa, Amer. Assoc. Petroleum Geologists, Studies in Geol. No. 2:1–39.

PURSER, B. H., (ED.), 1973. *The Persian Gulf*. Springer-Verlag, New York, 471 p.

PURSER, B. H., AND EVANS, G., 1973. Regional sedimentation along the Trucial Coast, SE Persian Gulf, *in* Purser, B. H. (ed.), *The Persian Gulf, Holocene Carbonate Sedimentation and Diagenesis in a Shallow Epicontinental Sea*. Springer-Verlag, New York, 211–231.

PURSER, B. H., AND SEIBOLD, E., 1973. The principal environmental factors influencing Holocene sedimentation and diagenesis in the Persian Gulf, *in* Purser, B. H. (ed.), *The Persian Gulf.* Springer-Verlag, New York, 1–10.

RAAF, J. F. M. DE, AND BOERSMA, J. R., 1971. Tidal deposits and their sedimentary structures (seven examples from western Europe). Geol. Mijnbouw, 50:479–504.

RAAF, J. F. M. DE, BOERSMA, J. R., AND GELDER, A. VAN, 1977. Wave-generated structures and sequences from a shallow marine succession, Lower Carboniferous, County Cork, Ireland. Sedimentology, 24:451–483.

RAMSDEN, J., AND WESTGATE, J. A., 1971. Evidence for reorientation of a till fabric in the Edmonton area, Alberta, *in* Goldthwait, R. P. (ed.), *Till: A Symposium.* Ohio State University Press, Columbus, Ohio, 335–344.

READING, H. G., 1978. *Sedimentary Environments and Facies.* Elsevier, New York, 557 p.

READING, H. G., AND WALKER, R. G., 1966. Sedimentation of Eocambrian tillites and associated sediments in Finnmark, northern Norway. Palaeogeog. Palaeoclimatol. Palaeoecol., 2:177–212.

REDFIELD, A. C., 1958. The influence of the continental shelf on the tides of the Atlantic coast of the United States. Jour. Mar. Res., 17:432–448.

REEVES, C. C., 1968. *Introduction to Paleolimnology.* Elsevier, New York, 228 p.

REID, G. K., 1961. *Ecology of Inland Waters and Estuaries.* Van Nostrand Reinhold, New York, 375 p.

REINECK, H.-E., 1955, Marken, Spuren und Fährten in den Waderner Schickten (ro) bei Martinstein/Nahe. Jeues Jahrb. Geol. Paläontol. Abh., 101:75–90.

REINECK, H.-E., 1960. Über Zeitlucken in rezenten Flachsee-Sedimenten. Geol. Rundschau., 49:149–161.

REINECK, H.-E., 1967. Layered sediments of tidal flats, beaches and shelf bottoms of the North Sea, *in* Lauff, G. H. (ed.), *Estuaries.* Amer. Assoc. Adv. Sci., Publ. No. 83, Washington, D.C., 191–206.

REINECK, H.-E., 1970, Marine Sandkörper, recent und fossil. Geol. Rundschau, 60:302–321.

REINECK, H.-E., 1972. Tidal flats, *in* Rigby, J. K., and Hamblin, W. K. (eds.), *Recognition of Ancient Sedimentary Environments.* Soc. Econ. Paleontologists and Mineralogists, Spec. Publ. No. 16, Tulsa, Okla., 146–159.

REINECK, H.-E., AND SINGH, I. B., 1980. *Depositional Sedimentary Environments.* 2nd edition, Springer-Verlag, New York, 549 p.

REINECK, H.-E., AND WUNDERLICH, F., 1968a. Classification and origin of flaser and lenticular bedding. Sedimentology 11:99–104.

REINECK, H.-E., AND WUNDERLICH, F., 1968b. Zeitmessungen und Bezeitens-chickten. Natur und Museum, 97:193–197.

REINSON, G. E., 1979. Barrier island systems, *in* Walker, R. G. (ed.), *Facies Models.* Geoscience Canada, Reprint Ser. 1, 57–74.

REX, R. W., AND MARTIN, B. D., 1966. Clay mineral formation in sea water by sub-marine weathering of K-feldspar, *in Clays and Clay Minerals,* v. 14. Pergamon Press, New York, 235–240.

RHOADS, D. C., 1972. Mass properties, stability and ecology of marine muds related to burrowing activity. Mar. Geol., 13:391–406.

RICCI-LUCCHI, F., 1975, Depositional cycles in two turbidite formations of northern Apennines (Italy). Jour. Sed. Petrology, 45:3–43.

RICHTER, R., 1936, Marken und Spuren im Hunsrückschiefer. II. Schichtung und Grundleben. Senkenbergiana 18:215–244.

RILEY, J. P., AND CHESTER, R., 1971. *Introduction to Marine Chemistry*. Academic Press, New York, 465 p.

ROBERTSON, A. H. F., AND HUDSON, J. D., 1974. Pelagic sediments in the Cretaceous and Tertiary history of the Troodos Massif, Cyprus, *in* Hsu, K. J., and Jenkyns, H. C. (eds.), *Pelagic Sediments: On Land and under the Sea*. Internat. Assoc. Sedimentologists, Spec. Publ. No. 1. Blackwell, London, 403–436.

ROBINSON, A. H. W., 1966. Residual currents in relation to sandy shoreline evolution of the East Anglian coast. Mar. Geol., 4:57–84.

ROSEN, P. S., 1979. Erosion susceptibility of the Virginia Chesapeake Bay shoreline. Mar. Geol., 34:45–59.

ROSS, D. A., 1980. *Applications and Uses of the Oceans*. Springer-Verlag, New York, 320 p.

ROSS, D. A.,1982, Introduction to Oceanography, 3rd Edition Prentice-Hall, Inc., Englewood Cliffs, New Jersey, 544 p.

ROUSE, H., 1937. Nomogram for settling velicity of spheres. Div. Geo. Exhibit D, Natl. Res. Counc., Rept. Comm. Sedimentation, Washington, D.C., 57–64.

RUPKE, N. A., 1977. Growth of an ancient deep-sea fan. Jour. Geol., 85:725–744.

RUSNAK, G. A., 1960. Sediments of Laguna Madre, *in* Shepard, F. P., Phleger, F. B., and Van Andel, T. H. (eds.), *Recent Sediments of the Northwest Gulf of Mexico*. Amer. Assoc. Petroleum Geologists, Tulsa, Okla., 153–196.

RUST, B. R., 1972a. Pebble orientation in fluvial sediments. Jour. Sed. Petrology, 42:384–388.

RUST, B. R., 1972b. Structure and process in a braided river. Sedimentology, 18:221–246.

RUST, B. R., 1976. Stratigraphic relationships of the Malbaie Formation (Devonian), Gaspe, Quebec. Can. Jour. Earth Sci., 13:1556–1559.

RUST, B. R., 1978. Depositional model for braided alluvium, *in* Miall, A. D. (ed.), *Fluvial Sedimentology*. Can. Soc. Petroleum Geologists, Mem. No. 5, Calgary, Alberta, 605–626.

RUST, B. R., 1979. Coarse alluvial deposits, *in* Walker, R. G. (ed.), *Facies Models*. Geoscience Canada, Reprint Ser. 1, 9–21.

RUST, B. R., AND GOSTIN, V. A., 1981, Fossil transverse ribs in Holocene alluvial fan deposits, Depot Creek, South Australia. Jour. Sed. Petrology, 51:441–44.

RYDER, R. T., FOUCH, T. D., AND ELISON, J. H., 1976. Early tertiary sedimentation in the western Uinta Basin, Utah. Geol. Soc. Amer. Bull., 87:496–512.

RYLER, J. M., 1972. Some aspects of morphometry of peraglacial alluvial fans in southcentral British Columbia. Can. Jour. Earth Sci., 8:1252–1264.

SANDBERG, P. A., 1975. New interpretations of Great Salt Lake ooids and ancient non-skeletal carbonate mineralogy. Sedimentology, 22:497–537.

SANDERS, J. E., 1960. Origin of convoluted laminae. Geol. Mag., 97:409–421.

SANDERS, J. E., 1968. Stratigraphy and primary sedimentary structures of fine-grained, well-bedded strata, inferred lake deposits, Upper Triassic, central and southern Connecticut, *in* Klein, G. D. (ed.), *Late Paleozoic and Mesozoic Continental Sedimentation, Northeastern North America*. Geol. Soc. Amer., Spec. Paper No. 106, 265–305.

SARJEANT, W. A. S., 1975. Fossil tracks and impressions of vertebrates, *in* Frey, R. W. (ed.), *The Study of Trace Fossils*. Springer-Verlag, New York, 283–324.

SCHAFER, A., AND STAPF, K. R. G., 1978. Permian Saar-Nahe Basin and recent Lake Constance (Germany): two environments of lacustrine algal carbonates, *in* Matter A., and Tucker, M. E. (eds.), *Modern and Ancient Lake Sediments*. Internat. Assoc. Sedimentologists, Spec. Publ. No. 2. Blackwell, London, 127–145.

SCHLAGER, W., AND CHERMAK, A., 1979. Sediment facies of platform-basin transition, tongue of the ocean, Bahamas, *in* Doyle, L. J., and Pilkey, O. H., Jr. (eds.), *Geology of Continental Slopes*. Soc. Econ. Paleontologists and Mineralogists, Spec. Publ. No. 27, Tulsa, Okla., 193–208.

SCHLEE, J., 1957. Fluvial gravel fabric. Jour. Sed. Petrology, 27:162–176.

SCHNEIDER, J. F., 1975. Recent tidal deposits, Abu Dhabi, UAE, Arabian Gulf, *in* Ginsburg, R. N. (ed.), *Tidal Deposits*. Springer-Verlag, New York, 209–214.

SCHÄFER, W., 1956. Wiskungen der Benthos-Organismen auf den jungen Schichtverband. Senkenbergiana Lethaea, 37:183–263.

SCHÄFER, W., 1972. *Ecology and Palaeocology of Marine Environments*. Oliver and Boyd, Edinburgh, 538 p.

SCHOLL, D. W., 1969. Modern coastal mangrove swamp stratigraphy and the ideal cyclothem, *in* Dapples, E. C., and Hopkins, M. E. (eds.), *Environments of Coal Deposition*. Geol. Soc. Amer., Spec. Paper 114, 37–61.

SCHOLLE, P. A., AND SPEARING, D., 1982. *Sandstone Depositional Environments*. Memoir No. 31, American Assoc. of Petroleum Geologists, Tulsa, Oklahoma, 410 p.

SCHOPF, T. J. M., 1980. *Paleoceanography*. Harvard University Press, Cambridge, Mass., 341 p.

SCHUBEL, J. R., 1971. *Estuarine Circulation and Sedimentation*. Amer. Geol. Inst. Short Course Lecture Notes, Washington, D.C., VI–1 to VI–7.

SCHUBEL, J. R., 1972. Distribution and transportation of suspended sediments in Upper Chesapeake Bay, *in* Nelson, B. W. (ed.), *Environmental Framework of Coastal Plain Estuaries*. Geol. Soc. Amer., Mem. No. 133, 151–168.

SCHULTZ, C. B., AND FRYE, J. C., (EDS.), 1968. *Loess and Related Eolian Deposits of the World*. Proc. 7th Cong., Internat. Assoc. Quart. Res. University of Nebraska Press, Lincoln, Nebr., 369 p.

SCHUMM, S. A., 1968. Speculations concerning paleohydrologic controls of terrestrial sedimentation. Geol. Soc. Amer. Bull., 79:1573–1588.

SCHUMM, S. A., 1977. *The Fluvial System*. John Wiley, New York, 338 p.

SCHWARTZ, M. L., 1971. The multiple casuality of barrier islands. Jour. Geol., 79:91–94.

SCHWARTZ, M. L., (ED.), 1973. *Barrier Islands*. Benchmark Papers in Geology. Dowden, Hutchinson & Ross, Stroudsburg, Pa., 451 p.

SCHWARTZ, R. K., 1975. *Nature and Genesis of Some Storm Washover Deposits*. U.S. Army, Corps of Engineers, Coastal Eng. Res. Center, Tech. Memo. No. 61, 69 p.

SCHWARZBACH, M., 1963. *Climates of the Past*. D. Van Nostrand, London, 328 p.

SCOTT, A. J., HOOVER, R. A., AND McGOWEN, J. H., 1969, Effects of Hurricane Beulah, 1967, on Texas coastal lagoons and barriers. Mem. Int. Symp. on Coastal Lagoons. UNAM-UNESCO, 221–236.

SCOTT, R. M., AND BIRDSALL, B. C., 1978. Physical and biogenic characteristics of sediments from Hueneme Submarine Canyon, California Coast, *in* Stanley, D. J., and Kelling, G. (eds.), *Sedimentation in Submarine Canyons, Fans, and Trenches*. Dowden, Hutchinson & Ross, Stroudsburg, Pa., 51–64.

SCRUTON, P. C., 1960. Delta building and the delta sequence, *in* Shepard, F. P., Phleger, F. B., and Van Andel, T. H. (eds.), *Recent Sediment of the Northwest Gulf of Mexico*. Amer. Assoc. Petroleum Geologists, Tulsa, Okla., 82–102.

SEILACHER, A., 1953. Die fossilen Ruhespuren (Cubichnia). Neues Jahrb. Geol. Palaeontol. Abhandl., 98:87–124.

SEILACHER, A., 1954. Die geologische Bedeutung fossiler Lebensspuren. Z. Deutsch. Geol. Ges., 105:214–227.

SEILACHER, A., 1964. Biogenic sedimentary structures, *in* Imbrie, J., and Newell, N. D. (eds.), *Approaches to Paleoecology*. John Wiley, New York, 296–316.

SELLEY, R. C., 1964. The penecontemporaneous deformation of heavy mineral bands in the Torridonian sandstone in northwest Scotland, *in* Van Straaten, L. M. J. U. (ed.), *Deltaic and Shallow Water Deposits*. Developments in Sedimentology. Elsevier, New York, 1464 p.

SELLEY, R. C., 1968. A classification of paleocurrent models. Jour. Geol., 76:99–110.

SELLEY, R. C., 1970, *Ancient Sedimentary Environments*, Chapman and Hall, Ltd., London, 237 p.

SELLWOOD, B. W., 1972. Tidal flat sedimentation in the Lower Jurassic of Bornholm, Denmark. Palaeogeog., Palaeoclimatol. and Palaeoecol., 11:93–106.

SELLWOOD, B. W., 1975. Lower Jurassic tidal-flat deposits, Bornholm, Denmark, *in* Ginsburg, R. N. (ed.), *Tidal Deposits*. Springer-Verlag, New York, 93–101.

SENGUPTA, S., 1966. Studies on orientation and imbrication of pebbles with respect to cross-stratification. Jour. Sed. Petrol., 36:362–369.

SESTINI, G., 1970, Vertical variation in flysch sequences; a review. Jour. Earth Sci., 8:15–30.

SHARP, R. P., 1963. Windripples. Jour. Geol., 71:617–636.

SHARP, R. P., 1960. *Glaciers*. University of Oregon Press, Eugene, Oreg., 78 p.

SHAVER, R. H., 1974. Silurian reefs of northern Indiana: reef and interreef macrofauna. Amer. Assoc. Petroleum Geologists Bull., 58:934–956.

SHAW, A. B., 1964, *Time in Stratigraphy*. New York, McGraw-Hill Book Co., 365 p.

SHAW, J., 1972. Sedimentation in the ice-contact environment with examples from Shropshire (England). Sedimentology, 18:23–62.

SHEPARD, F. P., 1973. *Submarine Geology*, 3rd ed. Harper & Row, New York, 551 p.

SHEPARD, F. P., 1979. Currents in submarine canyons and other types of sea valleys, *in* Doyle, L. J., and Pilkey, O. H., Jr. (eds.), *Geology of Continental Slopes*. Soc. Econ. Paleontologists and Mineralogists, Spec. Publ. No. 27, Tulsa, Okla., 85–94.

SHEPARD, F. P., AND DILL, R. F., 1966. *Submarine Canyons and Other Sea Valleys*. Rand McNally, Chicago, 381 p.

SHEPARD, F. P., AND INMAN, D. L., 1950. Nearshore water circulation related to bottom topography and wave refraction. Am. Geophys. Union Trans., 31:196–212.

SHEPARD, F. P. AND MOORE, D. G., 1960. Bays of central Texas coast, *in* Shepard, F. P., Phleger, F. B., and Van Andel, T. J. (eds.), *Recent Sediments of the Northwest Gulf of Mexico*. Amer. Assoc. Petroleum Geologists, Tulsa, Okla., 117–152.

SHEPARD F. P., EMERY, K. O., AND LA FOND, E. C., 1941. Rip currents: a process of geological importance. Jour. Geol., 49:337–369.

SHEPARD, F. P., DILL, R. F., AND VON RAD, U., 1969. Physiography and sed. processes of La Jolla submarine fan and fan-valley, California. Amer. Assoc. Petroleum Geologists Bull., 53:390–420.

SHINN, E. A., 1963. Spur and groove formation on the Florida reef tract. Jour. Sed. Petrology, 33:291–303.

SHINN, E. A., LLOYD, R. M., AND GINSBURG, R. N., 1969. Anatomy of a modern carbonate tidal flat, Andros Island, Bahamas. Jour. Sed. Petrology, 39:1202–1228.

SHIRLEY, M. L. (ED.), 1966. *Deltas in Their Geologic Framework*. Houston Geological Society, Houston, Tex., 251 p.

SHROCK, R. R., 1948. *Sequence in Layered Rocks*. McGraw-Hill, New York, 507 p.

SILVER, B. A., AND TODD, R. G., 1969. Permian cyclic strata, northern Midland and Delaware basins, West Texas and southeastern New Mexico. Amer. Assoc. Petroleum Geologists Bull., 53:2223–2251.

SIMONS, D. B., AND RICHARDSON, E. V., 1961. Forms of bed roughness in alluvial channels. Amer. Soc. Civil Eng., Proc., 88, Hydraulics Div., Jour. HY3, 87–105.

SIMONS, D. B., AND RICHARDSON, E. V., 1963, Forms of bed roughness in alluvial channels. Amer. Soc. Civil Engineers, Trans. 127:927–953.

SIMONS, D. B., AND RICHARDSON, E. V., 1966. *Resistence to Flow in Alluvial Channels*. U.S. Geol. Survey, Prof. Paper 422–J, 61 p.

SIMONS, D. B., AND SENTURK, F., 1977. *Sediment Transport Technology*. Water Resources Publications, Fort Collins, Colo., 807 p.

SIMONS, D. B., RICHARDSON, E. V., AND NORDIN, C. F., 1965. Sediment sorting in alluvial channels, *in* Middleton, G. V. (ed.), *Primary Sedimentary Structures and Their Hydrodynamic Interpretation*. Soc. Econ. Paleontologists and Mineralogists, Spec. Publ. No. 12, Tulsa, Okla., 34–52.

SINGH, I. B., 1972. On the bedding in the natural-levee and the point bar deposits of the Gomti River, India. Sed. Geol., 7:309–317.

SKIPPER, K., AND MIDDLETON, G. V., 1975. The sedimentary structures and depositional mechanics of certain Ordovician turbidites, Cloridorme Formation, Gaspe Peninsula, Quebec. Can. Jour. Earth Sci., 12:1934–1952.

SLY, P. G., 1978. Sedimentary processes in lakes, *in* Lerman, A. (ed.), *Lakes: Chemistry, Geology, Physics.* Springer-Verlag, New York, 65–68.

SMALLEY, I. J., (ED.), 1975. *Loess, Lithology and Genesis.* Dowden, Hutchinson & Ross, Stroudsburg, Pa., 429 p.

SMITH, A. J., 1968. Lakes, *in* Fairbridge, R. W. (ed.), *Encyclopedia of Geomorphology.* Reinhold, New York, 598–603.

SMITH, G. I., 1979. *Subsurface Stratigraphy and Geochemistry of Late Quaternary Evaporites, Searles Lake, California.* U.S. Geol. Survey, Prof. Paper 1043, 130 p.

SMITH, N. D., 1970. The braided stream, depositional environment: Comparison of the Platte River with some Silurian clastic rocks, north-central Appalachians. Geol. Soc. Amer. Bull., 82:2993–3014.

SMITH, N. D., 1971. Transverse bars and braiding in the lower Platte River, Nebraska. Geol. Soc. Amer. Bull., 82:3407–3420.

SMITH, N. D., 1972. Some sedimentological aspects of planar cross-stratification in a sandy braided river. Jour. Sed. Petrology, 42:624–634.

SMITH, N. D., 1974. Sedimentology and bar formation in the upper Kicking Horse River, a braided outwash stream. Jour. Geol., 82:205–223.

SMITH, N. D., 1978. Some comments on terminology for bars in shallow rivers, *in* Miall, A. D. (ed.), *Fluvial Sedimentology.* Can. Soc. Petroleum Geologists, Mem. No. 5, Calgary, Alberta, 85–88.

SMOOT, J. P., 1978. Origin of the carbonate sediments in the Wilken's Peak Member of the lacustrine Green River Formation (Eocene) Wyoming, USA, *in* Matter, A., and Tucker, M. E. (eds.), *Modern and Ancient Lake Sediments.* Internat. Assoc. Sedimentologists, Spec. Publ. No. 2. Blackwell, London, 107–127.

SOUTHARD, J. B., AND STANLEY, D. J., 1976. Shelf-break processes and sedimentation, *in* Stanley, D. J., and Swift, D. J. P. (eds.), *Marine Sediment Transport and Environmental Management.* John Wiley, New York, 351–378.

SPEARING, D. R., 1975. Shallow marine sands, *in* Harms, J. C., Southard, J. B., Spearing, D. R., and Walker, R. G. (eds.), *Depositional Environments As Interpreted from Primary Sedimentary Structures and Stratification Sequences.* Soc. Econ. Paleontologists and Mineralogists, Short Course No. 2, Lecture Notes, Tulsa, Okla., 103–132.

SPEARING, D. R., 1976. Upper Cretaceous Shannon sandstone: an offshore, shallow marine sand body. Wyoming Geol. Assoc. Guidebook, 28th Field Conf., 65–72.

SPENCER, A. M., 1975. Late Precambrian glaciation in the North Atlantic region, *in* Wright, A. E., and Moseley, F. (eds.), *Ice Ages; Ancient and Modern.* Seel House Press, Liverpool, England, 7–42.

STANLEY, D. J. (ED.), 1969. *New Concepts of Continental Margin Sedimentation*. Amer. Geol. Inst. Short Course Notes, Washington, D.C., 400 p.

STANLEY, D. J., AND SWIFT, D. J. P., 1976. *Marine Sediment Transport and Environmental Management*. John Wiley, New York, 602 p.

STANLEY, D. J., FENNER, P., AND KELLING, G., 1972. Currents and sediment transport at Wilmington Canyon shelf break, as observed by underwater television, *in* Swift, D. J. P., Duane, D. B., and Pilkey, O. H. (eds.), *Shelf Sediment Transport: Process and Pattern*. Dowden, Hutchinson & Ross, Stroudsburg, Pa., 621–644.

STAUFFER, P. H., 1967. Grain flow deposits and their implications, Santa Ynez Mountains, California. Jour. Sed. Petrology, 37:487–508.

STEINKER, D. C., AND STEINKER, C. A., 1972. The meaning of facies in stratigraphy. Compass, 49:45–53.

STERNBERG, R. N., AND LARSON, L. H., 1976, Frequency of sediment movement on the Washington continental shelf; a note. Mar. Geol., 21:M37–M47.

STOCKDALE, P. B., 1939. *Lower Mississippian Rocks of the East-Central Interior (United States)*. Geol. Soc. Amer., Spec. Paper No. 22, 248 p.

STOCKMAN, K. W., GINSBURG, R. N., AND SHINN, E. A., 1967. The production of lime mud by algae in south Florida. Jour. Sed. Petrology, 37:633–648.

STOKES, W. L., 1968. Multiple parallel-truncation bedding planes—a feature of wind deposited sandstone. Jour. Sed. Petrology, 38:510–515.

STRAATEN, L. M. U. U. VAN, AND KEUNEN, PH. H., 1958. Tidal action as a cause of clay accumulation. Jour. Sed. Petrology, 28:406–413.

STRUM, M., AND MATTER, A., 1978. Turbidites and varves in Lake Brienz (Switzerland): deposition of clastic detritis by density currents, *in* Matter, A., and Tucker, M. E. (eds.), *Modern and Ancient Lake Sediments*. Internat. Assoc. Sedimentologists, Spec. Publ. No. 2. Blackwell, London, 145–168.

STUBBLEFIELD, W. L., AND SWIFT, D. J. P., 1976. Ridge development as revealed by sub-bottom profiles on the central New Jersey shelf. Mar. Geol., 20:315–334.

STUBBLEFIELD, W. L., LAVELLE, J. W., SWIFT, D. J. P., AND MCKINNEY, T. F., 1975. Sediment response to the present hydraulic regime on the central New Jersey shelf. Jour. Sed. Petrology, 45:337–358.

SUGDEN, D. E., AND JOHN, B. S., 1976. *Glaciers and Landscapes*. Edward Arnold, London, 376 p.

SUNDBORG, A., 1956. The River Klaralven, a study of fluvial processes. Geog. Ann., 38:217–316.

SWIFT, D. J. P., 1969. Outer shelf sedimentation: processes and products, *in* Stanley, D. J. (ed.), *New Concepts of Continental Margin Sedimentation*. Amer. Geol. Inst. Short Course Notes, Washington, D.C., DS-5-1 to DS-5-26.

SWIFT, D. J. P., 1974. Continental shelf sedimentation, *in* Burk, C. A., and Drake, C. L. (eds.), *Geology of Continental Margins*. Springer-Verlag, New York, 117–135.

SWIFT, D. J. P., 1976. Continental shelf sedimentation, *in* Stanley, D. J., and Swift,

D. J. P. (eds.), *Marine Sediment Transport and Environment Management*. John Wiley, New York, 311–350.

SWIFT, D. J. P., AND MCMULLEN, R. M., 1968. Preliminary studies of intertidal sand bodies in the Minas Basin, Bay of Fundy, Nova Scotia. Can. Jour. Earth Sci., 5:175–183.

SWIFT, D. J. P., STANLEY, D. J., AND CURRAY, J. R., 1971. Relict sediments on continental shelves: a reconsideration. Jour. Geol., 79:322–346.

SWIFT, D. J. P., DUANE, D. E., AND MCKINNEY, T. F., 1973. Ridge and swale topography of the Middle Atlantic Bight, North America: secular response to the Holocene hydraulic regime. Mar. Geol., 15:227–247.

SWIFT, D. J. P., FREELAND, G. L., AND YOUNG, R. A., 1979, Time and space distribution of megaripples and associated bedforms. Middle Atlantic Bight, North American Atlantic Shelf. Sedimentology, 26:389–406.

SWIFT, D. J. P., HOLLIDAY, B., AVIGNONE, N., AND SHIDELER, G., 1972. Anatomy of a shoreface ridge system, False Cape, Virginia. Mar. Geol., 12:59–84.

SWINCHATT, J. P., 1965. Significance of constituent composition, texture, and skeletal breakdown in some recent carbonate sediments. Jour. Sed. Petrology, 35:71–90.

TANNER, W. F., 1965, Upper Jurassic paleogeography of the Four Corners region. Jour. Sed. Petrology, 35:564–574.

TANNER, W. F., 1967. Ripple mark indices and their uses. Sedimentology, 9:89–104.

TAYLOR, J. H., 1963. Sedimentary features of an ancient deltaic complex: the Wealden rocks of southeastern England. Sedimentology, 2:2–28.

TERWINDT, J. H. J., 1971. Litho-facies of inshore estuarine and tidal-inlet deposits. Geol. Mijnbouw, 51:515–526.

THOMPSON, R., AND KELTS, K., 1974. Holocene sediments and magnetic stratigraphy from Lakes Zug and Zurich, Switzerland. Sedimentology, 21:577–596.

THOMPSON, W. O., 1949. Lyons Sandstone of Colorado front range. Amer. Assoc. Petroleum Geologists Bull., 33:52–72.

TURNBULL, W. J., KRINITIZSKY, E. L., AND WEAVER, F. S., 1966. Bank erosion in soils of the lower Mississippi valley. Amer. Soc. Civil Eng., Soil Mech. and Foundations Sec., 92:121–131.

TWENHOFEL, W. H. (ED.), 1932. *Treatise on Sedimentation*. Williams & Wilkins, Baltimore, 926 p.

TWENHOFEL, W. H., 1950. *Principles of Sedimentation*. McGraw-Hill, New York, 673 p.

TWIDALE, C. R., 1972. Landform development in the Lake Eyre region, Australia. Geog. Rev., 62:40–70.

UDDEN, J. A., 1898. *The Mechanical Composition of Wind Deposits*. Augustana Library Publ. No. 1, 69 p.

VAN ANDEL, TJ. H., AND BUKRY, D., 1973. Basement ages and basement depths in the eastern equatorial Pacific. Geol. Soc. Amer. Bull., 84:2361–2370.

VAN ANDEL, TJ. H., HEATH, G. R., AND MOORE, T. C., 1975. *Cenozoic History and*

Paleoceanography of the Central Equitorial Pacific Ocean. Geol. Soc. Amer., Mem. No. 143, 134 p.

VAN DE GRAFF, W. J. E., 1975. Carboniferous deltas in the Pisuerga area, Cantabrian Mountains, Spain, *in* Broussard, M. L. (ed.), *Delta, Models for Exploration.* Houston Geological Society, Houston, Tex., 451–456.

VAN DE KAMP, P. C., HARPER, J. D., CONNIFF, J. J., AND MORRIS, D. A., 1974. Facies relations in the Eocene-Oligocene in the Santa Ynez Mountains, California. Jour. Geol. Soc. London, 130:545–565.

VAN HOUTEN, F. B., 1964. Cyclic lacustrine sedimentation, Upper Triassic Lockatong Formation, central New Jersey and adjacent Pennsylvania, *in* Merrian, D. F., (ed.), *Symposium on Cyclic Sedimentation.* Kansas Geol. Survey Bull., 169, 2:497–531.

VAN HOUTEN, F. B., 1965. Composition of Triassic Lockatong and associated formations of Newark Group, central New Jersey and adjacent Pennsylvania. Amer. Jour. Sci., 263:825–863.

VAN STRAATEN, L. M. J. U., 1954. Composition and structure of recent marine sediments in the Netherlands. Leidse Geol. Mededelin, 19:1–110.

VAN STRAATEN, L. M. J. U., AND KUENEN, P. H., 1957. Accumulation of fine-grained sediments in the Dutch Wadden Sea. Geol. Mijnbouw, 19:329–354.

VAN STRAATEN, L. M. J. U., AND KUENEN, P. H., 1958. Tidal action as a cause of clay accumulation. Jour. Sed. Petrology, 28:406–413.

VINE, F. J., AND MOORES, E. M., 1972. A model for gross structure, petrology and magnetic properties of oceanic crust. Geol. Soc. Amer., Mem. No. 132, 195–205.

VISHER, G. S., 1965. Use of a vertical profile in environmental reconstruction. Amer. Assoc. Petroleum Geologists Bull., 49:41–61.

VISHER, G. S., 1969. Grain size distributions and depositional process. Jour. Sed. Petrology, 39:1074–1106.

VON DER BORCH, C. C., 1976. Stratigraphy of stromatolite occurrences in carbonate lakes of the Coorong Lagoon area, South Australia, *in* Walter, M. R. (ed.), *Stromatolites.* Elsevier, New York, 413–420.

VOS, R. G., 1975. An alluvial plain and lacustrine model for the Precambrian Witwatersrand deposits of South Africa. Jour Sed. Petrology, 45:480–493.

WADELL, H. A., 1932. Volume, shape and roundness of rock particles. Jour. Geol., 40:443–451.

WAGNER, C. W., AND VAN DER TOGT, C., 1973. Holocene sediment types and their distribution in the southern Persian Gulf, *in* Purser, B. H. (ed.), *The Persian Gulf.* Springer-Verlag, New York, 123–156.

WALKER, K., AND LAPORTE, L. F., 1970. Congruent fossil communities from Ordovician and Devonian carbonates of New York. Jour. Paleontol., 44: 928–944.

WALKER, R. G., 1965. The origin and significance of the internal sedimentary structures of turbidites. Yorkshire Geol. Soc., 35:1–29.

WALKER, R. G., 1966a. Deep channels in turbidite-bearing formations. Amer. Assoc. Petroleum Geologists Bull., 50:1899–1917.

WALKER, R. G., 1966b. Shale Grit and Grindslow Shales: transition from turbidite to shallow water sediments in the Upper Carboniferous of northern England. Jour. Sed. Petrology, 36:90–114.

WALKER, R. G., 1973. Mopping up the turbidite mess, *in* Ginsburg, R. N. (ed.), *Evolving Concepts in Sedimentology*. Johns Hopkins University Press, Baltimore, 1–37.

WALKER, R. G., 1975. Generalized facies models for resedimented conglomerates of turbidite association. Geol. Soc. Amer. Bull., 86:737–749.

WALKER, R. G., 1976. Facies models 2. Turbidites and associated coarse clastic deposits. Geoscience Canada, 3:25–36.

WALKER, R. G., 1978. Deep water sandstone facies and ancient submarine fans: models for exploration for stratigraphic traps. Amer. Assoc. Petroleum Geologists Bull., 62:932–966.

WALKER, R. G. (ED.), 1979. *Facies Models*. Geoscience Canada, Reprint Ser. 1, Geol. Assoc. Canada, 211 p.

WALKER, R. G., AND CANT, D. J., 1979. Sandy fluvial systems, *in* Walker, R. G., (ed.), *Facies Models*. Geoscience Canada, Reprint Series 1, Geol. Assoc. Canada, 23–32.

WALKER, R. G., AND MUTTI, E., 1973. Turbidite facies and facies associations, *in* Middleton, G. V., and Bouma, A. H. (eds.), *Turbidites and Deep-water Sedimentation*. Soc. Econ. Paleontologists and Mineralogists, Pacific Section Short Course Notes, 119–157.

WALKER, T. R., 1967. Formation of red beds in ancient and modern deserts. Geol. Soc. Amer. Bull., 78:353–368.

WALKER, T. R., 1974. Formation of red beds in moist tropical climate: a hypothesis. Geol. Soc. Amer. Bull., 85:633–638.

WALKER T. R., AND HARMS, J. C., 1972. Eolian origin of flagstone beds, Lyons Sandstone (Permian), type area, Boulder County, Colorado. Mountain Geologist, 9:279–288.

WALLS, R. A., 1975. Late Devonian-Early Mississippian subaqueous facies in a portion of the southeastern Appalachian Basin, *in* Broussard, M. L. (ed.), *Deltas, Models for Exploration*. Houston Geological Society, Houston, Tex., 359–368.

WANLESS, H. R., BARROFFIO, J. R., AND TRESCOTT, P. C., 1969. Conditions of deposition of Pennsylvanian coal beds, *in* Dapples, E. C., and Hopkins, M. E. (eds.), *Environments of Coal Deposition*. Geol. Soc. Amer. Spec. Paper No. 114, 105–142.

WANLESS, H. R., TUBB, J. B., GEDNETZ, D. E., AND WEINER, J. L., 1963, Mapping sedimentary environments of Pennsylvanian cycles. Geol. Soc. America Bull., 74:437–486.

WANLESS, H. R., 1974. Mangrove sedimentation in geologic perspective, *in* Gleason, P. J. (ed.), *Environments of South Florida: Past and Present*. Miami Geol. Soc., Mem. No. 2, 190–200.

WARD, L. G., 1978. Suspended-material transport in marsh tidal channels, Kiawah Island, South Carolina, Mar. Geol., 40:139–154.

WARME, J. E., 1967. Graded bedding in the recent sediment of Mugu Lagoon, California. Jour. Sed. Petrology, 37:540–547.

WARME, J. E., 1969. Mugu Lagoon, coastal southern California: origin sediments and productivity, *in* Castanares, A. A., and Phleger, F. B. (eds.), *Coastal Lagoons—A Symposium*. Universidad Nacional Autónoma de México/ UNESCO, Mexico City, 137–154.

WARME, J. E., 1971. *Paleoecological Aspects of a Modern Coastal Lagoon*. Univ. Calif. Publ. Geol. Sci., Berkeley, Calif., 87, 131 p.

WARME, J. E., 1975. Borings as trace fossils, and the processes of marine bioerosion, *in* Frey, R. W. (ed.), *The Study of Trace Fossils*. Springer-Verlag, New York, 181–227.

WARME, J. E., DOUGLAS, R. G., AND WINTERER, E. L., (EDS.) 1981, *The Deep Sea Drilling Project: A Decade of Progress*. Soc. Econ. Paleontologists and Mineralogists, Spec. Publ. No. 32, Tulsa, Okla., 564 p.

WARME, J. E., SCANLAND, T. B., AND MARSHALL, N. F., 1971. Submarine canyon erosion; contribution of marine rock burrowers. Science, 173:1127–1129.

WARME, J. E., SANCHEZ-BARREDA, L. A., AND BIDDLE, K. T., 1977. Sedimentary patterns and processes in west coast lagoons, *in* Wiley, M. (ed.), *Estuarine Processes, v. 2: Circulation Sediments and Transfer of Materials in the Estuary*. Academic Press, New York, 167–181.

WEIMER, R. J., 1970. Late Cretaceous deltas, Rocky Mountain region, *in* Morgan, J. P., and Shaver, R. H. (eds.), *Deltaic Sedimentation*. Soc. Econ. Paleontologists and Mineralogists, Spec. Publ. No. 15, Tulsa, Okla., 270–291.

WIEMER, R. J., 1976. *Deltaic and Shallow Marine Sandstones: Sedimentation, Tectonics and Petroleum Occurrences*. Amer. Assoc. Petroleum Geologists, Cont. Educ. Course Notes, Ser. No. 2, Tulsa, Okla., 167 p.

WENTWORTH, C. K., 1919. A laboratory and field study of cobble abrasion. Jour. Geol., 27:507–521.

WENTWORTH, C. K., 1922. A scale of grade and class terms for clastic sediments. Jour. Geol., 30:377–392.

WHITAKER, J. H. M., 1962. The geology of the area around Leintwardine, Herefordshire. Geol. Soc. London, Quart. Jour., 118:319–347.

WHITAKER, J. H. M., 1974. Ancient submarine canyons and fan valleys, *in* Dott, R. H., Jr., and Shaver, R. H. (eds.), *Modern and Ancient Geosynclinal Sedimentation*. Soc. Econ. Paleontologists and Mineralogists, Spec. Publ. No. 19, Tulsa, Okla., 106–125.

WIEGEL, R. L., 1964. *Oceanographical Engineering*. Prentice-Hall, Englewood Cliffs, N.J., 532 p.

WILLIAMS, E. G., FERM, J. C., GUBER, A. L., AND BERGENBACK, R. E., 1964. Cyclic sedimentation in the Carboniferous of western Pennsylvania, *in* Guidebook, 29th

Ann. Field Conf. of Pennsylvanian Geologists, Pennsylvania State University, University Park, Pa., 35 p.

WILLIAMS, G., 1964. Some aspects of the eolian saltation load. Sedimentology, 3:257–287.

WILLIAMS, G. E., 1966. Palaeogeography of the Torridonian Applecross Group. Nature, 209:1303–1306.

WILLIAMS, G. E., 1969. Characteristics and origin of a Pre-Cambrian pediment. Jour. Geol., 77:183–207.

WILLIAMS, G. E., 1971. Flood deposits of the sand-bed ephemeral streams of central Australia. Sedimentology, 17:1–40.

WILLIAMS, P. F., AND RUST, B. R., 1969. The sedimentology of a braided river. Jour. Sed. Petrology, 39:649–679.

WILSON, I. G., 1972. Aeolian bedforms—their development and origin. Sedimentology, 19:173–210.

WILSON, I. G., 1973. Ergs. Sed. Geol., 10:77–106.

WILSON, J. L., 1975. *Carbonate Facies in Geologic History*. Springer-Verlag, New York, 471 p.

WOLMAN, M. G., AND LEOPOLD, L. B., 1957. *River Flood Plains: Some Observations on Their Formation*. U.S. Geol. Survey, Prof. Paper 282-C.

WRIGHT, A. E., AND MOSELEY, F. (EDS.), 1975. *Ice Ages: Ancient and Modern*. Seel House Press, Liverpool, England, 320 p.

WRIGHT, L. D., 1977. Sediment transport and deposition at river mouths: a synthesis. Geol. Soc. Amer. Bull., 88:837–868.

WRIGHT, L. D., 1978. River deltas, *in* Davis, R. A., Jr. (ed.), *Coastal Sedimentary Environments*. Springer-Verlag, New York, 5–68.

WRIGHT, L. D., AND COLEMAN, J. M., 1974. Mississippi River mouth processes: effluent dynamics and morphologic development. Jour. Geol., 82:751–778.

WRIGHT, L. D., COLEMAN, J. M., AND ERICKSON, M. W., 1974. *Analysis of major River Systems and Their Deltas: Morphologic and Process Comparisons*. Tech. Rept. 156, Louisiana State University, Coastal Studies Inst., Baton Rouge, La., 114 p.

WRIGHT, L. D., COLEMAN, J. M., AND THOM, B. G., 1975. Sediment transport and deposition in a macrotidal river channel, Ord River, Western Australia, *in* Cronin, L. E. (ed.), *Estuarine Research*, v. 2: *Geology and Engineering*. Academic Press, New York, 309–322.

WUNDERLICH, F., 1972. Georgia coastal region, Sapelo Island, U.S.A., Sedimentology and biology: III. Beach dynamics and beach development. Senckenbergiana Maritima, 4:47–79.

YALIN, M. S., 1977. *Mechanics of Sediment Transport*. Pergamon Press, Oxford, 290 p.

YEAKEL, L. S., 1962, Tuscarora, Juniata and Bald Eagle paleocurrents and paleogeography in the Central Appalachians. Geol. Soc. Amer. Bull., 73: 1515–1540.

ZEIGLER, J. M., WHITNEY, G. C., JR., AND HAYES, C. R., 1960. Woods Hole rapid sediment analyzer. Jour. Sed. Petrology, 30:490–495.

ZINGG, T., 1935. Beitrage zur Schatteranalyse. Min. Petrog. Mitt. Schweiz., 15:39–140.

Index